W9-COH-732

M.Greenberg
3/1/77

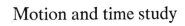

Motion and time study

THE IRWIN SERIES IN MANAGEMENT
AND
THE BEHAVIORAL SCIENCES

L. L. CUMMINGS and E. KIRBY WARREN
Consulting Editors

JOHN F. MEE *Advisory Editor*

Motion and time study

BENJAMIN W. NIEBEL

Head of Department and
Professor of Industrial Engineering
The Pennsylvania State University

 Sixth edition   1976

RICHARD D. IRWIN, INC.   Homewood, Illinois   60430

Irwin-Dorsey International   Arundel, Sussex   BN18 9AB
Irwin-Dorsey Limited   Georgetown, Ontario   L7G 4B3

© RICHARD D. IRWIN, INC., 1955, 1958, 1962, 1967, 1972, and 1976

All rights reserved. No part of this publication may be
reproduced, stored in a retrieval system, or transmitted,
in any form or by any means, electronic, mechanical,
photocopying, recording, or otherwise, without the prior
written permission of the publisher.

Sixth Edition

First Printing, May 1976

ISBN 0-256-01775-1
Library of Congress Catalog Card No. 75–43161

Printed in the United States of America

Preface

Inflation plus a prolonged worldwide recession have awakened almost every segment of industry, business, and government to the absolute necessity of improving productivity in every sector. The improvement of productivity, whether it be at the office, in a hospital, on a railroad, in the postal system, in any business or any industry, is made possible by the application of sound time standards, wage payment plans that are fair to both the worker and the company, and modern techniques of methods improvement and worker motivation.

This sixth edition has been written primarily for three reasons. First, to update the material and examples that have become obsolete because of inflation and technological change. Second, to broaden the application of motion and time study into such indirect and service areas as warehousing, maintenance, government, and hospitals. Third, to introduce new material that has proven successful and to refine some of the more traditional material that still has application in modern business and industry.

Considerable new material has been added in connection with macroscopic approaches to improvement that allow analysts to make improvements related to the immediate area of the work environment. This new material deals with not only the physical environment of the work station, but also physiological and psychological factors relating to both the operator and the work force.

New material has also been introduced on the application of the manufacturing progress function (learning curves); on fundamental motion data systems, including MTM–3; on group technology; and on wage payment systems.

New material has been added on the application of videotape facilities to motion and methods analysis, on the implications of the Occupational Safety and Health Act, on the use of travel charts in plant layout, and on work element sharing in conjunction with line balancing.

New problems have been added at the end of each chapter, and new lists of pertinent reference material give the text much broader application.

The suggestions of the many colleges and universities, technical institutes, industries, and labor organizations that have adopted the text have helped materially in the preparation of this sixth edition. It has the same objectives as the first five—that is, to provide a practical up-to-date college text in the area of methods, time study, and wage payment; and to give practicing analysts from both labor and management an authentic source of reference material.

The author wishes to acknowledge in particular the constructive criticisms of John Samuels and Emory Enscore, who have taught regularly from this text for the past several years.

April 1976   BENJAMIN W. NIEBEL

Contents

good housekeeping. Arrange for the disposal of irritating and harmful dusts, fumes, gases, vapors, and fogs. Provide guards at nip points and points of power transmission. Provide personal protective equipment. Sponsor and enforce a well-formulated first-aid program. OSHA. MATE-RIAL HANDLING: Reduce the time spent in picking up material. Reduce material handling by using mechanical equipment. Make better use of existing handling facilities. Handle material with greater care. PLANT LAYOUT: Layout types. Travel charts. Making the layout. PRINCIPLES OF MOTION ECONOMY: Both hands should work at the same time. Each hand should go through as few motions as possible. The workplace should be arranged to avoid long reaches. Avoid using the hand as a holding device.

Man and machine process charts. Construction of the man and machine process chart. Using the man and machine process chart. Gang process charts. Construction of the gang process chart. Using the gang process chart. Quantitative techniques for man and machine relationships. Line balancing. The operator process chart. Constructing the operator process chart. Using the operator process chart.

The fundamental motions. Definitions of basic divisions of accomplishment. Therblig summary. Principles of motion economy. EXPLANA-TION OF LAWS OF MOTION ECONOMY. THE PRACTICAL USE OF MOTION STUDY IN THE PLANNING STAGE: Motion analysis as applied in planning.

Operator selection for micromotion study. Micromotion study as a training aid. Micromotion equipment. The motion-picture camera. The light meter. Projection equipment. Videotape facilities. Taking the motion pictures. Analyzing the tape or film. Create an improved method. Teach and standardize the new method. Memomotion study.

Management and control of the physical environment. The visual environment. The impact of color. Noise. Vibration. Thermal conditions. Protection against heat. Low-temperature effects on performance. Radiation. Fundamentals of work physiology. Motor fitness, reaction time, and visual capacity. Memory. Physiological fatigue. Individual differences. The work regimen. Behavioral concepts. Safety and health concepts. Job factors leading to unsatisfactory performance. Indicator lights. Display information. Acoustic signals. Shape and size coding. Control size, displacement, and resistance criteria.

Development of standard time data. Calculation of cutting times. Drill press work. Lathe work. Milling machine work. Determining horsepower requirements. Plotting curves. The least squares method. Solving by regression line equations. Using standard data.

Definition of synthetic basic motion times. The necessity of synthetic basic motion times. Work-Factor. Detailed Work-Factor system. Simplified Work-Factor system. Abbreviated Work-Factor system. Ready Work-Factor. Methods-Time Measurement (MTM–1). MTM—General Purpose Data. MTM–2. MTM–3. Applying synthetic basic motion times. The development of standard data.

Application of formulas. Advantages and disadvantages of formulas. Characteristics of the usable formula. Steps to follow in formula construction. Analyzing the elements. Compute expressions for variables. Graphic solutions for more than one variable. Least squares and regression techniques. Multiple regression. Some precautions. Develop synthesis. Compute expression. Check for accuracy. Write formula report. Formula number. Part. Operation and work station. Normal time. Application. Analysis. Procedure. Time studies. Table of detail elements. Synthesis. Inspection, payment, and signatures. Representative formula. Use of the digital computer. Programming.

Illustrative example. Selling work sampling. Planning the work sampling study. Determining the observations needed. Determining the frequency of the observations. Designing the work sampling form. The use of control charts. Observing and recording the data. Using a random activity analysis camera. Work sampling for the establishment of allowances. Work sampling for the determination of machine utilization. Work sampling in establishing indirect and direct labor standards.

Automation. Methods improvements on indirect and expense work. Indirect labor standards. Factors affecting indirect and expense standards. Basic queuing theory equations. Monte Carlo simulation. Expense standards. Supervisory standards. Standard data on indirect and expense labor. Universal indirect standards. Advantages of work standards on indirect work.

Making the follow-up. METHODS OF ESTABLISHING STANDARDS: Stopwatch time study. Predetermined motion time data systems. Standard data, formulas, and queuing methods. Work sampling. PURPOSES OF STANDARDS: A basis for wage incentive plans. A common denominator in comparing various methods. A means for securing an efficient layout of the available space. A means for determining plant capacity. A basis for purchasing new equipment. A basis for balancing the working force with the available work. Improving production control. The accurate control and determination of labor costs. Requisite for standard cost methods. As a basis for budgetary control. As a basis for the supervisory bonus. Quality standards are enforced. Personnel standards are raised. Problems of management are simplified. Service to customers is bettered.

Data processing and the establishment of standards. Advantages of methods and standards automation. Approach to automation through data processing. The work measurement system. A computer program for indirect labor standards. Using the computer to calculate standards from standard data. Some typical computer programs.

Direct financial plans. Indirect financial plans. Plans other than financial. Classification of direct wage financial plans. Plans where the employee participates in all the gain above standard. Plans where the employee shares the gains with the employer. The unions' attitudes toward wage incentives. Prerequisites for a sound wage incentive plan. Design for a sound wage incentive plan. The motivation for incentive effort. Reasons for incentive plan failures. Administration of the wage incentive system.

Individual plant methods training programs. Training in methods and time study. Developing creativity. Decision-making methods and processes. Labor relations and work measurement. Union objectives. Employee reactions. The human approach. Research in methods, time study, and wage payment. Methods. Time study. Job evaluation.

1

Methods, time study, and wage payment today

The importance of productivity

The only way a business or enterprise can grow and increase its profitability is by increasing its productivity. And the fundamental tool that results in increased productivity is the tool of methods, time study, and wage payment. Of the total cost of the typical metal products manufacturing enterprise, 15 percent is direct labor, 40 percent direct material, and 45 percent overhead. It should be clearly understood that all aspects of a business or industry—sales, finance, production, engineering, cost, maintenance, and management—provide fertile areas for the application of methods, time study, and sound wage payment. Too often, only the production function is considered when applying methods, standards, and wage payment. Important as the production function is, it should be remembered that other aspects of the enterprise also contribute substantially to the cost of operation and are equally valid areas for the application of cost improvement techniques. In the field of sales, for example, modern information retrieval methods will usually introduce significant savings, product quotas for specific territories provide a base or standard that individual salesman will endeavor to exceed, and payment for results will always result in above-standard performance.

Since the field of production within manufacturing industries utilizes the greatest number of young men and women in methods, time study, and wage payment work, this text will treat that field in more detail than any other. However, it should be remembered that the philosophies and techniques of methods, time study, and wage payment are equally applicable in nonmanufacturing industries. For example, they may be readily employed in such service sectors as hospitals, government, and transportation. Wherever men, materials, and facilities interact to obtain some objective, productivity can be improved through the intelligent application of the principles of methods, time study, and wage payment.

The areas of opportunity existing in the field of production for students enrolled in engineering, industrial management, business administration, industrial psychology, and labor-management relations are: (1) work measurement, (2) work methods, (3) production engineering, (4) manufacturing analysis and control, (5) facilities planning, (6) wage administration, (7) safety, (8) production and inventory control, and (9) quality control. Other position areas, such as personnel or industrial relations, cost, and budgeting, are closely related to, and dependent upon, the production group. These areas of opportunity are not confined to manufacturing industries. They exist and are equally important in such enterprises as department stores, hotels, educational institutions, hospitals, and airlines.

The production section of an industry may well be called the heart of that industry, and once the activity of this section is interrupted, the whole industry ceases to be productive. The production department includes the methods engineering, time study, and wage payment activity, which offers the young technical graduate one of the most satisfying fields of endeavor.

It is in the production department that material to produce is requisitioned and controlled; the sequence of operations and methods determined; tools ordered; time values assigned; work scheduled, dispatched, and followed up; and customers kept satisfied. Training in this field demonstrates how production is accomplished, where it is done, when it is performed, and how long it takes to do it. A background including such training will prove invaluable, whether one's ultimate objective is in sales, production, or cost.

If the production department is considered the heart of an industrial enterprise, the *methods, time study, and wage payment* activity is the heart of the production group. Here, more than in any other place, it is determined whether a product is going to be produced on a competitive basis. Here is where initiative and ingenuity are used to develop efficient tooling, man and machine relationships, and work stations on new jobs in advance of production, thus assuring that the product will stand the test of facing stiff competition. Here is where creativity is continually used to improve existing methods and to help assure the company of leadership in its product line. In this activity good labor relations may be maintained through establishing fair labor standards, or may be impeded by setting one inequitable rate.

Methods, time study, and wage payment offer real challenges. Industries with competent engineers, business administrators, industrial relations personnel, specially trained supervisors, and psychologists carrying out methods, time study, and wage payment techniques are inevitably better able to meet competition and better equipped to operate profitably.

The objective of the manufacturing manager is to produce a quality product, on schedule, at the lowest possible cost, with a minimum of capital investment, and with a maximum of employee satisfaction. The quality control manager centers his or her objectives on the control of quality so that engineering specifications are maintained and customers are kept satisfied by the quality level. The production control manager is principally interested in establishing and maintaining production schedules with due regard both for customer needs and for the favorable economics obtainable with careful scheduling. The manager of methods, time study, and wage payment is mostly concerned with combining the lowest possible production cost and maximum employee satisfaction. The maintenance manager is primarily concerned with minimizing facility downtime because of unscheduled breakdowns and repairs. Figure 1–1 illustrates the relationship of the manager of the methods, time study, and wage payment department to the staff and line departments under the general manager.

FIGURE 1–1
Typical organization chart showing the influence of methods, time study, and wage payment on the operation of the enterprise

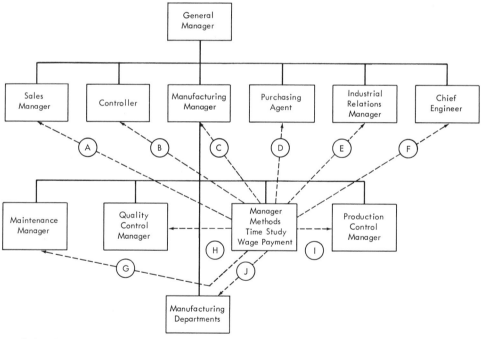

A–Cost is largely determined by manufacturing methods.
B–Time standards are the bases of standard costs.
C–Standards (direct and indirect) provide the bases for measuring the performance of production departments.
D–Time is a common denominator for comparing competitive equipment and supplies.
E–Good labor relations are maintained with equitable standards and fair base rates.
F–Methods and processes strongly influence product designs.
G–Standards provide the bases for preventive maintenance.
H–Standards enforce quality.
I–Scheduling is based on time standards.
J–Methods and standards provide how the work is to be done and how long it will take.

The scope of methods engineering and time study

The field of methods engineering and time study includes designing, creating, and selecting the best manufacturing methods, processes, tools, equipment, and skills to manufacture a product after working drawings have been released by the product engineering section. The best method must then be interfaced with the best skills available so that an efficient man-machine relationship exists. Once the complete method has been established, the responsibility of determining the time required to produce the product falls within the scope of this work. Also included is the responsibility of following through to see that predetermined standards are met and that workers are adequately compensated for their output.

This procedure includes defining the problem related to expected cost, breaking the job down into operations, analyzing each operation to determine the most economical manufacturing procedures for the quantity involved, applying proper time values, and then following through to assure that the prescribed method is put into operation. Figure 1–2 illustrates the opportunities for reducing manufacturing time through the application of methods engineering and time study.

FIGURE 1–2
Opportunities for savings through the application of methods engineering and time study

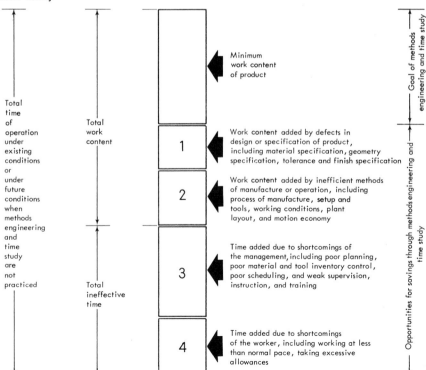

Methods engineering

The terms *operation analysis, work simplification,* and *methods engineering* are frequently used synonymously. In most cases the person is referring to a technique for increasing the production per unit of time and, consequently, reducing the unit cost. However, *methods engineering,* as defined in this text, entails analysis work at two different times during the history of a product. Initially, the methods engineer is responsible for designing and developing the various work centers where the product will be produced. Secondly, he or she continually restudies the work center to find a better way to produce the product. The more thorough the methods study made during the planning stages, the less the necessity for additional methods studies during the life of the product.

In developing the work center to produce the product, the methods engineer should follow a systematic procedure. This will include:

1. Gathering all the facts related to the design, such as drawings, quantity requirements, and delivery requirements.
2. Listing all facts in orderly form. The development of process charts is recommended.
3. Making an analysis. Consider the primary approaches to operation analysis and the principles of motion study.
4. Developing a method.
5. Presenting the method.
6. Installing the work center.
7. Developing a job analysis of the work center.
8. Establishing time standards at the work center.
9. Following up the method.

When methods studies are made to improve the existing method of operation, experience has shown that to achieve the maximum returns, a systematic procedure similar to that advocated for designing the initial work center should be followed. The Westinghouse Electric Corporation, in its Operation Analysis program, advocates the following steps for assuring the most favorable results:

1. Make a preliminary survey.
2. Determine the extent of analysis justified.
3. Develop process charts.
4. Investigate the approaches to operation analysis.
5. Make motion study when justified.
6. Compare the old and the new method.
7. Present the new method.
8. Check the installation of the new method.
9. Correct time values.
10. Follow up the new method.

Actually, methods engineering includes all of these steps.

Methods engineering can be defined as the systematic procedure for subjecting all direct and indirect operations to close scrutiny in order to introduce improvements that will make work easier to perform and will allow work to be done in less time and with less investment per unit. Thus, the real objective of methods engineering is profit improvement.

Time study

Time study involves the technique of establishing an allowed time standard to perform a given task, based upon measurement of the work content of the prescribed method, with due allowance for fatigue and for personal and unavoidable delays. The time study analyst has several techniques that can be used to establish a standard: stopwatch time study, standard data, fundamental motion data, work sampling, and estimates based upon historical data. Each of these techniques has application under certain conditions. The time study analyst must know when it is best to use a certain technique and must then use that technique judiciously and correctly.

A close association exists between the functions of the time study analyst and those of the methods engineer. Although the objectives of the two differ, a good time study analyst is a good methods engineer, since his or her position will include methods engineering as a basic component.

To be assured that the prescribed method is the best, the time study engineer frequently assumes the role of a methods engineer. In small industries these two activities are often handled by the same individual. Note that establishing time values is a step in the systematic procedure of developing new work centers and improving methods related to existing work centers.

Wage payment

The wage payment function, similarly, is closely associated with the time study and methods sections of the production activity. In many companies, and particularly in smaller enterprises, the wage payment activity is performed by the same group responsible for the methods and standards work. In general, the wage payment activity is performed by the same group responsible for conducting job evaluations and maintaining the wage payment plan or plans so that they function smoothly.

Job evaluation is a technique for equitably determining the relative worth of the different work assignments within an organization. It is this technique that results in establishing fair base rates for the different work assignments. In general, job evaluation methodologies give consideration to what the employee brings to the job in the form of education, experience, and special skills, and to what the job takes from him or her from the standpoint of physical and mental effort. Responsibility is a third important factor that is always considered in effective job evaluation.

Because of the nature of a given enterprise, it may be desirable to have two, or even three, entirely different wage payment plans in effect (daywork, piecework, group incentives), and the administration of these plans falls on the wage payment group.

Production control, plant layout, purchasing, cost accounting and control, and process and product design are additional areas closely related to both the methods and the standards functions. All of these areas depend on time and cost data, facts, and procedures of operation from the methods and standards department to operate effectively. These relationships are briefly discussed in Chapter 23.

Objectives of methods, time study, and wage payment

The principal objectives of methods, time study, and wage payment are to increase productivity and lower unit cost, thus allowing more goods to be produced for more people. The ability to produce more for less will result in more jobs for more people for a greater number of hours per year. It is only through the intelligent application of the principles of methods, time study, and wage payment that there can be more producers of goods and services while, at the same time, the purchasing potential of all consumers is increased. It is through the exercise of these principles that unemployment and relief rolls can be minimized, thus reducing the spiraling cost of economic support to nonproducers.

Corollaries that apply to the principal objectives are to:

1. Minimize the time required to perform tasks.
2. Conserve resources and minimize costs by specifying the most appropriate direct and indirect materials for the production of goods and services.
3. Produce with a concern for the availability of power.
4. Provide an increasingly reliable and high-quality product.
5. Maximize the safety, health, and well-being of all employees.
6. Produce with an increasing concern to protect our environment.
7. Follow a humane program of management.

TEXT QUESTIONS

1. What job opportunities exist in the general field of production?
2. What is the scope of methods engineering?
3. What activities are considered the key links to the production group within a manufacturing enterprise?
4. What is meant by the terms *operation analysis, work simplification,* and *methods engineering?*

5. What steps has the Westinghouse Electric Corporation advocated to assure real savings during a methods improvement program?

6. What is the function of the time study department?

7. Is it possible for one enterprise to have more than one type of wage payment plan? Explain.

8. What is the principal objective of methods engineering?

9. What four broad opportunities are there for savings through methods and time study?

10. Explain in detail what a "work center" encompasses.

11. What is the function of job evaluation?

12. What three considerations are included in a successful job evaluation plan?

13. What functions within a manufacturing organization depend on the methods, standards, and wage payment department for data? Explain.

GENERAL QUESTIONS

1. How do well-organized methods, time study, and wage payment procedures benefit the company?

2. Show the relationships between time study and methods engineering. Explain each fully.

3. Discuss the reasoning behind the statement, "A good time study man is a good methods engineer."

4. Comment on the general responsibility of the wage payment group.

5. Why does the purchasing department need data and information from the methods, time study, and wage payment department? Give several examples.

6. Why is job evaluation a part of the wage payment function?

7. What is meant by fundamental motion data?

8. Based on your reading of Chapter 1, what percentage savings do you estimate is possible in a hospital which has never practiced methods engineering and time study? Do you feel that this is a realistic estimate?

9. Explain why, on the average, in the typical metal manufacturing plant only 15 percent of total cost is direct labor cost.

10. Contact the following service industries in your community and find out what use is being made of professional competence in methods, standards, and wage payment:

    a. A hospital.
    b. A high school.
    c. A post office.
    d. A police department.
    e. A bus service.

PROBLEMS

1. In the XYZ hospital, management has been charging $90 per day for a semiprivate room. An analysis of present costs revealed the following:

Direct labor . . . . $5.50 per hour
Materials . . . . $1.50 per patient/day
Indirect costs . . . . $2.00 per 100 square feet/day
Hospital room occupancy . . . . . . 80 percent

The typical patient utilizes 8.2 hours of direct labor help per 24 hours. The average semiprivate room is 14 feet × 20 feet.

Was the $90 per day (for full occupancy) an adequate figure? What dollar charge would you recommend? Explain how you would initiate a cost reduction program.

2. In the Dorben Department Store, the local union and management entered into an agreement on the installation of work measurement standards and base rates determined by job evaluation. After the job evaluation was made, the following job classes and money rates were established, based upon the job points noted:

Job class	Job points	Money base rate per hour
A	100	$2.25
B	250	2.50
C	400	3.00
D	550	3.40
E	700	4.00
F	850	4.75

Because of the wide point range associated with each job class, the union representatives asked for the establishment of five additional job classes based on these job points: 175, 325, 475, 625, 775.

What money rates per hour should be assigned to the five new job classes? (Hint: The original plan plotting points against money was parabolic.)

SELECTED REFERENCES

Vaughn, Richard G. *Introduction to Industrial Engineering.* Ames Iowa: Iowa State University Press, 1967.

Ackoff, Russell L. *Redesigning the Future.* New York: John Wiley & Sons, Inc., 1974.

Blanchard, Benjamin S. *Logistics Engineering and Management.* Englewood Cliffs, N.J.: Prentice-Hall, Inc., 1974.

Riggs, James L., and Kalbaugh, A. James. *The Art of Management: Principles and Practices and Student Involvement Guide.* New York: McGraw-Hill Book Co., 1974.

Salvendy, Gauriel, and Seymour, W. Douglas. *Prediction and Development of Industrial Work Performance.* New York: John Wiley & Sons, Inc., 1973.

2

The work of Taylor

Frederick W. Taylor is generally conceded to be the father of modern time study in this country. However, time studies were conducted in Europe many years before the time of Taylor. In 1760, a Frenchman, Perronet, made extensive time studies on the manufacture of No. 6 common pins, and arrived at a standard of 494 per hour. Sixty years later an English economist, Charles Babbage, made time studies on No. 11 common pins, and as a result of these studies determined that one pound (5,546 pins) should be produced in 7.6892 hours.[1]

Taylor began his time study work in 1881 while associated with the Midvale Steel Company in Philadelphia. After 12 years' work, he evolved a system which was based upon the "task" idea. Here Taylor proposed that the work of each employee be planned out by the management at least one day in advance, and that each man receive complete written instructions describing his task in detail and noting the means to be used in accomplishing it. Each job was to have a standard time which was to be fixed after time studies had been made by experts. This time was to be based upon the work possibilities of a first-rate man who, after being instructed, was able to do the work regularly. In the timing process, Taylor advocated breaking up the work assignment into small divisions of effort known as "elements." These were timed individually, and their collective values were used to determine the allowed time of the task.

In June 1895, Taylor presented his findings and recommendations at a Detroit meeting of the American Society of Mechanical Engineers. His paper was received without enthusiasm because many of the engineers present interpreted his findings to be a new piece rate system rather than a technique for analyzing work and improving methods.

[1] Charles Babbage, *On the Economy of Machinery and Manufactures,* 1832.

The distaste for piecework that prevailed in the minds of many of the engineers of the time can well be appreciated. Piecework standards were then established by supervisors' estimates, and, at best, these were far from being accurate or consistent. Both management and employees were rightfully skeptical of piece rates based upon the foreman's guess. Management looked upon the rates with doubt, in view of the possibility that the foreman would make a conservative estimate so as to protect the performance of his department. The worker, because of unfortunate past experiences, was concerned over any rate established merely by judgment and guess, since the rate vitally affected his earnings.

Then, in June 1903, Taylor presented his famous paper, "Shop Management," at the Saratoga meeting of the A.S.M.E. In that paper he gave the elements of the mechanism of scientific management as follows:

Time study, with the implements and methods for properly making it.

Functional, or divided, foremanship, with its superiority to the old-fashioned single foreman.

The standardization of all tools and implements used in the plant, and also of the acts or movements of workers for each class of work.

The desirability of a planning room or department.

The "exception principle" in management.

The use of slide rules and similar timesaving implements.

Instruction cards for the worker.

The task idea in management, accompanied by a large bonus for the successful performance of the task.

The "differential rate."

Mnemonic systems for classifying manufactured products as well as the implements used in manufacturing.

A routing system.

A modern cost system.

Taylor's "Shop Management" technique was well received by many factory managers, and with modifications it resulted in many satisfactory installations.[2]

At this time the country was going through an unprecedented inflationary period. The word *efficiency* became passé, and most businesses and industries were looking for new ideas that would improve their performance. The railroad industry also felt the need to substantially increase shipping rates to cover general cost increases. Louis Brandeis, who at

---

[2] In 1917, C. Bertrand Thompson reported on the record of 113 plants which had installed "scientific management." Of these, 59 considered their installations completely successful, 20 partly successful, and 34 failures. C. Bertrand Thompson, *The Taylor System of Scientific Management* (Chicago: A. W. Shaw Co., 1917).

that time represented the eastern business associations, contended that the railroads did not deserve or, in fact, need the increase because they had been remiss in not introducing the new "science of management" into their industry. Brandeis claimed that the railroad companies could save $1 million a day by introducing the techniques advocated by Taylor. Thus, it was Brandeis and the Eastern Rate Case (as the hearing came to be known) that first introduced Taylor's concepts as "scientific management."

At this time many men without the qualifications of Taylor, Barth, Merrick, and other early pioneers, but eager to make a name for themselves in this new field, established themselves as "efficiency experts" and endeavored to install scientific management programs in industry. Here they encountered a natural resistance to change from employees, and since they were not equipped to handle problems of human relations, they met with great difficulty. Anxious to make a good showing and equipped with only a pseudoscientific knowledge, they generally established rates that were too difficult to meet. The situation would become so acute that management would be obliged to discontinue the whole program in order to continue operation.

In other instances, factory managers would allow the establishment of time standards by the foreman, and, as has been pointed out, this was seldom satisfactory.

Then, too, once standards were established, many factory managers of that time, interested primarly in the reduction of labor costs, would unscrupuously cut rates if some employee made what the employer felt was too much money. The result was harder work at the same, and sometimes less, take-home pay. Naturally, violent worker reaction resulted.

These developments spread in spite of the many favorable installations started by Taylor. At the Watertown Arsenal, labor objected to the new time study system to such an extent that in 1910 the Interstate Commerce Commission started an investigation of time study. Several derogatory reports on the subject influenced Congress in 1913 to add a rider to the government appropriation bill, stipulating that no part of the appropriation should be made available for the pay of any person engaged in time study work. This restriction applied to the government-operated plants where government funds were used to pay the employees.

The Military Establishment Appropriation Act, 1947 (Public Law 515, 79th Congress), and the Navy Department Appropriation Act, 1947 (Public Law 492, 79th Congress), provide as follows:

Sec. 2. No part of the appropriation made in this Act shall be available for the salary or pay of any officer, manager, superintendent, foreman or other person having charge of the work of any employee of the United States Government while making or causing to be made with a stopwatch, or other time-measuring device, a time study of any job of any such employee between the starting and completion thereof, or of the movements of any such empolyee while engaged upon such work; nor shall any part of the appropriation made in this Act be available to pay any premiums or bonus or cash reward to any employee in addition to his regular wages, except as may be otherwise authorized in this Act.

Finally, in July 1947, the House of Representatives passed a bill which allowed the War Department to use time study, and in 1949, the prohibition against using stopwatches was dropped from appropriation language, so that now no restriction of time study practice prevails.

Motion study

Frank B. Gilbreth was the founder of the modern motion study technique. This may be defined as the study of the body motions used in performing an operation, with the thought of improving the operation by eliminating unnecessary motions and simplifying necessary motions, and then establishing the most favorable motion sequence for maximum efficiency.

Gilbreth originally introduced his ideas and philosophies into the bricklayer's trade where he was employed. After introducing methods improvements through motion study and operator training, he was able to increase the average number of bricks laid to 350 per man per hour. Prior to Gilbreth's studies, 120 bricks per man per hour was considered a satisfactory rate of performance.

Frank Gilbreth and his wife, Lillian Gilbreth, more than anyone else, were responsible for industry's recognition of the importance of a minute study of body motions to its ability to increase production, reduce fatigue, and instruct operators in the best method of performing an operation.

Frank Gilbreth, with assistance from his wife, also developed the moving-picture technique for studying motions, which has since been applied in many walks of life. In industry, this technique is known as micromotion study, but the study of movements through the aid of the slow motion moving picture is by no means confined to industrial applications. The world of sports finds it invaluable as a training tool to show the development of form and skill.

The Gilbreths also developed the cyclegraphic and chronocyclegraphic analysis techniques for studying the motion paths made by an operator. The cyclegraphic method involves attaching a small electric light bulb to the finger or hand or part of the body being studied and then photographing the motion while the operator is performing the operation. The resulting picture gives a permanent record of the motion pattern employed and can be analyzed for possible improvement.

The chronocyclegraph is similar to the cyclegraph, but in its use the electric circuit is interrupted regularly, causing the light to flash. Thus, instead of showing solid lines of the motion patterns, as with the cyclegraph, the resulting photograph will show short dashes of light spaced in proportion to the speed of the body motion being photographed. Consequently, with the chronocyclegraph it is possible to compute velocity, acceleration, and deceleration as well as study body motions.

Early contemporaries

Carl G. Barth, an associate of Frederick W. Taylor, developed a production slide rule for determining the most efficient combination of speeds and feeds for cutting metals of various hardnesses, considering depth of cut, size of tool, and life of tool.

Barth is also noted for the work he did in determining allowances. He investigated the number of foot-pounds of work a man could do in a day. He then developed a rule that for a certain push or pull on a man's arms it is possible for him to be under load for a certain percentage of the day.

Harrington Emerson applied scientific methods to work on the Santa Fe Railroad and wrote a book, *Twelve Principles of Efficiency,* in which he made an effort to inform management of procedures for efficient operation. He reorganized the company, integrated its shop procedures, installed standard costs and a bonus plan, and transferred its accounting work to Hollerith tabulating machines. This effort resulted in annual savings in excess of $1.5 million.

It was Emerson who coined the term *efficiency engineering.* His ideal was efficiency everywhere and in everything. His philosophy of efficiency as a basis for operation in all fields of endeavor first appeared in 1908 in *Engineering Magazine.* When, in 1911, Emerson enlarged his ideas in *Twelve Principles of Efficiency,* this volume was perhaps the most comprehensive guide to good management.

In 1917, Henry Laurence Gantt developed simple graphs that would measure performance while visually showing projected schedules. This production control tool was enthusiastically adopted by the shipbuilding industry during World War I. This tool made it possible for the first time to compare actual performance against the original plan, and to adjust daily schedules in accordance with capacity, backlog, and customer requirements.

Gantt is also known for his invention of the task and bonus wage system. This was developed in 1901, after he spent six years as Taylor's right-hand man at the Midvale and Bethlehem Steel Companies. Gantt's wage payment system rewarded the worker for above-standard performance and eliminated any penalty for failure. Perhaps more important than the deviation from Taylor's advocacy of penalizing the below-standard operator was offering the foreman a bonus for every one of his workers who performed above standard. Gantt made it obvious that scientific management could and should be much more than an inhuman "speedup" of labor. Gantt proclaimed, "We do not approve of foremen 'cussing' their men, but we do approve of their showing the men how the work is to be done."

Morris L. Cooke, former director of the Department of Public Works in Philadelphia, made an effort to bring the principles of scientific management into city governments. In 1940, Cooke and Philip Murray, a past president of the CIO, published "Organized Labor and Production," in which they brought out that the goal of both labor and management is "Optimum Productivity." This they defined as "the highest possible balanced output of goods and services that management and labor skills can produce, equitably shared and consistent with a rational conservation of human and physical resources."

After Taylor retired, Dwight V. Merrick started a study of unit times, and these were published in the *American Machinist,* edited by L. P. Alford. Merrick, with the assistance of Carl Barth, developed a technique for determining allowances on a rational basis. Merrick is also known for his multiple piece rate wage payment plan in which he recommends three graded piece rates.

Motion and time study received added stimulus during World War II when Franklin D. Roosevelt, through the Department of Labor, advocated establishing standards, from which increases in production resulted. On November 11, 1945, Regional War Labor Board III (for Pennsylvania, southern New Jersey, Maryland, Delaware, and the District of Columbia) issued a memorandum stating the policy of the War Labor Board on incentive proposals. Sections I, II, and IV are reproduced since they contain matter pertinent to standards and wage incentives.

I—*General Considerations Applicable to All Incentive Proposals*

1. The expected effect of an incentive plan should be an increase in the present production per man-hour without increasing the unit labor cost in the plant, department, or job affected.
2. The proposal should not be in effect merely as a means of giving a general increase in wages, nor should it result in wage decreases.
3. The plan should offer more pay only for more output.
4. If a union has bargaining rights for workers affected, the plan in all of its details should be collectively bargained.
5. No incentive wage plan should be proposed as a substitute for carrying out the responsibilities of both the management and employees.
6. No incentive plan should be put into operation, even if the money is held rather than advanced to the workers, until the approval of the War Labor Board has been secured.

II—*Establishing Incentive Rates for a Specific Production Operation*

When incentive rates are proposed for a specific production job or job classification the following principles apply:
1. Where feasible the operation should be *time studied* carefully to set the production standard. Results of this time study in as much detail as possible should appear in the application. If a time study is impracticable, the application should show why.

2. If a time study is impracticable, the production standard may be based upon *records of past production* provided: (*a*) The records for an appropriate production period are submitted with the application; (*b*) The applicant establishes that the period is representative and that the present product, methods, prospective volume of work, and work force are comparable to those existing during the particular period of past production; (*c*) Exceptionally high or low production figures for short periods are satisfactorily explained in the application.

3. The production standard should be a quantity of output *higher than has been attained previously* by the average worker, or at least, higher than has been attained customarily. If the production standard is below an amount previously attained at some one or more times, the application should explain fully the reasons therefor.

IV—Plant-wide Incentive Plans

Because plant-wide incentive plans are comparatively new to American industry and because the effects of such plans on worker efficiency and production are difficult to predict, the Regional Board is not adopting a position on them at this time, but will consider each case on its individual merits. The general considerations set forth in I above are applicable here.

Organizations

Since 1911, there has been an organized effort to keep industry abreast of the latest developments in the techniques inaugurated by Taylor and Gilbreth. Technical organizations have contributed much toward bringing the science of time study, motion study, work simplification, and methods engineering up to present-day standards.

In 1911, the Conference of Scientific Management, under the leadership of Morris L. Cooke and Harlow S. Persons, was started at the Amos Tuck School of Dartmouth College.

In 1912, the Society to Promote the Science of Management was organized. It was renamed the Taylor Society in 1915.

The Society of Industrial Engineers was organized in 1917 by men interested in production methods.

The American Management Association was formed in 1922 by groups interested in training personnel through so-called Corporation Training Schools. Today its main purpose is "to advance the understanding of the principles, policies, practices, and purposes of modern management and of the method of creating and maintaining satisfactory relations in commerce and industry; to work toward the practical solution of current business problems and the development of the science of management." Annually the A.M.A. presents the Gantt Memorial Medal for the most distinguished contribution to industrial management as a service to the community.

The Society for the Advancement of Management (S.A.M.) was organized in 1936 by the merging of the Society of Industrial Engineers and the Taylor Society. This organization has continued to emphasize the importance of time study and methods and wage payment up to the present time. Annually it offers the Taylor key for the outstanding contribution to the advancement of the art and science of management as conceived by Frederick W. Taylor. Also awarded annually is the Gilbreth medal for noteworthy achievement in the field of motion, skill, and fatigue study.

The American Institute of Industrial Engineers has had a rapid growth since its founding at Columbus, Ohio, on September 9, 1948. It is a national technical society of industrial engineers. The purposes of the A.I.I.E. are to maintain the practice of industrial engineering on a professional level; to foster a high degree of integrity among the members of the industrial engineering profession; to encourage and assist education and research in areas of interest to the industrial engineer; to promote the interchange of ideas and information among members of the industrial engineering profession; to serve the public interest by identifying persons qualified to practice as industrial engineers; and to promote the professional registration of industrial engineers. Within the A.I.I.E. is the Work Measurement and Methods Engineering Division which is devoted to keeping the membership up-to-date on all facets of this area of work. This division annually gives the Phil Carroll achievement award, established in memory of the division's first director. The criteria for the award specifically state that the recipient's contribution to the profession must apply to work measurement and/or methods engineering.

Present trends

Time and motion study has steadily improved since the 1920s until today it is recognized as a necessary tool for the effective operation of business or industry. The practitioner of the art and science of time and motion study has come to realize the necessity of considering the "human element." No longer is the "cut-and-dried" procedure so characteristic of the "efficiency expert" acceptable. Today, through employee testing and training, consideration is given to the fact that individuals differ in performance potential. It is now recognized that such factors as sex, age, health and well-being, physical size and strength, aptitude, training attitudes and response to motivation have a direct bearing on output. Furthermore, the present-day analyst recognizes that workers object, and rightfully so, to being treated as machines. Workers tend to dislike and fear a purely scientific approach to methods, work measurement, and wage incentives. They inherently dislike any change from their present way of operation. This psychological reaction is not characteristic of factory workers only, but is the normal reaction of all people. Management frequently will reject worthwhile methods innovations because of its reluctance to change. In fact, in the experience of the writer, management has been harder to sell on new ideas than any other group within the plant. After all, it is responsible for the existing methods, and it frequently will defend those methods regardless of the potential savings through change.

Workers tend to fear methods and time study, for they see that this will result in an increase in productivity. To them, this means but one thing: less work and consequently less pay. They must be sold on the fact that they, as consumers, benefit from lower costs, and that broader markets result from lower costs, meaning more work for more people for more weeks of the year.

Some fear of time study today is without a doubt due to unpleasant experiences in the days of the efficiency expert. To many workers, motion and time study is synonymous with the *speedup* or the *stretch-out*. These terms denote using incentives to spur employees to higher levels of output, followed by establishing new levels as normal production, thus forcing the workers to still greater exertions to maintain even their previous earning power. In years past, undoubtedly some shortsighted and unscrupulous managements did resort to this practice.

Even today, most unions oppose the establishment of standards by measurement, the development of hourly base rates by job evaluation, and the application of incentive wage payment. It is the belief of these unions that the time allowed to perform a task and the amount that an employee should be paid represent issues that should be resolved by collective bargaining arrangements.

The practitioner of motion and time study today must use the "humane" approach. He must be well versed in the study of human behavior and be accomplished in the art of communication. He must be a good listener at all times, indicating that he respects the ideas and thinking of others, particularly the worker on the bench. He must give credit where credit is due. In fact, he should get in the habit of always giving the "other person" credit, even if there is some question of the other person's deserving it.

Regardless of his technical knowledge and ability, he will have little success in motion and time study work unless he is competent in dealing with the human element.

A great number of colleges and universities in their industrial engineering curricula are teaching the principles, techniques, and philosophies of this field. Most labor unions are training their representatives in the results and uses of motion and time study. Managements of both small and large industries are embarking on mass training programs, realizing the potentialities of a well-formulated program utilizing this tool.

The Industrial Management Society annually presents a Time and Motion Study and Management Clinic. Here leaders of government, labor, and industry, along with engineers, gather to discuss common problems and to increase their knowledge of what is being done in the area of methods, standards, and wage payment.

Industry, business, and government are in agreement that the untapped potential for increasing productivity is the best hope for dealing with inflation and competition. And the principal key to increased productivity is a continuing application of the principles of methods, standards, and wage payment. Only in this way can more output from men and machines be realized. American labor expects and has the bargaining strength to get a continuing increase in wage levels. American government has pledged itself to an increasingly paternalistic philosophy of providing for the disadvantaged—housing for the poor, medical care for the aged, jobs for the minorities, and so on. In order to accommodate the spiraling costs of labor and government taxes and still stay in business, we must get more from our productive elements—men and machines. The extended application of method, standards, and wage payment to all combinations of men, materials, and machines is assured.

TEXT QUESTIONS

1. Where were time studies originally made and who conducted them?
2. Explain Frederick W. Taylor's principle of functional foremanship.
3. What effect has Congress had on time study?
4. What is meant by motion study, and who is generally conceded to be the founder of the motion study technique?
5. What is Carl G. Barth primarily noted for in the production end of industry?
6. Which organizations are concerned with advancing the ideas of Taylor and the Gilbreths?
7. Was the skepticism of management and labor toward rates established by "efficiency experts" understandable? Why or why not?
8. What psychological reaction is characteristic of workers when methods changes are suggested?
9. Explain the importance of the humanistic approach in methods and time study work.
10. What contributions in the area of motion study were made by the Gilbreths?
11. What was Emerson's philosophy of efficiency?
12. What was Gantt's contribution to motion and time study?
13. Who was Louis Brandeis, and what part did he play in the introduction of scientific management?
14. How did the Military Establishment Appropriation Act, 1947, submerge the development of work measurement methodology?
15. Why have the CIO steelworkers endeavored to extend the application of incentives?

16. Who was the first director of the Work Measurement and Methods Engineering Division of the American Institute of Industrial Engineers?

GENERAL QUESTIONS

1. How is time and motion study used in industry today?
2. Why are labor unions training their representatives in time and motion study techniques?
3. Explain the function of the A.M.A. as compared to that of the A.I.I.E.
4. Why was the government restriction on the use of stopwatches carried through the World War II years?
5. Interview a group of five industrial workers and obtain their opinions on the necessity of motion and time study.

PROBLEMS

1. The industrial engineer of a local township is considering the location of a hospital on a plot 3.2 miles from borough A and 5.8 miles from borough B. If borough A is 4.7 miles from borough B, find the angle at which straight roads joining the two boroughs with the hospital will intersect.

   The population of borough A is three times that of borough B, and a second alternative is to place the new hospital 2.3 miles from borough A and 7.4 miles from borough B. If the cost of a hospital visit is computed at 0.20 per vehicle-mile, and hospital visitors are known to be 25 percent of the population per year, which location would be the most favorable from the standpoint of visitation cost?

SELECTED REFERENCES

Barnes, Ralph M. *Motion and Time Study: Design and Measurement of Work.* 6th ed. New York: John Wiley & Sons, Inc., 1968.

Mundel, Marvin E. *Motion and Time Study: Principles and Practices.* 4th ed. Englewood Cliffs, N.J.: Prentice-Hall, Inc., 1960.

Nadler, Gerald. *Work Design: A Systems Concept.* Rev. ed. Homewood, Ill.: Richard D. Irwin, Inc., 1970.

Salvendy, Gavriel, and Seymour, W. Douglas. *Prediction and Development of Industrial Work Performance.* New York: John Wiley & Sons, Inc., 1973.

3

Graphic tools of the methods analyst

Whether methods work is being used to design a new work center or to improve one already in operation, it is helpful to present in clear, logical form the factual information related to the process. In Chapter 1, it was brought out that after a preliminary survey indicates the desirability of going ahead with a methods study, the first step in methods work is to gather all the necessary facts related to the operation or process. Pertinent information—such as the quantity to be produced, delivery schedules, operational times, facilities, machine capacities, special materials, and special tools—may have an important bearing on the solution of the problem. This chapter will discuss the techniques that best present the factual data.

Every craftsman has the tools he needs to facilitate his performance. Just as the machinist has micrometers and calipers and the patternmaker has chisels and plane, so does the methods analyst have at his disposal tools to help him do a better job in a shorter time. One of the most important tools of the methods engineer is the process chart. A process chart is defined as a graphic presentation of any manufacturing or business process. In methods work, one usually uses eight different types of process charts, each of which has its specific applications. These are:

1. The operation process chart.
2. The flow process chart.
3. The flow diagram.
4. The man and machine process chart.
5. The gang process chart.
6. The operator process chart.
7. The travel chart.
8. The PERT chart.

The operation, flow, and PERT charts and the flow diagram are used principally to present a problem. Usually a problem cannot be adequately solved unless it is properly presented. Consequently, these charts will be discussed at this time. Man and machine, gang, and operator process charts are discussed in Chapter 6. The travel chart is discussed in Chapter 5.

The operation process chart

The operation process chart shows the chronological sequence of all operations, inspections, time allowances, and materials used in a manufacturing or business process—from the arrival of raw material to the packaging of the finished product. It depicts the entrance of all components and subassemblies to the main assembly. Just as a blueprint gives at a glance such design details as fits, tolerances, and specifications, so does an operation process chart give manufacturing and/or business details at a glance.

Before we can improve a design, we must first get a blueprint of the product as it is presently designed. Likewise, before we can improve a manufacturing process, it is well to construct an operation process chart so that we understand the problem fully and will be able to determine what areas afford the most possibilities for improvement. The operation process chart effectively states the problem, and if a problem cannot be stated, it usually cannot be solved.

Constructing the operation process chart

In constructing the operation process chart, two symbols are used: a small circle, usually ⅜ inch in diameter, denotes an operation, and a small square, usually ⅜ inch on a side, denotes an inspection.

An operation takes place when the part being studied is intentionally transformed or when it is being studied or planned prior to performing productive work on it.

An inspection takes place when the part is being examined to determine its conformity to standard.

Before beginning the actual construction of the operation process chart, the analyst should identify it with a title placed at the top of the paper. Usually the identifying information—which would include the part number, drawing number, process description, present or proposed method, date, and name of the person doing the charting—would be headed with the words "Operation Process Chart." Sometimes additional information may be added to identify completely the subject being charted. This may include such items as the chart number, plant, building, and department.

Vertical lines are used to indicate the general flow of the process as work is being accomplished, while horizontal lines feeding into the vertical flow lines are used to introduce material, either purchased or upon which work has been performed during the process. In general, the operation process chart should be constructed so that vertical flow lines and horizontal material lines do not cross. If for some reason it becomes necessary to cross a vertical and a horizontal line, the conventional practice to show that no juncture occurs is to draw a small semicircle in the horizontal line at the point where the vertical line crosses it (see Figure 3–1).

A typical completed operation process chart appears in Figure 3–2.

FIGURE 3–1
Conventional practice to show that no juncture occurs when vertical flow line crosses horizontal material line

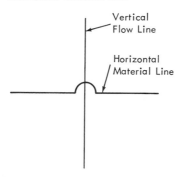

Using the operation process chart

After the analyst has completed the operation process chart, he is ready to use it. He should review each operation and inspection from the standpoint of the primary approaches to operation analysis (see Chapters 4 and 5). In particular, the following approaches apply when studying the operation process chart:

1. The purpose of the operation.
2. The design of the part.
3. Tolerances and specifications.
4. Materials.
5. The process of manufacture.
6. Setup and tools.
7. Working conditions.
8. Plant layout.

The procedure is for the analyst to adopt a questioning attitude on each of these eight criteria influencing the cost and output of the product under study.

The most important question that the analyst should ask when studying the events on the operation process chart is "Why?" Typical questions that should be asked are:

"Why is this operation necessary?"

"Why is this operation performed in this manner?"

"Why are these tolerances this close?"

"Why has this material been specified?"

"Why has this class of operator been assigned to do the work?"

Motion and time study

FIGURE 3–2
Operation process chart illustrating manufacture of telephone stands

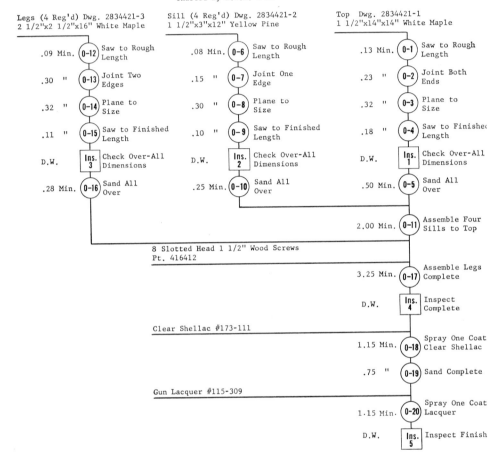

OPERATION PROCESS CHART

Manufacturing Type 2834421 Telephone Stands--Present Method
Part 2834421 Dwg. No. SK2834421
Charted By B.W.N. 4-12-

SUMMARY:

Event	Number	Time
Operations	20	17.58 minutes
Inspections	5	Day work

The analyst should take nothing for granted. He should ask these and other pertinent questions about all phases of the process and then proceed to gather the information to answer the questions so that a better way of doing the work may be introduced.

The question "Why" immediately suggests other questions, including

"What," "How," "Who," "Where," and "When." Thus, the analyst might ask:

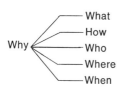

1. "*What* is the purpose of the operation?"
2. "*How* can the operation be performed better?"
3. "*Who* can best perform the operation?"
4. "*Where* could the operation be performed at a lower cost?"
5. "*When* should the operation be performed to give the least amount of material handling?"

For example, in the operation process chart shown in Figure 3–2, the analyst might ask the questions listed in Table 3–1 to determine the practicability of the methods improvements indicated.

By answering such questions, the analyst will become aware of other questions that may lead to improvement. Ideas seem to generate ideas, and the experienced analyst will always arrive at several possibilities for improvement. He must keep an open mind and not let previous disappointments discourage the trial of new ideas.

The completed operation process chart helps the analyst to visualize the present method with all its details, so that new and better procedures may be devised. It shows the analyst what effect a change on a given operation will have on the preceding and subsequent operations. The mere construction of the operation process chart will inevitably suggest possibilities for improvement to the alert analyst.

The operation process chart indicates the general flow of all components entering into a product, and since each step is shown in its proper chronological sequence, the chart is in itself an ideal plant layout. Consequently, methods analysts, plant layout engineers, and persons in related fields will find this tool extremely helpful in making new layouts and in improving existing ones.

The operation process chart is an aid in promoting and explaining a proposed method. Since it gives so much information so clearly, it provides an ideal comparison between two competing solutions.

The flow process chart

In general, the flow process chart contains considerably more detail than does the operation process chart. Consequently, it is not adapted to complicated assemblies as a whole. It is used primarily on one component of an assembly or a system at a time to effect maximum savings in manu-

TABLE 3-1

	Question	Method improvement
1.	Can fixed lengths of 1½" × 14" white maple be purchased at no extra square footage cost?	Eliminate waste ends from lengths that are not multiples of 14".
2.	Can purchased maple boards be secured with edges smooth and parallel?	Eliminate jointing of ends (operation 2).
3.	Can boards be purchased to thickness size and have at least one side planed smooth? If so, how much extra will this cost?	Eliminate planing to size.
4.	Why cannot two boards be stacked and sawed into 14" sections simultaneously?	Reduce time of 0.18 (operation 4).
5.	What percentage of rejects do we have at the first inspection station?	If the percentage is low, perhaps this inspection can be eliminated.
6.	Why should the top of the table be sanded all over?	Eliminate sanding of one side of top and reduce time (operation 5).
7.	Can fixed lengths of 1½" × 3" yellow pine be purchased at no extra square footage cost?	Eliminate waste ends from lengths that are not multiples of 12".
8.	Can purchased yellow pine boards be secured with edges smooth and parallel?	Eliminate jointing of one edge.
9.	Can sill boards be purchased to thickness size and have one side planed smooth? If so, how much extra will this cost?	Eliminate planing to size.
10.	Why cannot two or more boards be stacked and sawed into 14" sections simultaneously?	Reduce time of 0.10 (operation 9).
11.	What percentage of rejects do we have at the first inspection of the sills?	If the percentage is low, perhaps this inspection can be eliminated.
12.	Why is it necessary to sand the sills all over?	Eliminate some sanding and reduce time (operation 10).
13.	Can fixed lengths of 2½" × 2½" white maple be purchased at no extra square footage cost?	Eliminate waste ends from lengths that are not multiples of 16".
14.	Can a smaller size than 2½" × 2½" be used?	Reduce material cost.
15.	Can purchased white maple boards be secured with edges smooth and parallel?	Eliminate jointing of edges.
16.	Can leg boards be purchased to thickness size and have sides planed smooth? If so, how much extra will this cost?	Eliminate planing to size.
17.	Why cannot two or more boards be stacked and sawed into 14" sections simultaneously?	Reduce time (operation 15).
18.	What percentage of rejects do we have at the first inspection of the legs?	If the percentage is low, perhaps this inspection can be eliminated.
19.	Why is it necessary to sand the legs all over?	Eliminate some sanding and reduce time (operation 16).
20.	Could a fixture facilitate assembly of the sills to the top?	Reduce assembly time (operation 11).
21.	Can a sampling inspection be used on the first inspection of the assembly?	Reduce inspection time (operation 4).
22.	Is it necessary to sand after one coat of shellac?	Eliminate operation 19.

facturing or in the procedures applicable to a particular component or sequence of work. The flow process chart is especially valuable in recording hidden costs, such as distances traveled, delays, and temporary storages. Once these nonproductive periods are highlighted, the analyst can take steps for improvement.

In addition to recording operations and inspections, the flow process chart shows all the moves and delays in storage encountered by an item as it goes through the plant. In addition to the operation and inspection symbols used in the construction of operation process charts, several other symbols are used. A small arrow signifies a transportation, which can be defined as the moving of an object from one place to another, except when the movement takes place during the normal course of an operation or an inspection. A large capital *D* indicates a delay. A delay occurs when a part is not permitted to be immediately processed at the next work station. An equilateral triangle standing on its vertex signifies a storage, which occurs when a part is held and protected against unauthorized removal. When it becomes necessary to show a combined activity, such as one operator performing both an operation and an inspection at a work station, then the identifying symbol is a square ⅜ inch on a side with a small circle ⅜ inch in diameter inscribed within the square. Figure 3–3 illustrates how one company uses process chart symbols to identify industrial activity.

Constructing the flow process chart

Like the operation process chart, the flow process chart should be properly identified with a title appearing at the top. It is customary practice to head the identifying information with the words "Flow Process Chart." The identifying information usually shown includes the part number, drawing number, process description, present or proposed method, date and name of the person doing the charting.

Sometimes additional data is valuable to identify completely the job being charted. This may include the plant, building or department, chart number, quantity, and cost information.

Since the flowchart represents only one item rather than an assembly, a neat-appearing chart can be constructed by starting on the top central section of the paper. First, a horizontal material line is drawn, over which are shown the part number and the description as well as the material from which the part is processed. A short vertical flow line (about ¼ inch) is then drawn to the first event symbol, which may be an arrow indicating a transportation from the storeroom. Just to the right of the transportation symbol is recorded a brief description of the move, such as "moved to cutoff saw by material handler." Immediately under this is shown the type of material handling equipment used, if any. For example, "hand two-wheeled truck" or "gasoline-powered fork truck" would identify the equipment used. To the left of the symbol is shown the time required to perform

FIGURE 3–3
Examples of process chart symbols

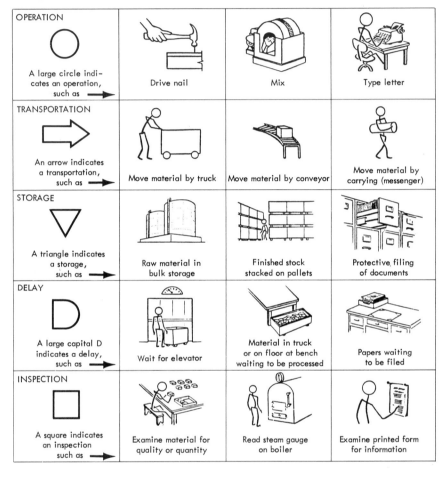

OPERATION			
A large circle indicates an operation, such as ➤	Drive nail	Mix	Type letter
TRANSPORTATION			
An arrow indicates a transportation, such as ➤	Move material by truck	Move material by conveyor	Move material by carrying (messenger)
STORAGE			
A triangle indicates a storage, such as ➤	Raw material in bulk storage	Finished stock stacked on pallets	Protective filing of documents
DELAY			
A large capital D indicates a delay, such as ➤	Wait for elevator	Material in truck or on floor at bench waiting to be processed	Papers waiting to be filed
INSPECTION			
A square indicates an inspection such as ➤	Examine material for quality or quantity	Read steam gauge on boiler	Examine printed form for information

the event, and about one inch still farther to the left, the distance in feet or meters moved is recorded.

This charting procedure is continued by recording all operations, inspections, moves, delays, permanent storages, and temporary storages that occur during the processing of the part. All events are numbered chronologically for reference purposes, using a separate series for each event class.

The transportation symbol is used to indicate the direction of the flow. Thus, when straight-line flow is taking place, the symbol is plotted with the arrow pointing to the right side of the paper. When the process reverses or backtracks, the change of direction is illustrated by plotting the arrow so that it points to the left. If a multifloor building is housing the process, an upward-pointing arrow indicates that the process is moving upward, and a downward-pointing arrow shows the flow of work to be descending.

To determine the distance moved, it is not necessary to measure each move accurately with a tape or a six-foot rule. Usually a sufficiently correct figure will result by counting the number of columns that the material is moved past and then multiplying this number, less one, by the span. Moves of five feet or less are usually not recorded; however, they may be if the analyst feels that they will materially affect the overall cost of the method being plotted.

It is important that all delay and storage times be included on the chart. It is not sufficient to indicate that a delay or storage takes place. Since the longer a part stays in storage or is delayed, the more cost it accumulates, it is important to know how much time it spends at each delay or storage.

The most economical method of determining the duration of delays and storages is to mark several parts with chalk indicating the exact time they went into a storage or were delayed. Then the section is checked periodically to see when the parts marked are brought back into production. By taking a number of cases and recording the elapsed time and then averaging the results, the analyst will obtain sufficiently accurate time values.

Using the flow process chart

The flow process chart, like the operation process chart, is not an end in itself, but merely a means to an end. It is used as a tool of analysis for eliminating the hidden costs of a component. Since the flowchart clearly shows all transportations, delays, and storages, it is helpful in reducing either the quantity or the duration of these elements.

Once the analyst has constructed the flow process chart, he or she uses the questioning approach based on the considerations primary to operation analysis. With the flow process chart, special consideration will be given to:

1. Material handling.
2. Plant layout.
3. Delay time.
4. Storage time.

In all probability, the analyst has already constructed and analyzed an operation process chart of the assembly of which the part under study in the flowchart is a component. The flowchart was constructed of the components of the particular assembly where it was thought that further study of the hidden costs would be practical. If the operations and inspections performed on the component have already been studied, then the analyst would not spend a great deal of time restudying them when analyzing the flowchart. He or she would be more concerned with studying the distance parts must be moved from operation to operation, and with the delays that occur. Of course, if the flow process chart was constructed initially, then all of the primary approaches to operation analysis should be used for studying the events shown on it. The analyst is principally interested in improving the following: first, the time for each operation, inspection, move, delay, and storage; and second, the distance in feet each time the item being plotted is transported.

To eliminate or minimize delay and storage time in order to improve deliveries to customers, and to reduce cost, the analyst should consider these check questions in studying the job:

1. How often is the full amount of material not delivered to the operation?
2. What can be done to schedule materials so that they come in more even quantities?
3. What is the most efficient batch or lot size, or manufacturing quantity?
4. How can schedules be rearranged to provide longer runs?
5. What is the best sequence for scheduling orders to allow for the type of operation, tools required, colors, and so on?
6. What can be done to group similar operations so that they are performed at the same time?
7. How much can downtime and overtime be reduced by improved scheduling?
8. What is the cause of emergency maintenance and rush orders?
9. How much delay and storage time can be saved by running certain products on certain days in order to make schedules more regular?
10. What alternate schedules can be developed in order to use materials most efficiently?
11. Would it be worthwhile to accumulate pickups, deliveries, and shipments?
12. What is the proper department to do the job so that it will be done with the same class of work and save a move, delay, or storage?
13. How much would be saved by doing the job on another shift? At another plant?
14. What is the best and most economical time to run tests and experiments?
15. What information is lacking on orders issued to the factory that may cause a delay or storage?
16. How much time is lost by changing shifts at different hours in related departments?
17. What are the frequent interruptions to the job, and how should they be eliminated?
18. How much time does the employee lose by waiting for or not receiving the proper instructions, blueprints, and specifications?
19. How many holdups are caused by congested aisles?
20. What improvements can be made in locating doorways and aisles and in making aisles that will reduce delays?

Specific check questions that the analyst should use to shorten the distances traveled and reduce material handling time include the following:

1. Is product group technology being practiced to reduce the number of setups and to allow for larger production runs? (Product group technology is the classification of different products into similar geometric configurations and sizes so as to take advantage of the economics in manufacture brought about by large quantities.)
2. Can a facility be economically relocated to reduce the distances traveled?
3. What can be done to reduce the handling of materials?
4. What is the correct equipment for handling materials?
5. How much time is lost in getting materials to and from the work station?
6. Should product grouping be considered rather than process grouping?
7. What can be done to increase the size of the unit of material handled in order to reduce handling, scrap, and downtime?
8. How can elevator service be improved?
9. What can be done about runways and roadways to speed up transportation?
10. What is the proper position in which to place material in order to reduce the amount of handling required by the operator?
11. What use can be made of gravity delivery?

A study of the completed flowchart (Figure 3–4) will familiarize the analyst with all the pertinent details related to the direct and indirect costs of a manufacturing process so that they can be analyzed for improvement. Unless all the facts relating to a method are known, it is difficult to improve that method. Casual inspection of an operation will not provide the information needed to do a thorough job of methods improvement. Since distances are recorded on the flow process chart, the chart is exceptionally valuable in showing how the layout of a plant can be improved. Intelligent use of the flow process chart will result in improvements.

The flow diagram

Although the flow process chart gives most of the pertinent information relative to a manufacturing process, it does not show a pictorial plan of the flow of work. Sometimes this added information is helpful in developing a new method. For example, before a transportation can be shortened, it is necessary to see or visualize where room can be provided to add a facility so that the transportation distance can be diminished. Likewise, it is helpful to visualize potential temporary and permanent storage areas, inspection stations, and work points. The best way to provide this information is to take an existing drawing of the plant layout of the areas involved, and then to sketch in the flow lines indicating the movement of the material from one activity to the next. A pictorial representation of the layout of floors and buildings which shows the location of all activities on the flow process chart is known as a flow diagram.

FIGURE 3–4
Flow process chart

FLOW PROCESS CHART

SUBJECT CHARTED___Shower head face_____CHART NO.__1128_____

DRAWING NO.__BA-14782_____PART NO.__B-14782-2_____CHART OF METHOD__Present____

CHART BEGINS__Bar stock storage_____CHARTED BY____E. Dunnick_____

CHART ENDS___Assembly Department Storeroom_____DATE_9-7__SHEET_1_OF__2__

DIST. IN FEET	UNIT TIME IN MIN.	CHART SYMBOLS	PROCESS DESCRIPTION	DIST. IN FEET	UNIT TIME IN MIN.	CHART SYMBOLS	PROCESS DESCRIPTION
		▽1	In bar stock storage until requisitioned.	60		□6	Waiting press operator.
20	.02	○1	Bars loaded on truck upon receipt of requisition.	100		⇒5	To Bliss 74 1/2 press #16 by operator.
600	.05	⇒1	Extruded rod to air saw #72.		.075	○6	Pierce 6 holes.
15	.02	○2	Bars removed from truck and stored on rack near machine.		120	□7	Waiting drill press operator.
120		□1	Waiting for operation to begin.	50		⇒6	To drill press by operator.
	.077	○3	Saw slug on air saw.		.334	○7	Rough ream and chamfer L. & G. drill press #19.
	30	□2	Waiting for move--man.		30	□8	Waiting drill press operator.
70	.03	⇒2	Slugs to Nat. Maxi-press #8.	20		⇒7	To Avery drill press #21 by operator.
	15	□3	Awaiting forging operation.		.152	○8	Drill three 13/64" holes Avery press #21.
	.234	▢1	Forge (3-man operation) and inspect.	20		○9	Awaiting turret lathe operator.
	10	□4	Waiting for press operator.	60		⇒8	To turret lathe section by operator.
30		⇒3	To press by operator.		.522	○9	Turn stem and face #3 W. & S.
	.061	○4	Trim Flash Bliss 74 1/2 press #16.	60		□10	Awaiting turret lathe operator.
	30	□5	Waiting for pickling operator.	30		⇒9	To adjacent turret lathe operator.
100		⇒4	To pickling tanks by operator.		.648	○10	From outside diameter and face back.
	.007	○5	Pickle (HCL tank).	15		□11	Waiting press operator.

When constructing a flow diagram, the analyst should identify each activity by symbols and numbers corresponding to those appearing on the flow process chart. The direction of flow is indicated by placing small arrows periodically along the flow lines. If it is desirable to show the flow of more than one part, a different color can be used for each part.

FIGURE 3–4 (continued)

FLOW PROCESS CHART

SUBJECT CHARTED __Shower head face_____ CHART NO. __1128__

DRAWING NO. __BA-14782____ PART NO. __B-14782-2_____ CHART OF METHOD __Present____

CHART BEGINS __Bar stock storage_____ CHARTED BY __E. Dunnick____

CHART ENDS __Assembly Department Storeroom_____ DATE __9-7__ SHEET __2__ OF __2__

DIST. IN FEET	UNIT TIME IN MIN.	CHART SYMBOLS	PROCESS DESCRIPTION	DIST. IN FEET	UNIT TIME IN MIN.	CHART SYMBOLS	PROCESS DESCRIPTION
60		▷10	To Bliss 20B press by operator.				
	.097	◯11	Stamp identification Bliss 20B press #21.				
	15	▢12	Waiting next press operator.				
	.167	◯12	Broach six holes to size Bliss 20B press #22				
	60	▢13	Waiting for move--man.				
350	.012	▷11	To inspection by move--man.				
	20	▢14	Waiting for inspector.				
	.05	▢1	Inspect complete (10% spot check).				
75		▷12	To storeroom by inspector.				
		▽2	Stored until requisitioned.				

SUMMARY

EVENT	NUMBER	TIME	DISTANCE
OPERATIONS	13	2.414 min.	
INSPECTIONS	1	.050 min.	
COMBINED ACTIVITY	1	.234 min.	
TRANSPORTATIONS	12		1580 ft.
STORAGES	2	undetermined	
DELAYS	14	605 min.	

FIGURE 3–5
Flow diagram of the old layout of a group of operations on the Garand rifle.
(Shaded section of plant represents the total floor space needed for the revised
layout [Figure 3–6]. This represented a 40 percent savings in floor space.)

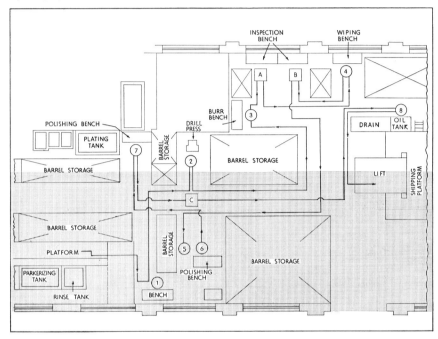

Figure 3–5 illustrates a flow diagram made in conjunction with a flow
process chart to improve the production of the Garand (M1) rifle at
Springfield Armory. This pictorial representation together with the flow
process chart resulted in savings that increased production from 500 rifle
barrels per shift to 3,600—with the same number of employees. Figure
3–6 illustrates the flow diagram of the revised layout.

FIGURE 3–6
Flow diagram of the revised layout of a group of operations on the Garand rifle

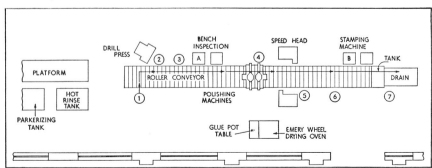

FIGURE 3–7
Network showing critical path (heavy line)

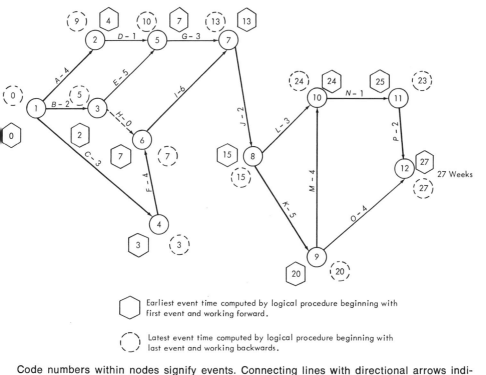

Earliest event time computed by logical procedure beginning with first event and working forward.

Latest event time computed by logical procedure beginning with last event and working backwards.

Code numbers within nodes signify events. Connecting lines with directional arrows indicate operations that are dependent upon prerequisite operations. Time values on the connecting lines represent normal duration in weeks. Hexagonals associated with events show the earliest event time. Dotted circles associated with events show the latest event time.

It can be seen that the flow diagram is a helpful supplement to the flow process chart because it shows up backtracking and areas of possible traffic congestion, and facilitates making an ideal plant layout.

PERT charting

PERT charting is a prognostic planning and control method that graphically portrays the optimum way to attain some predetermined objective, generally in terms of time. Usually the methods analyst is able to use PERT charting to improve scheduling from the standpoint of cost reduction and/or customer satisfaction.

In using PERT for scheduling, the analyst will generally provide two or three time estimates for each activity. If three time estimates are used, they are based upon the following questions:

1. What is the time in which you can expect to complete this activity if everything works out ideally (optimistic estimate)?
2. Under average conditions, what would be the most likely duration of this activity?
3. What is the time required to complete this activity if almost everything goes wrong (pessimistic estimate)?

With these estimates, a probability distribution of the time required to perform the activity can be made.

On the PERT chart, events—represented by nodes—are positions in time showing the start and/or the completion of a particular operation or group of operations. Each operation or group of operations in a department is referred to as an activity and is identified as an arc on the PERT chart. Each arc has attached to it a number representing the time (days, weeks, months) needed to complete the activity. Activities that utilize no time or cost, yet are necessary to maintain a correct sequence, are referred to as dummy activities and are plotted as dotted lines.

The minimum time needed to complete the entire project would correspond to the longest path from the initial node to the final node. In Figure 3–7 the minimum time needed to complete the project would be the longest path from node 1 to node 12. The longest path is termed the critical path since it is this path that establishes the minimum project time. There is always at least one such path through any project. However, more than one path can reflect the minimum time needed. This is the meaning behind the concept of critical paths.

It should be evident that activities which do not lie on the critical path

TABLE 3–2

Cost and time values to perform a variety of activities under normal and emergency conditions

	Normal		Emergency	
Activities	*Weeks*	*Dollars*	*Weeks*	*Dollars*
A.........	4	4,000	2	6,000
B.........	2	1,200	1	2,500
C.........	3	3,600	2	4,800
D.........	1	1,000	0.5	1,800
E.........	5	6,000	3	8,000
F.........	4	3,200	3	5,000
G.........	3	3,000	2	5,000
H.........	0	0	0	0
I.........	6	7,200	4	8,400
J.........	2	1,600	1	2,000
K.........	5	3,000	3	4,000
L.........	3	3,000	2	4,000
M.........	4	1,600	3	2,000
N.........	1	700	1	700
O.........	4	4,400	2	6,000
P.........	2	1,600	1	2,400

have a certain time flexibility. This time flexibility, or freedom, is referred to as "float." The amount of float is computed by subtracting the normal time from the time available. Thus, the float is the amount of time that a noncritical activity can be lengthened without delaying the project's completion date.

Figure 3–7 illustrates an elementary network portraying the critical path. This path, identified by a heavy line, would involve a duration of 27 weeks. Several methods can be used to shorten the project's duration. The cost of the various alternatives can be estimated. For example, let us assume that the following cost table has been developed and that a linear relation exists between the time and the cost per week.

The cost of various time alternatives can be readily computed:

27-week schedule—normal duration of project. . . . . cost = $22,500

26-week schedule—the least expensive way to gain one week would be to reduce activity M or J by one week for an additional cost of $400. . . . . cost = $22,900

25-week schedule—the least expensive way to gain two weeks would be to reduce activities M and J by one week each for an additional cost of $800. . . . . cost = $23,300

24-week schedule—the least expensive way to gain three weeks would be to reduce activities M, J, and K by one week each for an additional cost of $1,300. . . . . cost = $23,800

23-week schedule—the least expensive way to gain four weeks would be to reduce activities M and J by one week each and activity K by two weeks for an additional cost of $1,800. cost = $24,300

22-week schedule—the least expensive way to gain five weeks would be to reduce activities M and J by one week each, activity K by two weeks, and activity I by one week for an additional cost of $2,400. . . . . cost = $24,900

21-week schedule—the least expensive way to gain six weeks would be to reduce activities M and J by one week each and activities K and I by two weeks each for an additional cost of $3,000. . . . . cost = $25,500

20-week schedule—the least expensive way to gain seven weeks would be to reduce activities M, J, and P by one week each and activities K and I by two weeks each for an additional cost of $3,800.. . . . . cost = $26,300

19-week schedule—the least expensive way to gain eight weeks would be to reduce activities M, J, P, and C by one week each and activities K and I by two weeks each for an additional cost of $5,000.. . . . . cost = $27,500
(Note that a second critical path is now developed through nodes 1, 3, 5, and 7.)

18-week schedule—the least expensive way to gain nine weeks would be to reduce activities M, J, P, C, E, and F by one week each and activities K and I by two weeks each for an additional cost of $7,800. . . . . cost = $30,300
(Note that by shortening the time to 18 weeks, we develop a second critical path.)

Summary

The methods analyst should become familiar with the operation and flow process charts and the flow diagram so that he will be able to use these valuable tools in solving problems. Just as several types of tools are available for a particular job, so several designs of charts can be utilized in solving an engineering problem. However, in determining a specific solution, one chart usually has advantages over another. The analyst should know the specific functions of each of the process charts and only resort to those he needs to help solve his specific problems. In summary, their functions are as follows.

1. OPERATION PROCESS CHART. Used to analyze relations between operations. Good for studying operations and inspections on assemblies involving several components. Helpful for plant layout work.

2. FLOW PROCESS CHART. Used to analyze hidden or indirect costs, such as delay time, storage costs, and material handling costs. Best chart for complete analysis of the manufacture of one component part.

3. FLOW DIAGRAM. Used as a supplement to the flow process chart, especially where considerable floor space is involved in the process. Shows up backtracking and traffic congestion. Necessary tool in making revised plant layouts.

The operation and flow process charts and the flow diagram all have their place in developing improvements. Their correct use will aid in presenting the problem, solving the problem, selling the solution, and installing the solution.

4. PERT CHART. Used as a project scheduling tool. Especially desirable for use in major projects involving relatively long periods of time (six months or more).

TEXT QUESTIONS

1. Who uses process charts?
2. Who is credited as being the original designer of process charts?
3. What does the operation process chart show?
4. What symbols are used in constructing the operation process chart?
5. How are materials introduced into the general flow when constructing the operation process chart?
6. How does the flow process chart differ from the operation process chart?
7. What is the principal purpose of the flow process chart?
8. What symbols are used in constructing the flow process chart?
9. When would you advocate using the flow diagram?
10. How can the flow of several different products be shown on the flow diagram?
11. Why are the operation and flow process charts merely a means to an end?
12. In the construction of the flow process chart, what method can be used to estimate distances moved?
13. How can delay times be determined in the construction of the flow process chart? Storage times?
14. How can the concept of PERT charting save a company money?

GENERAL QUESTIONS

1. What are the limitations of the operation and flow process charts, and of the flow diagram?
2. What relation is there between the flow chart and material handling? Between the flow diagram and plant bottlenecks? Between the operation process chart and material specifications?
3. What is the connection between effective plant layout and the operation process chart?
4. Distinguish between process and product grouping.
5. What is meant by group technology?

PROBLEMS

1. In studying the flowchart of the activities in a bus station, the analyst is seeking to reduce the cost of energy. He has found that electrical energy is costing $0.12 for each kilowatt-hour. A television display of schedules utilizes 0.25 kilowatt. Twenty lamps in the station consume 0.075 kilowatt each. An intercommunication system employs an amplifier in which $\frac{1}{15}$ of the total power supplied to the amplifier is converted to audio power output, and 10 watts of audio output are obtained. Heating averages 5.25 kilowatts over a 12-month period. What is the estimated daily cost of power in this bus station?

SELECTED REFERENCES

Francis, Richard L., and White, John A. *Facility Layout and Location: An Analytical Approach.* Englewood Cliffs, N.J.: Prentice-Hall, Inc., 1974.

Moore, James M. *Plant Layout and Design.* New York: Macmillan Co., 1962.

Muther, Richard. *Systematic Layout Planning.* 2d ed. Boston: Industrial Education Institute, 1961.

Mallick, R. W., and Gaudreau, A. P. *Plant Layout: Planning and Practice.* New York: John Wiley & Sons, Inc., 1951.

4

Operation analysis

Operation analysis is a procedure used by the methods analyst to analyze all productive and nonproductive elements of an operation with the thought of improvement. Methods engineering in itself is concerned with devising methods to increase production per unit of time and reduce unit costs. Operation analysis is in reality a technique for accomplishing the goal of methods engineering.

The operation analysis procedure is just as effective in planning new work centers as it is in improving those already in operation. By using the questioning approach on all facets of the work station, dependent work stations, and the design of the product, an efficient work center will be developed.

Since the improvement of existing operations is a continuing process in industry, this chapter will deal primarily with that process, recognizing, however, that the principles discussed are equally valid and important in planning new work centers. Investigating the approaches to operation analysis is the step immediately following the presentation of facts in the form of the operation or flow process chart. This is the step in which analysis takes place and the various components of the method to be proposed are crystallized.

With foreign competition becoming increasingly keen at the same time that labor and material costs are spiraling, operation analysis has become increasingly important. It is a procedure that can never be regarded as complete. Competition usually necessitates the continuing study of a given product so that the manufacturing processes can be improved and a part of the gains passed on to the consumer in the form of a better product at a reduced selling price. Once this is done by a given plant, competitors invariably introduce similar improvement programs, and it is only a matter of time until they have produced a more salable product at a reduced price. This starts a new cycle in which the given plant again reviews its operations and improves its manufacturing processes, again necessitating improvements in competing plants. Conditions in industry cannot be static;

otherwise, bankruptcy would result. Progress is the only key to continued profitable operation.

Another basic economic law, that of supply and demand, must be considered when savings are effected. The volume of goods consumed is inversely proportional to the selling price. As improvements that inaugurate real savings are developed, the market is broadened through lower selling prices. This has been proven time and time again. It happened with electric lighting, refrigerators, radios, automobiles, television, home air conditioning, motorboats, and many other products. As progress allowed reduction in costs, more people could afford to purchase. The increased volume permitted further economies, which again resulted in lower prices, and lower prices further increased the volume. Each time the selling price of any commodity is reduced by as little as 5 percent, the product is immediately brought within reach of more people's pocketbooks.

Experience has proven that practically all operations can be improved if sufficient study is given to them. Since the procedure of systematic analysis is equally effective in large and small industries, in job shop and mass production, we can safely conclude that operation analysis is applicable in all areas of manufacturing, business, and government. When properly utilized, it can be expected to develop a better method of doing the work by simplifying operational procedures and material handling, and by utilizing equipment more effectively to increase output and reduce unit cost; to ensure quality and reduce defective workmanship; and to facilitate operator enthusiasm by improving working conditions, minimizing operator fatigue, and permitting higher operator earnings.

Approach to operation analysis

Probably one of the most common attitudes of management is that its problems are unique. Consequently, it feels that any new method will be impractical for it. Actually, all work, whether it be clerical, maintenance office, machine, assembly, or general labor, is much the same. The Gilbreths concluded that any work, whether productive or nonproductive, was done by using combinations of 17 basic movements that they called "therbligs." Regardless of what the operation is, when considered in the light of its basic divisions, it will be found to be quite similar to others. For example, the elements of work in driving a car are much like those required to operate a turret lathe; the basic motions employed in dealing a bridge hand are almost identical with certain manual inspection and machine loading elements. The fact that all work is similar in many respects verifies the principle that if methods can be improved in one plant, opportunities exist for methods improvements in all plants.

History repeats itself over and over in accounts of the unwillingness of people to accept something new. Iron ships, steamboats, automobiles, airplanes, radios, telephones, and television are just a few examples of products that are commonplace today; yet at one time all of them were considered impractical by most people. Today, the average person still scoffs at the idea of interplanetary travel and of atomic power plants for private vehicles.

Anyone engaged in methods work is well aware of the natural, inherent resistance to change prevailing in all people's minds, regardless of their level in an organization. Overcoming this resistance to change is one of the major obstacles in the path of an improvement program. Managers will continually make statements to the effect that "it might have worked at the _____ plant, but our operation is quite different." Supervisors will say, "It can't work here," and even the operator will bluntly declare, "It won't work." Selling the whole organization on "It will work here" is a never-ending job for the methods analyst. A person who is successful in this type of work never accepts anything as being right just because it exists today or has been done this way for years. Instead, he questions, probes, investigates, and finally, after all angles have been considered, he makes his decision for the moment. Always he is conscious that this method may be satisfactory today, but that it will not be tomorrow, for there is always a better way.

In order to reduce the resistance to change that the methods analyst knows is characteristic of all people, he will endeavor to establish an atmosphere of participation, understanding, and friendliness. He will recognize the knowledge other people have of their jobs and will solicit their assistance in making improvements. He will see to it that all communication channels are kept open and are utilized, so that all personnel who are affected by contemplated changes are kept informed. He will provide information freely, and will openly discuss facts related to the investigation; thus, he will obtain confidence rather than mistrust and suspicion. Above all else, he will maintain an enthusiastic attitude toward improvement.

Operation analysis procedure

As outlined in Chapter 1, the use of a systematic procedure is invaluable if real savings are to be realized. The first step is to get all information related to the anticipated volume of the work. In order to determine how much time and effort should be devoted toward improving the present method or planning the new job, it is necessary to determine the expected volume, chance of repeat business, life of the job, chance for design changes, and labor content of the job. If the job promises to be quite active, a more detailed study will be justified than would otherwise be the case.

Once an estimate is made of the quantity, job life, and labor content, then all factual manufacturing information should be collected. This information would include all operations, facilities used to perform the operations, and operational times; all moves or transportations, facilities used for transportation, and transportation distances; all inspections, inspection facilities, and inspection times; all storages, storage facilities, and time spent in storage; all vendor operations together with their prices; and all drawings and design specifications. After all this information affecting cost is gathered, it must be presented in a form suitable for study. One of the most effective ways of doing this is through the flow process chart (see Chapter 3). This chart graphically presents all manufacturing information, much as a blueprint shows all design information.

Upon completing the flow process chart, the analyst reviews the problem with thought toward improvement. Up to this point he has merely stated the problem, which is a necessary preliminary to solving it. One of the most common techniques used by methods men is to prepare a check sheet for asking questions on every activity shown on the flowchart. Typical questions are: Is this operation necessary? Can the operation be better performed another way? Can it be combined with some other operation? Are tolerances closer than necessary? Can a more economical material be used? Can better material handling be incorporated?

A typical checklist of pertinent questions is shown in Figure 4–1. The figure also illustrates how this form was used to make a cost reduction study on an electric blanket control knob shaft. By redesigning the shaft so that it could be economically produced as a die casting rather than a screw machine part, the factory cost was reduced from $68.75 per thousand pieces to $17.19 per thousand pieces.

The analyst asks and answers all related questions on the check sheet for all steps appearing on the flowchart. This procedure invariably evolves efficient ways of performing the work. As these ideas develop, it is well to record them immediately so that they will not be forgotten. It is also wise to include sketches at this time. Usually the analyst is surprised at the numerous inefficiencies that prevail and will have little trouble in compiling many improvement possibilities. One improvement usually leads to another. The analyst must have an open mind and creative ability if he or she is to be a real success in this type of work. The check sheet is also useful in giving methods training to factory foremen and superintendents. Thought-provoking questions, when intelligently used, help factory supervisors to develop constructive ideas. The check sheet serves as an outline which can be referred to by the discussion leader handling the methods training.

The ten primary approaches to operation analysis

There are 10 primary approaches to operation analysis which should be used when studying the flowchart of the prevailing method. These approaches are:

1. Purpose of the operation.
2. Design of the part.
3. Tolerances and specifications.
4. Material.
5. Process of manufacture.
6. Setup and tools.
7. Working conditions.
8. Material handling.
9. Plant layout.
10. Principles of motion economy.

FIGURE 4-1

Date _____ 9/15 _____ Dept. _____ 11 _____ Dwg. _____ 18-4612 _____ Sub. _____ 2 _____

Mould _____ Die _____ Style _____ Item _____ 2 _____

Pattern _____ Ins. Spec. _____ C _____ L. Spec. _____ Sub. _____

Part Description _____ Blanket control knob shaft _____

Operation _____ Turn, groove, drill, tap, knurl, thread, cut-off _____ Operator _____ Blazer _____

DETERMINE AND DESCRIBE

1. **PURPOSE OF OPERATION**

   To form contours of 3/8" S.A.E. 1112 rod on automatic screw machine
   to achieve drawing specifications.

2. **COMPLETE LIST OF ALL OPERATIONS PERFORMED ON PART**

No.	Description	Work Sta.	Dept.
1.	Turn, groove, drill, tap, knurl, thread, cut-off	B. & S.	11
2.	Burr	Bench	12
3.	Inspect 1%	Bench	18
4.			
5.			
6.			
7.			
8.			
9.			
10.			

3. **INSPECTION REQUIREMENTS**

   a—Of previous oprn.

   b—Of this oprn. Yes. Perhaps S.O.C. will reduce amount of inspection.

DETAILS OF ANALYSIS

Can purpose be accomplished
better otherwise?
Yes - by die casting

Can oprn. being analyzed
be eliminated? _No_

be combined with another?
_No_

be performed during idle
period of another? _Yes, by machine coupling._

Is sequence of oprns. best
possible? _Yes_

Should oprn. be done in another dept. to save cost or
handling? _Perhaps can be purchased outside at a savings._

Are tolerance, allowance,
finish and other requirements
necessary?

4. MATERIAL   Zinc base die cast metal would be less expensive.

Cutting compounds and other supply materials

Consider size, suitability, straightness, and condition.

Can cheaper material be substituted?

5. MATERIAL HANDLING

a—Brought by 4 wheel truck to automatics

b—Removed by hand 2 wheel trucks

c—Handled at work station by

Should crane, gravity conveyors, totepans, or special trucks be used?

Consider layout with respect to distance moved. *Perhaps gravity to burring station.*

6. SET-UP (Accompany description with sketches if necessary)

This is satisfactory as being done.

How are dwgs. and tools secured?

Can set-up be improved?

Trial pieces.

Machine Adjustments.

Tools

Suitable?
Provided?
  Ratchet Tools
  Power Tools
  Spl. Purpose Tools
  Jigs, Vises
  Special Clamps
Fixtures
  Multiple
  Duplicate

a—Tool Equipment
  Present

Suggestions   Redesign part to be made as zinc base die casting rather than S.A.E. 1112 screw machine part.

ANALYSIS SHEET

FIGURE 4-1 (continued)

7. CONSIDER THE FOLLOWING POSSIBILITIES.

1. Install gravity delivery chutes.
2. Use drop delivery
3. Compare methods if more than one operator is working on same job.
4. Provide correct chair for operator.
5. Improve jigs or fixtures by providing ejectors, quick-acting clamps, etc.
6. Use foot operated mechanisms.
7. Arrange for two handed operation.
8. Arrange tools and parts within normal working area.
9. Change layout to eliminate back tracking and to permit coupling of machines.
10. Utilize all improvements developed for other jobs.

8. WORKING CONDITIONS.

Generally satisfactory.

a—Other Conditions

9. METHOD (Accompany with sketches or Process Charts if necessary.)

a—Before Analysis and Motion Study.

	RECOMMENDED ACTION
7.	*Yes, to accumulate for tumbling.*
	___
	___
	___
	___
	___
8.	Light *O.K.*
	Heat *O.K.*
	Ventilation, Fumes *O.K.*
	Drinking Fountains *O.K.*
	Wash Rooms *O.K.*
	Safety Aspects *O.K.*
	Design of Part *O.K.*
	Clerical Work Required (to fill out time cards, etc.) *O.K.*
	Probability of Delays *O.K.*
	Probable Mfg. Quantities *O.K.*
9.	Arrangement of Work Area
	Placement of Tools.

Control knob shaft designed as screw machine part.

b—After Analysis and Motion Study

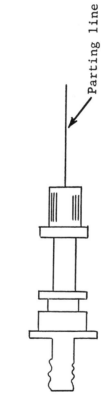

Parting line

Control knob redesigned as die cast part. Threads on left-hand extension cover only 50 percent of periphery; likewise knurl on right end extension on half of periphery thus allowing piece to be easily removed from die.

See Supplementary Report Entitled *Die cast.*
Control shaft.

Date

OBSERVER ____ R. Guild ____   APPROVED BY ____ R. Hussey ____

When all these approaches are used in studying each individual operation, attention is focused on the items most likely to produce improvement. All these approaches will not be applicable to each activity appearing on the flowchart, but usually more than one should be considered. The method of analysis recommended is to take each step in the present method individually and to analyze it with a specific approach toward improvement clearly in mind, considering all the key points of analysis. Then the same procedure should be followed on the succeeding operations, inspections, moves, storages, and so on, shown on the chart. After each element has been thus analyzed, it is well to consider the product being studied as a whole rather than in light of its elemental components, and to reconsider all points of analysis with thought toward overall improvement possibilities. There are usually unlimited opportunities for methods improvement in every plant. The best technique for developing the maximum savings is the careful study of individual and collective operations as outlined. Wherever this procedure has been followed by competent engineers, beneficial results have always been realized.

PURPOSE OF THE OPERATION

Probably the most important of the 10 points of operation analysis used to improve an existing method or plan a new job is the purpose of the operation. A cardinal rule that the analyst should observe is to try to eliminate or combine an operation before improving it. In my experience, as much as 25 percent of the operations being performed by American industry can be eliminated if sufficient study is given to the design and process.

Far too much unnecessary work is done today. In many instances, the job or the process should not be simplified or improved, but should be eliminated entirely. If a job can be eliminated, there is no need to spend money on installing an improved method. No interruption or delay is caused while an improved method is being developed, tested, and installed. It is not necessary to train new operators for the new method. The problem of resistance to change is minimized when a job or activity found to be unnecessary is eliminated. The best way to simplify an operation is to devise some way to get the same or better results at no cost at all.

Unnecessary operations are frequently a result of improper planning when the job was first set up. Once a standard routine is established, it is difficult to make a change, even if such a change would allow eliminating a portion of the work and make the job easier. When new jobs are planned, the planner usually will include an extra operation if there is any possibility that the product would be rejected without the extra work. For example, if there is some question of whether to take two or three cuts in turning a steel shaft to maintain a 40-microinch finish, invariably the planner will specify three cuts, even though the proper maintenance of cutting tools, supplemented by ideal feeds and speeds, would allow the job to be done with two cuts. Likewise, if there is some question of the ability of a drill press to hold a 0.005-inch tolerance on a ¼-inch drilled hole, the planner, when setting up the job, will call for a reaming operation. Actually, the drilling operation would be adequate if all variable factors (speed, feed, cutting lubricant, drill size) were controlled.

Many times, an unnecessary operation may develop because of the improper performance of a previous operation. A second operation must be done to touch up or make acceptable the work done by the first operation. In one plant, for example, armatures were previously spray painted in a fixture, making it impossible to cover the bottom of the armature with paint because the fixture shielded the bottom from the spray blast. It was necessary to touch up the armature bottoms after spray painting. A study of the job resulted in a redesigned fixture that held the armature and still allowed complete coverage. The new fixture permitted seven armatures to be spray painted simultaneously, while the old method called for spray painting one at a time. Thus, it was possible to eliminate the touch-up operation by considering that an unnecessary operation may develop because of the improper performance of a previous operation.

In manufacturing large gears, it was necessary to introduce a hand-scraping and lapping operation to remove waves in the teeth after they had been hobbed. An investigation disclosed that contraction and expansion brought about by temperature changes in the course of the day were responsible for the waviness in the teeth surface. By enclosing the whole unit and installing an air-conditioning system within the enclosure, proper temperature was maintained during the whole day. The waviness disappeared immediately, and it was no longer necessary to continue the hand-scraping and lapping operations.

Sometimes, unnecessary operations develop because an operation was introduced to facilitate an operation that followed. When wiring commutators, for example, it was thought necessary to twist each pair of wires to keep the correct pair of wires together and to increase the strength of the wires. However, it was found that the correct pair of wires could be placed in the proper slot without twisting. Also, observation revealed that twisting caused the wires to be unequal in length, and that the uneven tension in the twisted wires weakened rather than strengthened them. It was discovered that winding the wires was unnecessary when someone asked: "Can a change in assembly eliminate the need for a previous operation?"

In endeavoring to eliminate operations, the analyst should consider the question: Is an additional operation justified by savings that it will effect in a subsequent operation? For example, a brush holder was originally planned so that two holes were drilled and tapped in each holder.

FIGURE 4–2A
Painted armature as removed from old fixture and
as removed from improved fixture

FIGURE 4–2B
Armature in spray-painting fixture allowing complete
coverage of the armature bottom

One hole was drilled and tapped in the bottom; another was drilled and
tapped in the top. The brush holder was assembled so that a holder with a
tapped hole in the bottom and a holder with a tapped hole in the top
were used alternately. Since all parts had tapped holes in both the bottom
and the top, it was merely a matter of positioning the parts for assembly.
By questioning the need for having both tapped holes in each piece, it was
discovered to be more economical to have two parts—one with the hole
tapped in the top, the other with the hole tapped in the bottom. The two
parts were then assembled alternately. One drilled and tapped hole was

FIGURE 4-3
Large gear completely housed to maintain constant temperature during hobbing operation

eliminated from each brush holder because someone questioned the necessity of performing this operation.

Again, an unnecessary operation may develop because it was thought that it would give the product greater sales appeal. One company originally used cast brass nameplates for various lines of its products. Although the brass nameplate was attractive, it was found that the appearance could be kept attractive and the cost notably reduced by using an etched steel nameplate instead.

To produce a smooth joint in electric fan guards, it was thought necessary to perform a coining operation where the ends of the wire cage were joined by butt welding. By slightly modifying the design, it was possible to place the joined wire ends directly behind the crossbars of the fan guard, making it impossible to see the joined ends. Since the butt-welded ends could not be seen, it was no longer necessary to perform the coining operation. This design made it difficult to reach the butt-welded portion, so there was no chance of scratching hands on the rough welded area.

To eliminate, combine, or shorten each operation, the analyst should ask and answer the following question: "Does an outside supplier's tooling enable him to perform the operation more economically?" Ball bearings purchased from an outside vendor had to be packed in grease prior to

FIGURE 4–4
Twisting wires during commutator winding operation was found to be unnecessary because the correct pair of wires could be placed in the proper slot without twisting

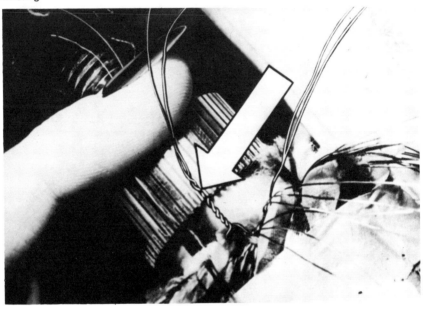

assembly. A study of bearing vendors revealed that "sealed-for-life" bearings could be purchased from another supplier at lower cost.

The above examples highlight the desirability of establishing the purpose of each operation before endeavoring to improve the operation. Once the necessity of the operation has been determined, then the remaining nine approaches to operation analysis should be considered to determine how it can be improved.

DESIGN OF THE PART

The methods engineer is often inclined to feel that once a design has been accepted his or her only recourse is to plan its manufacture in the most economical way. Granted, it is usually difficult to introduce even a slight change in design; still, a good methods analyst should review every design for possible improvements. Designs are not permanent—they can be changed; and if improvement is the result, and the activity of the job is significant, then the change should be made.

To improve the design, the analyst should keep in mind the following pointers for lower cost designs:

1. Reduce the number of parts, simplifying the design.
2. Reduce the number of operations and the length of travel in manu-

facturing by joining the parts better and by making the machining and assembly easier.

3. Utilize a better material.
4. Rely for accuracy upon "key" operations rather than upon series of closely held limits.

These general pointers should be kept in mind as consideration is given to design analysis on each component and each subassembly.

The General Electric Company has summarized the ideas shown in Table 4–1 to be kept in mind for developing minimum cost designs.

One manufacturer always used cast-iron brackets on its motors. A methods analyst questioned this design, and this led to the redesign of the bracket, making it from welded sheet steel. The new design was not only stronger, lighter, and more eye-appealing, but it was also less expensive to produce.

A similar design improvement was made in the construction of conduit boxes. Originally, they were built of cast iron, while the improved design, making a stronger, neater, lighter, and less expensive conduit box, was fabricated from sheet steel.

In another instance, a brass cam switch used in control equipment was made as a brass die casting. By slightly altering the design, the less expensive process of extruding was utilized. The extruded sections were cut to the desired length to produce the cam switch.

These methods improvements were brought about by considering a better material in an effort to improve the design.

Attaching nuts were originally arc welded to transformer cases. This proved to be a slow, costly operation. Furthermore, the resulting design had an unsightly appearance due to the overflow and spatter from the arc welding process. By projection welding nuts to the transformer case, not only were time and money saved, but the resulting design had considerably more sales appeal.

A similar improvement was brought about by changing from arc welding to spot welding to join mounting brackets to resistor tubes.

Design simplification through the better joining of parts was also used in assembling terminal clips to their mating conductors. It had been the practice to turn the end of the clip up to form a socket. This socket was filled with solder, and the wire conductor was then tinned, inserted into the solder-filled socket, and held there until the solder solidified. By altering the design to call for resistance welding the clip to the wire conductor, both the forming and the dipping operations were eliminated.

A motor cover thumbscrew was originally made of three components: head, pin, and screw. These components were assembled by joining the head to the screw with the pin. A much less costly thumbscrew was developed by redesigning the part for an automatic screw machine that was able to turn out the part complete with no secondary operations. A

TABLE 4-1

Castings
1. Eliminate dry sand (baked-sand) cores.
2. Minimize depth to obtain flatter castings.
3. Use minimum weight consistent with sufficient thickness to cast without chilling.
4. Choose simple forms.
5. Symmetrical forms produce uniform shrinkage.
6. Liberal radii—no sharp corners.
7. If surfaces are to be accurate with relation to each other, they should be in the same part of the pattern, if possible.
8. Locate parting lines so that they will not affect looks and utility, and need not be ground smooth.
9. Specify multiple patterns instead of single ones.
10. Metal patterns are preferable to wood.
11. Permanent molds instead of metal patterns.

Moldings
1. Eliminate inserts from parts.
2. Design molds with smallest number of parts.
3. Use simple shapes.
4. Locate flash lines so that the flash does not need to be filed and polished.
5. Minimum weight.

Punchings
1. Punched parts instead of molded, cast, machined, or fabricated parts.
2. "Nestable" punchings to economize on material.
3. Holes requiring accurate relation to each other to be made by the same die.
4. Design to use coil stock.
5. Punchings designed to have minimum sheared length and maximum die strength with fewest die moves.

Formed parts
1. Drawn parts instead of spun, welded, or forged parts.
2. Shallow draws if possible.
3. Liberal radii on corners.
4. Bent parts instead of drawn.
5. Parts formed of strip or wire instead of punched from sheet.

Fabricated parts
1. Self-tapping screws instead of standard screws.
2. Drive pins instead of standard screws.
3. Rivets instead of screws.
4. Hollow rivets instead of solid rivets.
5. Spot or projection welding instead of riveting.
6. Welding instead of brazing or soldering.
7. Use die castings or molded parts instead of fabricated construction requiring several parts.

Machined parts
1. Use rotary machining processes instead of shaping methods.
2. Use automatic or semiautomatic machining instead of hand-operated.
3. Reduce the number of shoulders.
4. Omit finishes where possible.
5. Use rough finish when satisfactory.
6. Dimension drawings from same point as used by factory in measuring and inspecting.
7. Use centerless grinding instead of between-center grinding.
8. Avoid tapers and formed contours.
9. Allow a radius or undercut at shoulders.

Screw machine parts
1. Eliminate second operation.
2. Use cold-rolled stock.
3. Design for header instead of screw machine.
4. Use rolled threads instead of cut threads.

Welded parts
1. Fabricated construction instead of castings or forgings.
2. Minimum sizes of welds.
3. Welds made in flat position rather than vertical or overhead.
4. Eliminate chamfering edges before welding.
5. Use "burnouts" (torch-cut contours) instead of machined contours.
6. Lay out parts to cut to best advantage from standard rectangular plates and avoid scrap.
7. Use intermittent instead of continuous weld.
8. Design for circular or straight-line welding to use automatic machines.

Treatments and finishes
1. Reduce baking time to minimum.
2. Use air drying instead of baking.
3. Use fewer or thinner coats.
4. Eliminate treatments and finishes entirely.

Assemblies
1. Make assemblies simple.
2. Make assemblies progressive.
3. Make only one assembly and eliminate trial assemblies.
4. Make component parts RIGHT in the first place so that fitting and adjusting will not be required in assembly.

    This means that drawings must be correct, with proper tolerances, and that parts must be made according to drawings.

General
1. Reduce number of parts.
2. Reduce number of operations.
3. Reduce length of travel in manufacturing.

(From A. M. _____, *American Machinist* reference sheets, 12th ed. (New York: McGraw-Hill Publishing Co.).

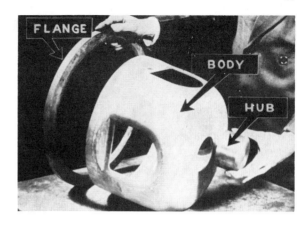

FIGURE 4–5
Improved design of
conduit box (new design
fabricated from sheet
steel)

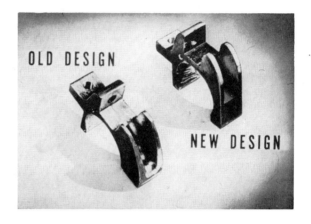

FIGURE 4–6A
Redesigned brass cam
switch allowing part to
be made from extrusion

FIGURE 4–6B
New design shown cut to
length from section of
brass extrusion

FIGURE 4–7
Projection welding of nuts to surface provides neat-appearing and inexpensive method of joining

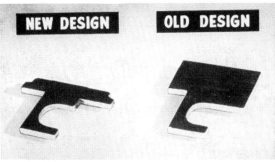

FIGURE 4–8A
Mounting bracket changed so that spot welding rather than arc welding may be used

FIGURE 4–8B
Illustration of bracket arc welded to resistor tube indicates the unsightly appearance necessitated by the old method

simplified design resulted in a less expensive part that still met all service and operating requirements.

Design simplification can be applied to the process as well as to the product. For example, Figure 4–11 illustrates a simple improvement in machine design which eliminated a division of work between operating and maintenance people.

The old design, bolted, required the services of a maintenance mechanic every time it was necessary to change the yarn guide because

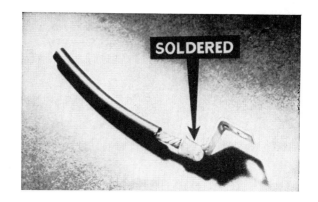

FIGURE 4–9A
Old design required
tinned and soldered
connection

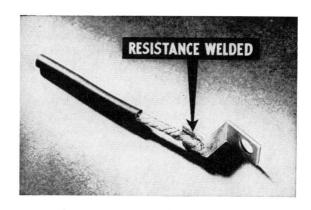

FIGURE 4–9B
Improved design
illustrating resistance-
welded clip to wire
conductor

FIGURE 4–10A
Old thumbscrew was
in three parts

FIGURE 4–10B
Improved one-piece
design of motor cover
thumbscrew

FIGURE 4–11
Old, bolted and new, snap-in design of rotating yarn guide

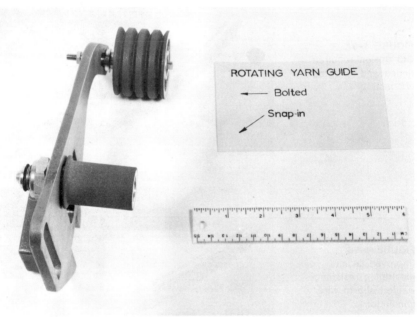

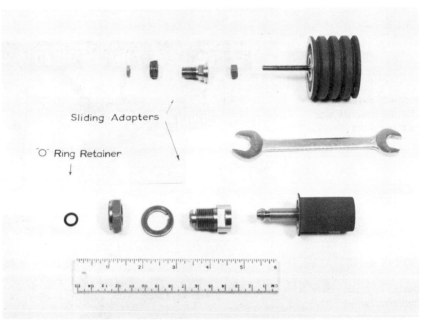

E. I. du Pont de Nemours & Co., Inc.

tools were required. The improved design, snap-in, permits the operator to pull out the guide and replace it with a new one. This new design permits more timely changing of guides; without tools or coordination with another group, the operator can change a guide as soon as it is found to be defective. In addition, it takes less time to make the yarn guide change.

The foregoing examples are characteristic of the possibilities for improvement when the design of the part is investigated. It is always wise to check the design for improvement because design changes can be valuable. To be able to recognize good design, the methods engineer should have had some training and practical experience in this area. Good designs do not just happen; they are the result of broad experience and creative thinking, tempered with an appreciation of cost.

TOLERANCES AND SPECIFICATIONS

The third of the 10 points of operation analysis to be considered in an improvement program is tolerances and specifications. Many times this point is considered in part when reviewing the design. This, however, is usually not adequate, and it is well to consider tolerances and specifications independent of the other approaches to operation analysis.

Designers have a natural tendency to incorporate more rigid specifications than necessary when developing a product. This is brought about by one or both of two reasons: (1) a lack of appreciation for the elements of cost, and (2) the feeling that it is necessary to specify closer tolerances and specifications than are actually needed to have the manufacturing departments produce to the required tolerance range.

The methods analyst should be versed in the aspects of cost and be fully aware of what unnecessarily close specifications can do to the selling price. Figure 4–12 illustrates the relationship between cost and machining tolerance. If it appears that designers are being needlessly "tight" in establishing tolerances and specifications, it may be wise for management to embark on a training program in which the economies of specifications are clearly presented.

The analyst must be alert for too liberal, as well as too restrictive, specifications. Closing up a tolerance often facilitates an assembly operation or some other subsequent step. This may be economically sound, even though it increases the time required to perform the present operation. In this connection, the analyst should recognize that the overall tolerance is equal to the square root of the sum of the squares of the individual tolerances comprised by the overall tolerance.

Beyond the principle of operating economies through correct tolerances and specifications is the consideration of establishing the ideal inspection procedure. Invariably, inspection is a verification of the quantity, quality, dimensions, and performance. Inspection in all these areas

Motion and time study

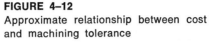

FIGURE 4–12
Approximate relationship between cost
and machining tolerance

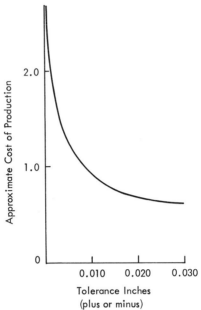

can usually be performed by numerous methods and techniques. One way is usually best, not only from the standpoint of quality control, but also from the time and cost consideration. The methods analyst should question the present way with thought toward improvement.

The possibilities for installing spot inspection, lot-by-lot inspection, or statistical quality control should be considered.

Spot inspection is a periodic check to assure that established standards are being realized. For example, a nonprecision blanking and piercing operation set up on a punch press should have a spot inspection to assure the maintenance of size and the absence of burrs. As the die begins to wear or deficiencies in the material being worked begin to show up, the spot inspection would catch the trouble in time to make the necessary changes without generating an appreciable number of rejects.

Lot-by-lot inspection is a sampling procedure in which a sample is examined to determine the quality of the production run or lot. The size of the sample selected depends on the allowable percentage defective and the size of the production lot under check.

Statistical quality control is an analytic tool employed to control the desired quality level of the process.

If a 100 percent inspection is being encountered, it is well to consider the possibility of spot inspection or lot-by-lot inspection. A 100 percent inspection is the process of inspecting every unit of product and rejecting the defective units. Experience has shown that this type of inspection does not assure a perfect product. The monotony of screening tends to create fatigue, thus lowering operation attention. There is always a good chance that the inspector will pass some defective parts as well as reject good parts. Because a perfect product is not assured under 100 percent inspection, acceptable quality may be realized from the considerably more economical methods of either lot-by-lot or spot inspection.

Usually an elaborate quality control procedure is not justified if the product does not require close tolerances, if its quality is easily checked, and if the generation of defective work is unlikely.

Just as there are several mechanical methods for checking a 0.500 to 0.502 inches reamed hole, so there are several overall policy procedures that can be adopted as a means of control. The methods analyst must be alert and well grounded in the various techniques so that he can make sound recommendations for improvement.

In one shop a certain automatic polishing operation was found to have a normal rejection quantity of 1 percent. It would have been quite expensive to subject each lot of polished goods to 100 percent inspection. It was decided, at an appreciable saving, to consider 1 percent the allowable percentage defective, even though this quantity of defective material would go through to plating and finishing only to be thrown out in the final inspection before shipment.

The analyst must always be aware that the reputation of and demand for his company's product depend on the care taken in establishing correct specifications and in maintaining them. Once quality standards are established, no deviations will be permitted. In general, tolerances and specifications can be investigated by asking these three questions: "Are they absolutely correct?" "Are ideal inspection methods and inspection procedures being used?" and "Are modern quality control techniques being exercised?"

One manufacturer's drawings called for a 0.0005-inch tolerance on a shoulder ring for a DC motor shaft. The original specifications called for a 1.8105 to 1.8110 inches tolerance on the inside diameter. It was thought necessary to hold this close tolerance because the shoulder ring was shrunk on the motor shaft.

An investigation revealed that a 0.003-inch tolerance was adequate for the shrink fit. The drawing was immediately changed to specify a 1.809 to 1.812 inches inside diameter. A reaming operation was saved because someone questioned the absolute necessity of a close tolerance.

In another instance, it was possible to introduce an automatic control on an external cylindrical grinder. Formerly, it was necessary to manually feed the grinding wheel to the required stop. Each piece had to be carefully inspected to assure the maintenance of tolerance on the outside diameter. With the automatic machine control, the feed is tripped and

FIGURE 4–13
Automatic control attached to external cylindrical grinder

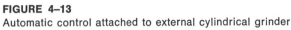

the piece released upon completion of the infeed. The automatic machine control freed the operator to do other work because some methods analyst endeavored to develop an ideal inspection procedure.

In an automatic screw machine shop, it was thought necessary to inspect 100 percent parts coming off the machine because of the critical tolerance requirements. However, it developed that adequate quality control would be maintained by inspecting every sixth piece. This sampling procedure allowed one inspector to service three machines rather than one machine.

By investigating tolerances and specifications and taking action when desirable, the costs of inspection will be reduced, scrap minimized, repair costs diminished, and quality kept high.

MATERIAL

One of the first questions an engineer considers when he is designing a new product is, "What material shall I use?" Since the ability to choose the right material is based upon the engineer's knowledge of materials, and since it is difficult to choose the correct material because of the great variety of materials available, many times it is possible and practical to incorporate a better and more economical material into an existing design.

The methods analyst should keep in mind six considerations relative

FIGURE 4–14A
Rectangular Micarta
bars were formerly used
to separate windings of
transformer coils

FIGURE 4–14B
Use of glass tubing
instead of bars provides
substantial savings

FIGURE 4–15A
Porcelain plate was
formerly used to hold
wire leads from
distribution transformers

FIGURE 4–15B
Fullerboard plate makes
economical and
effective substitute

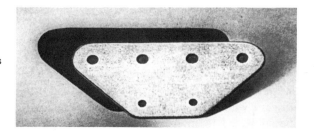

to the direct and indirect materials utilized in a process. These are (1) finding a less expensive material; (2) finding materials that are easier to process; (3) using materials more economically; (4) using salvage materials; (5) using supplies and tools more economically; and (6) standardizing on materials.

Finding a less expensive material

Prices of materials can be compared by their basic costs. Monthly publications available to all engineers summarize the approximate cost per pound of steel sheets, bars, and plates, and the cost of cast iron, cast steel, cast aluminum, cast bronze, and other basic materials (see Table 4–2). These costs can be used as anchor points from which to judge the application of new materials. Developments of new processes for producing and refining materials are continually taking place. Thus, a material that was not competitive in price yesterday may be so today.

In one company, Micarta spacer bars were used between windings of transformer coils. They were placed to separate the windings and permit the circulation of air between the windings. An investigation revealed that glass tubing could be substituted for the Micarta bars at a saving. Not only was the glass tubing less expensive, but it met service requirements better because the glass could withstand higher temperatures. Furthermore, the hollow tubing permitted more air circulation than did the solid Micarta bars.

A less expensive material that still met service requirements was also used in the production of distribution transformers. Originally, a porcelain plate was used to separate and hold the wire leads coming out of the transformers. A fullerboard plate was substituted which stood up just as well in service, yet was considerably less expensive.

The methods analyst should keep in mind that such items as valves, relays, air cylinders, transformers, pipe fittings, bearings, couplings, chains, hinges, hardware, and motors can usually be purchased at less cost than the cost at which they can be manufactured.

Today, many plastics are competing effectively with metals and wood. Aluminum alloys are competing with zinc as well as many ferrous materials in die casting. In the last five years, the price of some materials has more than doubled while that of other materials has risen only 20 percent.

Finding a material that is easier to process

One material is usually more readily processed than another. By referring to handbook data on the physical properties, it is usually easy to discern which material will react most favorably to the processes it must be subjected to in its conversion from raw material to finished product.

FIGURE 4–16
Packing Ivory bars before methods change

For example, machinability varies inversely with hardness, and hardness usually varies directly with strength.

By keeping in mind the thought of selecting a material that is easy to process, one methods analyst was able to show real savings when he changed the procedure for producing stainless steel bearing shells. Originally, they were made by drilling and boring to size cut lengths of stainless steel bar stock. By specifying stainless steel tubing as a material source, material was conserved, the rate of production was increased, and the cost of manufacture was reduced.

Using material more economically

A fertile field for analysis is the possibility of using material more economically. If the ratio of scrap material to that actually going into the product is high, then consideration should be given to greater utilization.

In the production of stampings from sheet, if the skeleton seems to

TABLE 4-2

Fabricated form	Raw material costs	Tool and die costs	Optimum lot sizes
Sand castings	9¢ per lb. for iron to 40¢ per lb. for magnesium (at the spout).	Pattern costs are low as compared to making dies and molds.	Wide range—from 100 to 10,000.
Permanent mold castings.	Medium cost—aluminum, copper, magnesium alloys most used—iron and steel sometimes used.	Moderate—less than for die casting but more than for other casting methods.	Large quantities—in the thousands.
Plaster mold castings.	Medium—28¢ aluminum; $1.10 bronze, etc.	Medium—between permanent mold and sand castings.	Short runs—100 to 2,000.
Die castings	Medium—about 15¢ per lb. for zinc; other materials somewhat higher; brass $1.90.	High—more than other casting methods. Typical dies cost $1,000 to $20,000.	Large quantities—1,000 to hundreds of thousands.
Precision castings . .	High generally—the process is best suited to special alloys which are usually costly.	Low if model is available; moderate if model must be made.	Wide range, although best for quantities of from 10 to 1,000.
Drop forgings	Low to moderate—most used steels 25¢ to 30¢ per lb., but alloys up to 70¢ per lb. or more are used; brass $1.50 per lb.	High—dies must be carefully made to withstand heat and impact—$700 to $4,000 or more.	Large quantities—10,000 to 100,000 or more, but used on small quantities when properties are required.
Press forgings	Low to moderate—compares with drop forgings.	High—but usually somewhat lower than dies for drop forgings.	Medium to high production lots.
Upset forgings.	Low to moderate—as with press and drop forgings.	High—often because of large number of impressions or difficult design.	Medium to high lot production best.
Cold headed parts. . .	Low to moderate—15¢ to 18¢ per lb. for steel wire, $1.20 per lb. for brass.	Medium—sometimes to a few hundred dollars.	Large lots are best.
Stampings, formed parts	Low to moderate, ranging from ordinary steel 12¢ to 17¢ per lb., to $1.25 per lb. for brass.	High—$1,500 to $5,000 is common for dies for small parts—more for large parts.	Large lots—over 10,000 best.
Spinnings.	Low to moderate. Steel 14¢ to 22¢ per lb., up to stainless steel sheet 60¢ to 80¢ per lb.	Low—form can cost from $100 up to $1,000.	Quantities under 1,000.

Screw machine parts	Low to medium—steel (12¢ to 16¢ per lb.) most common up to brasses ($2.00 to $2.25), with a few more expensive materials.	Medium—$200 to $1,000 common tooling cost range.	Large lots best—more than 1,000 essential—the higher the quantity the better.
Powder metal parts	Medium to high—powders run 35¢ to $1.50 per lb. for iron; 80 to 90¢ for copper, 80¢ to $1.50 for brass, $1.50 per lb. for nickel and stainless steel.	Medium—usually more than $400, range $400 to $5,000.	Large lots best to amortize die costs—10,000 pieces. However, small runs may be necessary.
Electroformed parts	Low to high, depending upon the material used: iron lowest, chromium highest.	Medium—master mold doesn't wear out. However, due to small lots die costs may be relatively high.	Best where small lots are involved.
Sectional tubing	High—invariably 40¢ per lb. or over; some tubing much higher.	Low—cutting is done with simple tools.	Wide range—good for small lots as well as large quantities.
Cut extrusions	High—usually over 45¢ per lb.; brass 75¢ per lb.	Low—cutting is done with simple tools.	Wide range—useful for large and small quantities.
Impact extrusions	Moderate—mostly used for aluminum, zinc, lead.	Medium—dies often cost less than $400.	Wide range—quantities can vary from thousands to millions.
Welded or brazed assemblies	Low cost materials can be used (carbon, steel, etc.).	Low to moderate—assembly jigs sometimes required.	Small quantities best, although small assemblies can be mass produced at low cost.
Molded plastics	Medium to high—material varies from 25¢ per lb. to over $2 per lb.	Medium to high—depending upon type of mold or die required; from $400 up, generally.	Large lots—over 5,000 pieces best.
Molded nonmetallic	Moderate to high.	Medium—molds are often rather simple.	Wide—because of need for parts of some materials, quantities must be low.

Materials and Methods

FIGURE 4–17
Packing Ivory bars after methods change (small end of box
requires 15 percent less boxboard because of smaller
overlapping flaps)

contain an undue amount of scrap material, it may be possible to go to
the next higher standard width of material and utilize a multiple die. If a
multiple die is used, care should be exercised in the arrangement of the
cuts to assure maximum utilization of material. Figure 4–18 illustrates how
the careful nesting of parts permits maximum utilization of flat stock.

At one time, the Procter & Gamble Company packed bars of Ivory soap
into boxes with the opening on the largest face of the cardboard box.
This operation is shown in Figure 4.16. Flaps on the four edges of the
open case fold over to form a double cover over that part of the box
during packing. As the result of a methods study, consideration was given
to using material more economically. It was found that the bars could be
packed satisfactorily, with no increase in time, by passing them through
an open end. This permitted redesign of the case so that the overlapping
of the flaps was at the smallest face of the case, using approximately 15

FIGURE 4–18
Method of torch cutting heavy gear case side plates (note nesting of the
point and the heel for most effective use of plate)

General Electric Co.

percent less boxboard. The end packing is shown in Figure 4–17. This
change resulted in a substantial reduction in the cost of the container.

Another example of economical use of material is in the compression
molding of plastic parts. By preweighing the amount of material put into
the mold, only the exact amount of material required to fill out the cavity
will be used and excessive flash will be eliminated.

Salvage materials

The possibility of salvaging materials that would otherwise be sold as
scrap should not be overlooked. Sometimes by-products resulting from
the unworked portion or scrap section offer real possibilities for savings.
One manufacturer of stainless steel cooling cabinets had sections of stainless steel four to eight inches wide left as cuttings on the shear. An analysis
brought out a by-product of electric light switch plate covers.

Another manufacturer, after salvaging the steel insert from defective
bonded rubber ringer rolls, was able to utilize the cylindrical hollow

rubber rolls as bumpers for protecting moored motor- and sailboats. Figure 4–19 a & b illustrates how one manufacturer used pieces of circular plate scrap to produce needed components.

If it is not possible to develop a by-product, then scrap materials should be segregated for top scrap prices. Separate bins should be provided in the shop for tool steel, steel, brass, copper, and aluminum; and chip-haulers and floor sweepers should be instructed to keep the scrap segregated.

It is usually wise to salvage such items as electric light bulbs if large quantities are used. The brass socket should be stored in one area, and after breaking and disposing of the glass bulb, the tungsten filament should be removed and stored separately for greatest residual value.

Wooden boxes from incoming shipments should be saved, and the boards sawed to standard lengths for use in making smaller boxes for outgoing shipments. This practice is always economical, and it is now being followed by many large industries as well as by service maintenance centers.

Full use of supplies and tools

Full use of all shop supplies should be encouraged. One manufacturer of dairy equipment introduced the policy that no new welding rod was to be distributed to workers without the return of old tips under two inches long. The cost of welding rods was reduced immediately by more than 15 percent.

It is usually economical to repair by brazing or welding expensive cutting tools, such as broaches, special form tools, and milling cutters. If it has been company practice to discard broken tools of this nature, it would be well to investigate the potential savings of a tool salvage program.

The unworn portion of grinding wheels, emery disks, and so forth, should be checked for possible use in the plant. Such items as gloves and rags should not be discarded once they are soiled. It is less expensive to store the dirty items in containers to await laundering than to replace them.

Waste of material benefits no one. The methods analyst can make a real contribution to his or her company by preventing wastes which today claim about one fifth of our material.

Standardize on materials

The methods analyst should always be alert to the possibility of standardizing materials. An effort should be made to minimize the sizes, shapes, grades, and so on of each material utilized in the production and assembly of products. The following typical economies result from reductions in the

FIGURE 4–19A
Circular piece is scrap from motor frame head fabrication part (this scrap is used to produce gusset support pieces for locomotive platform)

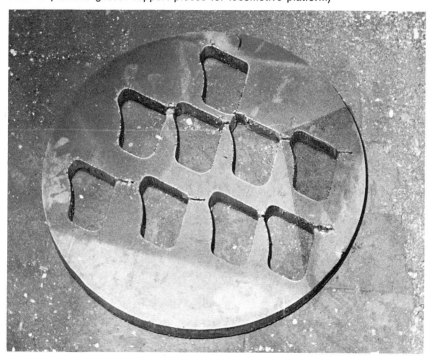

FIGURE 4–19B
Airco No. 50 Travo-graph 8 torch burner cutting gusset support pieces from scrap blank

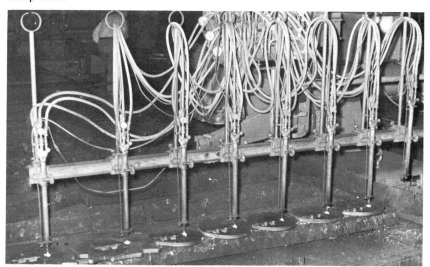

General Electric Co.

sizes and grades of the materials employed: purchase orders will be for larger amounts which are almost always less expensive; inventories will be smaller since less material must be maintained as a reserve; fewer entries will be made in stores records; fewer invoices will need to be paid; fewer spaces will be needed to house materials in the storeroom; sampling inspection will reduce the total number of parts inspected; and fewer price quotations and purchase orders will be prepared.

The standardization of materials, like other methods improvement techniques, is a continuing process. It requires the continual cooperation of members of the design, production planning, and purchasing departments.

TEXT QUESTIONS

1. Explain how design simplification can be applied to the manufacturing process.
2. How is operation analysis related to methods engineering?
3. Does increased competition submerge the necessity for operation analysis? Explain.
4. Explain the relationship between market price and volume as related to production.
5. What is the major obstacle in the path of the methods analyst?
6. How do unnecessary operations develop in an industry?
7. What four thoughts should the analyst keep in mind to improve design?
8. Explain why it may be desirable to "tighten up" tolerances and specifications.
9. What is meant by lot-by-lot inspection?
10. When is an elaborate quality control procedure not justified?
11. What six points should be considered when endeavoring to reduce material cost?
12. Explain why corrugated metal sheet is more rigid than flat sheet of the same material.
13. How does a changing labor and equipment situation affect the cost of purchased components?
14. The finish tolerance on a shaft was changed from 0.004 to 0.008. How much cost improvement resulted from this change?
15. What overall tolerance would be applied to three components making up the overall dimension if component 1 had a tolerance of 0.002; component 2, 0.004; and component 3, 0.005?

GENERAL QUESTIONS

1. Formulate a checklist that would be helpful in improving operations.
2. Explain how the conservation of welding rod can result in 20 percent material savings.

3. Investigate the operations required to convert waste sulfur dioxide to usable sulfur.

4. Explain why this country has lost a large share of its steel business to Japan since 1970.

PROBLEMS

1. A ceramic material is being considered as a possible mold material in conjunction with the die casting of 60–40 brass. A cylinder of the material 8 inches in diameter and 10 inches long was used to obtain a stress-strain relationship in compression. The material failed under a load of 265,000 pounds and a total strain of 0.012 inches. What was the material's fracture strength, percent contraction at fracture, and modulus of toughness?

2. The Dorben Company is designing a cast-iron part whose strength T is a known function of the carbon content C.

$$T = 2C^2 + \frac{3}{4}C - C^3 + k$$

In order to maximize strength, what carbon content should be specified?

3. In order to make a given part interchangeable, it was necessary to reduce the tolerance on the outside diameter from ± 0.010 to ± 0.005 at a resulting cost increase of 50 percent of the turning operation. The turning operation represented 20 percent of the total cost. By making the part interchangeable, the volume of this part could be increased by 30 percent. The increase in volume would permit production at 90 percent of the former cost. Should the methods engineer proceed with the tolerance change? Explain.

SELECTED REFERENCES

Fallon, Carlos. *Value Analysis to Improve Productivity.* New York: John Wiley & Sons, Inc., 1971.

Miles, Lawrence D. *Techniques of Value Analysis and Engineering.* 2d ed. New York: McGraw-Hill Book Co., 1961.

Mudge, Arthur E. *Value Engineering: A Systematic Approach.* New York: McGraw-Hill Book Co., 1971.

Ostwald, Philip F. *Cost Estimating for Engineering and Management.* Englewood Cliffs, N.J.: Prentice-Hall, Inc., 1974.

Peat, A. P. *Cost Reduction Charts for Designers and Production Engineers.* Brighton, England: Machinery Publishing Co., Ltd., 1968.

Rudd, Dale F., and Watson, Charles C. *Strategy of Process Engineering.* New York: John Wiley & Sons, Inc., 1968.

Trucks, H. E. *Designing for Economical Production.* Detroit: Society of Manufacturing Engineers, 1974.

5

Operation analysis (continued)

PROCESS OF MANUFACTURE

From the standpoint of improving the process of manufacture, investigation should be made in four ways: (1) when changing an operation, consider the possible effects on other operations; (2) mechanize manual operations; (3) utilize more efficient facilities on mechanical operations; and (4) operate mechanical facilities more efficiently.

Effects on subsequent operations resulting from changes in the present operation

Before changing any operation, it is wise to consider detrimental effects that may result on subsequent operations down the line. Reducing the costs of one operation can result in higher costs for other operations. For example, the following change in the manufacture of AC field coils resulted in higher costs and was, therefore, not practical. The field coils were made of heavy copper bands which were formed and then insulated with mica tape. The mica tape was hand wrapped on the already coiled parts. It was thought advantageous to machine wrap the copper bands prior to coiling. This did not prove practical as the forming of the coils cracked the mica tape, necessitating time-consuming repairs prior to acceptance.

Rearranging operations often results in savings. The flange of a motor conduit box required four holes to be drilled—one in each corner. Also, the base had to be smooth and flat. Originally, the job was planned by first grinding the base, then drilling the four holes in a drill jig. The drilling operation threw up burrs which had to be removed in the next step. By rearranging operations so that the holes were drilled first and the base then ground, it was possible to eliminate the deburring operation. The base-grinding operation automatically removed the burrs.

Costs can usually be reduced by combining operations. Formerly, the fan motor support and the outlet box of electric fans were fabricated,

FIGURE 5-1
Machine-wrapped
copper bands prove
unsatisfactory in
manufacture of AC
field coils

A. Method of forming
heavy copper bands

B. Hand-wrapped field
coils

C. Cracked insulation
in coils machine
wrapped prior to forming

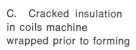

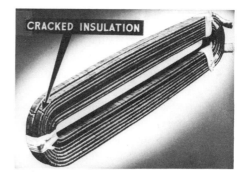

painted separately, and then riveted together. By riveting the outlet box to the fan motor support prior to painting, appreciable savings in time were effected on the painting operation.

Mechanize manual operations

Anytime heavy manual work is encountered, consideration should be given to possible mechanization. To clean insulation and dried varnish from armature slots, one company resorted to tedious hand filing. Questioning this process of manufacture led to the development of an end mill placed in a power air drill. This not only took most of the physical effort out of the job, but allowed a considerably higher rate of production.

The use of power assembly tools, such as power nut- and screw-drivers, electric or air hammers, and mechanical feeders, is often more economical than is the use of hand tool methods.

Utilize more efficient mechanical facilities

"Can a more efficient method of machining be used?" is a question that should be foremost in the analyst's mind. If an operation is done mechanically, there is always the possibility of a more efficient means of mechanization. Let us look at two examples. Turbine blade roots were machined by performing three separate milling operations. Both the cycle time and the costs were high. By means of external broaching, all three surfaces were finished at once. A pronounced saving was the result.

The possibility of utilizing press operation should never be overlooked. This process is one of the fastest for forming and sizing operations. A stamped bracket had four holes that were drilled after the bracket was formed. By designing a die to pierce the holes, the work was performed in a fraction of the drilling time.

Operate mechanical facilities more efficiently

A good slogan for the methods analyst to keep in mind is "Design for two at a time." Usually multiple-die operation in presswork is more economical than single-stage operation. Again, multiple cavities in die-casting, molding, and similar processes should always be considered when there is sufficient volume.

On machine operations the analyst should be sure that proper feeds and speeds are being used. He should investigate the grinding of cutting tools so that maximum performance will result. He should check to see whether the cutting tools are properly mounted, whether the right lubricant is being used, and whether the machine tool is in good condition and is being adequately maintained. Many machine tools are being operated at a fraction of their possible output. Always endeavoring to operate mechanical facilities more efficiently will pay dividends.

SETUP AND TOOLS

One of the most important elements applying to all forms of work holders, tools, and setups is the economic one. The amount of "tooling up" that proves most advantageous depends on (1) the quantity to be produced, (2) the chances for repeat business, (3) the amount of labor involved, (4) delivery requirements, and (5) the amount of capital required.

One of the most prevalent mistakes made by planners and toolmakers is to tie up money in fixtures that may show a large saving when in use but are seldom used. For example, a saving of 10 percent in direct labor cost on a job in constant use would probably justify greater expense in tools than an 80 or 90 percent saving on a small job that appears on the production schedule only a few times a year. The economic advantage of lower labor costs is the controlling factor in the determination of the tooling; consequently, jigs and fixtures may be desirable even where only small quantities are involved. Other considerations, such as improved interchangeability, increased accuracy, or reduction of labor trouble, may provide the dominant reason for elaborate tooling, although this is usually not the case.

For example, the production engineer in a machining department has under study a job being machined in the shop now. He has devised two alternate methods involving different tooling for this same job. Data on the present and proposed methods follow. Which method would be the more economical in view of the activity? The base pay rate is $2.40 per hour. The estimated activity is 10,000 pieces per year. The fixtures are capitalized and depreciated in five years.

Method	Time value (minutes)	Fixture cost	Tool cost	Average tool life
Present method	3.50 each	None	$ 6	10,000 pieces
Alternate 1.	2.80 each	$300	20	20,000 pieces
Alternate 2.	1.85 each	600	35	5,000 pieces

A cost analysis of the above would reveal that a unit total cost of $0.0930 represented by Alternate 2 is the most economical for the quantity anticipated. The elements of cost entering into this total are as follows:

Method	Unit direct labor cost	Unit fixture cost	Unit tool cost	Unit total cost	Annual cost
Present method	$0.140	None	$0.0006	$0.1406	$1,406
Alternate 1.	0.112	$0.006	0.0010	0.1190	1,190
Alternate 2.	0.074	0.012	0.0070	0.0930	930

FIGURE 5–2
Crossover (break-even) chart illustrating total costs for various
methods and quantities (note that at 10,000 pieces the total cost
of all three methods is about the same)

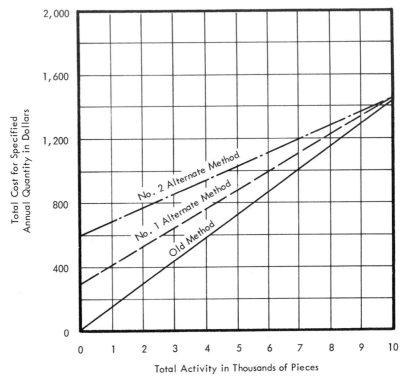

Total Activity in Thousands of Pieces

A crossover (break-even) chart, as illustrated in Figure 5–2, is a helpful
tool for deciding which method to use for given quantity requirements.

Once the needed amount of tooling has been determined, or if tooling
already exists, once the ideal amount needed has been determined, then
specific points should be kept in mind to produce the most favorable
designs. The following points should be considered.

Can the fixture be used to produce other similar designs to advantage?
Will the fixture be similar to some other that has been used to advantage?
If so, how can you improve upon it?
Has the part undergone any previous operations? If so, can you use any of
these points or surfaces to locate or master from?
Can any stock hardware be used for making the fixture?
Can the part be quickly placed in the fixture?
Can the part be quickly removed from the fixture?
Is the part held firmly so that it cannot work loose, spring, or chatter while
the cut is being made?

Always bear in mind that the cut should be against the solid part of the fixture and not against the clamp.

Can the output be increased by placing more than one part in the fixture?

Can the chips be readily removed from the fixture?

Are the clamps on the fixture strong enough to prevent them from buckling when they are tightened down on the work?

In using cams or wedges for binding or clamping the work, always bear in mind that through vibration or chatter they are apt to come loose and cause damage.

Must any special wrenches be designed to go with the fixture?

Is the work adequately supported so that the clamping force will not bend or distort it?

Can a gage be designed, or hardened pins added, to help the operator set the milling cutters or check up on the work?

Must special milling cutters, arbors, or collars be designed to go with the fixture?

Is there plenty of clearance for the arbor collars to pass over the work without striking?

If the fixture is of the rotary type, have you designed an accurate indexing arrangement?

Can the fixture be used on a standard rotary indexing head?

Can the fixture be made to handle more than one operation?

Have you, in designing the fixture, brought the work as close to the table of the miller as possible?

Can the part be milled in a standard miller vise by making up a set of special jaws and thus doing away with an expensive fixture?

If the part is to be milled at an angle, could the fixture be simplified by using a standard adjustable milling angle?

Can lugs be cast on the part to be machined to enable you to hold it?

How many different-sized wrenches must the operator have in order to tighten all clamps? Why will one not do?

Can the work be gaged in the fixture? Can a snap gage be used?

Can you use jack pins to help support the work while it is being milled?

Have you placed springs under all clamps?

Are all steel contact points, clamps, etc., hardened?

What kind or class of jigs are you going to design? Will any of the standard jig designs shown help you?

What takes the thrust of the drill? Can you use any jack pins or screws to support the work while it is being drilled?

Can you use a double or triple thread on the screw that holds the work in the jig, so that it will take fewer turns to get the screw out of the way in order to remove the part more quickly?

Have you made a note on the drawing, or have you stamped all loose parts with a symbol indicating the jig that they were made for, so that in case they are lost or misplaced they can be returned to the jig when found?

Are all necessary corners rounded?

Can the toolmaker make the jig?

Are your drill bushings so long that it will be necessary to make up extension drills?

Are the legs on the jig long enough to allow the drill, the reamer, or the pilot of the reamer to pass through the part a reasonable distance without striking the table of the drill press?

Are all clamps located in such a way as to resist or help resist the pressure of the drill?

Has the drill press the necessary speeds for drilling and reaming all holes?

Must the drill press have a tapping attachment?

Always remember that it is not practical to drill and ream several small holes and only one large one in the jig since quicker results can be obtained by drilling the small holes on a small drill press, while having only one large one would require the jig to be used on a large machine. The questions then arise, Is it cheaper to drill the large hole in another jig, and will the result of doing so be accurate enough?

Is the jig too heavy to handle?[1]

Is the jig identified with both a location number for storing and a part number that identifies the part or parts that the jig helps produce?

Setup ties in very closely with the tooling consideration in that the tooling of a job invariably determines the setup and teardown time. When we speak of setup time, we usually include such items as punching in on the job; procuring instructions, drawings, tools, and material; preparing work stations so that production can begin in the prescribed manner (setting up tools; adjusting stops; setting feeds, speeds, and depth of cut; and so on); tearing down the setup; and returning tools to the crib.

It can readily be seen that setup operations are extremely important in the job shop when production runs tend to be small. Even if this type of shop had modern facilities and high effort were put forth, it would still be difficult to meet competition if setups were long because of poor planning and inefficient tooling. When the ratio of setup time to production-run time is high, then the methods analyst will usually be able to develop possibilities for setup and tool improvement. A notable possibility here is to design and develop a system of group technology.

The essence of group technology is to classify the various components entering into a company's products so that parts similar in shape and processing sequence are numerically identified. Parts belonging to the same family group, such as rings, sleeves, discs, and collars, are scheduled for production over the same interval of time on a line of general purpose facilities arranged in the operation sequence optimum to most conditions. Since both the size and the shape of the family of parts will involve considerable variability, the line of facilities will usually be equipped with universal-type, quick-acting jigs and fixtures.

The resulting line can mean greater output, less setup time, greater machine utilization, less material handling, shorter cycle time, and better cost improvement. The design and development of universal-type jigs and fixtures will mean that less such equipment is required, with the added

---

[1] Abridged from F. H. Colvin and L. L. Haas, *Jigs and Fixtures*, 5th ed. (New York: McGraw-Hill Book Co., 1948), pp. 14–17.

advantage of reduction of such hidden costs as tool storage and obsolescence.

For example, Figure 5–3 illustrates a subdivision of a system grouping by nine classes of parts. Note the similarity of parts within a given vertical

FIGURE 5–3

	10	20	30	40	50	60	70	80	90
0 Without Subforms									
1 Set-Off or Shoulder on One Side									
2 Set-Offs or Shoulders on Two Sides									
3 With Flanges, Protuberances									
4 With Open or Closed Forking or Slotting									
5 With Hole									
6 With Hole and Threads									
7 With Slots or Knurling									
8 With Supplementary Extensions									

column. If we were machining a shaft with external threads and bored partially through at one end, the part would be identified as Class 207.

To develop better methods, the analyst should investigate the setup and tools in these three ways: (1) reduce setup time by better planning and production control; (2) design tooling to utilize the full capacity of the machine; and (3) introduce more efficient tooling.

Reduce setup time by better planning and production control

The time spent in requisitioning tools and materials, in preparing the work station for actual production, and in cleaning up the work station and returning the tools to the tool crib, is usually known as setup time. Since this time is often difficult to control, it is the portion of the work that is usually performed least efficiently. This time can often be reduced through effective production control. By making the dispatch section responsible for seeing that the tools, gages, instructions, and materials are provided at the correct time, and that the tools are returned to their respective cribs after the job has been completed, the need for the operator to leave his work area will be eliminated. Thus, the operator will have to perform only the actual setting up and tearing down of the machine. The clerical and routine function of providing drawings, instructions, and tools can be performed by those more familiar with this type of work. Thus, large numbers of requisitions for these requirements can be performed simultaneously, and setup time will be minimized. Here again, group technology can be advantageous.

Another function of production control that should be carefully reviewed for possible improvement is scheduling. Considerable setup time can be saved by scheduling like jobs in sequence. For example, if a one-inch round bar stock job is scheduled to a No. 3 Warner and Swasey turret lathe, from the standpoint of setup time, it would be economical to have other one-inch round bar stock work scheduled to immediately follow the first one-inch job. This would eliminate collet changing time, and would probably minimize the number of tool changes in the square and hex turrets.

Usually it is advisable to provide duplicate cutting tools for the operator rather than to make him responsible for sharpening his own tools. When it becomes necessary for the operator to get a new tool, the dull one should be turned in to the tool crib attendant and replaced with a sharp one. The benefits of standardization cannot be realized when tool sharpening is the responsibility of the operator.

To minimize downtime, there should be a constant backlog of work ahead of each operator. There should never be a question on what each operator's next work assignment will be. A technique frequently used to keep the work load apparent to the operator, supervisor, and superintendent is to provide a board over each production facility with three wire clips or pockets to receive work orders. The first clip contains all work orders scheduled ahead; the second clip holds the orders currently being worked on; and the last clip holds the completed orders. As the dispatcher issues work orders, they are placed in the "work ahead" station. At the same time, the dispatcher picks up all completed job tickets from the "work completed" station and delivers them to the scheduling department for record. It is obvious that this system assures the operator of a perpetual load in front of him and makes it unnecessary for him to go to his supervisor for his next work assignment.

By making a record of difficult recurring setups, considerable setup time can be saved when repeat business is received. Perhaps the simplest and yet most effective way to compile a record of a setup is to take a photograph of the setup once it is complete. The photograph should be either stapled to and filed with the production operation card, or placed in a plastic envelope and attached to the tooling prior to storage in the tool crib.

Design tooling to utilize the full capacity of the machine

In designing tooling to utilize the full capacity of the machine, the analyst should ask: "Can the work be held to permit all machining operations in one setup?" A careful review of many jobs will bring out possibilities for multiple cuts, thus utilizing a greater share of the machine's capacity. For example, it was possible to change a milling setup for a toggle lever so that the six faces were milled simultaneously by five cutters. The old setup required that the job be done in three steps. The part had to be placed in a separate fixture three different times. The new setup reduced the total machining time and increased the accuracy of the relationship between the six machined faces.

Another thought to keep in mind is the possibility of positioning one part while another is being machined. This opportunity exists on many milling machine jobs where it is possible to conventional mill on one stroke of the table and climb mill on the return stroke. While the operator is loading a fixture at one end of the machine table, a similar fixture is holding a piece being machined by power feed. As the table of the machine returns, the first piece is removed from the machine and the fixture is reloaded. While this internal work is taking place, the machine is cutting the piece in the second fixture.

In view of the ever-increasing cost of energy, it is becoming more and more important to utilize the most economical equipment to do the job. A few years ago, the cost of energy was such an insignificant proportion of total cost that little attention was given to utilization of the full capacity of machines. I have seen literally thousands of operations where but a fraction of machine capacity was utilized, with a resulting waste of electric power. In the metal trades industry today, the cost of power is over 2 percent of total cost, with strong indications that the present cost of power will at least double in the next decade. It is highly probable that careful planning to utilize a large proportion of the capacity of the machine selected to do the work can effect a 50 percent savings in power usage in many of our plants.

Introduce more efficient tooling

Everyone is aware of the vast number of screwdriver designs. Each design provides an efficient driver under a particular set of conditions. A screwdriver efficient under one set of conditions may be very inefficient under another set.

The analyst should investigate to see whether the proper hand tools are being provided and used.

More important than the proper hand tools are the proper cutting tools. Grinding wheels should be carefully checked to be sure that the right wheel is being used on the job. Excessive grinding wheel wear, poor finish, and a slow cutting rate are indications that the wrong wheel has been selected.

Carbide tools offer large savings over high-speed steel tools on many jobs. For example, a 60 percent saving was realized by changing the milling operation of a magnesium casting. Formerly, the base was milled complete in two operations, using high-speed steel milling cutters. An analysis resulted in employing three carbide-tipped fly cutters mounted in a special

FIGURE 5–4
Equipping this milling machine with three carbide-tipped fly cutters resulted in a 60 percent saving over the old milling method using high-speed steel cutters

holder to mill parts complete. Faster feeds and speeds were possible, and surface finish was not impaired.

Selection of the correct drill is important. Just as there are a variety of hand tools—each designed for a specific task, so there are broad varieties of cutting tools—each having its own limited range of application.

The old method for drilling a hole ½ inch in diameter and 7 inches deep in a hard steel shaft with a high-speed twist drill resulted in considerable difficulty. It was necessary to back out the drill about 20 times to remove the chips and cool the drill. It was not possible to get the coolant to the cutting edge because of the depth of the hole. By utilizing a carbide-tipped V-shaped gun drill with a hole in the center through which the coolant was pumped, it was possible to drill the hole in about one third the time.

Again, it was possible to greatly reduce the costs of die maintenance by considering the use of carbide as a die-cutting edge material. To produce a certain stamping, it was necessary to resharpen a die with tool steel cutting edges after every 50,000 pieces. A study of die-cutting edges resulted in an improved die with edges of tungsten carbide. Now, more than 600,000 pieces are produced before resharpening is necessary.

In the introduction of more efficient tooling, develop better methods

FIGURE 5–5
Carbide-tipped V-shaped gun drill with hole through center for coolant answers deep-hole drilling problem

FIGURE 5–6
More than 600,000 pieces are produced on this die before resharpening is necessary

for holding the work. Be sure that the work is held so that it can be positioned and removed quickly. A quick-acting bench vise has been developed, utilizing the feet to open and close the vise while the hands are doing useful work. The vise is opened by pressing one foot pedal and is closed by pressing a second foot pedal.

WORKING CONDITIONS

The methods analyst should accept as part of his or her responsibility the provision of good, safe, comfortable working conditions. Experience has proven conclusively that plants providing good working conditions will outproduce those maintaining poor conditions. The economic return provided through investment in an improved working environment is usually significant. Ideal working conditions will improve the safety record, reduce absenteeism and tardiness, raise employee morale, and improve public relations, in addition to increasing production.

Some common considerations for improving working conditions are as follows:

1. Improve lighting.
2. Control temperature.
3. Provide for adequate ventilation.
4. Control sound.
5. Promote orderliness, cleanliness, and good housekeeping.
6. Arrange for the immediate disposal of irritating and harmful dusts, fumes, gases, vapors, and fogs.
7. Provide guards at nip points and at points of power transmission.
8. Provide personal protective equipment.
9. Sponsor and enforce a well-formulated first-aid program.
10. Utilize work physiology principles.

Improve lighting

The intensity of light required depends primarily on what operations are being performed in an area. It is obvious that a toolmaker or an inspector requires greater intensity of light than would be needed in a storeroom. Glare, the quality of the light, the location of the light source, contrasts in color and brightness, and flickering and shadows must also be considered. Some ways to obtain good lighting are as follows:

1. Reduce glare by installing a large number of light sources to give the total required light output.
2. Enclose filament-type bulbs in opalescent bowls to reduce glare by spreading the light output over a greater surface.
3. Produce a satisfactory approximation to white light for most uses by using a filament-type bulb or a single white fluorescent unit. (White

light, or the composition of average sunlight, is generally considered ideal.)

4. Eliminate all shadow by providing the correct level of illumination at all points of the work station. In view of power costs, areas with too much illumination as well as areas with inadequate illumination should be identified.

Control temperature

The human body endeavors to maintain a constant temperature of about 98° F. When the body is exposed to unusually high temperatures, large amounts of perspiration evaporate from the skin. During the perspiration process, sodium chloride is carried through the pores of the skin and is left on the skin surface as a residue when evaporation takes place. This represents a direct loss to the system and may create a disturbance to the normal balance of fluids in the body. The result is heat fatigue and heat cramps, with an accompanying slowdown in production. The performance of a very good operator deteriorates as rapidly as the performance of the average and less-than-average operator. In clerical operations, such as typing, output not only suffers, but the number of errors also increases.

Conversely, detailed time studies have repeatedly brought to attention the loss in production when working conditions are unduly cold. Temperature should be controlled so that it will be between 65 and 75° F the year round. If this level can be maintained, losses and slowdowns from heat fatigue, heat cramps, and lack of manipulative dexterity will be kept to a minimum.

Provide for adequate ventilation

Ventilation also plays an important role in the control of accidents and operator fatigue. It has been found that disagreeable fumes, gases, dusts, and odors cause fatigue that definitely reduces the physical efficiency of the worker and often creates a mental tension. Laboratory findings indicate that the depressing influence of poor ventilation is associated with air movement as well as with temperature and humidity.

When humidity increases, evaporative cooling decreases rapidly, thus reducing the ability of the body to dissipate heat. Under these conditions high heart rates, high body temperatures, and slow recovery after work result in pronounced fatigue.

The New York State Commission on Ventilation disclosed that in heavy manual labor 15 percent less work was done at 75° F with 50 percent relative humidity than was done at 68° F with the same humidity, and that 28 percent less work was done at 86° F with 80 percent relative humidity. It also brought out that 9 percent less work was accomplished in stagnant air than in fresh air when the temperature and humidity remained constant. Further experiment showed a reduced work capacity of 17 percent at 75° F, and of 37 percent at 86° F, as compared with work done at 68° F.

Similar experiments made by the American Society of Heating and Ventilating Engineers showed that similar gains in production, safety, and employee morale follow when ideal ventilation is introduced to the production floor.

Control sound

Both loud and monotonous noises are conducive to worker fatigue. Constant and intermittent noise also tends to excite the worker emotionally, resulting in loss of temper and difficulty in doing precision work. Quarrels and poor conduct on the part of workers can often be attributed to disturbing noises. Tests have proven that irritating noise levels heighten the pulse rate and blood pressure and result in irregularities in heart rhythm. In order to overcome the effect of noise, the nervous system of the body is strained, resulting in neurasthenic states.

Promote orderliness, cleanliness, and good housekeeping

A good industrial housekeeping program will (1) diminish fire hazards, (2) reduce accidents, (3) conserve floor space, and (4) improve employee morale.

Industrial accident statistics indicate that a large percentage of accidents are the result of poor housekeeping. Many times the expression "A place for everything and everything in its place" has been cited as the basis for orderliness. This is true, and the methods analyst should be sure that a place is provided for everything and, if necessary, follow through with supervision to see that everything is in its place.

When the general layout of a plant shows management's and supervision's desire to have orderliness, cleanliness, and good housekeeping, then the employees will tend to follow the examples set for them, and they will practice good housekeeping themselves.

Arrange for the disposal of irritating and harmful dusts, fumes, gases, vapors, and fogs

Dusts, fumes, gases, vapors, and fogs generated by various industrial processes constitute one of the major dangers encountered by workers. The following classification of dusts, prepared by the National Safety Council, gives an indication of the problem:

1. Irritating dusts, such as metal and rock dusts.
2. Corrosive dusts, such as those from soda and lime.

3. Poisonous dusts, such as those from lead, arsenic, or mercury.
4. Dusts from fur, feathers, and hair may carry disease germs which may infect the worker.

All of these dangers can be eliminated by employing suitable methods, such as exhaust systems, complete enclosing of the process, wet or absorbing techniques, and complete protection of the operator through personal respiratory protective equipment. Probably the most effective measure for controlling dust and fumes is through local exhaust systems in which a hood is installed for collecting the substance to be removed right at the point of generation. A fan draws the contaminated air through metal pipes or ducts to the outside or to some properly provided place for disposal. The diameter of the exhaust pipe is an important detail that must be determined to insure a satisfactory installation. Generally, the larger the exhaust pipe, the more costly the initial installation. However, larger pipes allow increased efficiency of the motive system since they require less power for the exhaust of air (see Table 5–1). And, as already pointed out, the conservation of power is an important consideration today.

TABLE 5–1

Diameter of buffing wheels	Maximum grinding surface (sq. in.)	Minimum diameter of pipe (in.)
6 in. or less, not over 1 in. thick.	19	3
7 to 9 in., inclusive, not over 1½ in. thick	43	3½
10 to 16 in., inclusive, not over 2 in. thick.	101	4
17 to 19 in., inclusive, not over 3 in. thick.	180	4½
20 to 24 in., inclusive, not over 4 in. thick.	302	5
25 to 30 in., inclusive, not over 5 in. thick.	472	6

Source: National Safety Council Safe Practices pamphlet no. 32.

The analyst can consider the utilization of the dust or fumes collected. Perhaps wood dust can be used as a source of heat, as can blast furnace gases and other gases containing carbon monoxide. One manufacturer in the lead industry claims that it salvages enough lead annually to more than pay for the operation of its exhaust system.

Provide guards at nip points and at points of power transmission

In view of the Williams-Steiger occupational safety and health act (OSHA), the employer is legally responsible for properly guarding his facilities, so his employees may be protected. Guards must be designed correctly if they are to provide protection and still not hinder production. The general requisites of good guards are to:

1. Effectively protect the employee.
2. Permit normal operation of the facility at a pace equal to or greater than that used prior to the guard installation. (When an employee knows he is safe, he will tend to produce at a more effective pace.)
3. Permit normal maintenance to the facility.

Since the machine tool builder is in an ideal position to provide satisfactory guards, it is usually advantageous to purchase the necessary guards at the same time that the machines are procured. Although many home-made guards are doing an excellent job, they are not usually as efficient, attractive, or inexpensive as purchased units.

Production workers are fully aware that unguarded machinery is hazardous. If they are exposed to such working conditions, there will be a natural tendency to execute at low effort as a precautionary measure. Also, an employer who fails to spend any reasonable amount to remove a visible hazard cannot expect full cooperation from his employees.

Provide personal protective equipment

Because of the nature of the operation and/or economic considerations, it is not always possible to eliminate certain hazards by changing methods, equipment, and tools. When this is the case, the operator can often be fully protected by personal protective equipment. Representative personal protective equipment would include goggles, face shields, helmets, aprons, jackets, trousers, leggings, gloves, shoes, and respiratory equipment.

To be assured that operating personnel will conscientiously use protective equipment, it should either be furnished to the employee at cost or be provided at no expense. The policy of having the company absorb completely the cost of personal protective equipment is becoming more and more common. Innumerable cases can be cited where personal protective equipment has saved an eye, a hand, a foot, or a life. For example, one steel company reported that 20 fatalities were prevented in one year by the enforced wearing of company-provided helmets. A northwest lumber company reported, too, that six serious head injuries were prevented within a 20-day period through the use of protective hats.

Sponsor and enforce a well-formulated first-aid program

The most advanced program in industrial safety will never be able to completely eliminate all accidents and injuries. To care adequately for the injuries that do occur, a well-formulated first-aid program is essential. This will include training and publicity, so that the employees will be fully aware of the danger of infection and of the ease of avoiding infection through first aid. Also, a complete procedure to be followed in case of injury must be arranged, with proper instructions to all supervisory levels. A well-equipped first-aid room must be provided to care for injured and ill employees until professional medical aid is available.

FIGURE 5-7
Work center showing press operations (note that each person has two hand buttons spread well apart so that the operator must have both hands in a safe position when the press starts)

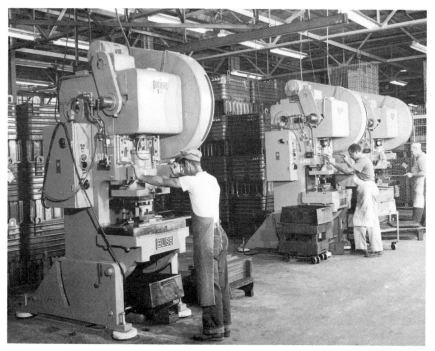

OSHA

The OSHA law and its provisions for unannounced plant inspections make it practically a necessity today to maintain good working conditions. This law gives an inspector the right to issue citations or levy fines for conditions he or she personally considers unsafe or unhealthful. Although there are some specific OSHA standards, a great deal of the evaluation is based on the working conditions the inspector sees as he or she visits the various departments of an enterprise. In a recent survey of the metal-working industry, it was determined that there was a 60 percent chance that either a fine or a citation would come about because of an inspector's visit. A majority of the fines or citations were issued because of:

1. Inadequate machine guarding.
2. Improper electrical grounding.
3. Unclearly marked aisles.
4. Generally poor housekeeping.
5. Improper ventilation in paint spray booths.
6. Excessive noise.
7. Excessive pressure on blowguns.

8. Missing handrails for steps.
9. A lack of proper fire extinguishers.

Working conditions should continually be improved to make plants clean, healthy, and safe. Good working conditions are reflected in health, output, the quality of work, and workers' morale. A better place to work results in better products at lower prices.

MATERIAL HANDLING

Good material handling provides for the delivery of an adequate inventory of material at the proper time and in the proper condition to the point of use at the least total cost. It is apparent that good material handling must act in concert with good material management. Thus, when the analyst considers the eighth primary approach to operation analysis, he should consider the following as an integrated system: inventory control, purchasing policy, receiving, inspection, storage, traffic control, pickup and delivery, layout, and facilities.

The tangible and intangible benefits of material handling can be reduced to four major objectives, as outlined by the American Material Handling Society. These are:

1. Reduction of handling costs.
   a. Reduction of labor costs.
   b. Reduction of material costs.
   c. Reduction of overhead costs.
2. Increase of capacity.
   a. Increase of production.
   b. Increase of storage capacity.
   c. Improved layout.
3. Improvement in working conditions.
   a. Increase in safety.
   b. Reduction of fatigue.
   c. Improved personnel comforts.
4. Better distribution.
   a. Improvement in handling system.
   b. Improvement in routing facilities.
   c. Strategic location of storage facilities.
   d. Improvement in user service.
   e. Increase in availability of product.

An axiom that the methods analyst should always keep in mind is that the best-handled part is the least manually handled part. Whether the distances of moves are large or small, the methods analyst should study them with thought toward improvement. By considering the following four points, it is possible to reduce the time and energy spent in handling material:

1. Reduce the time spent in picking up material.
2. Reduce material handling by using mechanical equipment.
3. Make better use of existing handling facilities.
4. Handle material with greater care.

A good example of the application of these four points is the evolution of warehousing so that the former storage center has become an automated distribution center. Today, the automated warehouse has computer control of material movement as well as information flow through data processing. In this type of mechanized warehouse, receiving, transporting, storing, retrieving, and inventory control are treated as an integrated function.

Reduce the time spent in picking up material

Many people think of material handling as only transportation, and neglect to consider positioning at the work station. This is equally important, and since it is often overlooked, it may offer even greater opportunities for savings than does transportation. Reducing the time spent in picking up material minimizes tiring, costly manual handling at the machine or the workplace. It gives the operator a chance to do his job faster and with less fatigue and greater safety.

Consider the possibility of avoiding loose piling on the floor. Perhaps the material can be stacked directly on pallets or skids after being processed at the work station. This can result in a substantial reduction of terminal transportation time (the time material handling equipment stands idle while loading and unloading take place). In fact, a good axiom to remember is, If terminal time in the transportation of any material is long, then an improved material handling installation is needed. A good example of reducing terminal time to a minimum is the utilization of an electromagnet on a crane to lift ferrous loads. The crane is tied up a negligible amount of time at the terminal points. Usually some type of conveyor or mechanical fingers can be used to bring material to the work station and thus avoid or reduce the time spent in picking up the material. Often gravity conveyors, used in conjunction with the automatic removal of finished parts, can be installed, thus minimizing material handling at the work station.

Several types of positioning equipment are available to reduce the time spent in handling material to and from the work station. Hydraulic tables that can place sheets of material at the proper feeding height for shears, presses, brakes, and other machines, as well as transport and bring to position dies and other heavy tools, often have application. The portable elevator or positioning truck is an elevating mechanism supported on a truck mast and base. The hoisting unit is either motorized or has a winch mechanism. Another well-known positioning aid is the welding positioner. The latest models provide for powered rotation and the elevation of materials to permit downhand welding motions.

FIGURE 5–8
Typical handling equipment used in industry today

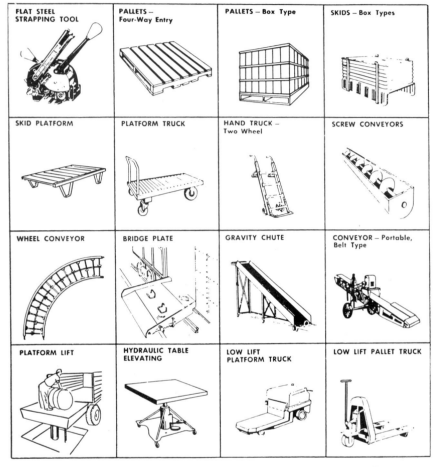

FLAT STEEL STRAPPING TOOL	PALLETS – Four-Way Entry	PALLETS – Box Type	SKIDS – Box Types
SKID PLATFORM	**PLATFORM TRUCK**	**HAND TRUCK –** Two Wheel	**SCREW CONVEYORS**
WHEEL CONVEYOR	**BRIDGE PLATE**	**GRAVITY CHUTE**	**CONVEYOR –** Portable, Belt Type
PLATFORM LIFT	**HYDRAULIC TABLE ELEVATING**	**LOW LIFT PLATFORM TRUCK**	**LOW LIFT PALLET TRUCK**

The Material Handling Institute

The Bethlehem Steel Corporation installed a strip-feeding table in its Lebanon, Pennsylvania, plant to position metal strip at the shear punch in its cold-rolled department,[2] in the expectation that the 24-foot table would pay for itself within four years. Again, by installing a positioning table, the Erie Malleable Iron Company of Erie, Pennsylvania, was able to reduce from 30 to 10 the man-minutes required to load a stack of annealing pots.[3]

---

[2] "Handling Shorts," *Factory Management and Maintenance,* Plant Operation Library no. 134, *Better Material Handling, a $2-Billion Target.*

[3] Donald W. Pennock, "Positioners: For Better Handling at the Workplace," *Factory Management and Maintenance,* Plant Operation Library no. 134, *Better Material Handling, a $2-Billion Target.*

With regard to the time spent in picking up material, the analyst should ask the following questions: "Can loose piling on the floor be avoided?" "Can material be weighed without picking it up?" and "Can a conveyor be used to avoid picking it up?"

Reduce material handling by using mechanical equipment

Mechanizing the handling of material will usually reduce labor costs, improve safety, alleviate fatigue, and increase production. However, care must be exercised on the proper selection of equipment and methods. Standardization of equipment is important because it simplifies operator training, provides interchangeability of equipment, and requires less stocking of repair parts.

FIGURE 5–9
Typical handling equipment used in industry today

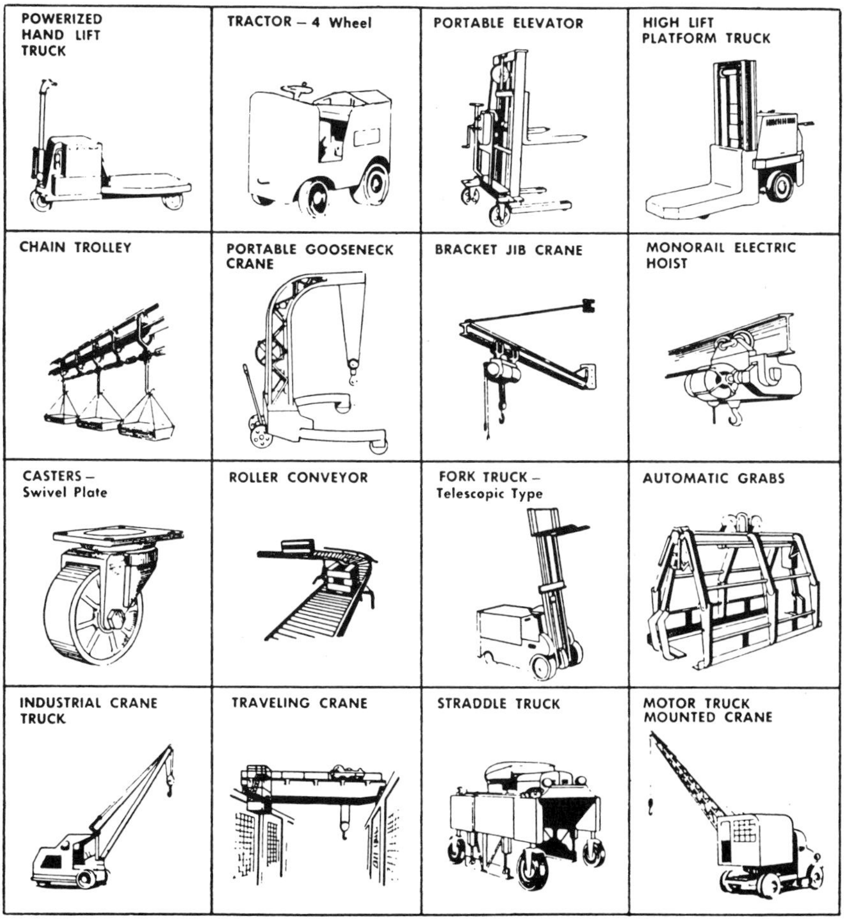

POWERIZED HAND LIFT TRUCK	TRACTOR – 4 Wheel	PORTABLE ELEVATOR	HIGH LIFT PLATFORM TRUCK
CHAIN TROLLEY	PORTABLE GOOSENECK CRANE	BRACKET JIB CRANE	MONORAIL ELECTRIC HOIST
CASTERS – Swivel Plate	ROLLER CONVEYOR	FORK TRUCK – Telescopic Type	AUTOMATIC GRABS
INDUSTRIAL CRANE TRUCK	TRAVELING CRANE	STRADDLE TRUCK	MOTOR TRUCK MOUNTED CRANE

The Material Handling Institute

Motion and time study

FIGURE 5–10
The method used previously to populate computer panels with Solid Logic
Technology circuit cards

1   2

3   4

The savings possible through the mechanization of material handling equipment are typified by the following examples. One plant installed a monorail hoist over two workplaces and a paint basin. Formerly, 25 tons of tools in process were transported weekly by hand. The return of the monorail installation investment was estimated at 200 percent in the first year. In another instance, a 100 percent investment return was realized by moving one heavy cutoff saw close to another cutoff saw, and then furnishing a jib crane to service the two saws.

At the outset of the 360 program, IBM considered several methods for populating panels with the Solid Logic Technology circuit cards used in the computers manufactured at its Edicott plant.

Under the previous method (see Figure 5–10), the circuit cards were stored in containers on shelving units in a storage crib area. The operator would go to the storage crib, select the correct cards required for a specific panel based upon its "plug" list, return to his workbench, and then proceed to insert the circuit cards into the panel in accordance with the plug list.

The improved method resulted in the installation of two automated vertical storage machines (see Figures 5–11 and 5–12). Both machines were located in an area under one of the "sawtooth" sections of the building roof, which literally uses storage space above the normal ceiling height. Each automated vertical storage unit has 10 carriers with four pullout

FIGURE 5–11
Automated vertical storage machine used in conjunction with the assembly of computer panels

FIGURE 5–12
Work area of vertical storage machine used in the assembly of computer panels

drawers per carrier; 20 stop positions were created so that two drawers are available at each stop position. The carriers move up and around in a system which is a compressed version of a Ferris wheel. With 20 possible stop positions on call, the unit always selects the closest route—either forward or backward—to bring the proper drawers to the opening in the shortest possible time.

The assembly station is arranged at the opening level, where an operator can select cards and place them in the proper position in the panel. From his seated position, the operator dials the correct stop. The drawer is pulled forward into a position exposing the needed cards, and the proper card is withdrawn and placed in the panel as designated on the "plug" list.

The improved method has reduced by approximately 50 percent the storage area required in valuable manufacturing floor space, has improved work station layout, and has substantially reduced populating errors by minimizing operator handling, decision making, and fatigue.

In considering the possible advantages of using mechanical equipment, the following questions should be asked: "Can heavy material be handled easier with mechanical equipment?" "Can parts be handled quicker with mechanical equipment?" and "Can material be stacked higher mechanically?"

Make better use of existing handling facilities

To assure the greatest return from material handling equipment, it must be used effectively. By palletizing material in temporary and permanent storage, greater quantities can be transported faster than when material is stored without the use of pallets. United Wallpaper, Inc., realized the following gross direct labor savings through palletizing and mechanical material handling:

1. Reduced labor cost of warehousing finished stock 66 percent.
2. Reduced labor cost of assembling and shipping finished goods orders 30 to 65 percent.
3. Reduced labor cost of unloading and warehousing raw stock 80 percent.
4. Reduced labor cost of unloading and warehousing other raw materials 40 percent.[4]

The maximum net load that can be safely handled can usually be easily computed. For example, the inch-pound rating of fork trucks is computed by multiplying the distance in inches from the center of the front axle to the center of the load (see Figure 5–13). If the distance from the center of the front axle to the front end of the fork truck is 18 inches, and the length of the pallet is 60 inches, then the maximum gross weight that a 200,000 inch-pounds fork truck should handle would be:

$$\frac{\text{inch-pounds}}{B} = \frac{200,000}{18 + \frac{60}{2}} = 4,167 \text{ pounds}$$

By planning the pallet size to take full use of the equipment, a greater return can be realized from the material handling equipment.

Sometimes it may be possible to design special racks so that the material can be handled in larger or more convenient units. When this is done, the compartments, hooks, pins, or supports for holding the work should be in multiples of 10 for ease of count during processing and final inspection.

If any material handling equipment is used only part of the time, consider the possibilities for putting it to use a greater share of the time. By relocating production facilities or adapting material handling equipment to diversified areas of work, greater utilization may be achieved.

---

[4] Philip F. Cannon, "Palletizing Makes New Warehousing System Really Work," *Factory Management and Maintenance,* Plant Operation Library no. 134.

Motion and time study

FIGURE 5–13
Typical forklift truck

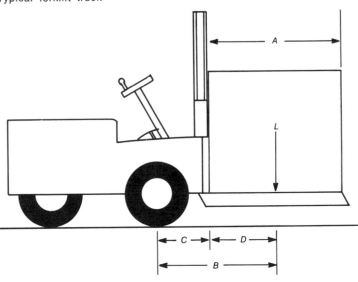

Here the analyst should ask and answer the following questions: "Can the material be handled in larger or more convenient units?" "Can auxiliary handling equipment facilitate service at the work station?" and "Can piece counting be combined with material handling?"

Handle material with greater care

Industrial surveys indicate that approximately 40 percent of plant accidents are a result of material handling operations. Of these, 25 percent are caused by lifting and shifting material. When care is exercised in handling material, and when the physical effort of material handling is transferred to mechanical mechanisms, fatigue and accidents are reduced. Records prove that the safe factory is also an efficient factory. Although it is factual knowledge that the greater the amount of mechanized material handling, the safer the factory, the analyst must consider the possibility of making the handling equipment safer, too. Safety guards at points of power transmission, safe operating practices, good lighting, and good housekeeping are essential to make material handling equipment safer.

It is often wise to consider the reduction of product damage by better handling. If the number of reject parts is at all significant in handling parts between work stations, then this area should be investigated. Usually damage to parts in handling can be kept to a minimum if specially designed racks or trays are fabricated to hold the parts immediately after processing. For example, one manufacturer of aircraft engine parts incurred a sizable number of damaged external threads on one component that was stored in metal tote pans after the completion of each operation. When the tote pans filled with parts were moved to the next work station by two-wheeled hand trucks, the machined forgings bumped against one another and against the sides of the metal pan to such an extent that they became badly damaged. Someone investigated the cause of the rejects and suggested making wooden racks with individual compartments to support the machined forgings. This prevented the parts from bumping against one another and against the sides of the metal tote pan. Production runs were also more easily controlled because of the faster counting of parts and rejects.

The introduction of a portable mechanized lift permitted much greater application of a Hubbard tank in connection with physical therapy treatment in a city hospital (see Figure 5–14). With this controllable material

FIGURE 5–14
Portable lift provides greater use of Hubbard tank in a physical therapy department

handling equipment, patients can be comfortably immersed in the tank while in either a sitting or a lying position.

Two typical questions that should be asked are: "Can the material be handled with greater safety?" and "Can product damage be reduced by better handling?"

FIGURE 5–15
Conveyorized final assembly of data processing machines

IBM

FIGURE 5–16
Individual work center typical of final assembly area

IBM

Summary: Material handling

The analyst should always be on the alert to eliminate inefficient handling of material. The following fundamental principles should be considered in doing a better job in material handling:

1. Material handling should be integrated with material management.
2. Gravity can frequently be used to move materials economically.
3. The terminal time of material handling equipment should be kept to a minimum.
4. Material handling costs per unit decrease with an increase in quantity up to plant capacity.
5. As the size of the handling unit increases, there is usually a corresponding decrease in the unit cost of material handling.
6. Flexible material handling equipment capable of a variety of uses or applications should be considered as an alternative to special purpose handling equipment.

7. Repairs and preventive maintenance should be well planned prior to the installation of material handling equipment.
8. It is usually best to move materials in a straight line.
9. Material handling equipment, like production equipment, becomes out-of-date. Modern handling equipment incorporates such productivity-increasing items as automatic couplers, nonfriction bearings, and rubber tires.

It may be well to reiterate that the predominant principle to be kept in mind is that the less a material is manually handled, the better it is handled.

The following example applies to several of the above-enumerated principles. In the Poughkeepsie, New York, Systems Manufacturing Division of IBM Corporation, data processing machines were moved from one assembly operation to the next on individual dollies. Subsequently, an operator-controlled conveyorized assembly system was introduced. Under the new installation, when an operator is ready to work on a unit, he actuates controls which transfer the unit from the main line to the spur. He then operates controls which transfer the unit from the spur to his work station. When the work is complete, the same controls operate in reverse to enable the operator to return the unit to the main line (see Figures 5–15 and 5–16).

This flexible, work station–controlled, conveyorized final assembly system has improved space use, reduced in-process inventory, and increased management control in the manufacture of punched card data processing machines.

PLANT LAYOUT

The principal objective of effective plant layout is to develop a production system that permits the manufacture of the desired number of products of the desired quality at the least cost. Thus, the physical layout is an important element of an entire production system that embraces operation cards, inventory control, material handling, scheduling, routing, and dispatching. All these elements must be carefully integrated in order to fulfill the stated objective.

The student may well ask, "Is there a type of layout that tends to be the best?" The answer is no. A given layout can be best in one set of conditions and yet be quite poor in a different set of conditions. And since work conditions are seldom static, the methods analyst often has the opportunity to make improvements in the layout. Paramount in dynamic conditions are: material handling systems, product mixes, process equipment, and process methods.

It should be apparent that the number of layout combinations is extremely large even in the relatively small plant or business. For example, 10 facilities will result in 10 factorial possibilities.

Softwear has been designed that can prove helpful to the analyst in developing realistic layout solutions rapidly and inexpensively. The CRAFT program (IBM share library No. SDA 3391) is one computer program that has been extensively used. It has the capacity for handling 40 activity centers. These could be departments or work centers within a department. Any one of these activity centers can be identified as being fixed, thus freezing it and allowing freedom of movement in those that can readily be moved. Input data includes the number and location of fixed work centers, material handling costs, interactivity center flow, and a representation of a block layout. The governing heuristic algorithm asks: What change in material handling costs would result if work centers were exchanged. The answer is stored, and the computer proceeds in an iterative manner until it converges on an optimal solution.

Another available program is known as CORELAP (Engineering Management Associates, Boston, Massachusetts). The input requirements for CORELAP are the number of departments, the departmental areas, the departmental relationships, and weights for these relationships. CORELAP then constructs layouts by locating departments, using rectangular-shaped areas.

ALDEP (IBM Corporation program order #360D–15.0.004) constructs plant layouts by randomly selecting a department and locating it in a given layout. The RELATION chart is then scanned, and a department that has a high closeness rating is introduced into the layout. This process continues until all departments are placed. A score is computed for the layout, and the process is repeated a specific number of times.

Although it is difficult and costly to make changes in arrangements that already exist, the methods analyst should be trained to review with a critical eye every portion of every layout that he comes in contact with. Poor plant layouts result in major costs. Unfortunately, most of these costs are hidden and, consequently, cannot be readily exposed. The indirect labor expense of long moves, backtracking, delays, and work stoppages due to bottlenecks are characteristic of a plant that has an antiquated plant layout.

Layout types

In general, all plant layouts represent one or a combination of two basic types of layout. These basic types are product, or straight-line, layouts and process, or functional, layouts. In the straight-line layout, the machinery is located so that the flow from one operation to the next is minimized for any product class. Thus, in an organization that utilizes this technique, it would not be unusual to see a surface grinder located between a milling machine and a turret lathe, with an assembly bench and plating tanks in the immediate area. This type of layout is quite popular for certain mass-production manufacture because material handling costs are lower than when the process grouping of facilities is practiced.

One of the principal advantages of group technology is that it utilizes a product grouping type of plant layout. Through group technology, a sufficient volume of work utilizing the same equipment in the same sequence permits this type of layout. Thus, a plant with, say, seven product groups would be laid out with seven lines of flow based upon product grouping for each of these seven product groups. The remainder of the plant may be planned on a process type of layout to accommodate all work not falling within any of the seven product groups.

There are some distinct disadvantages of product grouping that should be kept in mind before making any major change in plant layout. Since a broad variety of occupations are represented in a small area, employee discontent can be fostered. This is especially true when the different opportunities carry a significant money rate differential. Because unlike facilities are grouped together, operator training becomes more cumbersome, since no experienced employee on a given facility may be available in the immediate area to train the new operator. The problem of finding competent supervisors is also enhanced, due to the variety of facilities and jobs that must be supervised. Then, too, this type of layout invariably necessitates greater initial investment in view of the duplicate service lines required, such as air, water, gas, oil, and power lines. Another disadvantage of product grouping that can result in indirect costs is the fact that this arrangement of facilities tends to give the casual observer the impression that disorder and chaos prevail. With these conditions, it is often difficult to promote good housekeeping.

In general, the disadvantages of product grouping are more than offset by the advantages if production requirements are substantial.

Process or functional layout is the grouping of similar facilities. Thus, all turret lathes would be grouped in one section, department, or building. Milling machines, drill presses, and punch presses would also be grouped in their respective sections.

This type of arrangement gives a general appearance of neatness and orderliness, and tends to promote good housekeeping. Another advantage of functional layout is the ease with which a new operator can be trained. Since he is surrounded by experienced employees operating similar machines, he has the opportunity to learn from them. The problem of finding competent supervisors is lessened because the job demands are not as great with this type of grouping. Since a supervisor need be familiar with but one general type or class of facilities, his background does not have to be as extensive as that of supervisors in shops using product grouping.

Of course, the obvious disadvantage of process grouping is the chance for long moves and backtracking on jobs that require a series of operations on diversified machines. For example, if the operation card of a job specified a sequence of drill, turn, mill, ream, and grind, the movement of the material from one section to the next could prove extremely costly. Another major disadvantage of process grouping is the great volume of paperwork required to issue orders and control production between sections.

Usually, if production quantities of similar products are limited and the plant is "job" or special order in form, then a process type of layout will be more satisfactory.

No two plants have completely identical layouts, even though the nature of their operations is similar. Many times, a combination of process grouping and product grouping is in order. Regardless of what type of grouping is contemplated, the analyst should consider the following principal points for layout improvement:

1. For straight-line mass production—material laid aside should be in position for the next operation.
2. For diversified production—the layout should permit short moves and deliveries, and the material should be convenient to the operator.
3. For multiple-machine operations—the equipment should be grouped around the operator.
4. For efficient stacking—storage areas should be arranged to minimize searching and rehandling.
5. For better worker efficiency—service centers should be located close to production areas.

Travel charts

Before a new layout can be designed or an old one corrected, the methods analyst will have to accumulate all facts that directly and indirectly influence the layout. These will include many, if not all, of the following:

1. The present and anticipated sales volume of each product, line, or class.
2. The labor content of each operation on each product.
3. A complete inventory of existing machines and material handling equipment.
4. The status of existing facilities from the standpoint of condition and book value.
5. Possible changes in product design.
6. Drawings of the existing plant indicating the location of all service facilities, windows, doors, columns, and reinforced areas.
7. The amount of handling taking place between facilities.

Once all of these facts have been gathered, the analyst should construct a flow process chart (see Chapter 3), which in itself gives the general form the layout will take. In the construction of the flowchart, suggestions should be encouraged from operators, inspectors, material handlers, and line supervisors. These employees are closer to the production work than anyone else, and they will often provide valuable suggestions.

Another chart that can be helpful in connection with both plant layout and material handling work is the travel chart. This tool helps diagnose problems related to the arrangement of departments and services areas as well as to the location of equipment within a given sector of the plant. The travel chart presents in matrix form the magnitude of material handling that takes place between two facilities. The unit used to identify the amount of handling may be whatever seems most appropriate to the analyst making the study. It can be pounds, tons, frequency of handling, and so on. The reader should recognize that the travel chart would have application only in process-type layouts. Figure 5–17 illustrates a very elemen-

FIGURE 5–17
The travel chart is a useful tool in solving material handling and plant layout problems related to process-type layouts

Travel Chart									
		To							
		No. 4 W. & S. Turret Lathe	Delta 17" Drill Press	2-Spindle L. & G. Drill	No. 2 Cinn. Hor. Mill	No. 3 B. & S. Vertical Mill	Niagara 100-Ton Press	No. 2 Cinn. Centerless	No. 3 Excello Thd. Grinder
From	No. 4 W. & S. Turret Lathe		20	45	80	32	4	6	2
	Delta 17" Drill Press			6	8	4	22	2	3
	2-Spindle L. & G. Drill				22	14	18	4	4
	No. 2 Cinn. Hor. Mill	120				10	5	4	2
	No. 3 B. & S. Vertical Mill						6	3	1
	Niagara 100-Ton Press	60	12	2				0	1
	No. 2 Cinn. Centerless	15							15
	No. 3 Excello Thd. Grinder				15	8			

tary travel chart. Note that the volume of the material routed from the No. 4 W. & S. turret lathe to the No. 2 Cinn. Hor. Mill is considerably less than the volume of the material routed from the No. 2 Cinn. Hor. Mill to the No. 4 W. & S. turret lathe.

Making the layout

In making the proposed layout, templates of all facilities must be pre-pared. Templates are usually constructed to a scale of ¼ inch equals 1 foot unless the size of the project is quite large, when a scale of ⅛ inch equals 1 foot may be used. If an existing layout is available, a copy can be made of it, and all facilities can be cut out of the print for use as templates. If no layout exists, two-dimensional templates can be purchased in printed form, as illustrated in Figure 5–18. Of course, the analyst can

FIGURE 5–18
Standard templates of representative machine tools

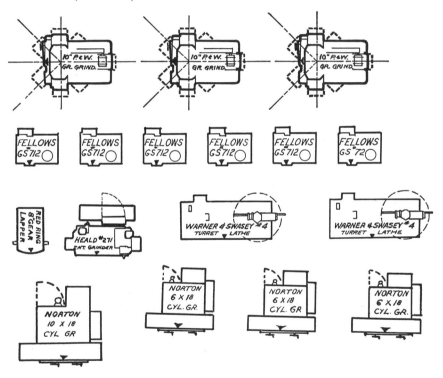

draw his or her own templates on a good grade of stiff cardboard and cut them out. The advantages of using stiff cardboard are apparent if the same templates are to be used several times.

Scale models give the third dimension to plant layouts, and are especially helpful to the analyst when he is endeavoring to sell his contemplated layout to a top executive who has neither the time nor the familiarity to grasp all details of the layout when it has been constructed on a two-dimensional basis (see Figure 5–19).

Once all the necessary templates have been made, a trial layout can be prepared. Good layouts will be evolved by giving consideration to the

FIGURE 5–19
Three-dimensional layout of a portion of a large industrial plant

Aluminum Co. of America

principal points for efficient layout and by providing adequate output capacity at each work station without introducing bottlenecks and interrupt-
ing the flow of production.

After an ideal layout has been designed, it is well to construct a flowchart of the proposed plan so that reduction in the distances traveled,
storages, delays, and overall costs will be highlighted. This will greatly
facilitate final approval of the design. A good technique for testing the
layout is to wind colored thread around the map tacks holding down the
templates, and with the thread to follow the flow of the product from its
raw material components to its transformation into a finished product. By
using a differently colored thread for every line of product produced, the
flow of all work can be visualized quite rapidly. This pictorial presentation,
supplemented with the flowchart, will bring to light most of the flaws of
the proposed method.

Summary: Plant layout

Possibilties for plant layout improvement are most likely to be uncovered if they are looked for systematically. Work stations and machines
should be arranged to permit the most efficient processing of a product
with a minimum of handling. A layout should not be changed until a care-

ful study of all the factors involved has been made. The methods analyst should learn to recognize poor layout and to present the facts to the plant engineer for his consideration. Computer programs (CRAFT, CORELAP, and so on) can rapidly provide layouts that represent a good first effort in the development of the recommended layout.

PRINCIPLES OF MOTION ECONOMY

The last of the 10 primary approaches to operation analysis has to do with improving the arrangement of parts at the workplace and of the motions required to perform the task. In Chapter 7, considerable detail is given to the laws of motion economy. At this point, consideration will be given only to the investigation of motion economy as practiced by the methods analyst. When studying work performed at a work station, the analyst should ask: "Are both hands working at the same time and in opposite symmetrical directions?" "Is each hand going through as few motions as possible?" "Is the workplace arranged so that long reaches are avoided?" and "Are both hands being used effectively and not as a holding device?"

If the answer to any of the above questions is no, then there are opportunities for improvement of the work station.

Both hands should work at the same time

The left hand, in right-handed people, can be used just as effectively as the right hand and should be given just as much consideration for use. It is common knowledge that a boxer learns to jab more effectively with his left hand than with his right hand. A stenographer will be just as proficient with one hand as the other. In a large number of instances, work stations can be designed to do "two at a time." This is a good slogan for the methods analyst to keep in mind. By providing dual fixtures to hold two components, it is often possible to have both hands working at the same time and making symmetrical moves in opposite directions (see Figure 5–20).

For example, a production increase of 100 percent was made possible by developing a fixture that utilized two-handed operation and permitted grinding two motor brushes at a time. The old method involved a one-handed operation, for only one piece was done during each grinding cycle.

Each hand should go through as few motions as possible

It is just common sense that the more motions the hands make while performing a task, the longer it will take to do the work. All hand motions are a series of reaches, moves, grasps, positions, and releases, and the more of these fundamental motions that can be eliminated or reduced, the more satisfactory the work station will be.

FIGURE 5-20
Work station and fixture designed to allow productive work with both hands

General Electric Co.

For example, by providing drop delivery and gravity chutes, it is possible to eliminate certain moves and positions and reduce release time. Likewise, installation of a belt conveyor to bring material to the work station and carry away the processed part will usually result in reduced "move" time. Position is usually a time-consuming basic division of accomplishment that in many instances can be minimized through well-designed fixtures. By using tapered channels and pilots, two mating parts can readily be assembled with considerable reduction or the complete elimination of the positioning element.

The workplace should be arranged to avoid long reaches

The time required to pick up an article depends to a great extent on the distance the hand must move. Likewise, move time is definitely related to distance. If at all possible, the workplace should be arranged so that all parts are within easy reach of the operator. If all components can be reached while both elbows are close to the body, then the work will be performed in the "normal" working area. This normal area represents the space within which work can be accomplished in a minimum time.

The work, tools, and parts should not be located beyond the reach of the hand when the arms are fully extended. The area encompassed with the arms fully extended in both the horizontal and the vertical plane represents the maximum working area, and the analyst should be on the alert for work stations that demand work performance beyond this area.

Avoid using the hand as a holding device

If either hand is ever used as a holding device during the processing cycle of a part, then useful work is not being performed by that hand. Invariably, a fixture can be designed to hold the work satisfactorily, thus allowing both hands to do useful work. Many times, foot-operated mechanisms can be introduced which permit utilizing both hands for productive

FIGURE 5–21
Ideal assembly work station with all components arranged to avoid long reaches and moves

Alden Systems Co.

work. The therblig "hold" represents an ineffective basic division of accomplishment and should be eliminated, if possible, from all work stations.

SUMMARY: 10 PRIMARY APPROACHES TO OPERATION ANALYSIS

The 10 primary approaches to operation analysis represent a systematic approach to analyzing the facts presented on the operation and flow process charts. To visualize how the various steps in methods engineering are integrated, a general flow of the entire analysis is illustrated in Figure 5–22. Other tools of analysis used in developing a method include the man and machine process chart, the operator process chart, and micromotion study. These subjects will be discussed in subsequent chapters.

Regardless of the nature of the work, whether continuous or intermittent, process or job shop, soft or hard goods, when systematic operation analysis is applied by competent personnel, real savings will result. It must be remembered that these principles are just as applicable in planning new work as in improving work already in production.

Increased output is the primary outcome of operation analysis, but operation analysis also distributes the benefits of improved production to all workers and helps develop better working conditions and methods, so that the worker can do more work at the plant, do a good job, and still have enough energy left to enjoy life.

A representative case history, which follows the method of analysis referred to, is that of the production of an Auto-Starter, a device for starting AC motors by reducing the voltage through a transformer. A subassembly of the Auto-Starter is the Arc Box. This part sits in the bottom of the Auto-Starter and acts as a barrier between the contacts so that there will be no short circuits. The present design consists of the following components:

Asbestos barriers with three drilled holes	6
Spacers of insulating tubing two inches long.	15
Steel rods threaded at both ends	3
Pieces of hardware	18
Pieces total	42

In assembling these components, the operator places a washer, lock washer, and nut on one end of each rod. The rods are then inserted through the three holes in the first barrier. Then one spacer is placed on each of the rods (three in all), and another barrier is added. This is repeated until six barriers are on the rods and separated by the tubing.

It was suggested that the six barriers be made with two slots—one at each end—and that two strips of asbestos for supporting the barriers be

FIGURE 5–22

The principal steps in a methods engineering program

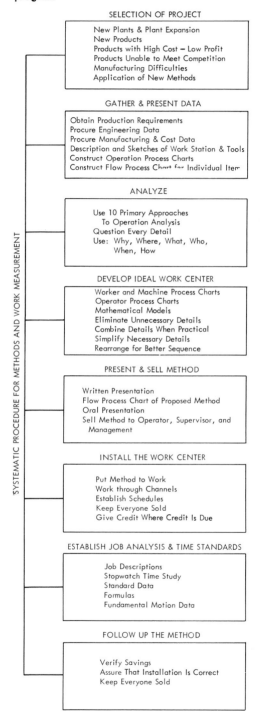

SELECTION OF PROJECT

New Plants & Plant Expansion
New Products
Products with High Cost — Low Profit
Products Unable to Meet Competition
Manufacturing Difficulties
Application of New Methods

GATHER & PRESENT DATA

Obtain Production Requirements
Procure Engineering Data
Procure Manufacturing & Cost Data
Description and Sketches of Work Station & Tools
Construct Operation Process Charts
Construct Flow Process Chart for Individual Item

ANALYZE

Use 10 Primary Approaches
  To Operation Analysis
Question Every Detail
Use: Why, Where, What, Who,
  When, How

DEVELOP IDEAL WORK CENTER

Worker and Machine Process Charts
Operator Process Charts
Mathematical Models
Eliminate Unnecessary Details
Combine Details When Practical
Simplify Necessary Details
Rearrange for Better Sequence

PRESENT & SELL METHOD

Written Presentation
Flow Process Chart of Proposed Method
Oral Presentation
Sell Method to Operator, Supervisor, and
  Management

INSTALL THE WORK CENTER

Put Method to Work
Work through Channels
Establish Schedules
Keep Everyone Sold
Give Credit Where Credit Is Due

ESTABLISH JOB ANALYSIS & TIME STANDARDS

Job Descriptions
Stopwatch Time Study
Standard Data
Formulas
Fundamental Motion Data

FOLLOW UP THE METHOD

Verify Savings
Assure That Installation Is Correct
Keep Everyone Sold

SYSTEMATIC PROCEDURE FOR METHODS AND WORK MEASUREMENT

made with six slots in each. These would be slipped together and placed in the bottom of the Auto-Starter as needed. The manner of assembly would be the same as that used in putting together the separator in an eggbox.

A total of 15 suggested improvements occurred after the analysis was completed, with the following results:

Old method	New method	Savings
42 parts	8	34
10 work stations	1	9
18 transportations	7	11
7,900 feet of travel	200	7,700
9 storages	4	5
0.45 hours time	0.11 hours time	0.34 hours
$1.55 costs	$0.60	$0.95

A program of operation analysis resulted in a 17,496-ton annual saving for one Ohio company. By forming a mill section into a ring and welding it, an original rough-forged ring weighing 2,198 pounds was replaced. The new mill section blank weighed only 740 pounds. The saving of 1,458 pounds of high-grade steel, amounting to twice the weight of the finished piece, was brought about by the simple procedure of reducing the excess material that had been cut away in chips.

An analytic laboratory in a New Jersey plant has applied the principles of operation analysis with gratifying results. New workbenches have been laid out in the form of a cross so that each chemist has an L-shaped work-table. This arrangement allows the chemist to reach any part of his work station by taking only one stride.

The new workbench has consolidated equipment, thus saving space and eliminating the duplication of facilities. One glassware cabinet services two chemists. A large four-place fume head allows multiple activity in this area that was formerly a bottleneck. All utility outlets are relocated for maximum efficiency.

In an effort to streamline its organization, a state government activity, developed a program of operation analysis that resulted in an estimated annual savings of more than 50,000 hours. This was brought about by combining, eliminating, and redesigning all paperwork activities; improving the plant layout; and developing paths of authority.

The application of methods improvement is as effective in office procedures as in production operations. One industrial engineering department of a Pennsylvania company[5] was given the problem of simplifying

---

[5] From a talk by Lynell Cooper, manufacturing engineer of Westinghouse Electric Corporation, to the central Pennsylvania chapter of the Society for Advancement of Management.

the paperwork necessary for shipping molded parts manufactured in one of its plants to an outlying plant for assembly. A new method was developed that reduced the average daily shipment of 45 orders from 552 sheets of paper forms to 50 sheets. The annual savings in paper alone brought about by this change was significant.

Methods improvement can and should be a continuing program. For example, in the early 1940s the average savings in the Procter & Gamble Company per member of management was approximately $500. In order to improve the situation and to maintain an acceptable level of cost reduction, a methods program was inaugurated, giving manufacturing management appreciation sessions in the concepts of operation analysis. Also, a special course was conducted for methods engineers, who were college graduates with one to five years of company experience. After completion of the operation analysis course, these engineers returned to their respective plants to work as methods specialists. The effect of this training was apparent in a very short time. Annual savings increased to approximately $700 per year per member of factory management. Since the program was a success, additional personnel were similarly trained and the program was extended to more plants.

By 1950, the position of the industrial engineer had changed from methods specialist to methods coordinator. It had been the practice for the engineer to suggest methods improvements to the line foreman. Since it was only natural for the foreman to take the suggested changes as criticism, frequently he did not participate actively.

Functioning as a methods coordinator, the engineer spent approximately two thirds of his time helping the factory supervisors on their projects and the other third working individually on special factory projects. Each member of plant management was assigned several selected projects where the cost was to be reduced, and each was encouraged to solicit assistance as required from the methods engineer.

The methods coordinator also periodically conducted training courses at his plant both for other staff members and for members of the line organization. With the active participation of plant management, the rate of savings had increased to $2,300 per year per member of factory management by 1950.

Subsequently, teams of four to eight employees, made up equally of line foremen and staff, were formed to work on selected projects having a high potential savings. More enthusiasm on the part of management resulted. Opportunities for the recognition of good work were increased. There was a friendly rivalry between teams for first place in plant standing. Display boards were set up in the front office to show team standings. Factory goals for cost reduction were established. By 1954, company-wide savings amounted to approximately $4,000 per member of management.

These results convinced top management that the cost reduction or profit improvement program should be plantwide and that all costs should be challenged. Consequently, the program was extended to all plants, and a summary sheet showing a comparison of the results of the program in each plant was circulated, which further stimulated the desire of individual plants to make a good showing.

The director of industrial engineering for Procter & Gamble made the following statement concerning the program:

Early in the program, plants tended to think that they had "skimmed the cream." The easy projects had all been picked off. Next year would be harder, they thought; consequently a lower goal was in order. We had to sell some people on the reasonableness of a higher goal each year.

Higher goals were reasonable because of increased experience at cost reduction. The growth in business was another factor. Getting the less active team members to do more offered real potential for increased savings.

Although we had concluded that the goal setting process should be democratic, some guides were useful. The teams were encouraged to compare themselves with others on a dollars-saved-per-member-of-management basis. Comparisons were also made in terms of goal as a percent of operating expense and as a percent of production value. The desire of plant management groups to show up well in all-plant comparisons was a strong factor in motivating teams to choose goals which required their best efforts for attainment.

The annual savings continued to increase each year, and by 1970 they amounted to $27,000 per member of management. This figure is conservative in that no credit is given for savings which continue beyond the first year. (See Figure 5-23).

Since 1970, the program has spread to involve people in all parts of

FIGURE 5–23
Progressive annual savings per member of management resulting from methods engineering in the Procter & Gamble Company

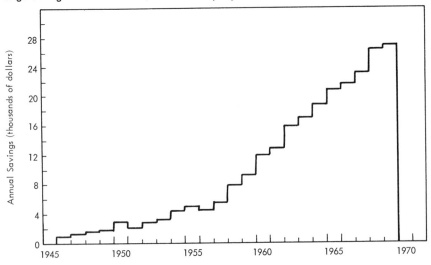

the business at both the nonmanagement and the management level, making dollars saved per member of management meaningless. By 1973, it was possible to adopt internally generated methods proposals which have benefited the company to the extent of over $100 million before tax annually.

TEXT QUESTIONS

1. In what four ways should an investigation to improve the process of manufacture be made?
2. Explain how rearranging operations can result in savings.
3. What process is usually considered the fastest for forming and sizing operations?
4. How should the analyst investigate the setup and tools to develop better methods?
5. Why should the methods analyst accept as part of his or her responsibility the provision of good working conditions?
6. Give the requisites of effective guarding.
7. What are the two general types of plant layout? Explain each in detail.
8. To what scale are templates usually constructed?
9. What is the best way to test a proposed layout?
10. What questions should the analyst ask himself when studying work performed at a work station?
11. For what purpose is operation analysis used?
12. Explain the advantage of using a checklist.
13. What are the primary approaches to operation analysis?
14. On what does the extent of tooling depend?
15. Do working conditions appreciably affect output? Explain.
16. How can planning and production control affect setup time?
17. How can you best handle a material?
18. Explain the effect of humidity on the operator.
19. Why does the travel chart have more application in process grouping than in product grouping?
20. Explain why a sound classification system is the first requisite of a successful group technology program.

GENERAL QUESTIONS

1. Where would you find application for a hydraulic elevating table?
2. What is the difference between a skid and a pallet?
3. Explain the significance of the colored code for stock templates.
4. When would you recommend using three-dimensional models in layout work?

5. What is the general flow of analysis procedure when it is applied to a product that has never been manufactured?

6. In a process like operation analysis, is it necessary to determine the point of diminishing returns? Why or why not?

7. Select five work stations and measure the footcandle intensity at these stations. How do these values compare with recommended practice?

8. Develop a classification system suitable for group technology where we have five product lines utilizing components with 13 geometric configurations and involving three types of thermosetting plastics. Make any assumptions that you feel are appropriate.

PROBLEMS

1. In the Dorben Company, the methods analyst was assigned the task of altering the work methods in the press department in order to meet OSHA standards relative to permissible noise exposures. She found the average sound level to be 100 dbA. The standard deviation was 10 dbA.

   The 20 operators in this department were provided with earplugs, and the power output from the public-address system was reduced from 30 watts to 20 watts. The deadening of the sound level by use of the earplugs was estimated to be 20 percent.

   What improvement resulted? Do you feel that this department is now in compliance with the law for 99 percent of its employees? Explain.

   $$\text{Hint: Decibel loss} = 10 \log \frac{\text{Power output original}}{\text{Power output planned}}$$

2. In the Dorben Company, the methods analyst designed a work station where the seeing task was difficult because of the size of the components going into the assembly. He established the desired brightness as being 100 footlamberts on the average, with a standard deviation of 10 footlamberts so as to accommodate 95 percent of the workers. The work station was painted a medium green having a reflectance of 50 percent.

   What would the required illumination in footcandles be at this work station in order to provide adequate illumination for 95 percent of the workers? Estimate what the required illumination would be if the work station were repainted with a light cream paint?

SELECTED REFERENCES

Buffa, Elwood S. *Modern Production Management*. 4th ed. New York: John Wiley & Sons, Inc., 1973.

Fogel, L. J. *Biotechnology: Concepts and Applications*. Englewood Cliffs, N.J.: Prentice-Hall, Inc., 1963.

Francis, Richard L., and White, John A. *Facility Layout and Location: An Analytical Approach*. Englewood Cliffs, N.J.: Prentice-Hall, Inc., 1974.

McCormick, E. J. *Human Factors Engineering*. 3d ed. New York: McGraw-Hill Book Co., 1970.

Miles, Lawrence D. *Techniques of Value Analysis and Engineering.* 2d ed. New York: McGraw-Hill Book Co., 1961.

Mudge, Arthur E. *Value Engineering: A Systematic Approach.* New York: McGraw-Hill Book Co., 1971.

Ostwald, Philip F. *Cost Estimating for Engineering and Management.* Englewood Cliffs, N.J.: Prentice-Hall, Inc., 1974.

Peat, A. P. *Cost Reduction Charts for Designers and Production Engineers.* Brighton, England: Machinery Publishing Co., Ltd., 1968.

Woodson, W. E. *Human Engineering Guide for Equipment Designers.* 2d ed., Berkeley: University of California Press, 1966.

6

Man and
machine relationships

Once an operation has been found necessary through analysis of the operation and flow process chart, it may frequently be improved through further analysis. The three process charts discussed in this chapter are helpful tools for the development of the ideal method. While the operation and flow process charts present facts and are used as tools of analysis, the man and machine process chart, gang process chart, and operator process chart are tools used to assist in the development of the ideal work center.

Man and machine process charts

While the operation and flow process charts are used primarily to explore a complete process or series of operations, the man and machine process chart is used to study, analyze, and improve only one work station at a time. This chart shows the exact relationship in time between the working cycle of the person and the operating cycle of his or her machine. With these facts clearly presented, possibilities exist for a fuller utilization of both man and machine time and a better balance of the work cycle.

Today, many of our machine tools are either completely automatic, such as the automatic screw machine, or are partially automatic, such as the turret lathe. In the operation of these types of facilities, the operator is often idle for a portion of the cycle. The utilization of this idle time can increase operator earnings and improve production efficiency.

The practice of having one employee operate more than one machine is known as "machine coupling." Machine coupling is not new. During the depression of the middle 1930s, one plant was unable to justify the amount of incentive earnings being turned in by the second shift in one of the machining departments. Investigation revealed that the operators were practicing machine coupling on their own initiative.

Today, some industries have encountered resistance to the practice of machine coupling from organized labor. The best way to sell machine coupling is to demonstrate the opportunity for added earnings. Since machine coupling will increase the percentage of "effort" time during the operating cycle, the opportunity for greater incentive earnings is enhanced if a company is on an incentive-type wage payment plan. Also, higher base rates will result when machine coupling is practiced, since the operator will have greater responsibility and will exercise more mental and physical effort under multiple-machine operation.

Construction of the man and machine process chart

In the construction of the man and machine process chart, the analyst should first identify the chart in the usual fashion by indicating at the top of the sheet "Man and Machine Process Chart." Immediately below this heading, the following information should be included: the part number, drawing number, description of the operation being charted, present or proposed method, date, and name of the person doing the charting.

Since man and machine charts are always drawn to scale, the analyst should next select a distance in inches to conform with a unit of time so that the chart can be neatly arranged on the sheet. The longer the cycle time of the operation being charted, the shorter the distance will be per decimal minute of time. Once exact values have been established for the distance in inches per unit of time, the analyst can begin the charting. On the left-hand side of the paper, the operations and time for the worker are shown, and to the right of the man time are shown graphically the working time and the idle time of the machine or machines, as the case may be. Working time on the part of the employee is represented by a solid line drawn vertically. A break in the vertical man-time line signifies idle time. Likewise, a solid vertical line under each machine heading indicates machine operating time, and a break in the vertical machine line designates idle machine time. Loading and unloading machine time is shown by a dotted line under the machine column, indicating that the machine is neither idle nor is production work being accomplished at the moment (see Figure 6–1).

All elements of occupied and idle time for both the worker and the machine he or she operates are charted until the termination of the cycle. At the foot of the chart are shown the employee's total working time and total idle time. Likewise, the total working time and idle time of each machine are posted. The productive time plus the idle time of the worker must equal the productive time plus the idle time of each machine he or she operates.

It will be noted that accurate elemental time values are necessary be-

FIGURE 6–1
Man and machine process chart for milling machine operation

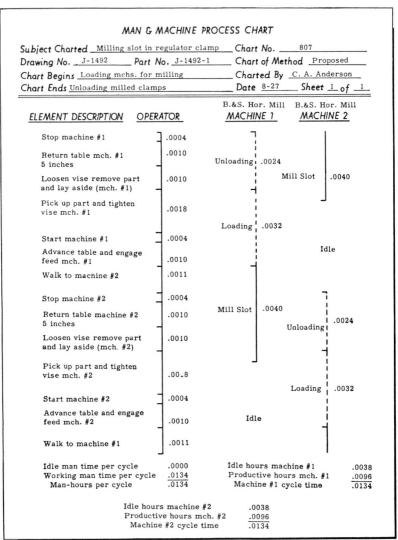

fore the man and machine chart can be constructed. These time values should represent standard times[1] which include an acceptable allowance to take care of fatigue, unavoidable delays, and personal delays. In no case should overall stopwatch readings be used in the construction of the chart.

The completed man and machine process chart clearly shows the areas

---

[1] See Chapter 17.

in which both idle machine time and man time occur. These areas are generally a good place to start effecting improvements. The analyst must be careful not to be deceived by what looks like an appreciable amount of idle man time. In many instances, it is far more economical to have a worker idle for a substantial portion of a cycle rather than chance having an expensive piece of equipment or process be idle for even a small portion of the cycle. In order to be sure that his proposal is the best solution, the analyst must know the cost of the idle facility as well as the cost of the idle worker. It is only when total cost

FIGURE 6–2
Man and machine process chart of thread-grinding operation with one man operating one machine

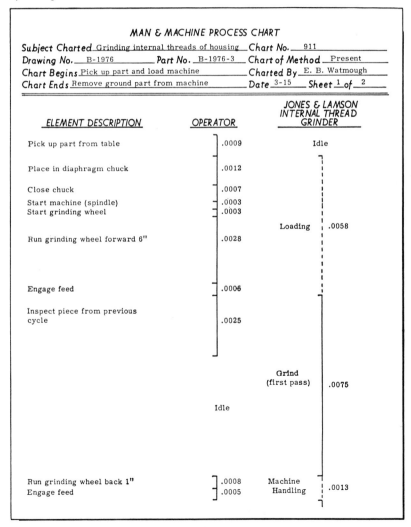

FIGURE 6–2 *(continued)*

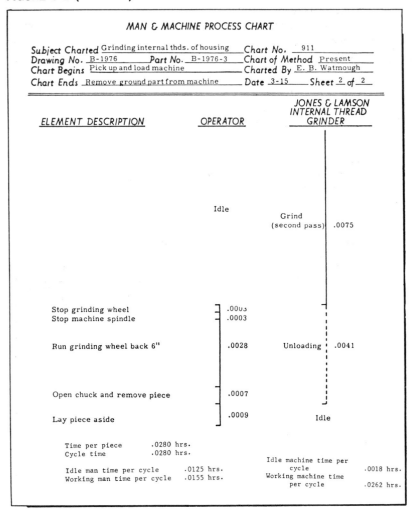

MAN & MACHINE PROCESS CHART

Subject Charted Grinding internal thds. of housing   Chart No. ___911___
Drawing No. _B-1976_____ Part No. _B-1976-3___ Chart of Method _Present___
Chart Begins _Pick up and load machine_____ Charted By _E. B. Watmough_
Chart Ends _Remove ground part from machine_____ Date _3-15_____ Sheet _2_ of _2_

| | | JONES & LAMSON INTERNAL THREAD |
| ELEMENT DESCRIPTION | OPERATOR | GRINDER |

Idle

Grind
(second pass) .0075

Stop grinding wheel .000ɔ
Stop machine spindle .0003

Run grinding wheel back 6"   .0028   Unloading .0041

Open chuck and remove piece   .0007

Lay piece aside   .0009   Idle

Time per piece   .0280 hrs.
Cycle time   .0280 hrs.

Idle man time per cycle   .0125 hrs.   Idle machine time per cycle   .0018 hrs.
Working man time per cycle .0155 hrs.   Working machine time per cycle .0262 hrs.

is considered that the analyst can safely recommend one method over another. Figure 6–3 illustrates a man and machine process chart that represents economies over the one shown in Figure 6–2.

Using the man and machine process chart

The analyst will construct a man and machine process chart when his preliminary investigation reveals that the working cycle of the operator is somewhat shorter than the operating cycle of the machine. Once he has completed the chart, the logical place to consider improvement possibilities is during the idle portion of the operator's cycle. Considering the

FIGURE 6–3

Man and machine process chart of thread-grinding operation with one man operating two machines

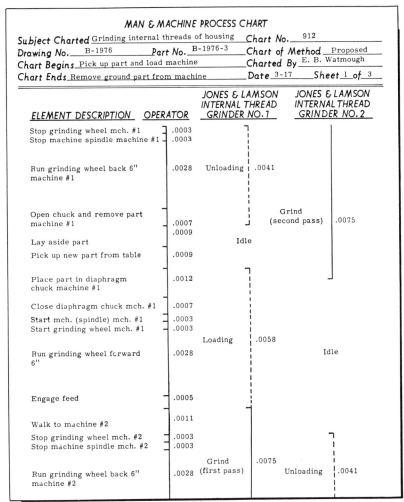

MAN & MACHINE PROCESS CHART

Subject Charted Grinding internal threads of housing   Chart No. 912

Drawing No. B-1976   Part No. B-1976-3   Chart of Method Proposed

Chart Begins Pick up part and load machine   Charted By E. B. Watmough

Chart Ends Remove ground part from machine   Date 3-17   Sheet 1 of 3

ELEMENT DESCRIPTION	OPERATOR	JONES & LAMSON INTERNAL THREAD GRINDER NO. 1		JONES & LAMSON INTERNAL THREAD GRINDER NO. 2	
Stop grinding wheel mch. #1	.0003				
Stop machine spindle machine #1	.0003				
Run grinding wheel back 6" machine #1	.0028	Unloading	.0041		
Open chuck and remove part machine #1	.0007		Grind (second pass)	.0075	
Lay aside part	.0009	Idle			
Pick up new part from table	.0009				
Place part in diaphragm chuck machine #1	.0012				
Close diaphragm chuck mch. #1	.0007				
Start mch. (spindle) mch. #1	.0003				
Start grinding wheel mch. #1	.0003	Loading	.0058	Idle	
Run grinding wheel forward 6"	.0028				
Engage feed	.0005				
Walk to machine #2	.0011				
Stop grinding wheel mch. #2	.0003				
Stop machine spindle mch. #2	.0003	Grind (first pass)	.0075		
Run grinding wheel back 6" machine #2	.0028			Unloading	.0041

amount of this time, he should investigate the possibility of assigning the worker the added responsibility of (1) operating a second machine during this idle time, and (2) performing some bench or manual operation, such as filing burrs or gaging parts, during the idle time.

Sometimes more available operator time can be obtained by reducing the speed and feed of the machine. This may permit machine coupling where it would not have been possible otherwise, and thus reduce total costs. As mentioned earlier, it is not always advisable to practice machine coupling, as the idle machine time introduced may more than offset the

FIGURE 6–3 *(continued)*

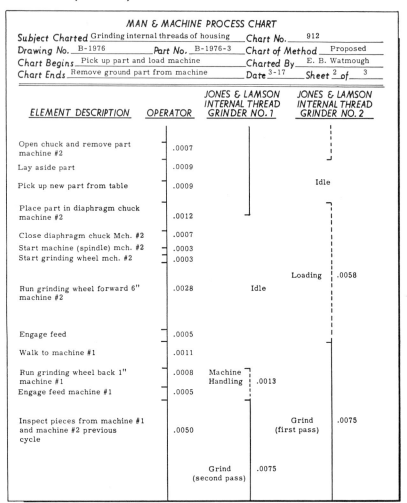

idle operator time saved. The only sure way to make the analysis in on a total cost basis.

Man and machine charts are effective in determining the extent of coupling justified in order to assure a "fair day's work for a fair day's pay." They are valuable for determining how idle machine time may be more fully utilized.

Gang process charts

The gang process chart is, in a sense, an adaptation of the man and machine chart. After completing a man and machine process chart, the analyst should be able to determine the most economical number of machines to be operated by one worker. However, several processes and

FIGURE 6–3 (concluded)

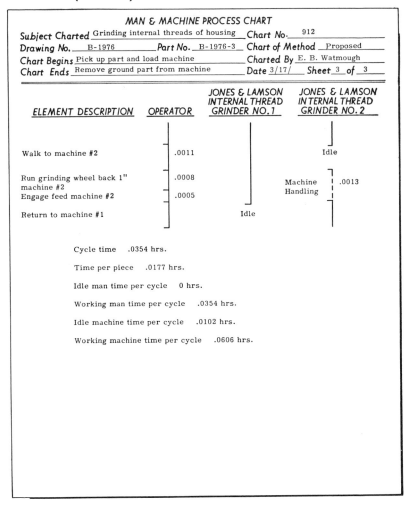

MAN & MACHINE PROCESS CHART
Subject Charted _Grinding internal threads of housing_ Chart No._912_
Drawing No. _B-1976_ Part No. _B-1976-3_ Chart of Method _Proposed_
Chart Begins _Pick up part and load machine_ Charted By _E. B. Watmough_
Chart Ends _Remove ground part from machine_ Date 3/17/ _Sheet 3 of 3_

ELEMENT DESCRIPTION	OPERATOR	JONES & LAMSON INTERNAL THREAD GRINDER NO. 1	JONES & LAMSON INTERNAL THREAD GRINDER NO. 2
Walk to machine #2	.0011		Idle
Run grinding wheel back 1" machine #2	.0008		Machine Handling .0013
Engage feed machine #2	.0005		
Return to machine #1		Idle	

Cycle time   .0354 hrs.

Time per piece   .0177 hrs.

Idle man time per cycle   0 hrs.

Working man time per cycle   .0354 hrs.

Idle machine time per cycle   .0102 hrs.

Working machine time per cycle   .0606 hrs.

facilities are of such magnitude that it is not a question of how many machines a worker should operate, but of how many workers it should take to operate one machine effectively. The gang process chart shows the exact relationship between the idle and operating cycle of the machine and the idle and operating time per cycle of the workers who service it. This chart shows clearly the possibilities for improvement by reducing both idle man time and idle machine time.

Construction of the gang process chart

After heading the chart "Gang Process Chart" and completely identifying the process being plotted with the part number, drawing number, description of the operation being charted, present or proposed method,

FIGURE 6-4

GANG PROCESS CHART OF PRESENT METHOD

HYDRAULIC EXTRUSION PRESS   DEPT. 11   BELLEFONTE PA. PLANT

CHARTED BY B.W.N.   4-15-   CHART NO. G-85

MACHINE	TIME	PRESS OPERATOR	TIME	ASSISTANT PRESS OPERATOR	TIME	FURNACE MAN	TIME	DUMMY KNOCKER	TIME	ASSISTANT DUMMY KNOCKER	TIME	PULL-OUT MAN	TIME
Elevate Billet	.07	Elevate Billet	.07	Grease Die & Position Back in Die Head	.12	Rearrange Billets in Furnace	.20	Position Shell on Small Press	.10	Move Away from Small Press and Lay Aside Tongs	.12	Pull Rod toward Cooling Rack	.20
Position Billet	.08	Position Billet	.08					Press Dummy Out of Shell	.12			Walk Back toward Press	.15
Position Dummy	.04	Position Dummy	.04					Dispose of Shell	.18				
Build Pressure	.05	Build Pressure	.05										
Extrude	.45	Extrude	.45	Idle Time	.68	Idle Time	.51	Dispose of Dummy and Lay Aside Tongs	.12	Idle Time	.68	Grab Rod with Tongs and Pull Out	.45
								Idle	.43				
Unlock Die	.06	Unlock Die	.06	Run Head & Shell Out	.11	Open Furnace Door & Remove Billet	.19			Guide Shell from Shear to Small Press	.20	Straighten Rod End with Mallet	.11
Loosen & Push Out Shell	.10	Loosen & Push Out Shell	.10	Shear Rod from Shell	.04			Grab Tongs & Move to Position	.05			Hold Rod while Die Removed at Press	.09
Withdraw Ram & Lock Die in Head	.15	Withdraw Ram & Lock Die in Head	.15	Pull Die Off End of Rod	.05	Ram Billet from Furnace & Close Furnace Door	.10						
WORKING TIME	1.00 MIN.		1.00 MIN.		.32 MIN.		.49 MIN.		.57 MIN.		.32 MIN.		1.00 MIN.
IDLE TIME	0 "		0 "		.68 "		.51 "		.43 "		.68 "		0 "

IDLE TIME = 2.30 MAN-MINUTES PER CYCLE = 18.4 MAN-HOURS PER EIGHT-HOUR DAY date, and name of the person doing the charting, the analyst should select a time scale that will give a neat-appearing chart on the paper being used. Like the man and machine process chart, the gang process chart is always drawn to scale.

On the left-hand side of the paper are shown the operations being performed on the machine or process. Immediately to the right of the operation description, the loading time, operating time, and idle time of the machine are represented graphically. Further to the right, the operating time and idle time of each operator manning the process are illustrated by flow lines in a vertical direction. A solid vertical line indicates that productive work is being done, while a dotted vertical line in connection with the facility shows that either loading or unloading operations are taking place. A void in the vertical flow line reveals idle time, and the length of the void determines the duration of the idle time. In the case of the operators, solid vertical lines show that work is being done, while breaks in the solid lines show idle time. Figure 6–4 illustrates a gang process chart, and it is apparent that a large number of idle man-hours exist. A better manning of the same process is shown on the gang process chart illustrated in Figure 6–5. The saving of 16 hours per shift was easily developed through the use of the gang process chart.

Using the gang process chart

The analyst usually constructs the gang process chart when his initial investigation of a given operation indicates that more workers than are necessary are being used to operate a facility or process. If he suspects this, he will find that the gang process chart is a useful tool for determining the exact number of operators needed to service a machine or process effectively. Once the chart has been constructed, the hours of idle man time can be analyzed to determine the possibility of utilizing one operator to perform the work elements currently performed by two or more.

For example, in the gang process chart shown in Figure 6–4, the company is employing two more operators than are needed. This is apparent when it is shown that under this process 18.4 idle man-hours are involved in every 8-hour turn.

By relocating some of the controls of the process, it was possible to reassign the elements of work so that four, rather than six, workers effectively operated the extrusion press.

Figure 6–5 illustrates a gang process chart of the proposed method using but four operators and so making a savings of 16 man-hours per shift. Without the gang process chart, this solution would have been quite difficult.

From the example given in Figures 6–4 and 6–5, it can be seen that the gang process chart aids in dividing the available work among the members

FIGURE 6–5

Gang process chart of the proposed method of operation of a hydraulic extrusion process

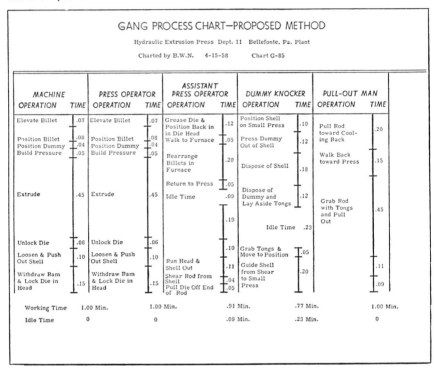

GANG PROCESS CHART—PROPOSED METHOD

Hydraulic Extrusion Press Dept. 11   Bellefonte, Pa. Plant

Charted by B.W.N.   4-15-58   Chart G-85

MACHINE OPERATION	TIME	PRESS OPERATOR OPERATION	TIME	ASSISTANT PRESS OPERATOR OPERATION	TIME	DUMMY KNOCKER OPERATION	TIME	PULL-OUT MAN OPERATION	TIME
Elevate Billet	.07	Elevate Billet	.07	Grease Die & Position Back in Die Head	.12	Position Shell on Small Press	.10	Pull Rod toward Cooling Rack	.20
Position Billet	.08	Position Billet	.08	Walk to Furnace	.05	Press Dummy Out of Shell	.12		
Position Dummy	.04	Position Dummy	.04						
Build Pressure	.05	Build Pressure	.05	Rearrange Billets in Furnace	.20	Dispose of Shell	.18	Walk Back toward Press	.15
				Return to Press	.05				
Extrude	.45	Extrude	.45	Idle Time	.09	Dispose of Dummy and Lay Aside Tongs	.12	Grab Rod with Tongs and Pull Out	.45
					.19	Idle Time	.23		
Unlock Die	.06	Unlock Die	.06		.10	Grab Tongs & Move to Position	.05		
Loosen & Push Out Shell	.10	Loosen & Push Out Shell	.10	Run Head & Shell Out	.11	Guide Shell from Shear to Small Press	.20		.11
Withdraw Ram & Lock Die in Head	.15	Withdraw Ram & Lock Die in Head	.15	Shear Rod from Shell	.04				.09
				Pull Die Off End of Rod	.05				
Working Time	1.00 Min.		1.00 Min.		.91 Min.		.77 Min.		1.00 Min.
Idle Time	0		0		.09 Min.		.23 Min.		0

of the team that operates the equipment, and then clearly determines the job assignments of all involved. Through the construction and use of the gang process chart, equipment will be operated to capacity, labor costs will be reduced, and employee morale will be improved as a result of the equitable distribution of work assignments.

Quantitative techniques for man and machine relationships

Although the man and machine process chart can be used to determine the number of facilities to be assigned to an operator, this can often be computed in much less time through the development of a mathematical model.

Man and machine relationships are usually one of three types: (1) synchronous servicing, (2) completely random servicing, and (3) a combination of synchronous and random servicing.

Assigning more than one machine to an operator seldom results in the ideal case where both the worker and the machine he or she is servicing are occupied the whole cycle. Ideal cases such as this are referred to as "synchronous servicing," and the number of machines to be assigned can be computed as:

$$N = \frac{l + m}{l}$$

where:

N = Number of machines the operator is assigned
l = Total operator servicing time per machine (loading and unloading)
m = Total machine running time (power feed)

For example, if the total operator servicing time was one minute, while the cycle time of the machine was four minutes, synchronous servicing would result in the assignment of five machines:

$$N = \frac{1 + 4}{1}$$
$$= 5$$

Graphically, this assignment would appear as shown below:

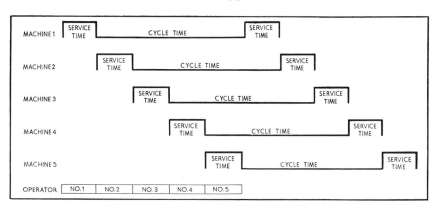

If the number of machines in the above example is increased, machine interference will take place, and we have a situation where one or more of the facilities are idle for a portion of the work cycle. If the number of machines is reduced to some figure less than five, then the operator will be idle for a portion of the cycle.

In such cases, the minimum total cost per piece usually represents the criterion for optimum operation. The best method will need to be established, considering the cost of each idle machine and the hourly rate of the operator. Quantitative techniques can be used to establish the best arrangement. The procedure is, first, to estimate the number of facilities that the operator should be assigned by establishing the lowest whole number from the equation:

$$N_1 \leqq \frac{l + m}{l + w}$$

where:

N_1 = Lowest whole number
w = Normal walking time to the next facility in hours

It can be seen from the above that the cycle time with the operator servicing N_1 machines is $l + m$ since in this case the operator will not be busy the whole cycle, while the facilities he or she is servicing will be occupied during the entire cycle.

Using N_1, the total expected cost may be computed as follows:

$$\text{T.E.C}_{N_1} = \frac{K_1(l + m) + N_1 K_2(l + m)}{N_1}$$

$$= \frac{(l + m)(K_1 + N_1 K_2)}{N_1}$$

where:

T.E.C. = Cost of production per cycle from one machine
K_1 = Operator rate in dollars per hour
K_2 = Cost of machine in dollars per hour

After this cost is computed, a cost should be calculated with $N_1 + 1$ machines assigned to the operator. In this case, the cycle time will be governed by the working cycle of the operator, since there will be some idle machine time. The cycle time will now be $(N_1 + 1)(l + w)$. Let $N_2 = N_1 + 1$. Then the total expected cost with N_2 facilities is:

$$\text{T.E.C}_{N_2} = \frac{(K_1)(N_2)(l + w) + (K_2)(N_2)(N_2)(l + w)}{(N_2)}$$

$$= [(l + w)][K_1 + K_2(N_2)]$$

The number of machines assigned will depend on whether N_1 or N_2 gives the lowest total expected cost per piece.

"Completely random servicing" situations refer to those cases in which it is not known when a facility will need to be serviced or how long servicing will take. Mean values are usually known or can be determined; with these averages, the laws of probability can provide a useful tool in determining the number of machines to assign the operator.

The successive terms of the binomial expansion will give a useful approximation of the probability of 0, 1, 2, 3, . . . n machines down (where n is relatively small), assuming that each machine is down at random times during the day and that the probability of running time is p and the probability of downtime is q.

For example, let us determine the minimum proportion of machine time lost for various numbers of turret lathes assigned to an operator where it has been estimated that on the average the machines operate 60 percent of the time unattended. Operator attention time at irregular intervals will be 40 percent on the average. The analyst estimates that three turret lathes should be assigned per operator on this class of work. Under this arrangement, the combinations of machine running (p) or down (q) expressed as probabilities would be:

$$(p + q)^n = (p + q)^3$$
$$= p^3 + 3p^2q + 3pq^2 + q^3$$
$$= (0.60)^3 + (3)(0.60)^2(0.40) + 3(0.60)(0.40)^2 + (0.40)^3$$
$$1.00 = 0.216 + 0.432 + 0.288 + 0.064$$

In tabular form, this would appear as follows:

Machine 1	Machine 2	Machine 3	Probability
	$R = 0.60$	$R = 0.60$	$(0.60)(0.60)(0.60) = 0.216$
		$D = 0.40$	$(0.60)(0.60)(0.40) = 0.144$
$R = 0.60$			
	$D = 0.40$	$R = 0.60$	$(0.60)(0.40)(0.60) = 0.144$
		$D = 0.40$	$(0.60)(0.40)(0.40) = 0.096$
	$R = 0.60$	$R = 0.60$	$(0.40)(0.60)(0.60) = 0.144$
		$D = 0.40$	$(0.40)(0.60)(0.40) = 0.096$
$D = 0.40$			
	$D = 0.40$	$R = 0.60$	$(0.40)(0.40)(0.60) = 0.096$
		$D = 0.40$	$(0.40)(0.40)(0.40) = \underline{0.064}$
			1.000

Thus, the proportion of time that some machines will be down may be determined, and the resulting lost time of one operator per three machines may be readily computed. In this example, we have:

No. of machines down	Probability	Machine hours lost per 8-hour day
0	0.216	0
1	0.432	0*
2	0.288	$(0.288)(8) = 2.304$
3	0.064	$(2)(0.064)(8) = \underline{1.024}$
	1.000	3.328

* Since only one machine is down at a time, the operator can be attending the down machine.

$$\text{Proportion of machine time lost} = \frac{3.328}{24.0} = 13.9 \text{ percent}$$

Similar computations can be made for more or less machine assignments to determine the assignment resulting in the least machine downtime. The most satisfactory arrangement is usually considered to be the arrangement showing the least Total Expected Cost per piece. The Total Expected Cost per piece of a given arrangement can be computed by the expression:

$$\text{T.E.C.} = \frac{K_1 + NK_2}{\text{Pieces from } N \text{ machines per hour}}$$

where:

$K_1 = $ Hourly rate of the operator
$K_2 = $ Hourly rate of the machine
$N = $ Number of machines assigned

The pieces per hour from N machines can be computed knowing the mean machine time required per piece, the average machine servicing time per piece, and the expected down or lost time per hour.

For example, under a five-machine assignment to one operator, it was determined that the machining time per piece was 0.82 hours, the machine serving time per piece 0.17 hours, and the machine downtime an average of 0.11 hours per machine per hour. Thus, each machine was available for production work only 0.89 hours each hour. The average time required to produce one piece per machine would be $\dfrac{0.82 + 0.17}{0.89} = 1.11$. Therefore, the five machines would produce 4.5 pieces per hour. With an operator hourly rate of $2.80 and a machine hourly rate of $6, we have a total expected cost per piece of:

$$\frac{\$2.80 + 5(\$6.00)}{4.5} = \$7.29$$

Combinations of synchronous and random servicing are perhaps the most common type of man and machine relationships. Here the servicing time is constant, but the machine downtime is random. Winding, coning, and quilling operations used in the textile industry are characteristic of this type of man and machine relationship. As in the former examples, algebra and probability can establish the mathematical model that will enhance a realistic solution, (see queuing theory, Chapter 22).

Line balancing

The problem of determining the ideal number of workers to be assigned to a production line is analogous to the problem of determining the number of workers to be assigned to a production facility where the use of the gang process chart was recommended. Perhaps the most elementary line balancing situation, yet one that is very often encountered, is a situation in which several operators, each performing consecutive operations, are working as a unit. In such a situation, it is obvious that the rate of production is dependent upon the slowest operator. For example, we may have a line of five operators assembling bonded rubber mountings prior to the curing process. The specific work assignments might be as follows: Operator 1, 0.52 minutes; Operator 2, 0.48 minutes; Operator 3, 0.65 minutes; Operator 4, 0.41 minutes; Operator 5, 0.55 minutes. Operator 3 establishes the pace, as is evidenced by the following:

Operator	Standard minutes to perform operation	Wait time based on slowest operator	Allowed standard minutes
1	0.52	0.13	0.65
2	0.48	0.17	0.65
3	0.65	–	0.65
4	0.41	0.24	0.65
5	0.55	0.10	0.65

The efficiency of this line can be computed as the ratio of the total standard minutes to the total allowed standard minutes, or:

$$E = \frac{\sum\limits_{1}^{5} S.M.}{\sum\limits_{1}^{5} A.M.} \times 100 = \frac{2.61}{3.25} \times 100 = 80 \text{ percent}$$

where:

E = Efficiency

$S.M.$ = Standard minutes per operation

$A.M.$ = Allowed standard minutes per operation

It is apparent that a real-life situation similar to the above example provides the opportunity for significant savings on the part of the methods analyst. If he can save 0.10 minute on Operator 3, the net savings per cycle is not 0.10 minute but 0.10×5 or 0.50 minutes.

Only in the most unusual situations would a line be perfectly balanced; that is, the standard minutes to perform an operation would be identical for each member of the team. The reader should recognize at this time that the "standard minutes to perform an operation" is really not a standard. It is only a standard in the eyes of the individual who established it. Thus, in our example, where Operator 3 was shown to have a standard time of 0.65 minutes to perform the first operation, it should be recognized that a different work measurement analyst might have allowed as little as 0.61 minutes or as much as 0.69 minutes. The range of standards established by different work measurement analysts on the same operation might even be greater than the range that has been suggested. The point is that whether the issued standard is 0.61, 0.65, or 0.69, the typical conscientious operator will have little difficulty in meeting the standard. In fact, he or she will probably better the standard in view of the performance of the operators on the line with less work content in their assignments. It should also be recognized that those operators who have a wait time based on the output of the slowest operator will seldom be observed as actually waiting. They will, instead, reduce the tempo of their movements so as to utilize the number of standard minutes established by the slowest operator.

The number of allowed standard minutes to produce one unit of the product will be equal to the summation of the standard minutes required times the reciprocal of the efficiency. Thus,

$$\Sigma A.M. = \Sigma S.M. \times \frac{1}{E}$$

It is apparent then that the number of operators needed is equal to the required rate of production times the total allowed minutes, where:

N = Number of operators needed in the line
R = Desired rate of production
$N = R \times \Sigma A.M.$

For example, let us assume that we have a new design for which we are establishing an assembly line. Eight distinct operations are involved. The line must produce 700 units per day, and since it is desirable to minimize storage, we do not want to produce many more than 700 units per day. The eight operations involve the following standard minutes based upon existing standard data: Operation 1, 1.25 minutes; Operation 2, 1.38 minutes; Operation 3, 2.58 minutes; Operation 4, 3.84 minutes; Operation 5, 1.27 minutes; Operation 6, 1.29 minutes; Operation 7, 2.48 minutes; and Operation 8, 1.28 minutes. The analyst wishes to plan this assembly line for the most economical setup. He is able to estimate the number of operators required at 100 percent efficiency as follows:

$$\sum_{1}^{8} S.M. = 15.37 \text{ minutes}$$

$$N = \frac{700}{480} \times \frac{15.37}{E} = \frac{22.4}{E}$$

If the analyst plans for 95 percent efficiency, he would estimate the number of operators to be: $\frac{22.4}{0.95} = 23.6$.

Since it is impossible to have six tenths of an operator, the analyst will endeavor to set up the line utilizing 24 operators.

The next step will be to estimate the number of operators to be utilized at each of the eight specific operations.

Since 700 units of work are required a day, it will be necessary to produce one unit in about 0.685 minutes $\left(\dfrac{480}{700}\right)$. The analyst can estimate how many operators will be needed on each operation by dividing the number of minutes in which it is necessary to have one piece produced into the standard minutes of each operation.

Operation	*Standard minutes*	*Standard minutes / Minutes/unit*	*No. of operators*
Operation 1	1.25	1.83	2
Operation 2	1.38	2.02	2
Operation 3	2.58	3.77	4
Operation 4	3.84	5.62	6
Operation 5	1.27	1.86	2
Operation 6	1.29	1.88	2
Operation 7	2.48	3.62	4
Operation 8	1.28	1.87	2
Total	15.37		24

In order to determine which is the slowest operation, the analyst divides the estimated number of operators into the standard minutes for each of the eight operations.

Operation 1	1.25/2 = 0.625
Operation 2	1.38/2 = 0.690
Operation 3	2.58/4 = 0.645
Operation 4	3.84/6 = 0.640
Operation 5	1.27/2 = 0.635
Operation 6	1.29/2 = 0.645
Operation 7	2.48/4 = 0.620
Operation 8	1.28/2 = 0.640

Thus, Operation 2 will determine the output from the line. In this case, it will be:

$$\frac{2 \text{ men} \times 60 \text{ min.}}{1.38 \text{ standard minutes}} = 87 \text{ pieces per hour, or } 696 \text{ pieces per day}$$

If this rate of production is inadequate, the analyst must increase the rate of production of Operator 2. This can be accomplished by:

1. Working one or both of the operators at the second operation overtime, thus accumulating a small inventory at this work station.

2. Utilizing the services of a third man (on a part-time basis) at the work station of Operation 2.
3. Reassigning some of the work of Operation 2 to Operation 1 or Operation 3. (It would be preferable to assign more work to Operation 1.)
4. Improving the method at Operation 2 so as to diminish the cycle time of this operation.

In the preceding example, the analyst has been given a cycle time and operation times and has determined the number of operators needed for each operation in order to meet a desired production schedule.

The production line work assignment problem can also be to minimize the number of work stations, given the desired cycle time; or, given the number of work stations, to assign work elements to the work stations, within the restrictions established, so as to minimize the cycle time.

A strategy that should not be overlooked in assembly line balancing is work element sharing. Thus, two or more operators whose work cycle includes some idle time may share the work of another station for the purpose of a more efficient line. For example, in Figure 6–6 an assembly line

FIGURE 6–6
Assembly line involving six work stations

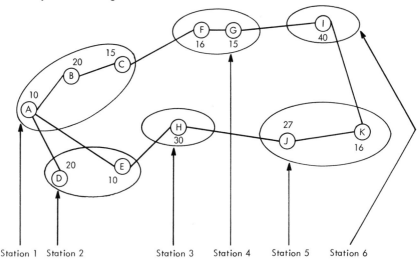

involving six work stations is shown. Station 1 has three work elements to accomplish—A, B, and C, for a total of 45 seconds. Note that work elements B, D, and E cannot begin until A is completed and that there is no precedence among B, D, and E. It may be possible to have element H shared by stations 2 and 4 with only a one-second increase in cycle time (from 45 seconds to 46 seconds) while saving 40 seconds per assembled unit. It should be recognized that element sharing may result in an increase in material handling since parts may have to be delivered to more than one location. In addition, element sharing may necessitate added costs for duplicate tooling.

A second possibility in improving the balance of an assembly line is dividing a work element. Referring again to Figure 6–6, it may be possible to divide element H rather than have half of the number of parts go to station 2 and the other half to station 4.

Many times it is not economical to divide an element. An example is driving home eight machine screws with a power screwdriver. Once the operator has located the part in a fixture, gained control of the power tool, and brought it to the work, it would usually be more advantageous to drive home all eight screws rather than drive only a portion home and have a different operator drive the rest home. Frequently, however, elements can be divided, and work stations may be better balanced as a result of the division.

The analyst should be aware that a different sequence of assembly may result in more favorable results. Product design generally dictates the assembly sequence. However, there are often alternatives that should not be overlooked. Balanced assembly lines are not only less costly, but they assist in maintaining worker morale, since in such lines little differential exists in the work content of the different workers.

The following procedure for assisting in the solution of the assembly line balancing problem is based on the General Electric Assembly Line Balancing publication. General Electric engineers have prepared a program, written for the GE–225 computer, for assigning work elements to stations on an assembly line.

The method established for the solution of the assembly line problem is based on the following:

1. Operators are not able to move from one work station to another in order to help maintain a uniform work load.
2. The work elements that have been established are of such magnitude that further division would substantially decrease the efficiency of performing the work element. (Once established, the work elements should be identified with a code.)

The first step in the solution of the problem is the determination of the sequence of the individual work elements. As can easily be recognized, the fewer the restrictions on the order in which the work elements can be done, the greater the probability of a favorable balance in the work assignments. In order to determine the sequence of the work elements, the analyst must ask and answer the question: "What other work elements, if any, must be completed before this work element can be started?"

It is recommended that a precedence chart be completed for the production line under study (see Figure 6–7). The analyst should recognize that not only functional design, but available production methods, floor space, and so on, can introduce contraints as far as the work element sequence is concerned.

FIGURE 6–7
Partially completed precedence chart

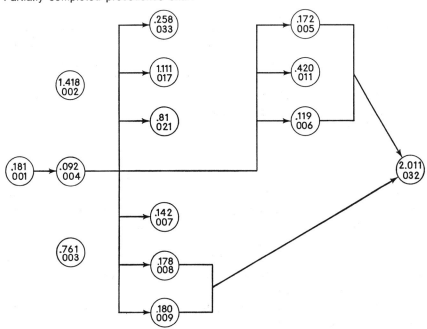

Note that work elements 002 and 003 may be done in any sequence with respect to any of the other work elements and that 032 cannot be started until 005, 006, 008, and 009 have been completed. Note also that after 004 has been finished we may start either 033, 017, 021, 007, 008, or 009.

A second consideration in the production line work assignment problem is the recognition of zoning restraints. A zone represents a subdivision, which may or may not be physically separated or identified from other zones in the system. Confining certain work elements to a given zone may be justified in order to congregate similar jobs, working conditions, or pay rates. Or again, it may be desirable to introduce zoning restraints in order to identify physically specific stages of a component, such as keeping it in a certain position while work elements are being performed. Thus, all work elements related to one side of a component may be performed in a certain zone before the component is turned over.

Obviously, the more the zoning restraints that are placed on the system, the fewer the combinational possibilities open to investigation. It is helpful for the analyst to make a sketch of the system and to code the applicable zones. Within each zone should be shown the work elements that may be done in this area.

The next step is to estimate the production rate. This is done with the expression:

$$\text{Production per day} = \frac{\text{Working minutes/day}}{\text{Cycle time of system (min./unit)}} \times \text{rating factor}$$

For example, assuming a 15 percent allowance, we would have 480 — 72, or 408, working minutes per day. The rating factor would be based on experience with the type of line under study. It could be either more or less than standard (100 percent). The cycle time of the system is the cycle time of the limiting station. Since we know the production requirements per day, we are able to compute the allowed cycle time of the limiting station.

The reader should recognize that the computer-generated assignment does not necessarily represent the optimum. It should be regarded as a good base from which the methods analyst can work and upon which he can hopefully improve. For example, the analyst may be able to divide an element and thus bring about a more efficient overall arrangement. Or again, he may see fit to modify zoning restrictions, or to alter precedence relationships.

The computer can be used to select the assignments for each operator, taking into account the cycle time of the system, precedence, and zoning.

In order to illustrate the logic of the computer routine, the following precedence graph will be described:

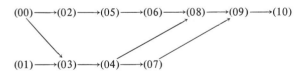

It can be seen from this precedence graph that work unit (00) must be completed before (02), (03), (05), (06), (04), (07), (08), (09), and (10); and that work unit (01) must be completed before (03), (04), (07), (08), (09), and (10). Either (00) or (01) can be done first, or they can be done concurrently. Work unit (03) cannot be started until work units (00) and (01) are completed, and so on.

In order to describe these relationships, the computer uses a precedence matrix, as illustrated in Figure 6–8. Here the numeral 1 signifies a "must precede" relationship. For example, work unit (00) must precede work units (02), (03), (04), (05), (06), (07), (08), (09), and (10). Also, work unit (09) must precede only work unit (10).

Now a "positional weight" must be computed for each work unit. This

FIGURE 6–8
A precedence matrix as used by a digital computer for a line-balancing problem

Estimated work unit time (minutes)	Work unit	Work unit										
		00	01	02	03	04	05	06	07	08	09	10
0.46	00			1	1	1	1	1	1	1	1	1
0.35	01				1	1		1	1	1	1	
0.25	02						1	1		1	1	1
0.22	03					1		1	1	1	1	
1.10	04							1	1	1	1	
0.87	05							1		1	1	1
0.28	06								1	1	1	
0.72	07									1	1	
1.32	08										1	1
0.49	09											1
0.55	10											
6.61												

is done by computing the summation of each work unit and all the work units that must follow it. Thus, the "positional weight" for work unit (00) would be: Σ 00, 02, 03, 04, 05, 06, 07, 08, 09, 10 = 0.46 + 0.25 + 0.22 + 1.10 + 0.87 + 0.28 + 0.72 + 1.32 + 0.49 + 0.55 = 6.26.

Listing the positional weights in decreasing order of magnitude gives the following:

Unsorted work elements	Sorted work elements	Positional weight	Immediate predecessors
00	00	6.26	—
01	01	4.75	—
02	03	4.40	(00), (01)
03	04	4.18	(03)
04	02	3.76	(00)
05	05	3.56	(02)
06	06	2.64	(05)
07	08	2.36	(04), (06)
08	07	1.76	(04)
09	09	1.04	(07), (08)
10	10	0.55	(09)

Work elements must now be assigned to the various work stations. This is based on the positional weights (those work elements with the highest positional weights are assigned first) and the cycle time of the system. Thus, the work element with the highest positional weight is assigned to the first work station. The unassigned time for this work station is determined by subtracting the sum of the assigned work element times from the estimated cycle time. If there is adequate unassigned time, then the work element with the next highest positional weight may be assigned, provided that the work elements in the "immediate predecessors" column have already been assigned.

The procedure is continued until all of the work elements have been assigned.

For example, let us assume that the required production per day was 300 units and that a 1.10 rating factor was anticipated. Then:

$$\text{Cycle time of system} = \frac{(480 - 72)(1.10)}{300}$$

$$= 1.50 \text{ minutes}$$

Work				Work element time	Station time		Remarks*
Station	Element	Positional weight	Immediate predecessors		Cumulative	Unassigned	
1	00	6.26	–	0.46	.46	1.04	–
1	01	4.75	–	0.35	.81	0.69	–
1	03	4.40	(00), (01)	0.22	1.03	0.47	–
1	~~04~~	~~4.18~~	~~(03)~~	~~1.10~~	~~2.13~~		N.A.
1	02	3.76	(00)	0.25	1.28	0.22	–
1	~~05~~	~~3.56~~	~~(02)~~	~~0.87~~	~~2.05~~		N.A.
2	04	4.18	(03)	1.10	1.10	0.40	–
2	~~05~~	~~3.56~~	~~(02)~~	~~0.87~~	~~1.97~~		N.A.
3	05	3.56	(02)	0.87	.87	0.63	–
3	06	2.64	(05)	0.28	1.15	0.35	–
3	~~08~~	~~2.36~~	~~(04), (06)~~	~~1.32~~	~~2.47~~		N.A.
4	08	2.36	(04), (06)	1.32	1.32	0.18	–
4	~~07~~	~~1.76~~	~~(04)~~	~~0.72~~	~~2.04~~		N.A.
5	07	1.76	(04)	0.72	.72	0.78	–
5	09	1.04	(07), (08)	0.49	1.21	0.29	–
5	~~10~~	~~.55~~	~~(09)~~	~~0.55~~	~~1.76~~		N.A.
6	10	.55	(09)	0.55	.55	0.95	–

* N.A. means not acceptable.

Under the arrangement illustrated, with six work stations we have a cycle time of 1.32 minutes (work station 4). This arrangement will more than meet the daily requirement of 300. It will produce:

$$\frac{(480 - 72)(1.10)}{1.32} = 341 \text{ units}$$

However, with six work stations we have considerable idle time. The idle time per cycle is:

$$\sum_{1}^{6} = 0.04 + 0.22 + 0.17 + 0 + 0.11 + 0.77 = 1.31 \text{ minutes}$$

For more favorable balancing the problem can be solved for cycle times of less than 1.50 minutes. This may result in more operators and in more production per day which may have to be stored. Another possibility includes operation of the line under a more efficient balancing for a limited number of hours per day.

The operator process chart

The operator process chart, sometimes referred to as a "left- and right-hand process chart" is, in effect, a tool of motion study. This chart shows all movements and delays made by both the right and the left hand, and the relationship between the relative basic divisions of accomplishment as performed by the hands. The purpose of the operator process chart is to present a given operation in sufficient detail so that the operation can be improved by means of an analysis. Usually it is not practical to make a detailed study through the operator process chart unless a highly repetitive manual operation is involved. Through the motion analysis of the operator chart, inefficient motion patterns become apparent, and violations of the laws of motion economy (see Chapter 7) are readily observed. This chart will facilitate changing a method so that a balanced two-handed operation can be achieved and ineffective motions either reduced or eliminated. The result will be a smoother, more rhythmic cycle which will keep both delays and operator fatigue to a minimum.

Constructing the operator process chart

Although Frank and Lillian Gilbreth stated that there are 17 fundamental motions, and that every operation consists of a combination of some of these elements, the methods analyst will find it practical when plotting operator process charts to use only eight basic divisions of accomplishment. These elemental motions with their symbols are:

Reach.	RE	Use	U
Grasp	G	Release	RL
Move	M	Delay	D
Position	P	Hold	H

See Chapter 7 for a detailed description of each of these basic divisions of human work.

Although there are several different delay types, such as unavoidable delay, avoidable delay, rest to overcome fatigue, and "balancing delay," in the operator process chart the methods analyst need only indicate a cessation of work with the identification "delay." Often, delays will not occur to both hands simultaneously. For example, an operator while hand-feeding an engine lathe would be performing "use" with his right hand while his left hand would be delayed. Other examples of delays would be interruptions of work due to a coughing spell of the operator or his sitting down to rest. In both of these examples, the right hand as well as the left hand would be shown as being delayed. Delays should be investigated with the thought of eliminating or minimizing them.

The chart should be headed "Operator Process Chart," and this should be followed by all necessary identifying information, including the part number, drawing number, operation or process description, present or proposed method, date, and name of the person doing the charting. Immediately below the identifying information should be a sketch of the work station, drawn to scale. The sketch will materially aid in presenting the method under study. A typical operator process chart form, with a section of the sheet laid out in coordinate form to facilitate sketching, is shown in Figure 6–9.

After the analyst has completely identified the operation and made a sketch of the work station showing dimensional relationships, he is ready to begin constructing the operator process chart. Since this chart is drawn to scale, the analyst should determine by observation the duration of the cycle. Then he can readily determine the amount of time represented by each $\frac{1}{4}$ inch of vertical space on the chart. For example, if it developed that the operation to be studied had a cycle time of 0.70 minutes and there were 7 vertical inches of available charting space, then each $\frac{1}{4}$ inch of chart space would equal 0.025 minutes.

It is usually less confusing to chart the activities of one hand completely, and then to chart all the basic divisions of accomplishment performed by the other hand. Although there is no fixed rule on what part of the work cycle should be used as a starting point, it is usually best to start plotting immediately after the "release" of the finished part. If this release is done with the right hand, the next movement that would normally take place would be the first motion shown on the operator process chart. This, for the right hand, would probably be "reach for new part." If the analyst observed that the "reach" element took about 0.025 minutes to perform, he would indicate this duration by drawing a horizontal line across the right-hand side of the paper $\frac{1}{4}$ inch from the top. Under the "Symbols" column, he would indicate a "RE" for "reach," and this would be shown in black pencil, indicating that an effective motion had been accomplished. Immediately to the right of the symbol would be a brief description of the event, such as "Reach 20 inches" for a $\frac{1}{2}$-inch nut.

FIGURE 6–9

Operator process chart of assembly of cable clamps

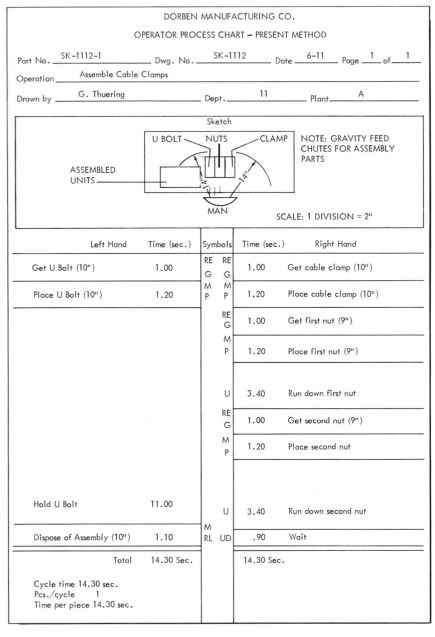

Immediately below, the next basic division would be shown, and so on until completion of the cycle. The analyst would then proceed to plot the left hand activities during the cycle. While plotting the left hand's activities, it is a good idea to verify that end points of the therbligs actually occur at the same point as indicated—a check for overall plotting errors.

It should be noted that elements must be large enough to be measured, since it is not possible in most instances to obtain with the stopwatch the time required for individual therbligs. For example, in Figure 6–9 the first element performed by the left hand was classified as "Get U-bolt." This element comprised the therbligs "reach" and "grasp." It would not have been possible with a stopwatch to determine the time required to perform either of these therbligs. Only through use of the motion-picture camera or videotape can time values as short as these be measured.

The observer in Figure 6–9 was using a decimal second watch, and by observing an element at a time, he could break the elements into one-second periods.

After the activities of both the right and the left hand have been charted, a summary should be shown at the bottom of the sheet, indicating the cycle time, pieces per cycle, and time per piece.

Using the operator process chart

After the operator process chart of an existing method has been completed, the analyst should see what improvements can be introduced. The "delays" and "holds" are good places to begin. For example, in Figure 6–9 it can be seen that the left hand was used as a holding device for almost the entire cycle. An analysis of this condition would suggest the development of a fixture to hold the U-bolt. Further consideration of how to get balanced motions of both hands would suggest that, when the fixture holds the U-bolts, then the left hand and the right hand would each assemble a cable clamp completely. Additional study of this chart might result in the introduction of an automatic ejector and gravity chute to eliminate the final cycle element "dispose of assembly."

The best way of doing a job is through systematic analysis of all the detailed elements constituting that job. The operator process chart clearly reveals the work done by each hand in performing an operation and shows the relative time and the relationships of all motions performed by the hands. The operator process chart is an effective tool to:

1. Balance the motions of both hands and reduce fatigue.
2. Eliminate and/or reduce nonproductive motions.
3. Shorten the duration of productive motions.
4. Train new operators in the ideal method.
5. Sell the proposed method.

The methods analyst should learn to make and use operator process charts to bring about improvements.

Summary: Man and machine, gang, and operator process charts

The methods analyst should know the specific functions of the man and machine, gang, and operator process charts so that he can select the appropriate one for improving operations. The operator process chart is adapted to all work where the operator goes through a series of manual motions. The man and machine and gang process charts are used only when machines or facilities are used in conjunction with the operator or operators. Frequently, where an operator is employed in running a machine, it will be worthwhile to make a study both from the standpoint of possible coupling and of motion pattern improvement. Thus, it may be helpful to construct both a man and machine and an operator process chart. In summary, the functions of these three charts are:

1. Man and machine chart: Used to analyze idle man time and idle machine time. Ideal for determining the amount of machine coupling to be practiced. Used as a training tool to show the relationships of work elements in a multimachine work center.
2. Gang process chart: Used to analyze idle facility time and idle time of operators servicing a facility or process. Ideal for determining the labor requirements of a production facility. Used as a training tool to show the work elements of several operators working with a production facility.
3. Operator process chart: Used to analyze work station for proper layout, proper operator motion patterns, and best sequence of elements. Best chart to use for improving repetitive manual motions.

Quantitative techniques can be used to determine the optimum arrangement of random servicing work centers. The analyst should be acquainted with sufficient algebra and probability theory to develop a mathematical model that will provide the best solution to the problem of machine or facility assignment.

TEXT QUESTIONS

1. When is it advisable to construct a man and machine process chart?
2. Why will higher base rates result when machine coupling is practiced?
3. How does the gang process chart differ from the man and machine process chart?
4. Explain how you would sell machine coupling to union officials strongly opposed to the technique.
5. In what way does an operator benefit through machine coupling?
6. What is the difference between synchronous and random servicing?
7. How many machines should be assigned to an operator when:
    a. Loading and unloading time one machine, 1.41 minutes.
    b. Walking time to next facility, 0.08 minutes.
    c. Machine time (power feed), 4.34 minutes.
    d. Operator rate, $4.40 per hour.
    e. Machine rate, $6.00 per hour.
8. What proportion of machine time would be lost in operating four machines when the machine operates 70 percent of the time unattended and the

operator attention time at irregular intervals averages 30 percent? Is this the best arrangement on the basis of minimizing the proportion of machine time lost?

9. In an assembly process involving six distinct operations, it is necessary to produce 250 units per eight-hour day. The measured operation times are as follows:

1. 7.56 minutes.   4. 1.58 minutes.
2. 4.25 minutes.   5. 3.72 minutes.
3. 12.11 minutes.   6. 8.44 minutes.

How many operators would be required at 80 percent efficiency?
How many operators will be utilized at each of the six operations?

GENERAL QUESTIONS

1. In the process-type plant, which of the following process charts has the greatest application: man and machine, gang, operation, flow? Why?

2. In the operation of some plant with which you are familiar, outline the "zoning restraints" and discuss how these influence the production line work assignment problem.

PROBLEMS

1. In the Dorben Company's automatic screw machine department, five machines are assigned to each operator. On a given job, the machining time per piece is 0.164 hours, the machine serving time 0.038 hours, and the average machine downtime 0.12 hours per machine per hour. With an operator hourly rate of $6.40 per hour and a machine rate of $7 per hour, calculate the expected cost per unit of output. Exclude material cost.

2. In the Dorben Company, an operator was assigned to operate three like facilities. Each of these facilities is down at random times during the day. A work sampling study indicated that on the average the machines operate 60 percent of the time unattended. Operator attention time at irregular intervals averages 40 percent. This arrangement results in the loss of about 14 percent of the available machine time due to machine interference. If the machine rate is $10 per hour and the operator rate is $6 per hour, what would be the most favorable number of machines (from an economic standpoint) that should be operated by one operator?

3. The analyst in the Dorben Company wishes to assign a number of like facilities to an operator based on minimizing the cost per unit of output. A detailed study of the facilities revealed the following:

$$\begin{aligned}
\text{Loading machine standard time} &= \ 0.34 \text{ minutes} \\
\text{Unloading machine standard time} &= \ 0.26 \text{ minutes} \\
\text{Walk time between two machines} &= \ 0.06 \text{ minutes} \\
\text{Operator rate} &= \$3.60 \text{ per hour} \\
\text{Machine rate (both idle and working)} &= \$4.20 \text{ per hour} \\
\text{Power feed time} &= \ 1.48 \text{ minutes}
\end{aligned}$$

How many of these machines should be assigned to each operator?

4. A work sampling study revealed that a group of three semiautomatic machines assigned to one operator operate 70 percent of the time unattended. Operator attention time at irregular intervals averaged 30 percent of the time on these three machines. What would be the estimated machine hours lost per eight-hour day because of lack of an operator?

SELECTED REFERENCES

Baker, Kenneth R. *Introduction to Sequencing and Scheduling.* New York: John Wiley & Sons, Inc., 1974.

Buffa, E. S., and Taubert, W. H. *Production-Inventory Systems: Planning and Control.* Rev. ed. Homewood, Ill.: Richard D. Irwin, Inc., 1972.

Graham, C. F. *Work Measurement and Cost Control.* Oxford, England: Pergamon Press, 1965.

Groff, G. K., and Muth, J. F. *Operations Management: Analysis for Decisions.* Homewood, Ill.: Richard D. Irwin, Inc., 1972.

Johnson, Lynwood A., and Montgomery, Douglas C. *Operations Research in Production Planning, Scheduling, and Inventory Control.* New York: John Wiley & Sons, Inc., 1974.

7

Motion study

Both visual motion study and micromotion study are used to analyze a given method and to help develop an efficient work center. These two techniques are used in conjunction with the principles of operation analysis when there is sufficient volume to justify the additional study and analysis required.

Motion study is the careful analysis of the various body motions employed in doing a job. Its purpose is to eliminate or reduce ineffective movements, and to facilitate and speed effective movements. Through motion study, the job is performed more easily and the rate of output is increased. The Gilbreths pioneered the study of manual motion and developed basic laws of motion economy that are still considered fundamental. They were also responsible for the motion-picture technique for making detailed motion studies, known as "micromotion studies," which have proved invaluable in studying highly repetitive manual operations.

Motion study, in the broad sense, covers two degrees of refinement that have wide industrial application. These are visual motion study and micromotion study.

Visual motion study has considerably broader application, because the activity of the work need not be as great to justify its use economically. This type of study involves a careful observation of the operation and construction of an operator process chart, and a probing analysis of the chart, considering the laws of motion economy.

The micromotion procedure (see Chapter 8), in view of its much higher cost, is usually practical only on extremely active jobs whose life and repetitiveness are great. The two types of studies may be compared to viewing a part under a magnifying glass and viewing it under a microscope. The added detail revealed by the microscope finds application only on the most productive jobs.

The fundamental motions

The basic division of accomplishment concept, developed by Frank Gilbreth in his early work, applied to all production work performed by the hands of the operator. Gilbreth called these fundamental motions "therbligs" (Gilbreth spelled backward), and concluded that any and all operations are made up of a series of these 17 basic divisions. The 17 fundamental hand motions, modified somewhat from Gilbreth's summary, and their symbols and color designations are shown in Table 7–1.

TABLE 7–1

Therblig Name	Symbol	Color Designation	Symbol
Search	S	Black	⌒⊃
Select	SE	Gray, Light	→
Grasp	G	Lake Red	∩
Reach	RE	Olive Green	‿
Move	M	Green	‿o‿
Hold	H	Gold Ocher	⌒o⌒
Release	RL	Carmine Red	⌒o⌒
Position	P	Blue	9
Pre–position	PP	Sky Blue	প
Inspect	I	Burnt Ocher	0
Assemble	A	Violet, Heavy	#
Disassemble	DA	Violet, Light	#
Use	U	Purple	U
Unavoidable Delay	UD	Yellow Ocher	⌒o
Avoidable Delay	AD	Lemon Yellow	∟o
Plan	PL	Brown	β
Rest to Overcome Fatigue	R	Orange	९

Definitions of basic divisions of accomplishment

In its "Glossary of Terms Used in Methods, Time Study, and Wage Incentives," the Management Research and Development Division of the Society for the Advancement of Management has provided definitions of the various therbligs. These definitions, in part, are included in the following summary.

1. SEARCH. Search is the basic operation element employed to locate an object. It is that part of the cycle during which the eyes or hands are

groping or feeling for the object. It begins the instant the eyes move in an effort to locate an object and ends the instant they are focused on the found object.

Search is a therblig that the analyst should always endeavor to eliminate. Well-planned work stations allow work to be performed continuously so that it is not necessary for the operator to perform this element. Providing an exact location for all tools and parts is the typical way to eliminate search from a work station.

A new employee, or one who is not familiar with his job, will find it necessary to use search periodically until his skill and proficiency develop.

The well-trained motion analyst will ask himself the following questions in order to eliminate or reduce "search" time:

1. Are articles properly identified? Perhaps labels or color could be utilized.
2. Can transparent containers be used?
3. Will a better layout of the work station eliminate searching?
4. Is proper lighting being used?
5. Can tools and parts be pre-positioned?

2. SELECT. Select is the therblig that takes place when the operator chooses one part over two or more analogous parts. This therblig usually follows search, but even with a detailed micromotion procedure it is difficult to determine the exact ending of search and the exact beginning of select. Select does occur without search when selective assembly is being encountered. In this case, it is usually preceded by inspect. Select can also be classified as an ineffective therblig and should be eliminated from the work cycle by better layout of the work station and better control of parts.

To eliminate this therblig, the analyst should ask:

1. Are common parts interchangeable?
2. Can tools be standardized?
3. Are parts and materials stored in the same bin?
4. Can a rack or tray be used so that parts will be pre-positioned?

3. GRASP. Grasp is the elemental hand motion of closing the fingers around a part in an operation. Grasp is an effective therblig and usually cannot be eliminated, but in many instances it can be improved. It occurs the instant the fingers of either or both hands begin to close around an object to maintain control of it, and it ends the moment control has been obtained. Grasp is usually preceded by reach and followed by move. Detailed studies have concluded that there are many types of grasp, some taking three times as much time to perform as others. The number of grasps occurring during the work cycle should be kept to a minimum, and the parts to be picked up should be arranged so that the simplest type of grasp can be used. This is done by having the object by itself in a fixed

location, and positioned so that there is no interference from the work-table, bin, or surrounding environment.

These therblig check questions may help improve the grasps performed during a cycle:

1. Would it be advisable for the operator to grasp more than one part or object at a time?
2. Can a contact grasp be used rather than a pickup grasp? In other words, can objects be slid instead of carried?
3. Will a lip on the front of bins simplify grasping small parts?
4. Can tools or parts be pre-positioned for easy grasp?
5. Can a vacuum, magnet, rubber fingertip, or other device be used to advantage?
6. Can a conveyor be used?
7. Has the jig been designed so that the part may easily be grasped when the operator is removing it?
8. Can the previous operator pre-position the tool or the work, thus simplifying grasp for the following operator?
9. Can tools be pre-positioned on a swinging bracket?

4. REACH.   Reach represents the motion of an empty hand, without resistance, toward or away from an object. The basic division reach was known as "transport empty" in Gilbreth's original summary. However, the shorter term is generally accepted by methods analysts today. Reach begins the instant the hand moves toward an object or general location, and it ends the instant hand motion stops upon arrival at the object or destination. Reach is usually followed by grasp and preceded by release. Obviously, the time required to perform a reach depends on the distance of the hand movement. The time to perform reach also depends to some extent upon the type of reach. Like grasp, reach can be classified as an objective therblig, and usually cannot be eliminated from the work cycle. However, it can be reduced by shortening the distances required for reaching and by providing fixed locations for objects that are reached for during the operator's work cycle. By keeping these elementary principles in mind, work stations can be developed that will keep reach time to a minimum.

5. MOVE.   Move is the basic division to signify a hand movement with a load. The load can be in the form of pressure. Move was originally known as "transport loaded." This therblig begins the instant the hand under load moves toward a general location, and it ends the instant motion stops upon arrival at the destination. Move is usually preceded by grasp and is usually followed by either release or position.

The time required to perform move depends on the distance, the weight to be moved, and the type of move. Move is an objective therblig, and it is difficult to eliminate this basic division from the work cycle. Nevertheless, the time to perform move can be diminished by reducing the distances to be moved, lightening the load, and improving the type of move by providing gravity chutes or a conveyor at the terminal point of the move, so that it will not be necessary to bring the object being moved to a specific location. Experience has proven that moves to a general location are performed more rapidly than moves to an exact location.

Both reach and move therbligs may be improved by asking and answering the following questions:

1. Can either of these therbligs be eliminated?
2. Can distances be shortened to advantage?
3. Are the best means being used? that is, conveyors, tongs, tweezers, and so on.
4. Is the correct body member being used? that is, fingers, wrist, forearm, shoulder.
5. Can a gravity chute be employed?
6. Can transports be effected through mechanization and foot-operated devices?
7. Will time be reduced by transporting in larger units?
8. Is time increased because of the nature of the material being moved or because of a subsequent delicate positioning?
9. Can abrupt changes in direction be eliminated?

6. HOLD. Hold is the basic division of accomplishment that occurs when either hand is supporting or maintaining control of an object while the other hand does useful work. Hold is an ineffective therblig and can usually be removed from the work cycle by designing a jig or fixture to hold the work rather than using the hand to do so. The hand is seldom considered an efficient holding device, and the methods analyst should always be on the alert to prevent hold from being a part of any work assignment.

Hold begins the instant one hand exercises control on the object, and it ends the instant the other hand completes its work on the object. A typical example of hold would occur when the left hand holds a stud while the right hand runs a nut on the stud. During the assembly of the nut to the stud, the left hand would be utilizing the therblig hold.

Hold can often be eliminated by asking and answering these questions:

1. Can a mechanical jig, such as a vise, pin, hook, rack, clip, or vacuum, be used?
2. Can friction be used?
3. Can a magnetic device be used?
4. Should a twin holding fixture be used?

7. RELEASE. Release is the basic division that occurs when the aim of the operator is to relinquish control of the object. Release takes the

least amount of time of all the therbligs, and little can be done to influence the time required for this objective therblig.

Release begins the instant the fingers begin to move away from the part held, and it ends the instant all fingers are clear of the part. This therblig is usually preceded by move or position and is usually followed by reach.

Release time may be eliminated or improved by asking:

1. Can the release be made in transit?
2. Can a mechanical ejector be used?
3. Are the bins that will contain the part after release of the proper size and design?
4. At the end of the therblig release, are the hands in the most advantageous position for the next therblig?
5. Can multiple units be released?

8. POSITION. Position is an element of work that consists of locating an object so that it will be properly oriented in a specific place.

The therblig position occurs as a hesitation while the hand or hands are endeavoring to place the part so that further work may be more readily performed. Actually, position may be a combination of several very rapid motions. Locating a contoured piece in a die would be a typical example of position. Position is usually preceded by move and followed by release. Position begins the instant the controlling hand or hands begin to agitate, twist, turn, or slide the part to orient it to the correct place, and it ends as soon as the hand begins to move away from the part.

Position frequently can be eliminated or improved through answering these and similar check questions:

1. Can such devices as a guide, funnel, bushing, stop, swinging bracket, locating pin, recess, key, pilot, or chamfer be used?
2. Can tolerances be changed?
3. Can the hole be counterbored or countersunk?
4. Can a template be used?
5. Are burrs increasing the problem of positioning?
6. Can the article be pointed to act as a pilot?

9. PRE-POSITION. Pre-position is an element of work that consists of positioning an object in a predetermined place so that it may be grasped in the position in which it is to be held when needed.

Pre-position often occurs in conjunction with other therbligs, one of which is usually move. It is the basic division that arranges a part so that it can be conveniently placed upon arrival. The time required to pre-position is difficult to measure since the therblig itself can seldom be isolated. Pre-position takes place if a screwdriver is being aligned while being moved to the screw it is going to drive.

These check questions will assist the analyst in studying pre-position therbligs.

1. Can a holding device be used at the work station so that tools may be kept in proper position and the handles kept in upright position?
2. Can tools be suspended?
3. Can a guide be used?
4. Can a magazine feed be used?
5. Can a stacking device be used?
6. Can a rotating fixture be used?

10. INSPECT. Inspect is an element included in an operation to assure acceptable quality through a regular check by the employee performing the operation.

Inspect takes place when the predominant purpose is to compare some object with a standard. It is usually not difficult to detect when inspect is encountered, since the eyes are focused upon the object, and a delay between motions is noted while the mind decides to accept or reject the piece in question. The time taken for inspect is determined primarily by the severity of the standard and the amount of deviation of the part in question. Thus, if an operator were sorting out all blue marbles from a bin, little time would be consumed in deciding what to do with a red marble. However, if a purple marble were picked up, there would be a longer hesitation for the mind to evaluate the marble and decide whether it should be accepted or rejected.

These questions, when answered by the analyst, may result in improvements on inspect therbligs:

1. Can inspection be eliminated or combined with another operation or therblig?
2. Can multiple gages or tests be used?
3. Will inspection time be reduced by increasing the illumination?
4. Are the articles being inspected at the correct distance from the worker's eyes?
5. Will a shadowgraph facilitate inspection?
6. Does an electric eye have application?
7. Does the volume justify automatic electronic inspection?
8. Would a magnifying glass facilitate the inspection of small parts?
9. Is the best inspection method being used? Has consideration been given to polarized light, template gages, sound tests, performance tests, and so on?

11. ASSEMBLE. Assemble is the basic division that occurs when two mating parts are brought together. This is another objective therblig, and it can more readily be improved than eliminated. Assemble is usually preceded either by position or move, and it is usually followed by release. It begins the instant the two mating parts come in contact with each other, and it ends upon completion of the union.

12. DISASSEMBLE. Disassemble is just the reverse of assemble. It occurs when two mating parts are disunited. The basic division is usually preceded by grasp and is usually followed by either move or release. Disassemble is objective in nature, and improvement possibilities are more likely than is complete elimination of the therblig. Disassemble begins the moment either or both hands have control of the object after grasping it, and it ends as soon as the disassembly has been completed, usually evidenced by the beginning of a move or a release.

13. USE. Use is a completely objective therblig that occurs when either or both hands have control of an object during that part of the cycle when productive work is being performed. When both hands are holding a casting against a grinding wheel, use would be the therblig that indicates the action of the hands. After a screwdriver has been positioned in the slot of the screw, use would occur as the screw is being driven home. The duration of this therblig depends on the operation as well as on the performance of the operator. Use is quite easily detected, since this therblig always advances the operation toward the ultimate objective.

In studying the three objective therbligs, assemble, disassemble, and use, thought should be given to the following questions:

1. Can a jig or fixture be used?
2. Does the activity justify automated equipment?
3. Would it be practical to make the assembly in multiple units?
4. Can a more efficient tool be used?
5. Can stops be used?
6. Is the tool being operated at the most efficient feeds and speeds?
7. Should a power tool be employed?

14. UNAVOIDABLE DELAY. Unavoidable delay is an interruption beyond the control of an operator in the continuity of an operation. It represents idle time in the work cycle experienced by either or both hands because of the nature of the process. Thus, while an operator is hand-feeding a drill with his right hand to a part held in a jig, the left hand would be unavoidably delayed. Since the operator has no control over unavoidable delays, to remove them from the cycle the process has to be changed in some manner.

15. AVOIDABLE DELAY. Any idle time that occurs during the cycle for which the operator is solely responsible, either intentionally or unintentionally, is classified as an avoidable delay. Thus, if an operator developed a coughing spell during a work cycle, this delay would be avoidable, for normally it would not appear in the work cycle. A majority of the avoidable delays encountered can be eliminated by the operator without changing the process or method of doing the work.

16. PLAN. The therblig plan is the mental process that occurs when the operator pauses to determine the next action. Plan may take place during any part of the work cycle, and it is usually readily detected as a hesi-

tation after all components have been located. This therblig is characteristic of new employees and can usually be removed from the work cycle through proper operator training.

17. REST TO OVERCOME FATIGUE. Usually this delay does not appear in every cycle but is evidenced periodically. The duration of rest to overcome fatigue will vary not only with the class of work but also with the individual performing the work.

To reduce the number of occurrences of the therblig rest, the analyst should consider:

1. Is the best order-of-muscles classification being used?
2. Are temperature, humidity, ventilation, noise, light, and other working conditions satisfactory?
3. Are benches of the proper height?
4. Can the operator alternately sit and stand while performing his or her work?
5. Does the operator have a comfortable chair of the right height?
6. Are mechanical means being used for heavy loads?
7. Is the operator aware of his or her average intake requirements in calories per day? The approximate number of calories required for sedentary activities is 2,400; for light manual labor, 2,700; for medium labor, 3,000; for heavy manual labor, 3,600.

Therblig summary

The 17 basic divisions can be classified as either effective or ineffective therbligs. Effective therbligs are those that directly advance the progress of the work. These therbligs can frequently be shortened, but it is difficult to eliminate them completely. Ineffective therbligs do not advance the progress of the work and should be eliminated by applying the principles of operation analysis and motion study.

A further classification breaks the therbligs into physical, semimental or mental, objective, and delay groups. Ideally, a work center should comprise only physical and objective therbligs.

A. Effective.
   1. Physical basic divisions.
      a. Reach.
      b. Move.
      c. Grasp.
      d. Release.
      e. Pre-position.
   2. Objective basic division.
      a. Use.
      b. Assemble.
      c. Disassemble.

B. Ineffective.
  1. Mental or semimental basic divisions.
     a. Search.
     b. Select.
     c. Position.
     d. Inspect.
     e. Plan.
  2. Delay.
     a. Unavoidable delay.
     b. Avoidable delay.
     c. Rest to overcome fatigue.
     d. Hold.

Principles of motion economy

Beyond the basic division of accomplishment concept, as first set forth by the Gilbreths, are the principles of motion economy developed by them and added to by others, notably Ralph M. Barnes. These principles are not all applicable to every job, and several find application only through the tool of micromotion study. However, those that apply to visual motion study, as well as to the micromotion technique, that should always be considered are broken down into three basic subdivisions: (1) the use of the human body, (2) the arrangement and conditions of the workplace, and (3) the design of tools and equipment.

The methods analyst should become familiar with the visual principles of motion economy so that he will be able to detect inefficiencies in the method by briefly inspecting the workplace and the operation. These basic principles under their respective divisions are as follows:

A. The use of the human body.
  1. Both hands should begin and end their basic divisions of accomplishment simultaneously and should not be idle at the same instant, except during rest periods.
  2. The motions made by the hands should be made symmetrically and simultaneously away from and toward the center of the body.
  3. Momentum should be employed to assist the worker wherever possible, and it should be reduced to a minimum if it must be overcome by muscular effort.
  4. Continuous curved motions are preferable to straight-line motions involving sudden and sharp changes in direction.
  5. The least number of basic divisions should be used, and these should be confined to the lowest practicable classifications. These classifications, summarized in ascending order of the time and fatigue expended in their performance, are:

      a. Finger motions.
      b. Finger and wrist motions.
      c. Finger, wrist, and lower arm motions.
      d. Finger, wrist, lower arm, and upper arm motions.
      e. Finger, wrist, lower arm, upper arm, and body motions.

6. Work that can be done by the feet should be arranged so that it is done simultaneously with work being done by the hands. It should be recognized, however, that it is difficult to move the hand and foot simultaneously.

7. The middle finger and the thumb are the strongest working fingers. The index finger, fourth finger, and little finger are not capable of handling heavy loads over extended periods.

8. The feet are not capable of efficiently operating pedals when the operator is in a standing position.

9. Twisting motions should be performed with the elbows bent.

10. To grip tools, the segments of the fingers closest to the palm of the hand should be used.

B. The arrangement and conditions of the workplace.

1. Fixed locations should be provided for all tools and material so as to permit the best sequence and to eliminate or reduce the therbligs search and select.

2. Gravity bins and drop delivery should be used to reduce reach and move times; also, wherever possible, ejectors to remove finished parts automatically should be provided.

3. All materials and tools should be located within the normal working area in both the vertical and the horizontal plane.

4. A comfortable chair should be provided for the operator and the height so arranged that the work can be efficiently performed by the operator alternately standing and sitting.

5. Proper illumination, ventilation, and temperature should be provided.

6. The visual requirements of the workplace should be considered so that eye fixation demands are minimized.

7. Rhythm is essential to the smooth and automatic performance of an operation, and the work should be arranged to permit an easy and natural rhythm wherever possible.

C. The design of tools and equipment.

1. Multiple cuts should be taken whenever possible by combining two or more tools in one, or by arranging simultaneous cuts from both feeding devices, if available (cross slide and hex turret).

2. All levers, handles, wheels, and other control devices should be readily accessible to the operator and should be designed so as to give the best possible mechanical advantage and to utilize the strongest available muscle group.

3. Parts should be held in position by fixtures.
4. The possibility of using powered or semiautomatic tools, such as power nut- and screwdrivers and speed wrenches, should always be investigated.

EXPLANATION OF THE LAWS OF MOTION ECONOMY

Both hands should begin and end their basic divisions of accomplishments simultaneously and should not be idle at the same instant, except during rest periods.

When the right hand is working in the normal area to the right of the body and the left hand is working in the normal area to the left of the body, there is a feeling of balance that tends to induce a rhythm in the operator's performance that leads to maximum productivity. When one hand is working under load and the other hand is idle, the body exerts an effort to put itself in balance. This usually results in greater fatigue than if both hands are doing useful work. This law can readily be demonstrated by reaching out with the right hand about 14 inches, picking up an object weighing about a half pound, and moving it 10 inches toward the body before releasing it. The operation should then be immediately repeated, only this time moving the part away from the body. Repeat the cycle about 200 times and note the discomfort in the body induced by "balance fatigue." Now repeat the operation using both hands simultaneously. Have the left hand reach out radially to the left of the body and the right hand reach out radially to the right of the body, grasp the objects, move both objects toward the body, and release them simultaneously. Repeat the cycle 200 times and note that the body feels less fatigued, even though twice as much load has been handled.

The motions made by the hands should be made symmetrically and simultaneously away from and toward the center of the body.

It is natural for the hands to move in symmetrical patterns: deviations from symmetry in a two-handed work station result in slow, awkward movements of the operator. The difficulty of patting the stomach with the left hand while rubbing the top of the head with the right hand is familiar to many. Another experiment that can readily be tried to illustrate the difficulty of performing nonsymmetrical operations is to draw a circle with the left hand while the right hand is drawing a square. Figure 7–1 illustrates an ideal work station that allows the operator to assemble two products by going through a series of symmetrical motions made simultaneously away from and toward the center of the body.

Momentum should be employed to assist the worker wherever possible, and it should be reduced to a minimum if it must be overcome by muscular effort.

As the hands progress through the elements of work constituting the operation, momentum will be developed during reach and move therbligs

FIGURE 7–1
An ideal work station that permits the operator to assemble two products by going through a series of symmetrical motions made simultaneously away from and toward the center of the body

General Electric Co.

and will be overcome during position and release therbligs. To make full use of the momentum built up, work stations should be designed so that a finished part can be released in a delivery area while the hands are on their way to get component parts or tools to begin the next work cycle. This allows the hands to perform their reaches with the aid of momentum and makes the therblig easier and faster to perform.

Detailed studies have proven conclusively that both reaches and moves are performed faster if the hand is in motion at the beginning of the therbligs.

Continuous curved motions are preferable to straight-line motions involving sudden and sharp changes in direction.

This law is very easily demonstrated by moving either hand in a rectangular pattern, and then moving it in a circular pattern of about the same magnitude. The greater amount of time required to make the abrupt 90° directional changes is quite apparent. To make a directional change, the hand must decelerate, change direction, and accelerate until it is time to decelerate preparatory to the next directional change. Continuous curved motions do not require deceleration, and consequently are performed faster per unit of distance.

The least number of basic divisions should be used, and these should be confined to the lowest practicable classifications.

To become fully aware of the significance of this fundamental law of motion economy, it is first necessary to be able to identify the various classifications of motions.

1. Finger motions are the fastest of the five motion classes and are readily recognized, for they are made by moving the finger or fingers while the remainder of the arm is kept stationary. Typical finger motions are running a nut down on a stud, depressing the keys of a typewriter, and grasping a small part. Usually there is a significant difference in the time required to perform finger motions with the various fingers. In most cases, the index finger will be able to move considerably faster than the other fingers, and in designing work stations this should be taken into consideration. Although with practice the fingers of the left hand (in right-handed people) can be trained to move with the same rapidity as the fingers of the right hand, detailed studies have shown that the fingers of the left hand will move somewhat slower. R. E. Hoke made a study of the "universal" keyboard used on typewriters and found that 88.9 taps were made with the fingers of the left hand for every 100 made with the right hand.[1]

The analyst should recognize that finger motions are the weakest of the five motion classes. Consequently, care should be exercised in designing work stations involving great manual effort so that higher classifications than finger motions may be employed.

2. Finger and wrist motions are made by movements of the wrist and fingers while the forearm and upper arm are stationary. In the majority of cases, finger and wrist motions consume more time than do strictly finger motions. Typical finger and wrist motions occur when a part is positioned in a jig or fixture or when two mating parts are assembled. Reach and move therbligs usually cannot be performed by motions of the second class unless the transport distances are very short.

3. Finger, wrist, and lower arm motions are commonly referred to as "forearm motions" and include those movements made by the arm below the elbow while the upper arm is stationary. Since the forearm includes a strong muscle, such motions are usually considered efficient, since they are not fatiguing. The time required to make forearm motions for a given operator depends on the distance moved and the amount of resistance overcome during the movement. By designing work stations so that these third-class motions, rather than fourth-class motions, will be used to perform the transport therbligs, the analyst will minimize cycle times.

---

[1] R. E. Hoke, *The Improvement of Speed and Accuracy in Typewriting*, The Johns Hopkins University Studies in Education no. 7 (Baltimore: Johns Hopkins Press, 1922).

4. Finger, wrist, lower arm, and upper arm motions, commonly known as "fourth-class" or "shoulder motions," are probably used more than any other motion class. The fourth-class motion for a given distance takes considerably more time than do the three lower classes just described. Fourth-class motions are required to perform transport therbligs of parts that cannot be reached without extending the arm. The time required to perform fourth-class motions depends primarily on the distance of the move and the resistance to the move.

5. Fifth-class motions include body motions, and those, of course, are the most time consuming. Body motions include movements of the ankle, knee, and thigh as well as movements of the trunk.

It will be noted that first-class motions require the least amount of effort and time, while fifth-class motions are considered the least efficient. Therefore, the analyst should always endeavor to utilize the lowest practicable motion classification with which to perform the work properly. This will involve careful consideration of the location of tools and materials so that ideal motion patterns can be arranged.

Work that can be done by the feet should be arranged so that it is done simultaneously with work being done by the hands.

Since the major part of work cycles is performed by the hands, it follows that it is economical to relieve the hands of work that can be done by the feet if this work is performed while the hands are occupied. Since the hands are more dexterous than the feet, it would be folly to have the feet perform elements while the hands are idle. Foot pedal devices allowing clamping, the ejection of parts, feeding, and so on, can often be arranged, thus freeing the hands for useful work and consequently reducing the cycle time (see Figure 7–2). When the hands are moving, the feet should not be moving, although they can be applying pressure, such as on a foot pedal. In the development of work stations involving the coordination of the hands and feet, care should be exercised to see that simultaneous movements of the hands and feet are not required.

The middle finger and the thumb are the strongest working fingers. The index finger, fourth finger, and little finger are not capable of handling heavy loads over extended periods.

Although the index finger is usually the finger that is capable of moving the fastest, it is not the strongest finger. Where a relatively heavy load is involved, it will usually be more efficient to use the middle finger or a combination of the middle finger and the index finger.

Twisting motions should be performed with the elbows bent.

When the elbow is extended, tendons and muscles in the arm are stretched. If in this position the arm is obliged to perform twisting motions, there can be an overstressing of muscle groups and tendons.

To grip tools, the segments of the fingers closest to the palm of the hand should be used.

Not only are the segments of the fingers closest to the palm stronger than the other segments, but, being closer to the load held in the hand,

FIGURE 7–2
This foot-operated press permits the operator's hands to procure parts in preparation for the next cycle while the press is in operation

General Electric Co.

they do not induce as great a bending moment as do the more remote segments.

Fixed locations should be provided for all tools and materials so as to permit the best sequence and to eliminate or reduce the therbligs search and select.

In driving an automobile, we are all familiar with the shortness of time required to apply the foot brake. The reason is obvious: since the brake pedal is in a fixed location, no time is required to decide where the brake is located. The body responds instinctively and applies pressure to the area where the driver knows the foot pedal will be. If the location of the brake foot pedal varied from time to time, considerably more time would be needed to apply braking to the car. Providing fixed locations for all tools and materials at the work station will eliminate, or at least minimize, the short hesitations required to search and select the various objects needed to do the work (see Figure 7–3).

Gravity bins and drop delivery should be used to reduce reach and move times.

The time required to perform both of the transport therbligs, reach and move, is proportional to the distance that the hands must move in per-

FIGURE 7–3

Fixed locations for all materials and tools are provided in this work station, minimizing search and select hesitations

General Electric Co.

forming these therbligs. By utilizing gravity bins, components can be continuously brought to the normal working area, thus eliminating long reaches to get supplies of parts. Likewise, gravity chutes allow the disposal of parts within the normal area, thus eliminating the necessity for long moves to dispose of the completed part or parts. Gravity chutes make possible a clean workplace area, as finished material will be carried away from the work area rather than be stacked up all around the workplace (see Figure 7–4).

All materials and tools should be located within the normal working area in both the vertical and the horizontal plane.

The normal working area in the horizontal plane of the right hand includes the area generated by the arm below the elbow when moving in an arc pivoted at the elbow. This area will represent the most convenient zone within which motions may be made by that hand with a normal expenditure of energy.

The normal area of the left hand may be established in a similar manner.

Since movements are made in the third dimension as well as in the

FIGURE 7–4
A work station utilizing gravity bins and a belt conveyor to reduce reach and move times. Note the conveyor in the background carrying other parts past this particular work station. The operator is feeding the conveyor from under the platform by merely dropping assembled parts onto the feeder belt.

Alden Systems Co.

horizontal plane, the normal working area also applies to the vertical plane. The normal area relative to height for the right hand includes the area made by the lower arm in an upright position hinged at the elbow moving in an arc. There is a like normal area in the vertical plane for the left hand (see Figures 7–5 and 7–6).

The maximum working area represents that portion of the workplace within which all tools and materials should be located and within which work may be performed without excessive fatigue. This area is formed by drawing arcs with the arms fully extended and, as with the normal working area, both the horizontal and the vertical plane are considered.

In the design of both facilities and work stations, consideration should be given to such factors as arm reach, leg clearance, and body support, since these human dimensions are important criteria in developing a worker environment that is comfortable and efficient.

A comfortable chair should be provided for the operator and the height

FIGURE 7–5
Normal and maximum working areas in the horizontal plane for women (for men, multiply by 1.09)

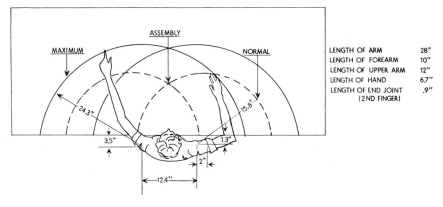

FIGURE 7–6
Normal and maximum working areas in the vertical plane for women (for men, multiply by 1.09)

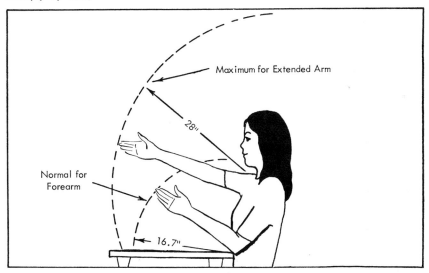

so arranged that the work can be efficiently performed by the operator alternately standing and sitting.

At the workplace, the operator should be seated if possible. Workplaces that require the operator to stand a significant portion of the day are conducive to high fatigue.

In order to reduce operator fatigue, the stool or chair used by the operator should receive careful attention. It should accommodate the 1st to 98th percentile of workers. This is equivalent to providing a work-surface height of from 36.5 inches to 44.7 inches. In general, the chair or stool seats should be broad and long enough to support the thighs, yet not so long as to cut into the back of the knees of shorter operators. The contour of the seat should be slightly saddle shaped, and the front end should be chamfered. Sixteen inches by 16 inches, with the curved portion beginning about 3 inches back from the front edge, is considered a good size. The seat should be lightly padded and aerated. Whenever possible, backs should be provided for all chairs, and these should be designed so that they do not interfere with movements of the arms. The backrest should not result in undue pressure against the pelvis or ribs, nor should it cause interference with shoulder blade movement. It should be slightly curved, with dimensions of approximately 5 inches in height and 10 inches in width. Some padding on the backrest is desirable in order to eliminate sharp edges. A tilting feature is desirable.

Good chair design should permit several effective working postures. The seats should be adjustable in height within the range of 15 to 21 inches. Half-inch increments of height adjustment should be provided. If the operator is working at benches more than 30 inches in height, chair adjustment should permit a range of seat heights of from 18 inches to 27 inches. Chair manufacturers supply industrial chairs that are adjustable in height from the floor to the top edge of the seat. In recent years, both industrial and medical experts have collected data that in many instances prove that production costs have been lowered by using chairs and benches of proper height.

If the height of the work station and chair is such that the operator can work alternately in a standing and a sitting position, fatigue and job monotony will be reduced substantially. Monotony is definitely an important factor in producing fatigue, and with the present-day tendency toward specialization and resulting increases in fatigue accidents, everything possible should be done to reduce monotony.

If it is not practicable to have the operator alternately in a standing and a sitting position, it has been found desirable to have a seat which can tilt the body slightly forward. Ideally, after the operator is comfortably seated with his feet on the floor, the workbench is positioned at the appropriate height to accommodate the operation. Thus, the work station too needs to be adjustable.

Proper illumination, ventilation, and temperature should be provided.

As explained in Chapter 5, good, comfortable working conditions are fundamental for peak production and operator satisfaction. Defective illumination has long been recognized as an important factor in causing operator fatigue, poor quality of product, and low production output. The National Safety Council, in its *Safe Practices Pamphlet No. 50*, mentions that eyes are used in "serious" work about 70 percent of the time and that failure to provide proper lighting will increase the consumption of body energy (see Figure 7–7).

Proper ventilation and temperature also help maintain good working conditions by controlling fatigue and reducing the causes of accidents.

FIGURE 7–7
Better illumination releases energy for useful work

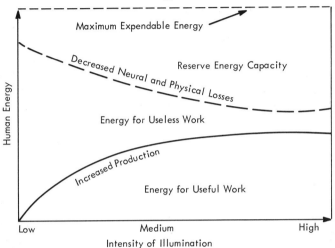

Source: From *National Safety Council Bulletin No. 50, p. 9.*

Laboratory data recorded in diversified industries are in agreement that atmospheric conditions exert an appreciable influence upon physical activity.

Members of the medical profession, industrialists, and psychologists have realized that colors are effective in stimulating or depressing. Having physical surroundings that relieve eyestrain and create a more cheerful atmosphere will result in fewer injuries and less absenteeism.

The visual requirements of the workplace should be considered so that eye fixation demands are minimized.

Certain visual requirements are characteristic of all work centers. Some equipment or controls may be scanned from nearby or remote points. Other areas require more concentrated attention. Pre-positioning components that require more concentrated observation, such as instruments and dials, will lessen eye fixation time as well as eye fatigue.

Rhythm is essential to the smooth and automatic performance of an operation, and the work should be arranged to permit an easy and natural rhythm wherever possible.

If the sequence of the basic motions performed can be arranged so that there is a regular recurrence of like therbligs, or a regular alternation in the therbligs, the hands will instinctively work rhythmically. When work is performed with such a regularity or flow of movement, the operator appears to be working effortlessly, yet invariably production is high and operator fatigue is low.

Multiple cuts should be taken whenever possible by combining two or more tools in one, or by arranging simultaneous cuts from both feeding devices, if available (cross side and hex turret).

Production planning in advance for the most efficient manufacture will include taking multiple cuts with combination tools and simultaneous cuts with different tools. Of course, the type of work to be processed and the number of parts to be produced will determine the desirability of combining cuts, such as cuts from both the square turret and the hexagon turret. Figures 7–8 and 7–9 illustrate typical combined and multiple cuts that can be utilized in turret lathe work (see also Figure 7–10).

FIGURE 7–8
4AC single-spindle automatic chucking machine

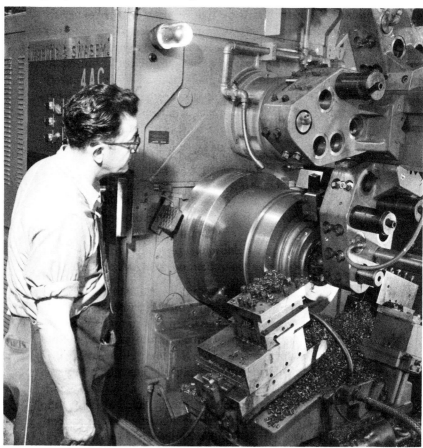

Warner & Swasey Co.

FIGURE 7–9
Forming cut and drilling operation on a single-spindle automatic chucking machine

Warner & Swasey Co.

Figure 7–8 shows combined cuts being made from the cross slide and the multiple turning head on the pentagon turret of a Warner & Swasey 4AC single-spindle automatic chucking machine. Two overhead turning tools and one boring tool are cutting from the multiple turning head while a facing cut is being made from the cross slide.

Figure 7–9 shows a forming cut being made from the cross slide while a drilling operation is being performed from the turret.

In Figure 7–10, combined cuts are being made from the square turret cross slide and the overhead turning tool mounted in a multiple turning head with an overhead pilot bar for extra support. Also, an internal boring operation is being made simultaneously.

All levers, handles, wheels, and other control devices should be readily

FIGURE 7–10
Combined cuts and an internal boring operation being made simultaneously

Warner & Swasey Co.

accessible to the operator and should be designed so as to give the best possible mechanical advantage and to utilize the strongest available muscle group.

Many of our machine tools and other devices are mechanically perfect, yet are incapable of effective operation because the designer of the facility overlooked various human factors in the operation of the equipment. Handwheels, cranks, and levers should be of such size and placed in such positions that the operator can manipulate them with maximum proficiency and minimum fatigue.

The controls that are used most often should be placed between elbow and shoulder height. Seated operators can apply maximum force to levers located at elbow level; standing operators, to levers located at shoulder height. Handwheel and crank diameters depend on both the torque to be expected and the mounting position. Maximum diameters of handgrips depend on the forces to be exerted. For example, if a 10- to 15-pound force is required, the diameter should be no less than ¼ inch and preferably larger; for 15 to 25 pounds, a minimum of ½ inch should be used; and for 25 or more pounds, a minimum of ¾ inch.

Diameters should not exceed 1½ inches, and the grip length should be at least 3¾ inches in order to accommodate the breadth of the hand.

Guidelines for crank and handwheel radii are: light loads, radii of 3 to 5 inches; medium to heavy loads, radii of 4 to 7 inches; very heavy loads, radii of more than 8 inches but not in excess of 20 inches.

Knob diameters of ½ to 2 inches are usually satisfactory. It should be recognized that the diameters of knobs should be increased as greater torques are needed.

The hand is seldom an efficient holding device because, when it is occupied in holding the work, it cannot be free to do useful work. Parts that have to be held in position while they are being worked on, should be supported by a fixture, freeing the hands for productive motions. Fixtures not only save time in processing parts, but permit better quality, in that the work can be held more accurately and firmly.

The possibility of using powered or semiautomatic tools, such as power nut- and screwdrivers and speed wrenches, should always be investigated.

Power hand tools will not only perform work faster than manual tools, but will do the work with considerably less operator fatigue. Greater uniformity of product can be expected when power hand tools are used. For example, a power nut driver will drive nuts consistently to a predetermined tightness in inch-pounds, while a manual nut driver cannot be expected to maintain constant driving pressure in view of operator fatigue.

THE PRACTICAL USE OF MOTION STUDY IN THE PLANNING STAGE

Motion analysis as applied in planning

Production personnel in general agree that it is better to concentrate thinking about methods improvement in the planning stage, rather than depend entirely on trying to correct manufacturing methods after they have been introduced.

Insufficient volume may make it impossible to consider many improvement proposals that might have been introduced in the planning stage with substantial saving over existing methods.

For example, let us take an operation that was done on a drill press in which a ½-inch hole was reamed to the tolerance of 0.500 to 0.502 inches. The activity of the job was estimated to be 100,000 pieces. The time study department established a standard of 8.33 hours per thousand to perform the reaming operation, and the reaming fixture cost $1,000. Since a base rate of $3.60 per hour was in effect, the money rate per thousand pieces was $30.

Now let us assume that a methods analyst suggests broaching the inside diameter because his calculations reveal that the part can be broached at the rate of five hours per thousand. This would be a saving of 3.33 hours per thousand pieces, or a total saving of 333 hours. At our $3.60 base rate, this would mean a direct labor savings of $1198.80. However, it would not be practical to go ahead with this idea, since the tool cost for broaching is

$1,400. Thus, the change would not be sound unless the labor savings can be increased to $1,400 to offset the cost of the new broaching tools.

Since the labor savings in a new broaching setup would be 3.33 × $3.60 per thousand, it can be calculated that 116,800 pieces would have to be ordered before the change in tooling would be justified.

$$\frac{\$1,400 \times 1,000}{\$3.60 \times 3.33} = 116,800 \text{ pieces}$$

However, if the broaching method had been used originally instead of the reaming procedure, it would have paid for itself in

$$\frac{\$1,400 - \$1,000}{\$3.60 \times 3.33/M} = 33,400 \text{ pieces}$$

With production requirements of 100,000 pieces, 3.33 × $3.60 × 66.6

FIGURE 7–11
Crossover chart illustrating the fixed and variable costs of two competing methods

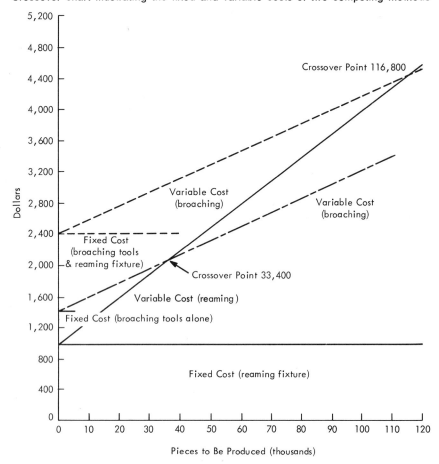

Pieces to Be Produced (thousands)

thousand (the difference between 100,000 and 33,400) = $796 in labor that would have been saved over the present reaming method. Had a motion analysis been made in the planning stage, this saving might have been realized. Figure 7–11 illustrates these relationships with the customary crossover (break-even) chart.

By keeping the principles of motion economy in mind, and breaking the proposed method into its basic divisions, it is possible to develop a right-hand, left-hand analysis prior to starting production. Then, by assigning synthetic time values to the various elements of the operator process chart (see Chapter 19), the practicability of the proposed method can be determined. Let us see how this technique is practiced on a simple manufacturing job. The job involves (1) a light blanking operation, (2) low annual production requirements (200,000 pieces per year), and (3) a competitive price.

With these requirements in mind, the methods analyst proposes performing the job on an arbor press, with an accompanying die to do the

FIGURE 7–12
Right- and left-hand analysis of a proposed method for a blanking operation

THE AIRFELT MFG. CORP.
RIGHT AND LEFT HAND ANALYSIS

Operation __Blank supporting strip on hand arbor press__

Part No. __P-1107-7__ Dwg. No. __PB-1107__ Date __1-12-__

Drawn By __W. Eitele__ Dept. __13__ Plant __Bellefonte__ Sheet __1__ of __1__

Sketch

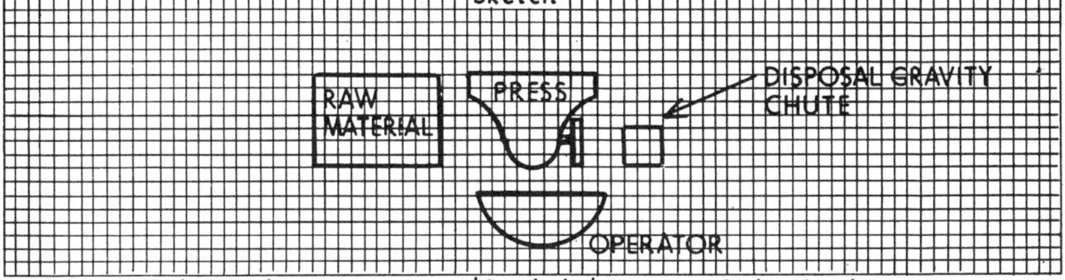

Left Hand	Symbols		Right Hand	
1. Pick up part from hopper			Pull arbor press handle down	1.
Reach for part	Re	H	Hold handle	
Grasp part	G	H	Hold handle	
Move part adjacent to die	M	M	Move press handle down	
2. Wait for right hand			Raise arbor press handle	2.
Unavoidable delay	UD	M	Move arbor press handle back	
Unavoidable delay	UD	R1	Release press handle	
			Get part from die	3.
3. Wait for right hand		Re	Reach for part in die	
Unavoidable delay	UD	G	Grasp part	
Unavoidable delay	UD			
			Aside part to chute	4.
4. Place part in fixture		M	Move part to chute	
Move part to fixture	M	R1	Release part	
Position part in fixture	P	UD	Wait for left hand	
Release part	R1			
5. Wait			Get arbor press handle	5.
Unavoidable delay	UD	Re	Reach for handle	
Unavoidable delay	UD	G	Grasp handle	

trimming. He then breaks the job down in the form of a right- and left-hand process chart, as shown in Figure 7–12.

A review of this operator process chart reveals that several of the principles of motion economy have been violated. Motions of the right and left hands are not balanced; unavoidable delays occur in the motion pattern of the left hand; and both hands do not finish their work simultaneously.

Analysis of the poor setup discloses that the first wait of the left hand occurs during the period that the right hand raises the handle of the arbor press. The left hand is then kept idle while the right hand removes the part from the die, and finally, it is unavoidably delayed while the right hand reaches out to gain control of the arbor press handle.

It can also readily be seen that the right hand must grasp the arbor press handle every cycle as a result of relinquishing it earlier in the cycle to remove the finished part from the die. Therefore, if the operator did not have to remove the finished piece from the die, he would also be freed of the element "grasp arbor press handle." The improved method in the form of an operator process chart is shown in Figure 7–13.

FIGURE 7–13
Right- and left-hand analysis of a proposed method for a blanking operation in which unavoidable delays have been omitted

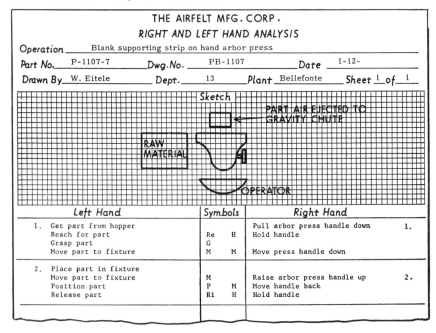

If this method were put into operation, we would have a balanced setup with considerably shorter cycle time. It is possible to develop a fixture to include leaf springs that will lift the part clear of the die on the return stroke of the arbor press. The piece can then be automatically ejected through the back of the press by an air blast actuated from the ram of the press.

This motion analysis in the planning stage allows the part to be blanked in the most economical manner.

Let us take another example and see how motion study in the planning stage helps determine the ideal method. The operation under study is the assembly of the components going into the upper jaw of a pipe vise. The parts involved are the jaw, brace, two lock washers, and two machine screws (see Figure 7–14). The production schedule calls for assembly

FIGURE 7–14
Components of upper jaw of pipe vise—upper panel shows parts in order of assembly (left to right); lower panel shows assembled parts

of 10,000 of these units per year, and the item must be priced to meet stiff competition.

Since the parts are all small and can be readily controlled with either hand, the methods analyst may first consider hand assembly at a bench, with the operator seated and all parts fed to the normal work area by

FIGURE 7–15

Right- and left-hand analysis of the assembly of a pipe-vise jaw

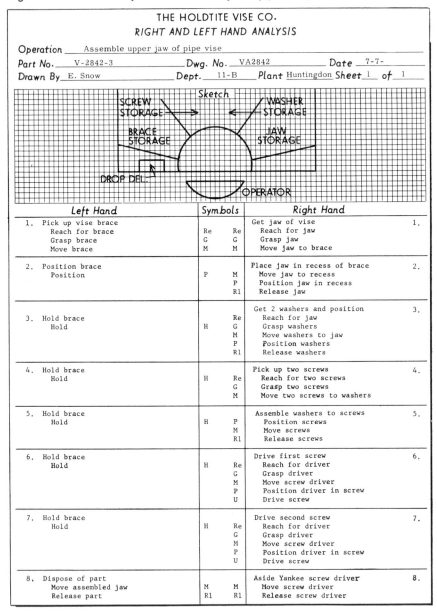

THE HOLDTITE VISE CO.

RIGHT AND LEFT HAND ANALYSIS

Operation ___Assemble upper jaw of pipe vise___

Part No.___V-2842-3___ Dwg. No.___VA2842___ Date ___7-7-___

Drawn By___E. Snow___ Dept.___11-B___ Plant ___Huntingdon___ Sheet__1__ of__1__

Sketch

Left Hand	Symbols		Right Hand	
1. Pick up vise brace			Get jaw of vise	1.
Reach for brace	Re	Re	Reach for jaw	
Grasp brace	G	G	Grasp jaw	
Move brace	M	M	Move jaw to brace	
2. Position brace			Place jaw in recess of brace	2.
Position	P	M	Move jaw to recess	
		P	Position jaw in recess	
		R1	Release jaw	
			Get 2 washers and position	3.
3. Hold brace		Re	Reach for jaw	
Hold	H	G	Grasp washers	
		M	Move washers to jaw	
		P	Position washers	
		R1	Release washers	
4. Hold brace			Pick up two screws	4.
Hold	H	Re	Reach for two screws	
		G	Grasp two screws	
		M	Move two screws to washers	
5. Hold brace			Assemble washers to screws	5.
Hold	H	P	Position screws	
		M	Move screws	
		R1	Release screws	
6. Hold brace			Drive first screw	6.
Hold	H	Re	Reach for driver	
		G	Grasp driver	
		M	Move screw driver	
		P	Position driver in screw	
		U	Drive screw	
7. Hold brace			Drive second screw	7.
Hold	H	Re	Reach for driver	
		G	Grasp driver	
		M	Move screw driver	
		P	Position driver in screw	
		U	Drive screw	
8. Dispose of part			Aside Yankee screw driver	8.
Move assembled jaw	M	M	Move screw driver	
Release part	R1	R1	Release screw driver	

gravity bins. This method in the form of a right- and left-hand analysis would appear as shown in Figure 7–15.

A quick review of this method discloses that the left hand is ineffectively employed for the majority of the cycle, since it is occupied by the therblig hold in five distinct areas. To alleviate this condition, the analyst considers making a fixture to hold the part. To eliminate the element "dispose of part," he introduces an ejector pin actuated by a foot pedal (see Figure 7–16). This ejects the part into an accompanying gravity chute.

FIGURE 7–16
Details of guide and ejector pins for a foot-operated mechanism

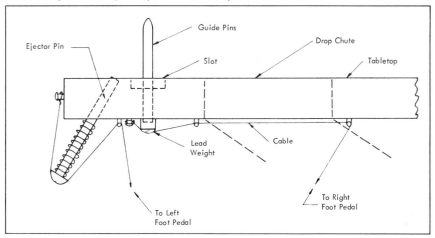

To dispose of the element "lay aside Yankee screwdriver," he suspends the tool overhead. To permit a balanced two-handed motion pattern throughout the cycle, an additional Yankee screwdriver is suspended overhead for the left hand. Thus, both screws can be driven home simultaneously with the aid of self-aligning sleeves incorporated into the fixture. Through the use of fundamental motion data, it can be estimated that the improved method will increase the productivity by at least 50 percent. A right- and left-hand analysis of this proposed method is shown in Figure 7–17.

Figure 7–18 shows a work station for a "doffing yarn" operation. Figure 7–19 illustrates the motion pattern that the operator was following. The operation in this case was not planned, and by means of stroboscopic photography it can be seen that the operator was going through an elaborate body motion pattern. The planned method shown in Figure 7–20 reduced the number and difficulty of the motions so that the operator was less fatigued and his production improved greatly.

The application of the laws of motion economy and an understanding of the therblig concept will prove valuable in establishing ideal methods in the planning stage as well as in improving existing methods.

FIGURE 7–17

Right- and left-hand analysis of a proposed setup for the assembly of a pipe-vise jaw

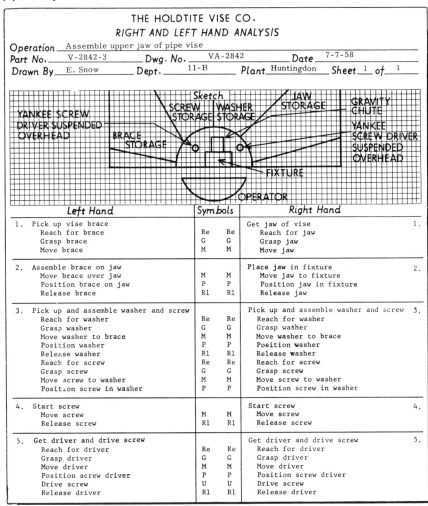

THE HOLDTITE VISE CO.

RIGHT AND LEFT HAND ANALYSIS

Operation __Assemble upper jaw of pipe vise__

Part No. __V-2842-3__ Dwg. No. __VA-2842__ Date __7-7-58__

Drawn By __E. Snow__ Dept. __11-B__ Plant __Huntingdon__ Sheet __1__ of __1__

Left Hand	Symbols		Right Hand
1. Pick up vise brace			Get jaw of vise 1.
Reach for brace	Re	Re	Reach for jaw
Grasp brace	G	G	Grasp jaw
Move brace	M	M	Move jaw
2. Assemble brace on jaw			Place jaw in fixture 2.
Move brace over jaw	M	M	Move jaw to fixture
Position brace on jaw	P	P	Position jaw in fixture
Release brace	Rl	Rl	Release jaw
3. Pick up and assemble washer and screw			Pick up and assemble washer and screw 3.
Reach for washer	Re	Re	Reach for washer
Grasp washer	G	G	Grasp washer
Move washer to brace	M	M	Move washer to brace
Position washer	P	P	Position washer
Release washer	Rl	Rl	Release washer
Reach for screw	Re	Re	Reach for screw
Grasp screw	G	G	Grasp screw
Move screw to washer	M	M	Move screw to washer
Position screw in washer	P	P	Position screw in washer
4. Start screw			Start screw 4.
Move screw	M	M	Move screw
Release screw	Rl	Rl	Release screw
5. Get driver and drive screw			Get driver and drive screw 5.
Reach for driver	Re	Re	Reach for driver
Grasp driver	G	G	Grasp driver
Move driver	M	M	Move driver
Position screw driver	P	P	Position screw driver
Drive screw	U	U	Drive screw
Release driver	Rl	Rl	Release driver

FIGURE 7–18
A work station for doffing yarn

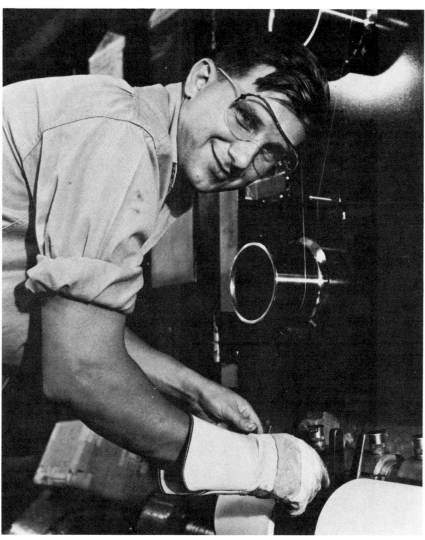

E. I. du Pont de Nemours & Co.

FIGURE 7-19
A stroboscopic photograph showing the motion pattern prior to motion planning on a yarn doffing operation

E. I. du Pont de Nemours & Co.

FIGURE 7-20
A stroboscopic photograph showing the motion pattern after motion planning on a yarn doffing operation

E. I. du Pont de Nemours & Co.

TEXT QUESTIONS

1. When is visual motion study practical?
2. Define and give examples of the 17 fundamental motions, or therbligs.
3. How may the basic motion "search" be eliminated from the work cycle?
4. What basic motion generally precedes "reach"?
5. What three variables affect the time for the basic motion "move"?
6. How does the analyst or observer determine when the operator is performing the element "inspect"?
7. Explain the difference between avoidable and unavoidable delays.
8. Why should fixed locations be provided at the work station for all tools and materials?
9. Which of the five classes of motions is the one most used by industrial workers?
10. Why is it desirable to have the feet working only when the hands are occupied?
11. Explain the significance of human dimensions.
12. What is the approximate number of calories required by the typical operator per day for sedentary activities? For light manual labor? For medium labor? For heavy manual labor?
13. Outline the principal guidelines in the design of operators' handgrips, handwheels, and knobs.

GENERAL QUESTIONS

1. Which of the 17 therbligs are classed as effective, and usually cannot be removed from the work cycle?
2. Take an operation such as dressing out of your daily routine, analyze it through the operator process chart, and try to improve the efficiency of your movements.
3. Explain why effective visual areas are reduced where concentrated attention will be required.
4. Why does a broach usually cost more than a reamer to perform the same sizing operation?

PROBLEMS

1. A given design has fixed costs of $5,000 and variable costs of $15,000 at 100 percent of company capacity that yields total sales of $30,000. How much more of the plant's capacity must be utilized in order to break even if the fixed costs rise to $8,000?
2. Compute the range of work-surface heights that would accommodate all workers through the 98th percentile.
3. What would be the effective visual surface area (in square inches) of an average employee whose work station was centered 26 inches from the center point between his eyes?

SELECTED REFERENCES

Bailey, G. B., and Presgrave, Ralph. *Basic Motion Time-Study.* New York: McGraw-Hill Book Co., 1957.

Barnes, Ralph M. *Motion and Time-Study: Design and Measurement of Work.* 6th ed. New York: John Wiley & Sons, Inc., 1968.

Maynard, H. B., Stegemerten, G. J., and Schwab, J. L. *Methods-Time Measurement.* New York: McGraw-Hill Book Co., 1948.

Mundel, M. E. *Motion and Time-Study.* 3d ed. Englewood Cliffs, N.J.: Prentice-Hall, Inc., 1960.

Nadler, G. *Work Design: A Systems Concept.* Rev. ed. Homewood, Ill.: Richard D. Irwin, Inc., 1970.

8

Micromotion study

Micromotion study is the most refined technique for the analysis of an existing work center. The cost of conducting a micromotion study of an operation is approximately four times as great as the cost of conducting a visual motion study of the same operation. Consequently, it is only when the activity of a specific job or class of work is high that it is economically sound to utilize videotape or moving pictures for motion study. The term *micromotion study* is used to signify the making of a detailed motion study employing either videotapes or motion pictures. Here, each space occupied by a single picture, known as a frame, is projected and studied independently, and then collectively with successive frames.

The basic division of motion, or therblig, concept is usually more important in making a micromotion study than in conducting a visual motion study, in that a work assignment can more readily be broken down into basic elements with the frame-by-frame analysis than with visual motion studies. Since the object of the micromotion procedure is to uncover every possible opportunity for improvement, down to the performance of each therblig, it is fundamental that the analyst to be able to identify each basic division as performed.

Several important corollaries to the principles of motion economy previously cited that have application to micromotion study include:

1. The best sequences of therbligs should be established.
2. Any substantial variation in the time required for a given therblig should be investigated and the cause determined.
3. Hesitations should be carefully examined and analyzed so as to determine and then to eliminate their causes.
4. Cycles and portions of cycles completed in the least amount of time should be used as a goal to attain. Deviations from these minimum times should be studied in order to determine the cause.

Operator selection for micromotion study

In making a micromotion study, it is wise to study either the best operator or, preferably, the best two operators. This procedure is quite different

189

from time study, where an average operator is usually selected for study (see Chapter 14). This is not always possible because the operation might be performed by only one person. In such instances, if past performance indicates that the individual is only a fair or a poor operator, then a competent, skilled, cooperative employee should be trained in the operation before the pictures are taken. Only first-class, proficient operators should be selected for filming. This is fundamental for a number of reasons: the proficient employee is usually a dexterous individual who will instinctively follow the principles of motion economy related to the use of the human body; this type of operator is usually cooperative and willing to be photographed; extra effort put forth by such an operator will show greater results than similar effort employed by the mediocre operator.

If the best two operators have been studied, analysis will indicate proficiency in different areas of the cycle for the two operators. This may lead to greater improvement than if only one individual is studied.

The people who are being studied should be given at least a day's notice prior to taking the motion picture. This allows them to make any personal preparations they may wish and also permits them to choose wearing apparel that will be conducive to making clear pictures.

To secure the cooperation of the foreman, it is essential that he be advised of the plans several days in advance. This is necessary so that he will be able to adjust his personnel to assure that his projected production schedules will be maintained. Interruptions brought about by film analysis work may cause a given section to lose several valuable man-hours, and if the foreman is not given advance notice of contemplated motion study work involving his section, he can hardly be expected to be cooperative.

Micromotion study as a training aid

In addition to being used as a methods improvement tool, micromotion study is finding increasing use as a training aid. The world of sports has used this tool for many years, to develop athletes' timing of movement, rhythm, and smoothness of performance. Motion pictures are made of outstanding performers in a given sport, and then the pictures are enlarged several times and projected onto the screen to facilitate detailed analysis of their motions. Less proficient performers are then able to pattern their efforts after the experts.

Industry is finding that it can attain the same results accomplished in the area of athletics. By filming highly skilled workers and showing the pictures in slow motion and highly enlarged, it is possible to train new employees in a minimum of time to follow the ideal motion-method pattern.

One Pennsylvania concern engaged in the manufacture of loose-leaf notebooks, writing paper, envelopes, tablets, and similar paper items continually takes videotapes for training purposes, not only in its own plant but in affiliated plants. By continually exchanging motion pictures taken at the different plants, it can take advantage of methods improvements introduced throughout the corporation, and is able to train its employees in following the new methods in a minimum of time.

Management should take full advantage of the industrial movies once a micromotion program has been launched. By showing all the pictures taken of the various operations to the operators involved and to their fellow workers, immeasurable goodwill and individual interest are established throughout an organization. Once operators understand the usefulness of micromotion study, their assistance can be enlisted in working out improved methods.

Micromotion equipment

To be able to do an acceptable job in micromotion study, it will be necessary to budget a minimum of $5,000 if motion pictures are used and of $35,000 if videotapes are used. These figures are only rough approximations based on current prices and can fluctuate appreciably as market conditions change.

Videotape equipment provides the distinct advantage of instantaneous replay. Immediately after taking pictures on videotape, the analyst can observe the operation under study. Videotapes may be used over and over again for different micromotion studies. Figure 8–1 illustrates videotape equipment that permits real-time projection speeds, slow motion, and frame-by-frame analysis.

Using motion pictures has the obvious advantage of requiring less

FIGURE 8–1
Videotape equipment that permits immediate playback at real-time speeds, in slow motion, or frame by frame, where the time between frames is $\frac{1}{120}$ second

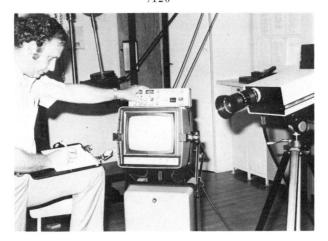

capital equipment than is required when using videotape. However, there is also the disadvantage of having to wait several days for the development of the film before frame-by-frame analysis can begin.

The most important equipment when using film for micromotion study is the camera. In selecting a camera, it is desirable to choose one that accommodates three lenses. This allows a standard lens, a wide-angle lens to provide additional viewing area, and a telephoto lens to provide a great amount of detail in a limited viewing area.

It should be recognized that the larger a lens is in relation to its focal length, which is the distance from the lens to the film, the greater the amount of light that will be admitted during a given interval and that the greater the amount of light per interval, the greater the scope that can be given to the camera.

The motion-picture camera

Several acceptable cameras are on the market today, each having certain desirable features. Some cameras are equipped with spring drive, others with electric motor drive, and still others provide both of these drives. Straight spring-driven cameras are on the average about 30 percent lower in cost than are constant speed, motor-driven cameras and offer the additional advantage of not requiring a nearby power source. In general, the film capacity of the spring-driven camera is as large as or larger than that of its motor-driven companion.

The principal disadvantage of the spring-driven camera is that rewinding is a periodic necessity. The operator may forget to do this and

FIGURE 8–2
Bell & Howell 70 TMR Time and Motion Study Camera

have to stop and rewind at a critical point in the job being studied. A decade ago, these cameras sometimes failed to maintain a sufficiently constant speed, but the better spring-driven cameras on the market today offer no problems in this respect.

Electric, synchronous, motor-driven cameras assure constant speed exposure, and thus it is not necessary to include a microchronometer or other timing device in the background of the pictures being taken. Standard models are available that allow speeds of either 1 frame per second, 10 frames per second, or 1,000 frames per minute. The last, which is the most popular for motion study work, is slightly faster than the normal camera speed of 960 frames per minute (see Figure 8–2).

The light meter

Even though a broad latitude is given in XX reversible film, judgment should not be relied upon to determine the intensity of the light being cast upon the subject to be photographed. A light meter is the only tool that assures an accurate recording of the light intensity. Since the light that strikes the film in the camera is reflected light, it is necessary to measure this light rather than incident light, which is the light cast upon the subject.

It must be remembered that the exposure should be constant for the film being used. Since exposure equals light times time, it is essential that the quantity of light be known. If the quantity of light is small, the time will be long; conversely, if the quantity of light is large, the time will be short.

A photoelectric cell exposure meter is shown in Figure 8–3. A light

FIGURE 8–3
Exposure meter for measuring the intensity of reflected light

Weston Electrical Instrument Corp.

reading is made by aiming the meter toward the scene and pointing the instrument downward. It is important that the meter not be pointed toward a floodlight or, for outdoor pictures, toward the sky. If this is done, an inflated reading will result in improper setting of the *f*-stop number and consequent underexposure of the film. Reflected brightness from the scene energizes the light-sensitive photoelectric cell, and intensity is read on the scale in candles per square foot.

Projection equipment

Many styles of projection equipment are available today, and care should be exercised to select a projector that will allow (1) counting

FIGURE 8–4
16-mm projector equipped with a stop action, slow motion (1–12 frames per second), illuminated frame counter and a self-contained control box for forward, reverse, cine-stop, or pulse control

Traid Corporation

frames, (2) stop action, (3) variable speed projecting, and (4) clear pictures.

If the projector is equipped with a frame counter, it will save the analyst the laborious, monotonous operation of counting frames while analyzing elemental motions.

Stop action is necessary so that each frame may be projected as a still slide for extensive analysis.

Variable speed is a characteristic that makes the projector more versatile. When this feature is incorporated into the projector, it will be useful for training operators in following the correct method. Also, it will be helpful in training time study men to performance rate or level (see Chapter 15). Figure 8–4 illustrates a 16-mm projector with all the fundamental characteristics required to do a good job of projecting for both motion study and micromotion study.

Videotape facilities

Videotape equipment developed in recent years offers the motion and time study analyst several advantages over the conventional motion-picture facilities. This equipment provides the problem-solving capabilities of both normal and high-speed film with long recording capacity and instant replay features.

Videotape systems can record up to one hour of black-and-white videotape on a single eight-inch reel of one-inch magnetic tape that may be erased, edited, or stored for future reference. This compares favorably with 16-mm motion-picture film, which permits less than four minutes of exposure when using 100-foot rolls of film and exposing at normal speed.

The videotape model illustrated in Figure 8–1 records at 120 frames per second. This speed provides the added advantage of slow-motion playback, which facilitates the study of work station methods. Of course, recorded tapes may be played back in real time as well as in a range of slow-motion times. Stop action analysis may also be accomplished with this equipment.

Videotape facilities can incorporate two cameras and a dual camera control unit for the added flexibility of split screen. This feature permits the simultaneous observation and study of two competing methods. It also provides an excellent training device in "performance rating," as will be explained in Chapter 15.

Videotape equipment is available with scene and frame identification features which preclude the necessity of including a microchronometer or some other frame-identifying device in the scene.

In the equipment illustrated, the scene and frame identification consists of two separate sets of numbers that appear at the top of each picture. The scene number is a three-digit number which is followed by a six-digit number for frame identification. The frame identification number starts at zero count at the beginning of the record mode, and increases by one count for each frame.

The principal disadvantage of videotape facilities is the first cost of the equipment, which is today about seven times the cost of motion-picture facilities.

Taking the motion pictures

In general, the fundamental steps in making a micromotion study, in chronological sequence, are:

1. Take videotape or moving pictures of the two best operators.
2. Analyze the film on a frame-by-frame basis, plotting the results on a simo (simultaneous motion) chart.
3. Consider the laws of motion economy and their corollaries and create an improved method.
4. Teach and standardize the new method.
5. Take videotape or moving pictures of the new method.

Several factors should be checked before taking motion pictures on the production floor or in the laboratory. First, be sure that the selected operators have had reasonable advance notice that pictures are to be taken. This is important so that they can be properly groomed and attired. It is also necessary to be sure that clearance for the use of the operators has been obtained from the departmental foreman. In most plants, it is necessary to obtain clearance from the union before proceeding with the picture taking. Even if this is not required, it is always wise, in the interests of healthy labor relations, to keep the union officers completely informed as to the purposes of all phases of the motion-picture procedure. This is important, since some plants have misused the data developed by the micromotion technique to the detriment of organized labor, and thus have caused suspicion and resentment toward the filming technique. Every effort should be made to avoid creating the same reaction to the micromotion procedure as that which developed toward the stopwatch during the 1920s.

Then, make sure that there is an adequate inventory of material so that it will not be necessary to replenish various supplies during the picture-taking process.

Upon the completion of several practice cycles, conducted to relieve the operator of any nervousness he may have developed while being "put under the spotlight," the operator is requested to give his best performance, and then the tape or film is exposed. Several cycles, the number depending on their duration, should be taken for analysis. Often, the first cycle is unsatisfactory, as the operator may develop a case of nerves when he hears the camera begin to operate. This reaction is usually brief, and he will soon acquire his normal feel for the work. Usually, operators enjoy being photographed, and little difficulty is encountered in getting excellent co-operation from them.

Analyzing the tape or film

Before analyzing the film or tape, the analyst should run the pictures through several times in order to determine the cycle that represents the best performance. The shortest cycle is usually the best one to study.

Once the cycle that is to be studied has been determined, the analyst can begin the frame-by-frame analysis. Although the study can be made at any point in the cycle, it is customary to begin the analysis at the frame after "release of finished part" of the previous cycle. Thus, the first basic operation employed in most instances would be the reach made in order to pick up a part or tool so as to begin production on the cycle being studied.

Once the starting point of the cycle has been determined, the analyst should set the frame counter to zero so as to facilitate recording elapsed elemental times.

After the therblig being employed by the operator has been noted, the film should be advanced a frame at a time until termination of the basic division. The number of frames exposed is then read off the frame counter and recorded on the micromotion data sheet.

Reach and move motions are considered as beginning in the frame preceding the frame in which noticeable movement occurs, and as ending in the frame in which noticeable movement ceases. Where only minor movements occur, such as grasp, release, and position, the beginning point is considered to be the frame in which the preceding motion ended, and the ending point is the frame preceding the one in which the following motion was noticeable.

In reviewing the tape or film, the analyst carefully observes the motion class being used while performing the basic division, and records this information on his data sheet. At all times, the analyst must have an open and probing mind so as to determine the possibilities of either eliminating or improving each basic division.

After briefly describing the basic division in the space provided (such as "move to fixture" in Figure 8–5), the analyst draws a horizontal line across the portion of the paper representing the body member being studied. For the most part, only the arms and hands of the operator are studied, although sometimes it is desirable to analyze foot motions and body motions.

The motion symbol should be recorded in the space provided, and in combined motions, symbols of the elements occurring simultaneously should be noted. Thus, a pre-position may occur during a move, such as when the hand aligns a bolt while transporting it. In this case, the ele-

Motion and time study

FIGURE 8–5
Film analysis of vise clamp assembly

| DATE: JANUARY 9 | DEPT. I.E. 320 | DRAWING: SEE REPORT | PART NO. 114 |

PART DESCRIPTION: VISE CLAMP

OPERATION: ASSEMBLY   FILM SPEED: 16 FRAMES/SEC.

CLOCK READINGS	LEFT HAND	SYMBOL	MOTION CL.	FINGER	WRIST	FOREARM	SHOULDER	BODY	TIME (IN WINKS)	BODY	SHOULDER	FOREARM	WRIST	FINGER	MOTION CL.	SYMBOL	RIGHT HAND	CLOCK READINGS
2059																		2059
2089	MOVE TO JAW BIN 14"	RE	3						0030							3	RE	MOVE TO BRACE BIN 14"	2089
2105	PICK UP JAW	G	1						0016										2105
2125	MOVE TO FIXTURE 8"	M	3						0020 0036								UD	IDLE	2125
2184	PLACE IN FIXTURE	U	1						0059							1	G	PICK UP BRACE	2184
									0038							3	M	MOVE TO FIXTURE 8"	2222
2451	MOVE TO FIXTURE 12"	M	3						0037							3	M	MOVE TO FIXTURE 12"	2451
2579	SCREW IN SCREWS	U	1						0128							1	U	SCREW IN SCREWS	2579

ment description would be "Move bolt _____ inches," and the symbols recorded would be "P.P. and M."

In the motion class columns, it is helpful to differentiate the productive elements from the nonproductive elements. This is usually done through color; thus, the elements reach, grasp, move, use, and assemble would be shown as solid black, and the remainder of the therbligs would be shown in red or by cross-hatching if the charts are to be duplicated. Delay times are signified as Class 1 motions in red. All other therbligs are identified on the chart by the motion class utilized in performing the operation.

The time scale selected should provide enough room to identify clearly the shortest basic divisions so as to give a neat-appearing chart that can be easily studied and diagnosed for improvement. Release is the shortest of all the therbligs, taking but one frame at normal camera speed. Thus, if each division of the time scale is equated to 0.002 of a minute, a neat-appearing chart would result.

After the first basic division of one hand has been analyzed, and the method recorded, the film should be advanced slowly, and the next basic division should likewise be analyzed, and so on until completion of the cycle. To avoid confusion, one hand should be completely studied before the study of the other hand is begun. As the second hand is being considered, a periodic check should be made to verify that the motions recorded are occurring in the same relationship with the other hand as is indicated in the analysis sheet. The analyst must remember that when the projected picture is reviewed the left arm is on the right side of the picture. In fact, all directions are reversed.

After both hands have been completely charted, a summary should be included at the bottom of the chart showing the cycle time, pieces made per cycle, productive time per cycle, and nonproductive time per cycle. Sometimes a sketch of the workplace is helpful, and this can be included at the bottom of the chart. The sketch is usually not necessary, since the film will give a pictorial presentation of the work area.

Create an improved method

After the simo chart has been completed, the next step is to use it. The nonproductive sections of the chart represent a good place to start. These sections will include the therbligs hold, search, select, position, pre-position, inspect, and plan, and all the delays. The more of these that can be eliminated, the better the proposed method will be. However, the analysis should not be confined to the red sections of the chart, since possibilities for improvement exist in the productive portions as well. For example, the productive element "reach 24" can be improved by reducing the distance and, consequently, lowering the motion class.

A micromotion checklist helps the analyst to avoid overlooking any possibility of improvement. Table 8–1 illustrates a checklist that covers

TABLE 8–1
Checklist for the simo chart

---

1. Are both hands beginning their therbligs simultaneously?
2. Are both hands ending their therbligs simultaneously?
3. Can delay or idle times be removed from the cycle?
4. Are the motion patterns of the arms symmetrical?
5. Are motions made radially from the body in opposite directions?
6. Can a fixture be incorporated to eliminate hold?
7. Are all materials and tools within the normal work area in both the horizontal and the vertical plane?
8. Does the chair and work station allow the operator to get close enough to the work?
9. Has the overhead space been utilized to suspend tools and store inventory components?
10. Are motions confined to the lowest possible classifications?
11. Are tools and materials located to permit the best sequence of motions?
12. Can foot-operated mechanisms be introduced advantageously?
13. Have gravity chutes and drop delivery been provided for?
14. Are all tools pre-positioned to minimize grasp time and reduce the search and select therbligs?
15. Has an automatic ejector been incorporated into the fixture?
16. Can position times be reduced by providing "pilots" or other devices for mating parts?

---

most of the considerations that apply to the principles of motion economy and their corollaries.

Figure 8–6 illustrates a completed simo chart of the assembly of a variable resistor. This chart was constructed after a new fixture was developed to facilitate assembly and represents a method considerably improved over the original method of assembly. Yet an analysis of the chart will disclose several areas for further improvement. The right- and left-hand movements are not balanced in many parts of the cycle, and 10 ineffective motions are performed by the left hand, and 11 ineffective motions by the right hand. Several of the moves and reaches appear to be unduly long, suggesting the shortening of distanes at the work station layout.

The micromotion technique should be used to uncover every inefficiency, regardless of its apparent insignificance. Sufficient numbers of minute improvements will in total result in an appreciable annual savings.

Teach and standardize the new method

To be assured that full benefits are realized from the micromotion study, it is important that the improved method be put into practice as soon as possible and that it be followed in exact detail by all operators to whom it applies. Verbal explanations of the motion pattern to be followed are usually inadequate: both time and effort will be saved if an instruction chart is used to give specific information on the "how" of the improved method.

In form, the typical micromotion instruction sheet is similar to the

FIGURE 8–6
Simo chart assembly of a variable resistor

MICROMOTION DATA

NAME OF PART. VARISTOR   NO. 30 A   FILM NO. 5209

OPERATION. ASSEMBLY OF VARISTOR   OPERATOR. MAYHEW

TIME SCALE	E.T.	NO.	LEFT-HAND DESCRIPTION				RIGHT-HAND DESCRIPTION	NO.	E.T.	TIME SCALE
0163	WINKS	NO.			RL	RELEASE ASSEMBLY	NO.	WINKS	0163
0170	7	1	MOVING TOWARD BINS	RE		RE	MOVE TO BIN	1	10	0173
0182	12	2	GRASP SPRING WASHER	G		G	GRASP BOLT	2	9	0182
0201	18	3	MOVE TO FIXTURE	M		M	MOVE TO FIXTURE POSITION BOLT	3	19	0201
0250	49	4	WAIT FOR RIGHT HAND	UD		P		4	49	0250
0254	4	5	POSITION WASHER	P		RL	RELEASE BOLT	5	4	0254
0272	18	6	PLACE ON BOLT	A		UD	WAIT FOR LEFT HAND	6	19	0273
0276	4	7	RELEASE WASHER	RL		RE	MOVE TO BIN	7	18	0291
0295	19	8	MOVE TO BINS	RE		G	GRASP BRASS WASHER	8	11	0302
0310	15	9	GRASP TUBE	G						
0340	30	10	MOVE TO FIXTURE	M						
0343	3	11	POSITION	P		M	MOVE TO FIXTURE	9	43	0345
0349	6	12	PLACE TUBE ON BOLT	A		P	POSITION WASHER	10	4	0349
0354		13	HOLD	H						
0356	2	14	RELEASE TUBE	RL						
			MOVE TO BIN AND	RE		A	PLACE WASHER ON BOLT	11	30	0379
0385	29	15	GRASP FIBER WASHER	G		RL	RELEASE WASHER	12	2	0381
0402	17	16	MOVE TO FIXTURE	M		RE	MOVE TO BIN	13	19	0400
						G	GRASP CONTACT	14	6	0406
						M	MOVE TO FIXTURE	42	21	1122
						P	POSITION DRIVER	43	5	1127
1184	100	40	HOLD POSITIONER UP	H		U	TIGHTEN NUT	44	57	1184
1186	2	41	RELEASE POSITIONER	RL		RL	RELEASE DRIVER	45	2	1186
						RE	MOVE TO FIXTURE	46	11	1197
						G	GRASP ASSEMBLY	47	4	1201
1205	19	42	MOVE TO BINS	RE		M	MOVE TOWARD BINS	48	11	1212

TIME PER PIECE
1049 WINKS
OR
0.5245 MIN.

R. E. Call

operator process chart. A layout of the work area drawn to scale heads the instruction sheet, and following this is shown the motion sequence of both the right hand and the left hand. These elements appear in relationship to one another. Special tools, such as form cutters, gages, jigs, and fixtures, are identified after the element in which they are used. Speeds, feeds, depth of cut, and other pertinent manufacturing information are also shown. Usually a time summary will accompany the instruction sheet, so that the operator will be able to measure his performance against standard while being trained in the new method. A typical instruction sheet appears in Figure 8–7.

It is important that the foreman or line supervisor be well versed in the proposed method so that he will be able to assist in training the workers. Usually instruction sheets are run off in Ditto or some other

FIGURE 8–7
Instruction sheet

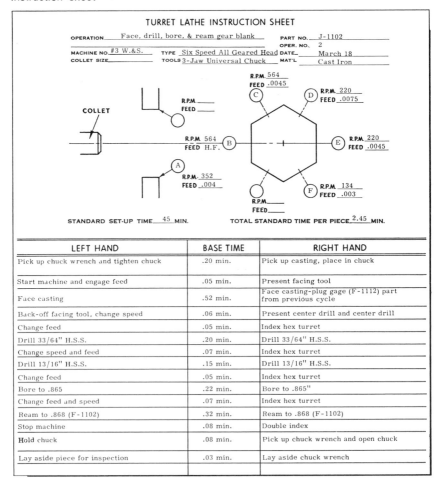

TURRET LATHE INSTRUCTION SHEET

OPERATION Face, drill, bore, & ream gear blank   PART NO. J-1102
                                                    OPER. NO. 2
MACHINE NO. #3 W.&S.   TYPE Six Speed All Geared Head DATE March 18
COLLET SIZE   TOOLS 3-Jaw Universal Chuck   MAT'L Cast Iron

STANDARD SET-UP TIME 45 MIN.   TOTAL STANDARD TIME PER PIECE 2.45 MIN.

LEFT HAND	BASE TIME	RIGHT HAND
Pick up chuck wrench and tighten chuck	.20 min.	Pick up casting, place in chuck
Start machine and engage feed	.05 min.	Present facing tool
Face casting	.52 min.	Face casting-plug gage (F-1112) part from previous cycle
Back-off facing tool, change speed	.06 min.	Present center drill and center drill
Change feed	.05 min.	Index hex turret
Drill 33/64" H.S.S.	.20 min.	Drill 33/64" H.S.S.
Change speed and feed	.07 min.	Index hex turret
Drill 13/16" H.S.S.	.15 min.	Drill 13/16" H.S.S.
Change feed	.05 min.	Index hex turret
Bore to .865	.22 min.	Bore to .865"
Change feed and speed	.07 min.	Index hex turret
Ream to .868 (F-1102)	.32 min.	Ream to .868 (F-1102)
Stop machine	.08 min.	Double index
Hold chuck	.08 min.	Pick up chuck wrench and open chuck
Lay aside piece for inspection	.03 min.	Lay aside chuck wrench

rapid duplicating process so that copies can be filed in the methods section, time study section, and foremen's office, as well as given to each operator performing the operation. A hanger for the instruction sheet should be provided at the operator's work station so that it can readily be referred to at all times during the operating cycle. If a clear plastic envelope is provided to hold the instruction sheet, it will remain readable for a considerably longer period.

The foreman, with the methods engineer, should periodically check each operation on the production floor to be sure that the new method is being followed, and to answer any question that may come up on the new procedure. Installation and follow-up are two very important phases in micromotion methods improvement. Often it is difficult for the average operator to justify the new method because the changes may appear so insignificant. Consequently, it will be necessary for both the foreman and the methods engineer to do a real job of selling the improved system.

Unless periodic checks are made for at least several weeks, operators may go back to their old way of doing the job and completely disregard the new technique. One of the added benefits of the written instruction sheet is that it helps the operator to conform to the established instructions.

As soon as one of the better operators has become skilled in the new method, motion pictures can be taken and a simo chart constructed to illustrate the improvements resulting from the study. The same operators selected in making pictures of the unimproved method should be used. This will give a better comparison for evaluating the savings than if different operators are used in the final pictures. Once the new pictures have been taken, they should be used as a training tool for all other operators performing the same operation. Seeing how the operation should be done is much more helpful than reading or hearing how it should be done. The psychological impact of seeing a fellow worker perform the new method is a force in breaking down the inherent human resistance to change.

Comparing the completed simo chart of the improved method to the one drawn of the old method will clearly present the savings. Figure 8–8 illustrates a method proposed after a complete micromotion study. Note that this tool not only shows the amount of savings but vividly explains how these results were made possible.

Memomotion study

One other important motion study technique gives more detail than visual motion study and less than micromotion study, and has application under certain conditions. This is the technique referred to as "memomotion study."

Memomotion study is a moving-picture technique developed by Marvin Mundel several years ago to analyze the principal movements in an

FIGURE 8–8

Simo chart of improved method for the assembly of lace finger in textile machinery

SIMO CHART

OPERATOR: Ken Reisch
DATE: May 21,
OPERATION: Assembly
PART: Lace Finger
METHOD: Proposed
CHART BY: Joseph Riley

TIME SCALE (winks)	ELEMENT TIME	LEFT-HAND DESCRIPTION	SYMBOL	MOTION CLASS	SYMBOL	RIGHT-HAND DESCRIPTION	ELEMENT TIME	TIME SCALE (winks)
4548	12	Reach for finger	RE		RE	Reach for finger	12	4548
4560	19	Grasp finger	G		G	Grasp finger	19	4560
4579	31	Move finger	M		M	Move finger	31	4579
4610	75	Position and release finger	P RL		P RL	Position and release finger	75	4610
4685	15	Reach for clamp	RE		RE	Reach for clamp	15	4685
4700	15	Grasp clamp	G		G	Grasp clamp	15	4700
4715	12	Grasp assembly	G		G	Grasp assembly	12	4715
7541	18	Move and release assembly	M RL		M RL	Move and release assembly	18	7541
7559								7559

SUMMARY

%	TIME	LEFT HAND SUMMARY	SYM.	RIGHT HAND SUMMARY	TIME	%
8.56	249	Reach	RE	Reach	245	8.4
7.49	218	Grasp	G	Grasp	221	7.5
12.16	354	Move	M	Move	413	14.2
30.45	887	Position	P	Position	1124	38.7
39.3	1145	Use	U	Use	876	30.1
1.03	30	Idle	I	Idle	0	0.0
.96	28	Release	RL	Release	32	1.1
100.0	3011	TOTALS			3011	100.0

FIGURE 8-9
Portion of a memomotion study of four men operating a hydraulic extrusion press
(film exposed at rate of one frame per second)

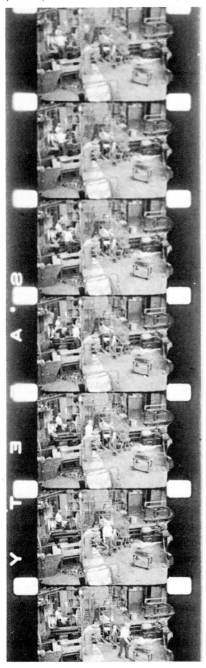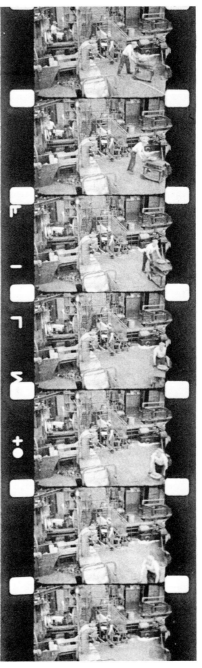

operation in order to improve methods, define problem areas, and establish standards.

Memomotion study usually uses a filming speed of 60 frames per minute with an accurate rate time interval between pictures to provide a record for frame-by-frame analysis. However, equipment is available to provide much greater flexibility in the rate of film exposure. For example, the equipment illustrated in Figure 8–10 allows exposures of from two frames per second to one frame every three seconds or even greater intervals.

I have standardized on 60 frames per minute, or 1 frame per second. This speed gives a measurement of elapsed elements that is precise enough for establishing standards, and permits significant economies over the regular camera speed of 960 frames per minute for 16-mm silent film. Thus, under memomotion, pictures are being exposed just $\frac{1}{16}$ as fast as under the normal speed (see Figure 8–9). For example, in the study of a three-minute cycle, the exposure and the analysis of only 4.5 feet of film are required, as compared with 72 feet when the micromotion procedure is used.

$$\frac{3 \text{ min.} \times 60 \text{ frames/min.}}{40 \text{ frames/foot}} = 4.5 \text{ feet}$$

Frequently, it is advantageous to employ memomotion study in the following areas of work:

1. Multiman activities.
2. Multifacility operations.
3. Irregular cycle studies.
4. Long-cycle operations.
5. Plant layout.
6. Recorded picture histories.
7. Work sampling studies.

Memomotion finds application in the above areas of study because:

1. It provides a more accurate record than do visual means for analysis and film appraisal.
2. It records interrelated events more accurately than do visual techniques.
3. It provides a basis for work measurement, since the camera operates at a constant speed.
4. It focuses attention on the major movements taking place at a work center.
5. It reduces the cost of purchasing and analyzing the film to about 6 percent of the cost incurred when filming at normal speed.

Memomotion is particularly helpful in the study of crew activities. Terminal points in the elements of various operators within a group and the relation of terminal points to one another are almost impossible to recognize by visual methods. Through frame-by-frame analysis, the relationships of the various work elements performed by the members of the crew are easily determined. Consequently, memomotion study lends

FIGURE 8–10
This time-lapse controller permits exposures of from two frames per second to one frame every three seconds or even greater intervals

Si-Con Signal Control, Inc.

itself to such uses as crew balance studies and gang process chart work (see Chapter 6).

Multifacility operation studies are quite similar to crew studies from the standpoint of adaptability to memomotion study work. Here it is desirable to follow not only the activities of the operator but also the facilities he or she services. Such factors as interference delays and machine breakdowns are very difficult to study by visual means, and yet they are easily observed through the frame-by-frame procedure. It is quite difficult to study irregular cycles, such as the periodic replenishing of stock supplies, sharpening tools, and moving completed work, by visual means. To begin with, the analyst does not know when they are going to take place, and consequently he is obliged to record the element and measure its duration while it is taking place. This is most difficult. The memomotion procedure records all the facts pictorially—the method employed and the time consumed—so that an analysis of the facts may be conveniently and leisurely made through frame-by-frame analysis.

Long-cycle operations employing many short elements are usually too costly to study by micromotion techniques and too difficult to study in adequate detail by visual methods. Memomotion provides the ideal alternative to both of these procedures.

Memomotion studies of a work area will highlight the major moves of all personnel coming into that area. Long moves, such as backtracking and trips for tools, raw materials, and supplies, are emphasized, and inefficient areas of the layout are pointed out. Frequently, a work area that appears satisfactory under a conventional micromotion film will indicate areas for improvement under memomotion.

Filing a permanent record of a method is frequently helpful. The permanent record simplifies setting up for the same job at a later date; it is also useful for later methods improvement studies, for settling labor disputes, and for operator training.

Another important use of the memomotion camera is in preparing work sampling studies (see Chapter 21). By using memomotion equipment, unbiased work sampling studies may be made. To the typical memomotion camera is attached a timer that permits taking random observations during the eight-hour day. The "observation times" can be preset by placing small metal clips around the circumference of the timer dial. A random numbers table may be used in locating the clips. Also, the timer can be set to take pictures at regular intervals during the day if this is desired. A motor-driven timer operates the synchronous motor that drives the camera.

Another mechanism that the author developed (see Figure 8–11) provides a means for randomly triggering the camera. Here random punchings on a constant-speed tape energize the switching mechanism. The roll of tape can allow operation of the equipment for 160 working hours, based upon a minimum of 10 seconds between random observa-

FIGURE 8–11
Triggering equipment for randomly exposing a predetermined number of frames of 16-mm film

tions. A timer can be preset to allow the exposure of 1 to 10 frames at each triggering of the camera. Thus, at random times during the work period, observations of 1 to 10 seconds are taken of the work area.

By using memomotion equipment adapted as described above, the operator can make work sampling studies that will give unbiased results. Work sampling studies made by an observer recording data at random intervals tend to be biased for the following reasons:

1. The mere arrival of the observer at the work scene will influence the operator to become productively engaged in work which he or she would not normally be doing at the time of the observation.
2. There is a natural tendency for the observer to record what has just happened or what will be happening at the exact moment of the observation.

The use of memomotion to make work sampling studies saves labor in the collection of the data, since it is only necessary to remove the film at the end of the work period rather than go to the work scene periodically in order to record data.

The reader should recognize that, for the most part, micromotion

FIGURE 8–12
Time-lapse TV system shown studying a group activity

Timelapse, Inc.

study will allow the same analysis that memomotion study does. However, memomotion offers the kinds of economies that open the door to the frame-by-frame analysis of activities that normally would not justify the cost of a micromotion study.

Time-lapse TV that provides the added feature of instantaneous replay is now available. After taking pictures on videotape, the analyst can observe immediately the operation under study. Tapes may be used over and over again for different time-lapse studies. Figure 8–12 illustrates equipment that is available in compression ratios of from 15 hours of activity to 1 hour of viewing to 2 hours of activity to 1 hour of showing.

TEXT QUESTIONS

1. How does micromotion study differ from motion study?
2. For micromotion study, what additional considerations should be given to "the use of the human body" and "the arrangement and conditions of the workplace"?
3. Why is the best operator usually selected in the micromotion procedure?
4. What consideration should be given the foreman relative to taking micromotion pictures in his department?
5. How can micromotion study be helpful in training?
6. Why is "stop action" an important feature of projection equipment?
7. What "extra" devices should be included with a projector to make it ideal for film analysis work?
8. What is meant by the "speed" of the lens?
9. How fast would an $f/1.8$ lens be as compared to an $f/5.6$ lens?
10. Would an $f/16$ lens be considered a fast lens? Why or why not?
11. What are the important steps in making a micromotion study?
12. How does the simo chart differ from the operator process chart?
13. Which motions are classified as ineffective on the simo chart?
14. What class of motions is signified on the simo chart when plotting a delay?
15. What is the purpose of the micromotion instruction sheet?
16. In film analysis, why is it inadvisable to analyze both hands simultaneously?
17. What advantages does memomotion study have over micromotion study?
18. How much film would be required to study a cycle 13.5 minutes in duration using a memomotion speed of 50 frames per minute?
19. What filming speed is generally used in making memomotion studies?
20. What range of speeds is possible in memomotion study?
21. Explain when you would advocate the use of time-lapse TV facilities.

GENERAL QUESTIONS

1. Make a sketch of your desk or work station and show an arrangement and conditions that would agree with the principles of motion economy and their corollaries.
2. Interview at least two factory workers and get their reaction to micromotion study through the film analysis technique.
3. What activity within an industrial organization other than methods analysis might include the use of micromotion equipment? How?

4. Discuss the relative merits of the telephoto lens and the wide-angle lens.
5. Why do unions require permission before motion pictures are taken on the production floor?
6. In general, what is the operator's attitude to being filmed?
7. How can industrial relations be improved by micromotion studies?
8. Explain what impact, if any, today's philosophy of throwaway type designs has on the use and application of micromotion study.
9. What influence does international competition have on the application of micromotion study?

PROBLEMS

1. In the assembly of small tube–type flashlights in the XYZ Company, two size C (2.5 cm diameter by 4.5 cm length) batteries provide the power. A breakdown of the work elements at each of the 10 assembly stations was as follows:

   a. Reach 44 cm with left hand and pick up tube-type housing.
   b. Simultaneously with (*a*), reach 50 cm with right hand and pick up two size C batteries.
   c. Palm one battery and assemble one battery in housing. Place second battery in housing.
   d. Reach 40 cm and pick up assembled head. Screw assembled head onto tube-type housing.
   e. Reach 40 cm and pick up spring-loaded end. Assemble end to tube-type housing.
   f. Place assembled flashlight in cardboard carton with left hand. This move averages 50 cm.

   Design an improved work station based upon the principles of motion economy. Estimate the hourly savings of your improved layout if 10 million flashlights are produced.

SELECTED REFERENCES

Barnes, Ralph M. *Motion and Time Study: Design and Measurement of Work.* 6th ed. New York: John Wiley & Sons, Inc., 1968.

Mundel, Marvin E. *Motion and Time Study: Principles and Practices.* 4th ed. New York: Prentice-Hall, Inc., 1960.

Nadler, G. *Work Design: A Systems Concept.* Rev. ed. Homewood, Ill.: Richard D. Irwin, Inc., 1970.

9

Macroscopic approaches to improvement

For the most part, operation analysis, motion study, and micromotion study have concentrated on the improvement of the work station. The principal objectives of these techniques, as has already been pointed out, are to:

1. Optimize physical work.
2. Minimize the time required to perform tasks.
3. Maximize the product quality per dollar of cost.
4. Maximize the welfare of the employee from the standpoint of earnings, job security, health, and comfort.
5. Maximize the profit of the business or enterprise.

Work station improvements utilizing the three aforementioned techniques are based primarily on:

1. Newton's laws of motion.
2. The biomechanics of the human body and the limitations of employees.
3. Optimization methodologies.

In order to better achieve the improvement of existing methods and to more thoroughly plan projected work, it is desirable for the analyst to be cognizant of some of the fundamentals related to the macroscopic approach to work improvement. The areas of study relating to the macroscopic approach include the physical environment of the work station and physiological and psychological factors relating to both the operator and the work force.

Thus, before implementation of the new or improved method resulting from operation analysis is inaugurated, there should be an understanding of more important basic principles that relate to a somewhat broader spectrum of the methods engineering program.

Measurement and control of the physical environment

The immediate physical environment has a significant impact not only upon the operator and his supervisor's performance, but also upon the reliability of the process. The principal environmental factors that influence the productivity of the working personnel and the dependability of the process include the visual environment, noise, vibration, humidity, temperature, and atmospheric contaminants.

The visual environment

The efficient performance of almost all tasks, whether industrial, office, business, service, or professional, depends to some degree on adequate vision. Adequate lighting is just as important for the dentist who is repairing a molar as it is for the toolmaker who is polishing contours in a mold for producing plastic components.

The principal criteria that govern the visual environment are the amount of lighting, the contrasts between the immediate surroundings and the specific task being carried out, and the presence or absence of glare.

Although much research has been conducted regarding the amount of light required for a specific job, the exact amounts needed are still controversial. The ability to see increases as the logarithm of the illumination, so that a point is soon reached at which large increases in illumination will result in very small increases in the worker's efficiency. In 1932, Lythgoe observed that the relationship between visual acuity and the logarithm of the luminance started to depart from linearity at approximately 10 footlamberts (see Figure 9–1).

FIGURE 9–1

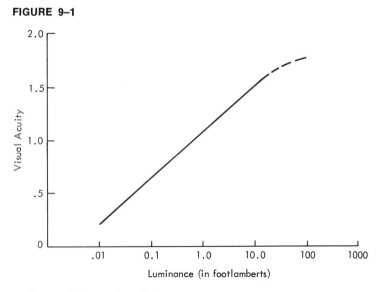

Luminance (in footlamberts)

Source: Redrawn from Lythgoe.

The quantity of light that is needed to perform a task satisfactorily is affected by several independent factors. Notable among these are:

1. The contrast between the object being viewed and the immediate surroundings. Colors also have a significant impact on contrast.
2. The reflectivity of the immediate surroundings.
3. The physical dimensions of the object being viewed.
4. The viewing distance.
5. The time permitted for seeing.

It is apparent that the methods analyst will have a certain amount of control over all of these factors except for the physical dimensions of the object being viewed. He may even exercise some control over this if he plans the work so that it is exposed dimensionally in its most advantageous position from the standpoint of the operator's visual perception.

The contrast between the object being viewed and the immediate surroundings can be thought of as brightness differences and may be expressed in terms of intensity. Contrast may then be expressed as:

$$C = \frac{B_L - B_D}{B_L} \times 100$$

where:

C = Contrast in percent

B_L = Luminance[1] of the lighter surface

B_D = Luminance of the darker surface

In many work situations, a second form of contrast is evident. This is color contrast. Thus, two or more colors within a work environment may reflect approximately the same amount of incident light, yet be significantly distinguishable because of their difference in hue. Where color is involved, the analyst has the opportunity to enhance contrast by selecting colors which are not only widely separated on the color circle, but which also have widely different reflectivities.

The luminance of an object depends on the amount of the incident light which is reflected. Reflectivity is the percentage of the incident flux which is reflected by a surface. The relationship between reflectivity, luminance, and illumination can be equated as follows:

$$R = \frac{B}{E} \times 100$$

where:

R = Reflectivity (%)

B = Luminance (quantity of reflected light in lumens per square cm)

E = Illumination (amount of light in lumens per square cm which falls upon object)

---

[1] The basic unit of luminance is the lambert, which can be defined as the brightness of a perfectly diffusing surface emitting one lumen/cm^2.

TABLE 9–1

Color or finish	Percent of reflected light	Color or finish	Percent of reflected light
White	85	Medium blue	35
Light cream	75	Dark gray	30
Light gray	75	Dark red	13
Light yellow	75	Dark brown	10
Light buff	70	Dark blue	8
Light green	65	Dark green	7
Light blue	55	Maple	42
Medium yellow	65	Satinwood	34
Medium buff	63	Walnut	16
Medium gray	55	Mahogany	12
Medium green	52		

Table 9–1 provides reflectance factors of typical paint and natural wood finishes.

The size of the object has a notable impact on its ability to be seen. The smaller the size, and consequently the less the visual angle, the more difficult it will be to see the part. Murrell has classified small work into six categories related to an 18-inch viewing distance, as shown in Table 9–2.[2]

TABLE 9–2

Category	Visual angle (minutes and seconds)	Size (inches)	Example
I	Less than 23 seconds	<0.002	Fine wires (about 0.001 inch in diameter)
II	23–45 seconds	0.002–0.004	Fine scribed line or human hair
III	45 seconds–1 minute 30 seconds	0.004–0.008	Lines on slide rule or micrometer
IV	1 minute 30 seconds–3 minutes	0.008–0.015	Size 40 sewing cotton or thickness of 6-point print
V	3–6 minutes	0.015–0.030	Thickness of 12-point print
VI	Over 6 minutes	>0.030	Any large object

When an object is reasonably large (over 0.030 inch when viewed from 18 inches) and the contrast is high, the object can be seen in a very short period of time. However, if the object is quite small and the contrast is low, the time required to see the object will be relatively high. It has been proven experimentally that the time required to see a small object can be increased by a factor of four if the contrast is altered by 20 percent (for example from 70 percent to 50 percent). This is an important factor in quality control, when an inspector is searching for minute flaws in work that is passing by rapidly on a conveyor.

---

[2] K. F. H. Murrell, *Ergonomics* (London, England: Chapman and Hall, 1969).

As a general guide to the analyst, light recommendations in foot-lamberts are shown in Table 9–3, and footcandle requirements for different seeing tasks where the reflectance and contrast are relatively high are shown in Table 9–4.

TABLE 9–3
Guide brightness for various categories of seeing at specified footcandles and reflectance*

Category of seeing task	Guide brightness (footlamberts)	Footcandles for specified reflectance conditions		
		90%	50%	10%
A. Easy	Below 18	Below 20	Below 36	Below 180
B. Ordinary	18–42	20–95	36–84	180–420
C. Difficult.	42–100	45–153	84–240	420–1,200
D. Very difficult. . . .	120–420	133–455	240–840	1,200–4,200
E. Most difficult. . . .	420 up	455	840	4,200

* From *Human Factors Engineering* by Ernest J. McCormick (New York: McGraw-Hill Book Co., 1964).

TABLE 9–4
Current light recommendations for different seeing tasks

Tasks	Footcandles in service, i.e., on task or 30 inches above floor
Most difficult seeing tasks	200–1,000
Finest precision work involving: Finest detail; poor contrasts; long periods of time. Extratime assembly; precision grading; extrafine finishing.	
Very difficult seeing tasks	100
Precision work involving: Fine detail; fair contrasts; long periods of time. Fine assembly; high-speed work; fine finishing.	
Difficult and critical seeing tasks	50
Prolonged work involving: Fine detail; moderate contrasts; long periods of time. Ordinary benchwork and assembly; machine shop work; finishing of medium-to-fine parts; office work.	
Ordinary seeing tasks.	30
Involving: Moderately fine detail; normal contrasts; intermittent periods of time. Automatic machine operation; rough grinding; garage work areas; switchboards; continuous processes; conference and file rooms; packing and shipping.	
Casual seeing tasks	10
Such as: Stairways; reception rooms; washrooms and other service areas; active storage.	
Rough seeing tasks	5
Such as: Hallways; corridors; passageways; inactive storage.	

Source: International Labour Office, Geneva, Switzerland.

The impact of color

Both color and texture have psychological effects on people. For example, yellow is the accepted color of butter; therefore, margarine must be made yellow in order to appeal to the appetite. Beefsteak cooked in

45 seconds on an electronic grill did not appeal to the customer because it lacked a seared, brown, "appetizing" surface. A special attachment had to be designed to sear the beefsteak. Employees in an air-conditioned midwestern plant complained of feeling cold, although the temperature was maintained at 72° F. When the white walls of the plant were re-painted in a warm coral color, all complaints ceased. Workers in a factory complained that boxes were too heavy until the plant engineer had the old boxes repainted a light green. The next day several workers said to the foreman, "Say, those new lightweight boxes sure make a difference."

Perhaps the most important use of color is to improve the environmental conditions of the workers by providing more visual comfort. Colors may be used to reduce sharp contrasts, to increase reflectance, to highlight hazards, and to call attention to features of the work environment that need to be highlighted.

Sales are conditioned by colors. People recognize a company's products instantly by the pattern of colors used on packages, trademarks, letterheads, trucks, and buildings. Some research has indicated that color preferences are influenced by nationality, location, and climate. Sales of a product formerly made in one color increased when several colors suited

TABLE 9–5
The emotional and psychological significance of the principal colors

Color	*Characteristics*
Yellow	Has the highest visibility of any color under practically all lighting conditions. It tends to instill a feeling of freshness and dryness. It can give the sensation of wealth and glory, yet can also suggest cowardice and sickness.
Orange	Tends to combine the high visibility of yellow and the vitality and intensity characteristic of red. It attracts more attention than any other color in the spectrum. It gives a feeling of warmth, and frequently has a stimulating or cheering effect.
Red	A high-visibility color having intensity and vitality. It is the physical color associated with blood. It suggests heat, stimulation, and action.
Blue	A low-visibility color. It tends to lead the mind to thoughtfulness and deliberation. It tends to be a soothing color, although it can promote a depressed mood.
Green	A low-visibility color. It imparts a feeling of restfulness, coolness, and stability.
Purple and violet	Low-visibility colors. They are associated with pain, passion, suffering, heroism, and so on. They tend to bring a feeling of fragility, limpness, and dullness.

to the differences in customer demands were supplied. Table 9–5 illustrates the typical emotional effects and psychological significance of the principal colors. Figure 9–2 illustrates pairs of colors that will give harmonious hues; the figure also shows complementary colors.

FIGURE 9–2
Color wheel based on the Munsell color
system.

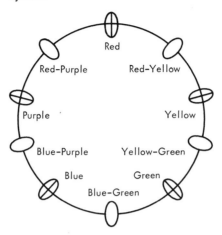

Five primary and five secondary colors
or hues are indicated. Pairs of adjacent
colors on this wheel are known tech-
nically as harmonious hues. Colors opposite to each other are complementary.

In a visual sense, texture, refers to the patterns of contrasts in the
light reflections that identify a surface. The influence of surface texture
on customers is as significant as that of color. Marketing people have
been aware for some time of the consuming public's increasing desire
for mirror finishes, such as glass-fronted stores and buildings, Koroseal
fabrics, and chrome finishes on appliances and automobiles.

Other finishes also create interest and inspire the worker with respect
for and pride in his or her environment. For example, there are hammered
finishes, which simulate the mottled texture of hammered metals, and
Rigid-Tex, rigidized metals rolled in any of a number of textured patterns
and finished in a variety of colors.

Noise

From the practical analyst's point of view, noise is any unwanted
sound. Sound waves originate from the vibration of some object, which
in turn sets up a succession of waves of compression and expansion
through the transporting medium (air, water, and so on). Thus, sound
can be transmitted not only through air and liquids, but also through
solids, such as machine tool structures. We know that the velocity of

sound waves in air is approximately 340 m/sec (1,100 ft). In visco-elastic materials, such as lead and putty, sound energy is dissipated rapidly as viscous friction.

Sound can be defined in terms of the frequencies which determine its tone and quality and of the amplitudes which determine its intensity. Frequencies which are audible to the human ear range from approximately 20 to 20,000 cycles per second. The unit of cycles per second is now commonly called Hertz, abbreviated Hz. The fundamental equation of wave propagation is: $c = \lambda f$, where c = sound velocity in m/sec; f = frequency in Hz; and λ = wave length in m. Figure 9–3 shows the relation-

FIGURE 9–3
Relationship between the frequency and the wavelength of sound in room-temperature air

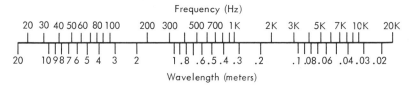

ship between the frequency and the wavelength of sound in room-temperature air. Sound intensity may be measured by means of a sound-level meter, and the unit of sound intensity is the decibel (db). The greater the amplitude of the sound waves, the greater the sound pressure which is measured on the decibel scale.

Because of the very large range (about 10^7 to 1) of sound intensities encountered in the normal human environment, the decibel scale has been chosen. In effect, it is the logarithmic ratio of the sound intensity to the sound intensity at the threshold of hearing of a young person. Thus, the sound pressure level L_P in decibels is given by:

$$L_P = 10 \log_{10} \frac{P^2_{\text{rms}}}{P^2_{\text{ref}}} \text{ db}$$

where

P_{rms} = Root-mean-square sound pressure in microbars (or dynes/cm²)
P_{ref} = sound pressure at the threshold of hearing of a young person at 1,000 Hz (0.0002 microbars)

Since sound-pressure levels are logarithmic quantities, the effect of the coexistence of two sound sources in one location requires that a logarithmic addition be performed. A graphic method is convenient in performing this operation. Figure 9–5 describes the method and gives examples of its application.

The A-weighted sound level used in Figure 9–4 is the most widely accepted measure of environmental noise. The A weighting recognizes that

FIGURE 9–4
Decibel values of typical sounds (dbA)

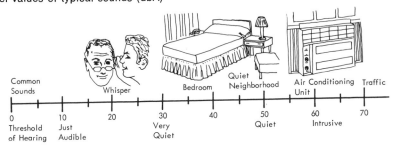

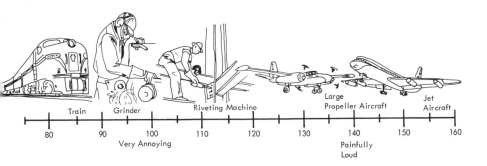

from both the psychological and the physiological points of view, the low frequencies (50–500 Hz) are far less annoying and harmful than are sounds in the critical frequency range of 1,000–4,000 Hz. Above frequencies of 10,000 Hz, hearing acuity (and, therefore, noise effects) again drops off.

The appropriate electronic network is built into sound-level meters to attenuate low and high frequencies, so that the sound-level meter can read in dbA units directly to correspond to the effect on the average human ear.

The chances of damage to the ear resulting in "conductive" deafness increase as the frequency approaches the 2,400 to 4,800 Hz range. This loss of hearing is a result of a loss of mechanical flexibility in the middle ear, so that it fails to transmit sound waves to the inner ear. Also, as the exposure time increases, especially where higher intensities are involved, there will eventually be an impairment in hearing. Nerve deafness is the

FIGURE 9-5
Chart for combining levels of uncorrelated noise signals*

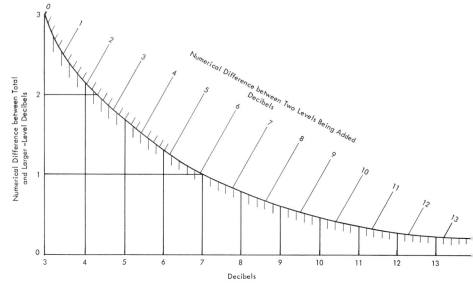

Numerical Difference between Total and Smaller Levels

To add levels, enter the chart with the "numerical difference between two levels bein added." Follow the line corresponding to this value to its intersection with the curve line, then left to read the "numerical difference between total and larger level." Add thi value to the larger level to determine the total.

Example: Combine 75 db and 80 db. The difference is 5 db. The 5-db line intersec the curved line at 1.2 db on the vertical scale. Thus, the total value is 80 + 1.2, or 81.2 db

To subtract levels, enter the chart with the "numerical difference between total an larger level." This value is less than 3 db. Enter the chart with the "numerical differenc between total and smaller levels" if this value is between 3 and 14 db. Follow the lin corresponding to this value to its intersection with the curved line, then either left or dow to read the "numerical difference between total and larger [smaller] levels." Subtract thi value from the total level to determine the unknown level.

Example: Subtract 81 db from 90 db. The difference is 9 db. The 9-db vertical line inte sects the curved line at 0.6 db on the vertical scale. Thus, the unknown level is 90 − 0.6 or 89.4 db.

* This chart is based on one developed by R. Musa.
Source: Reprinted by permission from ASHRAE Handbook of Fundamentals, 1967.

result of damage in the inner ear or in the auditory nerve itself. Both conductive and nerve deafness are most commonly due to excess exposure to noise, and one of their causes is occupational noise. Individuals vary widely in their susceptibility to noise-induced deafness.

In general, we can classify noise as being of two types: broadband noise and meaningful noise. Broadband noise is made up of frequencies covering a significant part of the sound spectrum. This type of noise can be either continuous or intermittent. Meaningful noise represents distracting information that will have an impact on the worker's efficiency.

In long-term situations broadband noise can result in deafness, and. in day-to-day operations it can result in reduced worker efficiency and ineffective communication.

Continuous broadband noise is typical of such industries as the textile industry and an automatic screw machine shop, where the noise level does not deviate significantly during the entire working day. Intermittent broadband noise is characteristic of the drop forge plant and the lumber mill. When a person is exposed to noise that exceeds the damage level, the initial effect is likely to be a temporary hearing loss from which there will be complete recovery within a few hours after leaving the work environment. If repeated exposure continues over a long time, then irreversible damage to hearing can result. The effects of excessive noise depend on the total energy which the ear has received during the work period. Thus, by reducing the time of exposure to excessive noise during the work shift, one is able to reduce the probability of permanent hearing impairment.

Both broadband and meaningful noise have proven to be distracting and annoying enough to result in decreased productivity and in an increase in employee fatigue. However, federal legislation was enacted primarily because of the possibility of permanent hearing damage. The *Federal Register,* vol. 34, no. 96 (May 20, 1969), in reference to the Walsh-Healy Act (the Occupational Safety and Health Act, or OSHA) on occupational noise exposure, provides the data contained in Table 9–6.

TABLE 9–6
Permissible noise exposures

Duration per day (hours)	Sound level (dA)
8	90
6	92
4	95
3	97
2	100
1.5	102
1	105
0.5	110
0.25 or less	115

The law (50–204.10 occupational noise exposure) states:

Protection against the effects of noise exposure shall be provided when the sound levels exceed those shown in Table I [Table 9–6] of this section when measured on the A scale of a standard sound level meter at slow response.

When noise levels are determined by octave band analysis, the equivalent A-weighted sound level may be determined as follows:

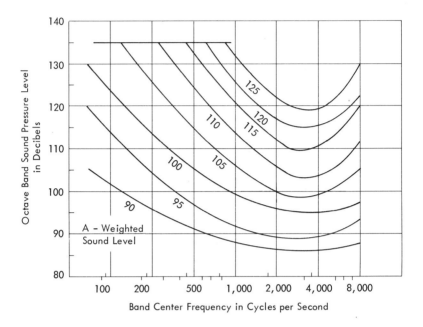

OSHA is currently preparing new noise standards for industry. The new standards are expected to be available by early 1977. It is anticipated that these new standards will set limitations on how much noise is tolerable at a workplace during an eight-hour period and will prescribe the steps that must be taken if noise exceeds these levels.

The control of the noise level at the ear may be accomplished in three ways. The best, and usually the most difficult, way is to reduce the noise level at its source. It would be very difficult to redesign such equipment as pneumatic hammers, steam forging presses, board drop hammers, and woodworking planers and jointers so that the efficiency of the equipment would be maintained and yet the noise level would be brought into a tolerable range. In some instances, however, more quietly operating facilities may be substituted for those operating at a high noise level. For example, a hydraulic riveter may be substituted for a pneumatic riveter, electrically operated apparatus for steam-operated apparatus, and an elastomer-lined tumble barrel for an unlined barrel.

If the noise cannot be controlled at its source, then the opportunity to isolate the equipment responsible for the noise should be investigated. The noise that emanates from a machine may be controlled by housing all or a substantial portion of the facility within an insulating enclosure. This has frequently been done in connection with power presses having automatic feeds. Ambient noise can frequently be reduced by isolating the noise source from the remainder of the structure, thus preventing a sounding board effect. This can be done by mounting the facility on a shear-type elastomer, thus damping the telegraphing of noise.

In situations where enclosing the facility would not prevent operation and accessibility, the analyst should take the following steps in order to assure the most satisfactory enclosure design:

1. Clearly establish the design goals and determine the acoustical performance that will be required of the enclosure. The analyst should establish octave band criteria at one meter (three feet) from the major machine surfaces. This distance is consistent with that recommended by NMTBA, GAGI, and so on.
2. Make actual measurements of the octave band noise levels of the equipment to be enclosed at the locations recommended in step 1.
3. Determine the accumulation of noise and then the net noise when multiple facilities are being used (see Figure 9–5).
4. Determine the required spectral attenuation of each enclosure. This is the difference between the design criteria determined in step 1 and the net noise level determined in step 3.
5. Select the acoustical panels and wall configuration for the enclosure. Table 9–7 provides several materials that are popular for relatively small enclosures. It is suggested that a viscolastic damping material be applied if any of these materials (with the exception of lead) are used. This can provide an additional attenuation of 3 to 5 db.

TABLE 9–7
Octave band noise reduction of single-layer materials commonly used for enclosures

Octave band center frequency	125	250	500	1,000	2,000	4,000
16-gauge steel	15	23	31	31	35	41
7-mm steel	25	38	41	45	41	48
7-mm plywood 0.32 kg/0.1 square meter	11	15	20	24	29	30
¾-in. plywood 0.9 kg/0.1 square meter	19	24	27	30	33	35
14-mm gypsum board 1 kg/0.1 square meter	14	20	30	35	38	37
7-mm fiberglass 0.23 kg/0.1 square meter	5	15	23	24	32	33
0.2 mm 0.45 kg/0.1 square meter	19	19	24	28	33	38
0.4-mm lead 0.9 kg/0.1 square meter	23	24	29	33	40	43

If the noise cannot be reduced at its source, and if the noisemaking source cannot be acoustically isolated, then acoustic absorption can provide beneficial results. The purpose of installing acoustical materials on

FIGURE 9–6
Spectrum of noise before and after acoustical treatment

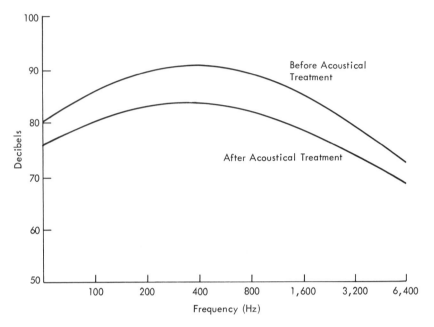

Source: Redrawn from McCormick, 1970.

walls, ceilings, and floors is to reduce reverberation. Figure 9–6 illustrates the amount of noise reduction typically possible through acoustical treatment.

Lastly, the personnel in the area can wear personal protective equipment, though in most cases OSHA will accept this as only a temporary solution. Personal protective equipment can include various types of earplugs, some of which are able to attenuate noises in all frequencies up to sound-pressure levels of 110 db or more. Also available are earmuffs which will attenuate noises to 125 db above 600 Hz and up to 115 db below this frequency.

Vibration

Vibration can cause detrimental effects on human performance. Vibrations of high amplitude and low frequency have especially undesirable effects on body organs and tissue. The parameters of vibration are frequency, amplitude, velocity, acceleration, and jerk. For sinusoidal vibrations, amplitude and its derivations with respect to time are:

Amplitude (s) = maximum displacement from static position in inches

Maximum velocity $\dfrac{ds}{dt}$ = $2\pi(s)(f)$ in./sec

Maximum acceleration $\dfrac{d^2s}{dt^2}$ = $4\pi^2(s)(f^2)$ in./sec

Maximum jerk $\dfrac{d^3s}{dt^3}$ = $8\pi^2(s)(f^3)$ in./sec

where:

f = Frequency
s = Displacement amplitude

Displacement and maximum acceleration are the principal parameters used to characterize the intensity of vibration.

There are three classifications of exposure to vibration:

1. Circumstances in which the whole or a major portion of the body surface is affected; for example, when high-intensity sound in air or water excites vibration.
2. Cases in which vibrations are transmitted to the body through a supporting area; for example, through the buttocks of a man driving a truck or through the feet of a man standing by a shakeout facility in a foundry.
3. Instances in which vibrations are applied to a localized body area, for example, to the hand when a power tool is being operated.

Of these three classifications, the second is of the most concern to the methods analyst. This category has the greatest effects on working efficiency and on the health, safety, and comfort of the working force.

Research in this area has suggested that the most sensitive frequency ranges are 4 Hz to 8 Hz for vertical vibration and below 2 Hz for horizontal vibration. Various parts of the body resonate at specific frequencies, causing disturbances. One should also recognize that human tolerance of vibration decreases as the exposure time increases. Thus, the tolerable acceleration level increases with decreasing exposure time. Low-frequency, high-amplitude vibrations are the principal cause of the motion sickness, that some people experience in sea and air travel. Workers experience fatigue much more rapidly when they are exposed to vibrations in the range from 1 to 250 Hz. Early symptoms of vibration fatigue are headache, loss of appetite, and loss of interest. Vibrations experienced in this range are often characteristic of the trucking industry. The vertical vibrations of many rubber-tired trucks when traveling at typical speeds over ordinary roads range from 3 Hz to about 7 Hz.

Protection against vibration may be accomplished in several ways. The applied forces responsible for initiating the vibration may be reduced. The body position may be altered so as to result in a lessening of the disturbing vibratory forces. Lastly, supports may be introduced that will cushion the body and thus damp higher amplitude vibrations. Seat suspension systems involving hydraulic shock absorbers, coil or leaf springs, rubber shear-type mountings, or torsion bars may be used.

Thermal conditions

Although man is able to function within a broad range of thermal conditions, his performance will deviate substantially if he is subjected to temperatures that vary from those characterized as "normal." When the analyst considers the temperature of the work environment, he should be cognizant of the following:

1.   The environmental temperature is the temperature actually experienced by the man or woman in a given environment. This temperature is the sum of the convective heat exchange; conduction from hot or cold floors or tools; the radiation exchange from the walls, floor, and ceiling; and any solar radiation that is transmitted or reflected to the occupant through the transparent areas of the work environment.

2.   Effective temperature is an experimentally determined index, which includes the temperature, air movement, and humidity. The normal range is from 18.3° C (65° F) to 22.8°C (73° F), with a relative humidity of 20 to 60 percent (see Figure 9–7).

3.   The operative temperature is the body temperature of the employee. It is determined by the cumulative effects of all heat sources and sinks. For an individual to maintain a comfortable skin temperature of approximately 32° C (90° F), a convective heat removal commensurate with the operative temperature needs to be effected. This may be expressed as:

$$E = M \pm R \pm C$$

where:

E = Heat loss by evaporation
M = Heat gain from metabolism
R = Heat exchange by radiation
C = Heat exchange by convection

If the environment is of the hot-dry type where the principal source of heat is radiant, then the problem is fundamentally one of heat gain, and it is solved by reducing the heat load. If the environment is hot-moist, the problem centers on heat loss, which is best solved by ventilation and by the dehydration of the ambient air.

The methods analyst should recognize that the maximum rise in body temperature should be about 1° C or 2° F. Conditions that cause greater change than this can result in heat stress.

In order to estimate the length of time that a person can be exposed to a particular working environment, it is necessary to estimate or measure the heat load. Measurements can be made upon both the environment and the person. When taking measurements of the person, we use one or more of three factors. These are the heart rate, oxygen consumption, and the body temperature.

The heart rate will increase with the increase in the presence of heat while a particular job is being done. This differential in the heart rate is a measure of the heat load on the employee. A general rule that the analyst may use is that the heat load should not cause an increase in the pulse rate of more than 15 beats per minute. A related benchmark is that the combination of the work load and the environment should not cause an increase in the pulse rate of more than 45 beats per minute above the normal rest pulse count. Obviously, conditioning and accommodation to an environment can make a great difference in the effects of the work load and the environment on an employee.

Measures of oxygen consumption alone will not provide a satisfactory estimate of the heat load. However, the normal relationship between the heart rate and oxygen consumption will not be maintained when work is performed in the presence of excess heat, and this variation can be used in an estimation of the heat load.

Body temperature, which is normally approximately 37.1° C (98.6° F), should not be allowed to rise to as much as 38.6° C (101.5° F). Allowing body temperatures to rise to this level can result in the physical collapse of the worker.

In measuring the variables that affect the environment, we are concerned with the dry-bulb temperature, humidity, air velocity, and radiation from the immediate surroundings.

Dry-bulb temperatures are usually measured with an ordinary liquid-in-glass thermometer. Thermocouples, thermistors, and resistance thermometers may also be used. The analyst should recognize that, to obtain reliable results, the sensing element must be shielded against radiation from all nearby surfaces that are significantly hotter or colder than the surrounding air.

The instrument that is most commonly used to determine the humidity of the air is probably the sling psychrometer. This instrument whirls a wet-bulb and a dry-bulb thermometer simultaneously. The whirling is interrupted at intervals to read the thermometers, and is continued until both thermometer readings stabilize. Then, by use of a table, the relative humidity can be established.

If the environment in which the relative humidity is being measured is subject to considerable radiant heat, a properly shielded and aspirated psychrometer should be used instead of the sling psychrometer.

A globe thermometer is used to measure the mean radiant temperature. In this instrument, a thermocouple, thermistor, or thermometer bulb is fixed at the center of a six-inch hollow copper sphere painted on the outside and inside with a matte black finish. Both the air temperature and the air velocity around the globe must be determined in order to compute the mean radiant temperature, which may be approximated by the following equation:

$$\text{MRT} = t_g + cv(t_g - t_a)$$

where:

MRT = Mean radiant temperature (F)
c = Convention coefficient = 0.17 for a 6-in. globe
v = Air velocity at the globe (ft/min)
t_g = Globe temperature (F)
t_a = Ambient air temperature (F)

A period of from 15 to 25 minutes is required for the air inside the globe to reach equilibrium.

Thermal anemometers of the heated-thermocouple, hot-wire, or heated-thermometer types are best suited for the measurement of the low-velocity, random-direction air movements characteristic of most work centers.

An index has been suggested that combines the effects of temperature, humidity, radiant heat, and air movement. The index can be computed by combining dry-bulb temperatures with wet-bulb temperatures and conventional dry-globe temperature as follows:

$$WBGT_i = 0.7(WBT) + 0.3(GBT)$$
$$WBGT_o = 0.7(WBT) + 0.2(GBT) + 0.1(DBT)$$

where:

$WBGT_i$ = Wet bulb–globe thermometer for indoor use
$WBGT_o$ = Wet bulb–globe thermometer for outdoor use
WBT = Wet-bulb temperature
GBT = Dry-globe temperature
DBT = Dry-bulb temperature

Although this index has not received wide industrial use, it can provide a measure of the thermal conditions related to a work center. When the $WBGT_o$ value reaches a 82° F to 85° F level, the U.S. Marine Corps posts a green flag, thus alerting personnel to the possible curtailment of drill. A similar value at a work station would suggest a need for improvement.

Another index is the effective temperature, which has already been defined. This can be estimated by the use of a nomogram originally developed by a research team within the American Society of Heating, Refrigerating, and Air-Conditioning Engineers, Inc. (see Figure 9–7).

FIGURE 9–7

Chart for determining the effective temperature for fully clothed worker performing light work from measurements of the dry-bulb temperature, wet-bulb temperature, and air velocity

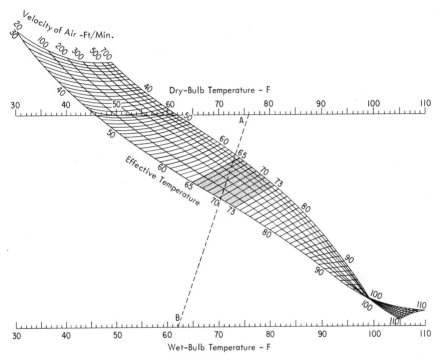

To use the chart, draw a line (for example, A–B) throught the measured dry-bulb and wet-bulb temperatures. Read the effective temperature or velocity at the desired intersections.

Source: Redrawn from data provided through the courtesy of the American Society of Heating, Refrigerating, and Air-Conditioning Engineers, Inc.

Protection against heat

Many industrial activities involve exposure to intense heat from which the worker needs protection. Typical examples include the hot forging of large work, tending a furnace for the production of glass or steel, and tapping a cupola in a foundry. For workers engaged in some of these activities, an air-conditioned enclosure with suitable windows assures protection while permitting them to perform effectively. For example, operators of overhead cranes, mechanical equipment, and the like, can be housed in such air-conditioned cubicles.

If the worker needs to get exceptionally close to a source of radiant heat, personal protective equipment will be needed. Air-conditioned suits are now available. This type of clothing is usually aluminized on the exterior for further protection. The fabric is light in weight and allows an envelope of circulating cool air.

Under less severe thermal exposure, gloves, protective clothing, and a face covering may prove adequate.

Low-temperature effects on performance

Relatively few assignments in modern business and industry require the personnel involved to be exposed to cold environments for long periods of time. The principal jobs that do result in such exposure are such outdoor jobs as winter work in the building trades and the police force, and work in cold storage warehouses, such as those for meat and other perishable commodities.

Research has indicated that performance deteriorates as temperature declines. One study indicated a fall of approximately 40 percent in performance as temperature declined from 30° F to −40° F. For the operator to maintain thermal balance under low-temperature conditions, there must be a close relationship between the operator's physical activity (heat production) and the body insulation provided by protective clothing. Figure 9–8 illustrates this relationship. Here, a clo unit represents the insulation necessary to maintain comfort for a person sitting where the relative humidity is 50 percent, the air movement 20 feet/minute, and the environmental temperature 21° C (70° F).

Radiation

Although all types of ionizing radiation can damage tissue, beta and alpha radiation are so easy to shield that most attention is given today to gamma ray, X-ray, and neutron radiation. It should be noted that high-energy electron beams impinging on metal in vacuum equipment can produce very penetrating X rays that may require much more shielding than does the electron beam itself.

Absorbed dose is the amount of energy imparted by ionizing radiation to a given mass of material. The unit of absorbed dose is the rad, which is equivalent to the absorption of 0.01 joules per kilogram (100 ergs per gram). Dose equivalent is a way of correcting for the differences in the biological effect on humans of different types of ionizing radiation. The unit of dose equivalent is called the rem, which produces a biological effect essentially the same as that of one rad of absorbed dose of X or gamma radiation. The roentgen (R) is a unit of exposure which measures the amount of ionization produced in air by X or gamma radiation. Tissue located at a point where the exposure is one roentgen will receive an absorbed dose of approximately one rad.

Very large doses of ionizing radiation—100 rads or more—received over a short time span by the entire body can cause radiation sickness.

FIGURE 9–8

Prediction of the total insulation required as a function of the ambient temperature (M = heat production in cal/m²/hr)

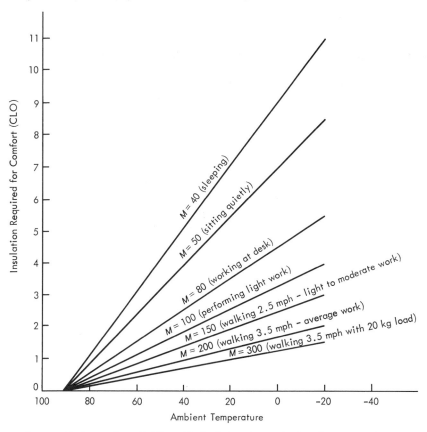

Source: Redrawn from Belding, H. S. Index for evaluating heat stress in terms of physiological strains. Heat, Pipe, and Air Cond., 27, 129

Small doses received over a long period of time may increase the probability of contracting various types of cancers or other diseases. Persons working in areas where access is controlled for the purpose of radiation protection are generally limited to a dose equivalent of five rem per year. The limit in uncontrolled areas is usually $\frac{1}{10}$ of that value. Working within these limits should have no significant effect on the health of the individuals involved.

Fundamentals of work physiology

In order to design a work station that will result in high productivity over a period of time during which different workers will be engaged, it is important that the analyst have a background in the fundamentals of work physiology. In many instances, the work personnel will differ in many significant respects, such as age, sex, background, physical and mental characteristics, and health.

Motor fitness, reaction time, and visual capacity

The motor fitness elements of strength, endurance, speed of movement, and reaching distance, coupled with visual capacity and the speed and accuracy of response to events, have a significant collective impact on both the rate of productivity and the total productivity over time of most manual operations.

Three factors influence the accuracy of control movements—the number of muscle fibers controlled by each motor nerve ending being utilized, the position of the body members, and neural stimuli. Arms have considerably more motor nerve endings and, consequently, much greater accuracy of control than do the legs. Also, the closer the position of a limb to the body, the more accurately it can be moved. Consequently, controls that are operated by a worker's hands should be located so that it is not necessary for the worker to extend his or her arms in order to manipulate them.

FIGURE 9–9
Effective visual areas for the average employee

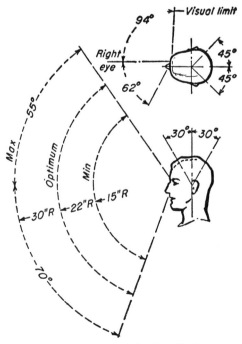

Product Engineering

The average person normally moves the head to the right or left only about 55 degrees. When eye movement is added, the total angle is substantially larger, averaging somewhere between 90 and 100 degrees for the normal person (see Figure 9–9).

The shoulder joint (ball and socket) permits more range of movement than does any other joint in the human body. Also, the shoulder is capable of applying about 30 percent more force than the elbow. The torque which can be applied by right-handed people is about 20 percent greater for the right hand than for the left hand, and is at a maximum when the elbow is at a right angle. Similarly, position influences the maximum weight that can be lifted. For example, if the lifting movement is performed with the palm of the hand facing upward (supine), the typical male adult will lift about 55 pounds. However, if the palm of the hand is palm downward (prone), the load that can be handled will be about 30 percent smaller.

Although there is little difference in the magnitude of the force that the typical operator can exert by either pulling or pushing, the analyst should be cognizant of the fact that the maximum force capable of being exerted decreases as the hands are brought nearer to the body. Also, this maximum push or pull can be achieved over only about three inches of travel.

The usual method of applying force with either leg is by pushing. If a worker is seated with a satisfactory backrest, he will be able to produce forces of up to about 300 kilograms for thrusts of short duration in the horizontal plane.

Response time is another important ingredient of overall performance. Generally, response time can be thought of as being made up of:

1. The time involved in sensing a signal.
2. The time for the decision process as to the nature of the response.
3. The time to perform the physical movement.

Reaction time, which is a combination of steps 1 and 2 above, is affected by several factors. The size and clarity of the signal affects the time needed to sense a signal; the larger the signal, the faster will be the reaction time up to some particular point. Also, the reaction time for a visual signal is faster when the signal is observed by the center or near the center of the eye rather than by the periphery of the eye. The type of stimuli—whether visual, auditory, or tactile—also affects reaction time.

The different body members have different response times. The right hand (in right-handed people) has the fastest response time, followed by the left hand, the right foot, and the left foot.

The average response time increases with age and is longer for women than for men, though individual differences in these respects are large.

Like response time, reading speed varies greatly for different individuals. An important factor in reading speed is eye fixation. To increase reading speed, one must decrease the number of fixations or the time of each fixation. Typically, a line of print 100 millimeters long requires about six eye fixations if one has had training comparable to that of the university graduate. A youngster in first grade will average about 18 fixations per line.

The number of fixations will vary with the number of difficulties encountered. Thus, more eye fixations will take place when more unfamiliar words appear in the reading material. Color has little effect on reading speed; however, contrast has a great effect. If the brightness contrast is low, then reading speed will also be slow.

Memory

It has been estimated that the storage capacity of the human memory system is between 10^8 and 10^{15} bits. The memory of humans appears to be of two types, which can be classified as static and dynamic. In the static, or long-term, memory we store relevant information, which we call upon and use from time to time. For example, the value of π, the sine of 30 degrees, and the modulus of elasticity of steel are typical values that most engineers maintain in their static memory bank. In the dynamic, or short-term, memory we store information or data that are needed for immediate use. For example, a telephone number looked up just prior to dialing represents data that are stored in the dynamic memory. After the data are used, they are no longer retained.

There is considerable variation in the memory or recall capacity of different people. This variation is characteristic of both static and dynamic memory. Memory—in particular, dynamic memory—is influenced by age. As an individual becomes older, his short- and long-term memory tend to decline.

Physiological fatigue

Everyone is familiar with the effects of physiological fatigue. When brush painting a ceiling, the arm, held over the head, soon tires because of the insufficient flow of blood into the contracted arm muscle. One must stop periodically in order to relax the muscle and allow the flow of blood through it.

The oxygen that is used by the body in doing work comes either from the blood or from chemical compounds within the muscle fibers. If one's ability to supply oxygen to working muscles is sufficient to prevent a build-up of the by-products of metabolism in the body during the course of the workday, the job assignment is referred to as "aerobic." If the job assignment is such that performing it depletes the reserve of oxygen in a muscle or muscles, the assignment is referred to as "anaerobic." The symptoms of this oxygen depletion are muscle pain and an accompanying physiological fatigue or muscle weakness.

It has been estimated that the basal metabolic rate averages 1,700 calories per day. This represents the number of calories needed to maintain the body in an inactive state. Thus, a worker requires additional calories to handle the duties and responsibilities that go with his or her job. Otherwise, a portion of the work will necessarily be anaerobic. Some estimates of human energy expenditures are shown in Table 9–8. Anaero-

TABLE 9–8
Energy expenditure and the corresponding oxygen consumption

Level of work	Energy expenditure (calories per minute)	Oxygen consumption (liters per minute)
Unduly heavy (carrying mortar upstairs at 54 feet per minute)	Over 12.5	Over 2.5
Very heavy (shoveling charge into furnace)	Over 10.0	Over 2.0
Heavy (carrying load upstairs at 27 feet per minute).	Over 7.5	Over 1.5
Moderate (wheelbarrowing sand at 220 feet per minute)	Over 5.0	Over 1.0
Light (light assembly work)	Over 2.5	Over 0.5

bic work is related more to the velocity of movement than to the duration of the work. Work of long duration will exhaust energy supplies in the muscles (glycogen) rather than cause the operator to suffer from lack of oxygen.

Individual differences

Human beings are variable in their performance. This variation represents one of the most important considerations in the design of man and machine systems. Not only is there considerable difference between the performance of different individuals, but even that of the same individual will vary from moment to moment, from one period of the day to the next, and from day to day. Even when carrying out the simplest functions, the individual will vary considerably in his or her performance.

The variations in performance among members of the work force are due in large part to variations in both static and dynamic anthropometry. Figures 9–10 to 9–11 illustrate some of the principal static body dimensions.

Then, too, body dimensions vary with age. From birth to about the middle 20s, most static body dimensions increase. From about the age of 60 on, there is a slight decrease in height and body weight, and often in the dynamic dimensions, such as functional arm reach. Table

FIGURE 9–10
Seating depends on reach, vision, and the work load on the operator (the dimensions shown are based on the small man, which will accommodate the reach limitations of 95 percent of all men)

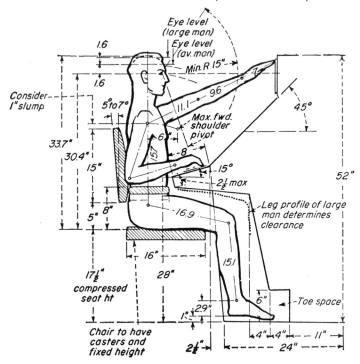

Product Engineering

9–9 provides data as to the mean height and weight, and the standard deviations therefrom, of white American men and women at various ages. A study of these data will give the reader a feel for the variability of human body dimensions in the work force.

Both age and sex have a bearing on the time required for response. Response time differences among people will increase as the work becomes more precise, more rushed, or more difficult to perform. Also, as the work environment becomes more adverse, there will be an increase in the differences in reaction time among the various workers.

As already mentioned, women have a slightly longer mean reaction time to both light and sound than do men (see Figure 9–12). Also, the reaction time of both men and women will increase with age after they are about 30 years old.

The differences among people as to visual capacity and memory have already been discussed. Similarly, there is considerable variation in the muscle strength and motor skills of the work force. Strength usually

FIGURE 9–11
The dimensions of the average adult male based on the 2½
to 97½ percentile range of the subjects measured

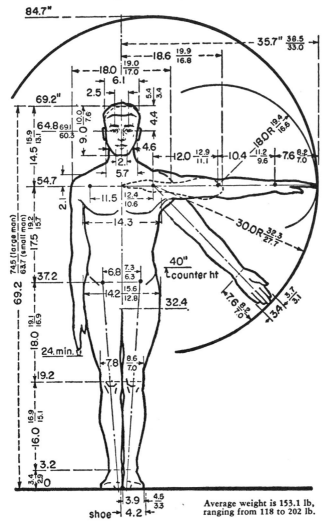

Average weight is 153.1 lb,
ranging from 118 to 202 lb.

Product Engineering

reaches a maximum at about age 25. After around age 30, strength declines
until, at age 60, it is only about three-fourths of what it was at maximum.
Not only is there considerable variation in the strength of a given worker
due to biological, environmental, and occupational reasons, but there is
considerable variation among workers. For example, it has been estimated
that the standard deviation of the maximum force that can be exerted by
a two-handed vertical pull on a horizontal bar 28 inches above floor level

TABLE 9–9

Age	Male				Female			
	Height (inches)		Weight (pounds)		Height (inches)		Weight (pounds)	
	Mean	Standard deviation	Mean	Standard deviation	Mean	Standard deviation	Mean	Standard deviation
3	37.8	1.3	32	3	37.5	1.4	31	4
6	46.1	2.1	47	6	45.7	1.9	45	5
9	52.8	2.4	66	8	52.1	2.3	64	11
12	58.3	2.9	87	12	59.6	2.7	93	18
15	66.3	3.1	128	16	63.4	2.4	117	20
18	68.5	2.6	149	20	64.1	2.3	123	17
22	68.7	2.6	158	23	64.0	2.4	125	19
26	68.7	2.6	163	24	63.7	2.5	127	21
30	68.5	2.6	165	25	63.6	2.4	130	24
34	68.5	2.6	165	25	63.6	2.4	130	24
38	68.4	2.6	166	25	63.4	2.4	136	25
48	68.0	2.6	167	25	63.2	2.4	142	27
58	67.3	2.6	165	25	62.8	2.4	148	28
68	66.8	2.4	162	24	62.2	2.4	146	28
78	66.5	2.2	157	24	61.8	2.2	144	27

Source: Adapted from Stoudt et al., 1960.

by male workers is about 68 pounds. Among female workers, the standard deviation for this backlift force is about 40 pounds.

Because of the wide variability in the performance of the population as a whole, task forces should be based on the research results obtained from subjects taken from a population representative of the particular job under study.

The work regimen

Today, the eight-hour day and the five-day week are generally thought of as being standard in American business and industry. However, some interesting experiments are being made with the 10-hour day and the four-day week. Reports from companies experimenting with this type of schedule are not consistent. Some reports state that 10 hours is too long for an employee to work in one day, and that the added productivity of the last two hours is considerably less, proportionately, than the productivity experienced over the first eight hours. Other objections to the 10-hour day, four-day week, stem from members of management who state that they are obliged to be on the job not only 10 hours for four days but for at least eight hours on the fifth day. Other reports indicate satisfaction with having the plant closed three days every week.

Experience has proven that the typical worker today responds well,

FIGURE 9–12
Mean reaction time to light and sound for men and women at various ages

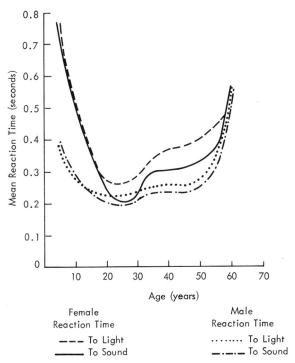

Source: Redrawn from J. Kennedy.

both psychologically and physiologically, to the 40-hour workweek, provided that the environmental conditions are satisfactory and that he receives adequate recognition (both monetary and nonmonetary) for his work. Thus, it appears that the 40-hour, or approximately 40-hour, week will continue to be the basis of the work schedule in most U.S. industries for many years into the future.

In most businesses, industries, and service organizations, the vast majority of work that is done may be classified as "light" from a physiological standpoint. Of course, some heavy physical work is done by the labor force. Typical heavy work exists in many of our coal mines, in bench molding in the foundry industry, and in connection with "pick and shovel" work related to building and highway construction.

As almost everyone is aware, after any activity has continued for a period of time, the worker feels the need to take a short break from the routine. If this cessation in work is not taken, there will be a pronounced and progressive falloff in productivity, even from highly motivated employees, until there is a forced break. This forced break could be a lunch period or a work stoppage due to equipment difficulty. The falloff in productivity can take place in two ways: first, the cycle time may increase, and second, there may be more rejects or poorer quality due to human errors. That period of time up to the point at which productivity begins to decline markedly is known as the actile period. At the end of this period, a rest break should take place. The length of the actile period is dependent upon both the work and the worker. Typically, the length of this period in most light work is about one hour. The minimum break that provides satisfactory recovery is usually about five minutes. Most workers will take this break periodically (about three times in the morning and three times in the afternoon), whether or not it is scheduled. The results of the periodic break are almost always positive. With regularly scheduled breaks, there will be more productivity with fewer rejects than if the worker forces himself to work continuously for four hours in the morning and four hours in the afternoon.

It is general practice to schedule a department or plant shutdown for 10 to 15 minutes in the middle of the morning and a similar work break in the middle of the afternoon. These forced breaks in the workday help assure that employees do not overextend the actile period. The employee does not really know the length of his actile period, and often he will continue to work for a period that exceeds it, to the detriment of his productivity. The conscientious employee, in particular, will overextend his actile period.

As will be pointed out later, in the development of work standards for light work, it is customary to provide at least a 10 percent allowance for personal delays and fatigue. This usually amounts to 24 minutes in the morning and a like amount of time in the afternoon. The alert employee will utilize this allowed time periodically, say, eight minutes after every hour's work, in order to avoid the decline in productivity that takes place at the end of the actile period.

In the performance of heavy work, the actile period will be shorter because of muscular exertion. Thus, for optimum productivity, the operator will need to take more breaks during the day. For recovery, the length of the break seldom needs to be more than about five minutes, as is also the case for light work.

A recent case history will illustrate the impact of the actile period. A 16-day study was made of capsule sorting involving three different sorting machines. Four operators were randomly assigned to the different machines, so that each operator spent a total of four days on each of the three machines. The information recorded for each sorted can of capsules included: the time used to sort, amount of scag removed, number of capsules per can, time of day, number of good capsules sorted out as scag, capsule size, inspection sample size, and transparency, Efficiency scores were computed for cans of work as well as for each total day of work. These efficiency scores were based upon the ratio of the actual performance to the expected performance, based upon work measurement techniques. The overall average daily efficiency was 95.9 percent. The overall average of the daily efficiencies adjusted for good capsules inadvertently sorted out as scag was 86.3 percent. The overall average number of good capsules sorted out as scag per day was 5,377 per machine. The overall average number of good capsules sorted and passed through inspection per day was 479,498 per machine.

Figure 9–13 illustrates the effect of monotony and boredom on the

IGURE 9–13

typical relationship between worker efficiency and the time of day under daywork operation.

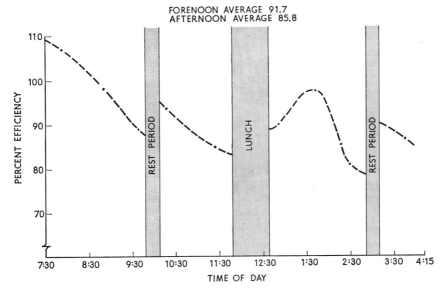

FORENOON AVERAGE 91.7
AFTERNOON AVERAGE 85.8

average performance of all the operators. It also illustrates the immediate increase in efficiency that occurs after rest periods. Figure 9–14 compares the effects of monotony and boredom on the operation of the three different machines.

It is interesting that this study revealed that there are significant differences between machines with fluorescent lighting and those with filament lighting. This study also revealed that boredom or vigilance factors associated with the time of day are apparently a normal (or expected) human fatigue characteristic. The decline in performance, regardless of the machine type used, suggests this conclusion.

FIGURE 9–14
The effects of monotony and boredom on the operation of three different designs c
machines

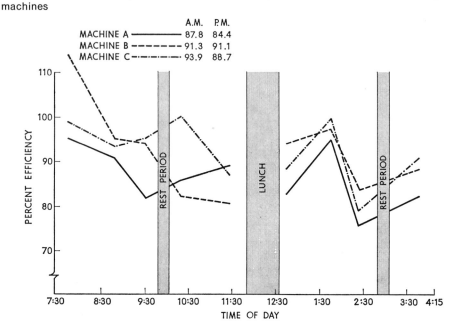

Although all four operators were similar in physical capabilities, and all had adequate experience with the work, there were significant differences among them, as is shown in Figure 9–15.

The results of the study indicated that the greatest improvements in work of this type can be made by providing: (1) more frequent rest periods of shorter duration; (2) work stations and work procedures that give the operator maximum opportunity to see imperfections throughout the day; and (3) test procedures that help select people who are more endowed with the visual ability and the mental qualifications that minimize the effect of boredom. It was recommended that rotating sorting operators to different types of sorting stations, or perhaps regularly rotating these operators with operators having entirely different work requirements, might help reduce the effects of monotony or vigilance.

When performing heavy work, the worker should be so conditioned that he is capable of expending about five calories per minute throughout the workday. If the employee is not capable of expending energy at this moderate rate, he will be obliged to take more frequent rest breaks, and his overall productivity will be limited. Figure 9–16 illustrates the well-conditioned and not-so-well-conditioned worker from the standpoint of

FIGURE 9–15

Operator differences among four typical female employees (the connecting lines indicate significant differences)

	Helen	Joy	Judy	Mary
Overall Efficiency for 16 Days	95.2	90.5	87.1	108.1
Average of Daily Efficiencies	94.6	90.7	87.8	108.9
Average of Daily Efficiencies—Adjusted	92.0	82.9	81.1	93.0
Average Daily Number of Good Capsules in Scag	1,900	4,700	4,700	10,900
Average Daily Number of Capsules Passed	554,700	501,200	445,000	418,100
Years of Sorting Experience	19	14	1	2
Age	48	37	20	21
Wear Eyeglasses	Yes	No	No	Sometimes
Preferred Hand	RH	RH	RH	RH

FIGURE 9–16

The relationship between a well-conditioned operator and a not-so-well-conditioned operator as to capacity to perform work

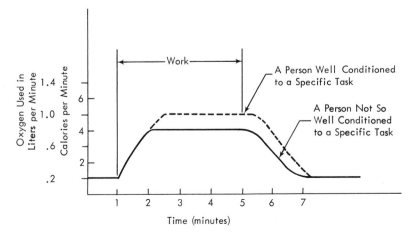

the ability to use up calories, or the capacity to perform work. Conditioning reflects the following differences among individuals:

1. Muscle tone: the degree to which the muscle reflexes are conditioned to a specific task.
2. Endurance: the degree to which fuel is made available, fuel is stored, and oxygen is made abundant, owing to a more adequate circulation of blood through active muscle.
3. Transmission: the facility of transmission of nerve impulses across the motor end plate in the muscle fiber.
4. Anaerobic efficiency: the efficiency (the ratio of the work done, in calories, to the net energy used, in calories) of the body during very heavy work.
5. Aerobic efficiency: the efficiency of the body during moderate work, where oxygen intake and demand are balanced.
6. Body health: the degree to which the physiological processes function normally.
7. Physical fitness: the degree of ability to execute a specific physical task under specific ambient conditions; long-term ability to perform (no aches, pains, strains, excessive tiredness, or "hangover").

These factors are interrelated and affect one another. They cannot be thought of as completely separate entities.

Human capacity to perform a specific task is not determined by efficiency alone, but is determined by the combination of all the physiological processes (and some psychological processes) and their interaction with one another. The degree of training and practice influences the conditioning of a worker for a given assignment, and the importance of this should be recognized in defining the production requirements of the job. Thus, an operator who continually works on an assignment involving heavy muscular demands will condition himself physically and otherwise in preparation for this class of work. His output will be considerably greater than that of an operator who performs the same operation at infrequent intervals and who, consequently, has not been conditioned for this type of work.

Behavioral concepts

The now-famous Hawthorne studies, published in 1939, were among the earliest studies to point out that social behavior is a reality of life that is continually taking place in every work environment. The Hawthorne studies demonstrated that, to a certain extent, the behavior of people in the work environment is conditioned by their social needs. The studies also clearly delineated the effect of informal leaders on the behavior of work groups as distinct from that of formally assigned leaders.

Other studies have identified a hierarchy of needs that typically apply to all people. These needs, in preferential order, are needs at the lowest level involving the basic necessities of life; then security; then the need for belonging to a group; then the need for status and esteem in some social system; and finally the need for personal self-actualization. The notion is that one tends to progress up this hierarchy of needs, and that as one's needs are filled in any of the more basic areas, the higher level needs are evoked, and the individual's behavior can be explained less by the satisfaction of lower level needs. Certainly, this idea of a hierarchy of needs seems to apply today. Until one's needs are satisfied at lower, more primitive levels, one will not move up the scale to more sophisticated requirements.

In order to apply these behavioral concepts in the effective operation of business and industry, it is important that the work be designed so that the principles of operation analysis incorporate the fundamental psychological and behavioral concepts of the work force. In the design of the job, there sometimes needs to be an enlargement of the work in certain situations. This enlargement may be characterized by multijob assignments, diversity, and flexibility or worker control. Perhaps as much as 15 to 20 percent of the jobs in industry would benefit through training personnel to perform a broader spectrum of the functions. This broadening provides an overall perception of the whole work process within an organization and some overall picture of that particular organization's goals.

The reader should be cognizant that just as there are a number of jobs that would benefit substantially through broadening, there are probably many more (this author would estimate up to 80 percent) that should not be broadened or enlarged. In other words, from the standpoint of the job itself, most of the work force is satisfied with the way it is. What is needed in the majority of work situations is a clear concept by the entire work force of what the organization's goals are and of how the individual operator's efforts can help fulfill those goals. This communication from management to each and every individual worker can go a long way toward fulfilling the hierarchy of needs of all employees. For long-term success, however, careful placement and advancement procedures must still be practiced.

Safety and health concepts

Certainly, one of the objectives of any progressive management team is to provide a safe and healthful workplace for the employees. In order to have this, there must be control over the physical environment of the business or operation. Most injuries are the result of accidents which are caused by an unsafe condition or an unsafe act, or a combination of the two. The unsafe condition is related to the physical environment. This involves the equipment used and all of the physical conditions surrounding the workplace. For example, hazards can stem from lack of guarding or inadequate guarding of the equipment, the location of machines, the condition of storage areas, and the condition of the building.

Some general safety considerations related to the building include adequate floor-loading capacity. This is especially important in storage areas, where overloading has caused many serious accidents every year. The danger signs of overloading include cracks in walls or ceilings, excessive vibration, and the displacement of structural members.

Aisles, stairs, and other walkways should be investigated periodically to assure that they are free of obstacles, are not uneven, and do not have oil or other material that could lead to slips and falls. In many old buildings, stairs should be inspected, since they are the cause of numerous lost time accidents. Stairs should have a slope of 30 to 35 degrees, with tread widths of approximately 9½ inches. Riser heights should not exceed 8 inches. All stairways should be equipped with a handrail, and should have at least 10 footcandles of illumination and be painted in light colors.

There should be at least two exits on all floors of a building, and the size of the exits should be in accordance with the Life Safety Code of the National Fire Protection Association. This code gives consideration to the occupancy and relative fire hazard that the exit area is servicing. Adequate fire protection should be incorporated, based upon both OSHA standards and specific local regulations. Thus, the building should contain adequate fire extinguishers, sprinkler systems, and standpipe and hose.

Aisles should be plainly marked and straight, with well-rounded corners or diagonals at turn points. If aisles are to accommodate vehicle travel, they should be at least three feet wider than twice the width of the broadest vehicle. When traffic is only one way, then two feet wider than the broadest vehicle is adequate. In general, aisles should have at least five footcandles of illumination. The initial installation of sufficient fixtures does not assure adequate illumination. There needs to be a continuing maintenance effort so as to assure that periodic cleaning of fixtures and replacement of blown lights take place.

Color should be used throughout to identify hazardous conditions. The color recommendations shown in Table 9–10 are in compliance with OSHA standards.

The majority of machine tools can be satisfactorily guarded so that the probability of being injured while operating the facility is remote. The problem is that there are many instances in which a facility can be guarded but isn't. These are the instances in which the analyst should take immediate action to see that a guard is provided and that it is workable and routinely used. There are, of course, exceptions, such as a jointer or a circular cutoff saw, where the process does not lend itself to foolproof guarding. In such cases, partial guarding is easily attainable, but complete guarding is excessively expensive or impossible because it interferes with the operator's manipulations. In such cases, the analyst has several alternatives. In some instances, the process may be automated, thus completely

TABLE 9–10

Color	Used for	Examples
Red	Fire protection equipment, danger, and as a stop signal	Fire alarm boxes, location of fire extinguishers and fire hose, sprinkler piping, safety cans for flammables, danger signs, emergency stop buttons
Orange	Dangerous parts of machines, other hazards	Inside of movable guards, safety starting buttons, edges of exposed parts of moving equipment
Yellow	Designating caution, physical hazards	Construction and material handling equipment, corner markings, edges of platforms, pits, stair treads, projections. Black stripes or checks may be used in conjunction with yellow
Green	Safety	Location of first-aid equipment, gas masks, safety deluge showers
Blue	Designating caution against starting or using equipment	Warning flags at starting point of machines, electrical controls, valves about tanks and boilers
Purple	Radiation hazards	Container for radioactive materials or sources
Black and white	Traffic and housekeeping markings	Location of aisles, direction signs, clear floor areas around emergency equipment

freeing the operator from the "nip" point. In other instances, a robot manipulator can be used in place of an operator, or the method can be planned and the operator trained to use manual feeders or devices to keep the hands and other portions of the body away from danger points.

In addition to making provision for guarding the facility at the nip point, the analyst should see that the operator has adequate protection from potential accidents resulting from the use of the tool. To control such accidents, management needs to take steps to:

1. Train operators in the correct and safe use of tools.
2. Provide the correct tool for the job.
3. Maintain the tool so that it is always in a safe condition.
4. Insure the use and maintenance of the necessary guards and safety devices or practices.

A system of quality control and maintenance should be incorporated within the tool room and the tool cribs, so that only reliable tools in good working condition are released to the worker. Examples of unsafe tools that should not be released to the operator include: power tools with broken insulation, electrically driven power tools lacking grounding plugs or wires, poorly sharpened tools, hammers with mushroomed heads, cracked grinding wheels, grinding wheels without guards, and tools with split handles or sprung jaws.

In addition to being cognizant of the work environment, the equipment, and the tools, in order to initiate and maintain a program of safety and employee health, the analyst should also be cognizant of the potential danger of certain materials. A large segment of our business and manufacturing enterprises uses some potentially dangerous chemicals. As a matter of company policy, the composition of every chemical compound used by a concern should be ascertained, its hazards determined, and control measures established to protect the employees. The long-term health effect of many materials is still unknown, and new procedures are being initiated as hazards are identified by OSHA and other groups.

Materials that are known to cause health and/or safety problems may be classified as falling into one of three categories: corrosive materials, toxic or irritant materials, and flammable materials.

Corrosive materials include a vast variety of acids and caustics that can burn and destroy human tissue upon contact. The chemical action of corrosive materials can take place by direct contact with the skin or through inhalation of their fumes or vapors. In order to avoid the potential danger resulting from the use of corrosive materials, the analyst should consider the following measures:

1. Be sure that the methods of material handling are completely foolproof.
2. Assure that the process will not result in any spilling or spatting, especially during initial delivery processes.
3. Be sure that operators who are exposed to corrosive materials have and are using correctly designed personal protective equipment and waste disposal procedures.
4. Assure that the dispensary or the first-aid area is equipped with the necessary emergency provisions, including deluge showers and eye baths.

Toxic or irritating materials include gases, liquids, or solids that poison the body or disrupt normal processes upon ingestion, absorption through the skin, or inhalation.

In order to control toxic materials, the following methods are used:

1. Completely isolate the process from the worker.
2. Provide adequate exhaust ventilation.
3. Provide the worker with reliable personal protective equipment.
4. Substitute a nontoxic or nonirritating material.

Flammable materials and strong oxidizing agents present problems of fire and explosion. The spontaneous ignition of combustible materials can take place when there is insufficient ventilation to remove the heat from a process of slow oxidation. To prevent such fires, combustible materials need to be stored in a well-ventilated, cool, dry area. Small quantities should be stored in covered metal containers.

An explosion can result when combustible dusts (of which some, such as sawdust, are not ordinarily known to be explosive) or flammable vapors or gases are present in the air in such proportions that ignition at any point is carried out at a rapid rate throughout an entire volume. For both gases and dusts, there are limiting concentrations in air below which and above which explosions do not occur. For light dusts, the generally accepted lower explosive limit is 0.015 ounce per cubic foot, and for heavy dusts, 0.5 ounce per cubic foot. Vapors and gases have a wider range over which an explosion is liable to take place. Concentrations in air of 0.5 percent by volume are frequently listed as lower limits. An increase in temperature will depress the lower limit.

In order to avoid explosions, the analyst will need to prevent ignition and to provide adequate ventilation-exhaust systems. He should also adequately control the processes involved so as to minimize the generation of dusts and the liberation of gases and vapors.

Gases and vapors may be removed from gas streams by absorption in liquids or solids, adsorption on solids, condensation, and catalytic combustion and incineration. In absorption, the gas or vapor becomes distributed in the collecting liquid or solid. Equipment for absorption includes absorption towers, such as bubble-cap plate columns, packed towers, spray towers, and wet-cell washers.

For the adsorption of gases and vapors, a variety of solid adsorbents with an affinity for certain substances have been used. Charcoal, for example, will adsorb many different substances, including benzene, carbon tetrachloride, chloroform, nitrous oxide, and acetaldehyde.

The catalytic combustion process uses a platinum alloy–alumina catalyst to burn hydrocarbons. The minimum catalytic ignition temperature varies from 350° F to 600° F. In catalytic combustion, gases and vapors go through a low-temperature oxidation process and are converted to odor- and color-free gases. The presence of the catalyst merely provides an activated surface on which the reaction proceeds more readily.

Job factors leading to unsatisfactory performance

In this chapter, we have discussed the principal considerations for macroscopic improvement, including the physical environment, physiological and psychological constraints, and sociological considerations. One additional aspect that needs to be studied by the analyst is those task factors that may lead to human error.

The facility, coupled with the worker's task of managing and operating a piece of equipment, may be too demanding of the operator so that he or she will have difficulty in operating it efficiently through a normal shift. At this time, we want to consider facility outputs from the standpoint of the operator's ability to grasp readily the meaning of the output and to respond promptly in the most effective manner.

One of the most common methods of providing information output of facilities is through visual displays. The principal forms of visual displays include: indicator lights, scales, counters, printers, graphic plotters, and cathode-ray tubes. To be effective, a display must be able to communicate information quickly, accurately, and efficiently. By efficiently, we mean that the eye as the sensory organ that gathers the information must be able to do so in a manner that is free from error. Thus, it should be possible to read the display quickly and accurately both from a position of straight on the display and at viewing angles of up to 45 degrees, as the job demands.

Simple indicator lights are typified by the "idiot" or pilot light in a car. Legend lights provide additional information. For example, a legend light in a modern car will advise that something is wrong, and in addition the legend "alt" informs you that the alternator is not producing sufficient output. Scales are used to display graduated information, such as air speed or the amount of fuel in a tank. Counters are used when a precise quantity or indication is desired in connection with the operation of the facility or process. A counter may be used on a press to inform the operator when to change a die. A printer is usually an electromechanical device for recording information. Thus, for later referral, it may be desirable to incorporate an output printer in connection with a desk calculator. The graphic plotter is similar to a printer, except that the output is in graphic form, which reflects continuous trends. Both plotters and printers are commonly used today in connection with computers so as to secure a permanent record of the material that is being processed by the computer. Cathode-ray tubes are used to present moving visual images, as in the TV screen. Also, cathode-ray tubes can be used in presenting alphanumeric information. This is done in connection with schedules, such as schedules of airline arrivals and departures and production schedules in the large job shop.

Indicator lights

Indicator or warning lights are probably the type of visual displays in greatest use. Several basic requirements should be incorporated into their use.

First, they should be designed so that they will immediately get the attention of the worker. They should also be designed so that the operator will know what is wrong and what action he or she should take.

Generally, only one warning light should be used in conjunction with a given system. Other lights that identify the cause and the action to be taken, and that operate in conjunction with the single warning light, may be located in less central positions. The warning light should remain on until the condition that caused it to be energized has been remedied.

If a flashing light is used (one flashing light will attract attention quickly, but several lose most of this ability), it should be planned to give four flashes per second. Immediately after the operator takes action, the flashing should stop and the light remain on until the improper condition has been completely remedied.

The warning light should be red or yellow and of sufficient size and intensity to be noticed immediately. A good rule is to make it twice the size and brightness of other panel indicators, and located not more than 30 degrees off the operator's expected line of sight.

Display information

Table 9–11 provides helpful information in connection with the relative advantages and disadvantages of using moving pointers, moving scales, and counters.

TABLE 9–11

Indicator	Service rendered			
	Quantitative reading	Qualitative reading	Setting	Tracking
Moving pointer	Fair	Good (changes are easily detected)	Good (easily discernible relation between setting knob and pointer	Good (pointer position is easily controlled and monitored
Moving scale	Fair	Poor (may be difficult to identify direction and magnitude)	Fair (may be difficult to identify relation between setting and motion	Fair (may have ambiguous relationship to manual-control motion
Counter.	Good (minimum time to read and results in minimum error)	Poor (position change may not indicate qualitative change)	Good (accurate method to monitor numerical setting)	Poor (not readily monitored)

Operator errors in reading display information will increase as the density of information per unit area of the display increases and as the operator time for reading the display and responding decreases. Coding is a method that improves the readability of the display and the operator's viewing efficiency. The best three coding methods are: color, alphanumerics (letters and digits), and shape (geometric figures). These three coding techniques require little space and allow easy identification, though some training may be required in their interpretation.

Colors have emotional and psychological significance. Table 9–5 provides some of these characteristics of the major colors. Universally in

Western culture, red is recognized as a stop situation, as is characterized by traffic control. Red also usually symbolizes danger. On the other hand, green is thought of as a proceed or go-ahead situation.

Yellow is usually thought of as a caution symbol. It is widely used in connection with hunting sportswear, and everyone is aware of its use to convey caution in connection with traffic lights. A recommended coding of simple indicator lights is shown in Table 9–12.

TABLE 9–12
Recommended coding of indicator lights

Diameter	State	Color			
		Red	*Yellow*	*Green*	*White*
12.5 mm	Steady	Failure; Stop action; Malfunction	Delay; Inspect	Circuit energized; Go ahead; Ready; Producing	Functional; In position; Normal (on)
25 mm or larger	Steady	System or subsystem in stop action	Caution	System or subsystem in go-ahead state	
25 mm or larger	Flashing	Emergency condition			

Alphanumeric coding provides many more combinations than does color coding. From an efficiency standpoint, it parallels color coding. In order to have good efficiency with alphanumeric coding, the analyst should consider the stroke width of the numerals and letters, the width-height ratio, and the type form, or "font."

Based on a viewing distance of up to 28 inches under a range of illuminating conditions, the letter or numeral height should be at least 0.20 inches, and the stroke width at least 0.04 inches, to give a width-height ratio of 1:5. A broader stroke is used with dark letters on a bright background, and a narrower stroke with bright letters on a dark background.

The font refers to the available type styles, such as Gothic, Futura, and Tempo. In general, capital, or uppercase, letters are more easily read for a few words than are lowercase letters. Consequently, uppercase letters are recommended, and their width-height ratio should be about 3:5.

Acoustic signals

In some instances, it is better to use auditory signals than visual presentations. For example, auditory signals are usually more efficient if the worker's job necessitates his or her continual movement about the plant or business, or if the person receiving the signal is located in a work area where it would be difficult to see a visual signal, such as a dark area or an excessively bright area. Short, simple messages are also usually better handled by auditory means.

The analyst should recognize the competence of the human auditory system. It can be alert continuously, and it can detect sources of different signals without orientation of the body, as is usually necessary with visual signals. Since hearing is omnidirectional, and reaction times to sounds are shorter than to visual indications, auditory messages are especially desirable in connection with warning signals. Of course, only acoustic means are satisfactory for speech. There are cases where auditory signals should not be considered as an alternative to visual signals but as an addition to

TABLE 9–13
Control size criteria

Control	Dimension	Control size Minimum (mm)	Control size Maximum (mm)
Pushbutton . . . . Fingertip	Diameter	13	*
Thumb/palm	Diameter	19	*
Foot	Diameter	8	*
Toggle switch	Tip diameter	3	25
	Lever arm length	13	50
Rotary selector	Length	25	*
	Width	*	25
	Depth	16	*
Continuous adjustment knob . . . . Finger/thumb	Depth	13	25
	Diameter	10	100
Hand/palm	Depth	19	*
	Diameter	38	75
Cranks . . . . For rate	Radius	13	113
For force	Radius	13	500
Handwheel.	Diameter	175	525
	Rim thickness	19	50
Thumbwheel	Diameter	38	*
	Width	*	*
	Protrusion from surface	3	*
Lever handle. . . . . . . Finger	Diameter	13	75
Hand	Diameter	38	75
Crank handle	Grasp area	75	*
Pedal	Length	88	†
	Width	25	†
Valve handle.	Diameter	75 inches per inch of valve size	

* No limit set by operator performance.
† Dependent upon space available.

them. In cases where the visual system of the operator may already be overburdened, it may be more efficient to add an auditory system.

Shape and size coding

Shape coding, where two- or three-dimensional geometric configurations are used, permits both tactual and visual identification. It finds most of its applications where redundant or double-quality identification is desirable, thus helping to minimize errors. Shape coding permits a relatively large number of discriminable shapes. However, if the operator must identify controls without vision, discrimination will be difficult and slow as the number of shapes increase. If the operator is obliged to wear gloves,

TABLE 9–14
Control displacement criteria

| | | Displacement | |
Control	Condition	Minimum	Maximum
Pushbutton	Thumb/fingertip operation	3 mm	25 mm
	Foot Normal	13 mm	–
	Heavy boot	25 mm	–
	Ankle flexion only	–	63 mm
	Leg movement	–	100 mm
Toggle switch	Between adjacent positions	30°	–
	Total	–	120°
Rotary selector	Between adjacent detents: Visual	15°	–
	Nonvisual	30°	–
	For facilitating performance	–	40°
	When special engineering is required	–	90°
Continuous adjustment knob	Determined by desired control/display ratio (mm. of control movement for each mm. of display movement)		
Crank	Determined by desired control/display ratio		
Handwheel	Determined by desired control/display ratio	90°– 120°†	
Thumbwheel	Determined by number of positions		
Lever handle.	Fore-aft movement	*	350 mm
	Lateral movement	*	950 mm
Pedals	Normal	13 mm	–
	Heavy boot	25 mm	–
	Ankle flexion (raising)	–	63 mm
	Leg movement	–	175 mm

* None established.
† Provided optimum control/display ratio is not hindered.

then shape coding is desirable only for visual discrimination or for the tactual discrimination of only two to four shapes.

Size coding, analogous to shape coding, permits both tactual and visual identification of controls. Size coding is used principally where the controls cannot be seen by the operator. Of course, as is the case with shape coding, size coding permits redundant coding, since controls can be discriminated both tactually and visually. In general, it is desirable to limit the number of size categories to three.

Control size, displacement, and resistance criteria

In both the micro- and macroscopic components of his work assignment, the worker is continually using various types and designs of controls. The

TABLE 9–15
Control resistance criteria

Control	Condition	Resistance Minimum (kg)	Maximum (kg)
Push button	Fingertip	0.17	1.14
	Foot: Normally off control	1.82	9.10
	Rested on control	4.55	9.10
Toggle switch	Finger operation	0.17	1.14
Rotary selector	Torque.	1 cm-kg	7 cm-kg
Continuous adjustment knob . .	Torque: Fingertip <1-in. dia	*	0.3 cm-kg
	Fingertip >1-in. dia	*	0.4 cm-kg
Crank	Rapid, steady turning: <3-in. radius	0.91	2.28
	5–8-in. radius	2.28	4.55
	Precise settings	1.14	3.64
Handwheel†	Precision operation: <3-in. radius	*	*
	5–8-in. radius	1.14	3.64
	Resistance at rim: One-hand	2.28	13.64
	Two-hand	2.28	22.73
Thumbwheel.	Torque	1 cm-kg	3 cm-kg
Lever handle	Finger grasp	0.34	1.14
	Hand grasp: One-hand	0.91	–
	Two-hands	1.82	–
	Fore-aft: Along median plane:		
	One-hand–10 in. forward SRP§	–	13.64
	–16–24 in. forward SRP	–	22.73
	Two-hand–10–19 in. forward SRP	–	45
	Lateral:		
	One-hand–10–19 in. forward SRP	–	9.09
	Two-hand–10–19 in. forward SRP	–	22.73
Pedal	Foot: Normally off control	1.82	–
	Rested on control	4.55	–
	Ankle flexion only	–	4.55
	Leg movement	–	80

* Not established.
† For valve handles/wheels: 25 ± cm-kg of torque/cm of valve size
 (8 cm-kg of torque/cm of handle diameter).
§ SRP = seat reference point.

three parameters that have a major impact on his performance are the control size, the control resistance when engaged, and the total displacement upon activation. A control that is either too small or too large cannot be actuated efficiently. Likewise, the amount of resistance and displacement will have an impact upon operator performance. The effect of both distance and resistance on performance time will become more apparent to the reader after he or she has read Chapter 19 on synthetic basic motion times. Tables 9–13, 9–14, and 9–15 provide helpful design information as to minimum and maximum dimensions for various control mechanisms.

TEXT QUESTIONS

1. What are the principal objectives of operation analysis, motion study, and micromotion study?
2. What areas of study relate to the macroscopic approach toward improvements?
3. What independent factors affect the quantity of light that is needed to perform a task satisfactorily?
4. What is the relationship between contrast and seeing time?
5. What footcandle intensity would you recommend 30 inches above the floor in the company washroom?
6. Explain how sales may be influenced by colors.
7. Would the combined colors of yellow and blue give a harmonious hue? Explain.
8. What color has the highest visibility?
9. How is sound energy dissipated in viscoelastic materials?
10. A frequency of 2,000 Hz would have approximately what wavelength in meters?
11. What would be the approximate decibel value of a grinder being used to grind a high-carbon steel?
12. Distinguish between broadband noise and meaningful noise.
13. According to the present OSHA law, how many continuous hours per day of a 100 dbA sound level would be permissible?
14. What three classifications have been identified from the standpoint of exposure to vibration?
15. In what ways can the worker be protected from vibration?
16. What is meant by the environmental temperature? The effective temperature? The operative temperature?
17. What is the maximum rise in body temperature that the analyst should allow?
18. How would you go about estimating the maximum length of time that a worker should be exposed to a particular heat environment?
19. With a dry-bulb temperature of 80° F, a wet-bulb temperature of 70° F, and an air velocity of 200 feet per minute, what would be the normal effective temperature?

20. What type of radiation is given the most attention by the safety engineer?
21. What is meant by absorbed dose of radiation? What is the unit of absorbed dose?
22. What is meant by the rem?
23. What insulation would be required for a stenographic pool of female operators if the ambient temperature were 40° F?
24. What three factors influence the accuracy of control movements?
25. What caloric intake would you recommend for an operator doing heavy work? Explain.
26. Explain the preferential order of needs that applies to most workers.
27. What color would you paint a container used for holding radioactive materials?

GENERAL QUESTIONS

1. What steps would you take to increase the amount of light in an assembly department by about 15 percent? The department currently uses fluorescent fixtures, and the walls and ceiling are painted a medium green. The assembly benches are a dark brown.
2. What color combination would you use to attract attention to a new product being displayed?
3. What is the relationship between the heartbeat rate and oxygen consumption?
4. When would you advocate that the company purchase aluminized clothing?
5. Are there any possible health hazards in conjunction with electron beam machining? With laser beam machining? Explain.
6. Explain why supine lifting permits heavier lifts than does prone lifting.
7. Explain why the effective visual areas for average employees is greater for the right eye than for the left eye.
8. Is there a satisfactory explanation for the deterioration of dynamic memory with age?

PROBLEMS

1. A work area has a reflectivity of 60 percent, based upon the color combinations of the work stations and the immediate environment. The seeing task of the assembly work could be classified as difficult. What would be your recommended illumination?
2. What would be the level of two uncorrelated noise signals of 86 and 96 decibels?
3. In the XYZ Company, the industrial engineer designed a work station where the seeing task was difficult because of the size of the components going into the assembly. He established the brightness desired was 100 footlamberts on the average, with a standard deviation of 10 footlamberts so as to accommodate 95 percent of the workers.

   The work station was painted a medium green having a reflectance of

50 percent. What illumination in footcandles would be required at this work station in order to provide adequate illumination for 95 percent of the workers? Estimate what the required illumination would be if you repainted the work station with a light cream paint.

4. In the XYZ Company, the industrial engineer was assigned the task of altering the work methods in the press department in order to meet OSHA standards relative to permissible noise exposures. He found that the sound level averaged 100 db and that the standard deviation was 10 db. The 20 operators in this department were provided with earplugs. Also, the power output from the public-address system was altered from 30 watts to 20 watts. The deadening of the sound level of the earplugs was estimated to be 20 percent effective. What improvement resulted? Do you feel that this department is now in compliance with the law for 99 percent of the employees? Explain.

SELECTED REFERENCES

McCormick, Ernest J. *Human Factors Engineering.* 3d ed. New York: McGraw-Hill Book Co., 1970.

Morgan, Clifford, T., Cook, Jesse S., Chapanis, Alphonse, and Lund, Max W. *Human Engineering Guide to Equipment Design.* New York: McGraw-Hill Book Co., 1963.

Murrell, K. F. H. *Ergonomics.* London, England: Chapman and Hall, 1969.

VanCott, H. P., and Kinkade, R. G. *Human Engineering Guide to Equipment Design.* Rev. ed. Washington, D.C.: U.S. Government Printing Office, 1970.

Woodson, W. E., and Conover, D. W. *Human Engineering Guide for Equipment Designers.* 2d ed. Los Angeles: University of California Press, 1964.

10

Presentation and installation of the proposed method

Selling the proposed method is the step following development in the systematic procedure for methods and work measurement. This step is as important as any of the preceding steps, since a method not sold will not be installed.

The analyst must keep in mind that it is natural for humans to resent the attempts of others to influence their thinking. When someone approaches us with a new idea, our instinctive reaction is to put up a defense against it. We feel that we must protect our own individuality—preserve the sanctity of our own ego. And all of us are just egotistical enough to convince ourselves that our ideas are better than those of anyone else. It is natural for us to react in this manner even if the new idea is for our own advantage. If the idea has merit, there is a tendency to resent it because we did not think of it first.

Some techniques that the successful analyst frequently uses to help sell his ideas include:

1. Introducing the idea into the other person's mind, so that he feels it is really his idea to a large extent. For example, begin by saying, "You recently gave me the idea that . . ."
2. Not appearing overly anxious to have an idea accepted. Introduce your thoughts with such statements as: "Do you think this idea has possibilities?" or "Have you considered this?"
3. Presenting objections to his own idea. This can get the other person arguing for the support of the idea.

The presentation of the proposed method should emphasize savings. Savings in material (both direct and indirect) and savings in direct and indirect labor should highlight the analyst's report. The second most important part of the presentation is that dealing with the recovery of capital investment.

Once the proposed method has been properly presented and sold, installation can take place. Installation, like presentation, requires sales ability. During installation, the successful analyst will continually be selling the proposed method to engineers and technicians on his or her own level, to subordinate executives and supervisors, and to labor and representatives of organized labor. The analyst's job during the presentation and installation of the proposed method will involve "selling up," "selling down," and "selling sideways."

The report on the proposed method

It is important that the analyst make his or her presentation in both written and oral form. Even if the company involved does not require a written report, it is good practice to make one for record purposes and for future applications. A well-written report is a major step in selling the proposed method.

The elements of a well-written report are:

1. Title page.   4. Summary.
2. Table of contents.   5. Body.
3. Letter of transmittal.   6. Appendix.

From the standpoint of presentation, the summary is the most important section of the report, since it will be the only part read by the busy executive. The summary is based upon the body of the report, and consequently it is usually not written until after the body has been completed. However, it is presented early in the report so that those who must pass or disapprove the proposal can obtain the facts quickly in order to make a decision.

The summary should contain three elements: an abstract explaining briefly the nature of the problem; conclusions outlining the results of the analysis; and recommendations setting forth the proposed method and summarizing estimated savings and the recovery of capital expenditures.

The body includes a section on the nature of the problem followed by details relative to the gathering of the data and the methods of analysis. This section contains the operation and flow process charts used to present the facts and the man and machine, gang, and operator process charts used in developing the proposed work center. This section also contains the reasons for the conclusions and recommendations given in the summary.

The entire report should have the qualities of correctness, clearness, conciseness, completeness, and accuracy. It should be prepared so that it can be easily read and studied.

Reporting the recovery of capital investment

The three most frequently used appraisal techniques for determining the desirability of investing in a proposed method are: (1) the return on sales method, (2) the return on investment method, and (3) the cash flow method.

The return on sales method involves a computation of the ratio of the average yearly profit brought about through using the method, to the average yearly sales or increase in dollar value added to the product, based upon the estimated pessimistic life of the product. Although this ratio provides information on the effectiveness of the method and of the resulting sales efforts, it does not consider the original investment required to get started on the proposed method.

The return on investment method gives the ratio of the average yearly profit brought about through using the method, based upon the estimated pessimistic life of the product, to the original investment. Of two proposed methods that would result in the same sales and profit potential, management would obviously prefer to use the one requiring the investment of the least capital.

The cash flow method computes the ratio of the present worth of cash flow, based upon a desired percentage return, to the original investment. This method introduces the rate of flow of money in and through the company and the time value of money. To compute the time value of money, we have the relationship:

$$S = P(1 + i)^n$$

where:

i = Interest rate for a given period
n = Number of interest periods
P = Present worth of principal
S = Worth of sum n periods later

An assumed return rate (i) is the basis of the cash flow computation. Then all cash flows following the initial investment for the new method are estimated and adjusted to their present worth, based upon the assumed return rate. The total estimated cash flows for the estimated pessimistic life of the product are then summed up as a profit or loss in terms of cash on hand today. This total is then compared with the initial investment.

An illustrative example will help clarify the use of these three appraisal methods for prognosticating the potential of a proposed method.

Investment for proposed method: $10,000.

Desired return on investment: 10 percent.

Salvage value of jigs, fixtures, and tools: $500.

Estimated life of the product for which the proposed method will be used: 10 years.

End of year	Increase in sales values due to proposed method	Cost of production with proposed method	Gross profit due to proposed method
1	$ 5,000	$ 2,000	$ 3,000
2	6,000	2,200	3,800
3	7,000	2,400	4,600
4	8,000	2,600	5,400
5	7,000	2,400	4,600
6	6,000	2,200	3,800
7	5,000	2,000	3,000
8	4,000	1,800	2,200
9	3,000	1,600	1,400
10	2,000	1,500	500
Totals.	$53,000	$20,700	$32,300
Average.	$ 5,300	$ 2,070	$ 3,230

Return on sales = $\dfrac{3,230}{5,300}$ = 61%   Return on investment = $\dfrac{3,230}{10,000}$ = 32.3%

Present worth of cash flow:

$$(3,000)(0.9091) = \$2,730 \qquad (3,800)(0.5645) = \quad 2,140$$
$$(3,800)(0.8264) = \quad 3,140 \qquad (3,000)(0.5132) = \quad 1,540$$
$$(4,600)(0.7513) = \quad 3,460 \qquad (2,200)(0.4665) = \quad 1,025$$
$$(5,400)(0.6830) = \quad 3,690 \qquad (1,400)(0.4241) = \quad 595$$
$$(4,600)(0.6209) = \quad 2,860 \qquad (\ 500)(0.3855) = \quad 193$$
$$\qquad\qquad\qquad\qquad\qquad\qquad\qquad\qquad\qquad \$21,373$$

Salvage value of tools:

$$(500)(0.3855) = \$193$$

Total present worth of anticipated gross profit and tool salvage value: $21,566. Ratio of present worth to original investment:

$$\frac{21,566}{10,000} = 2.16$$

The new method has satisfactorily passed all three of these appraisal methods. A 61 percent return on sales and a 32.3 percent return on capital investment certainly represent attractive returns. The cash flow analysis reveals that the original investment of $10,000 will be recovered in four years while earning 10 percent. During the 10-year anticipated life of the product, $11,566 more than the original investment will be earned.

The analyst must recognize that estimates of the product demand 10 years hence may deviate considerably from reality. Thus, the element of chance is introduced, and the probabilities of success tend to diminish with the increased length of the payoff period.

The oral presentation

Frequently, the analyst will be asked to present his method proposal orally. To help assure that approval is given the proposal, the analyst should be prepared to present the benefits and advantages accurately and forcefully. He should be prepared to give estimates of the resulting increase in productivity and/or decrease in cost. If quality will be improved or customer service enhanced, this information should be given.

It is important that the analyst plan his presentation in advance. He should have data relative to all the advantages of the proposed method, as well as cost information and facts giving the expected savings and the expected recovery of capital investment. It is a good idea to list in advance all the information that one's superior may ask for, and then be prepared to supply this information.

The analyst should also be prepared to answer objections raised to the proposal. These are usually centered on the initial cost, time to adopt the method, and inconvenience while installation is taking place. The analyst should point out that these objections have been carefully considered and that plans have been made to cope with them.

The oral presentation of the proposed method requires much study, preparation, and salesmanship. A good method will not sell itself; it must be sold.

Installation

After a proposed method has been approved, the next step is installation. Too frequently, the analyst does not stay close enough to the job during installation. He tends to feel that installation will take place automatically according to his proposal. Often this is not the case. The maintenance man, mechanic, or worker will make a slight change or modification on his own initiative without considering the consequences. This may mean that the proposed method will not give the results anticipated.

The analyst should stay with the job during installation to assure that all details are carried out in accordance with the proposed plan. During installation, he should verify that the work center being established is equipped with the facilities proposed, that the planned working conditions are provided, that the tooling is in accordance with his recommendations, and that the work is progressing satisfactorily.

During installation, the analyst will have an opportunity to sell the new method to the operator, foreman, setup man, and so on. Then, by the time the installation is complete, these employees will be readier to give the new method an enthusiastic try.

Once the new work center has been installed, the analyst should check all aspects to see whether they conform to the specifications established. In particular, he should verify that the "reach" and "move" distances the operator must perform are the correct length; that the tools are correctly sharpened; that the mechanisms function soundly; that stickiness and sluggishness have been worked out; that safety features are operative; that material will be available in the quantities planned; that working conditions associated with the work center are as anticipated; and that all parties have been informed of the new method.

After the analyst is sure that every aspect of the method is ready for operation, he should have the foreman assign the operator who will be working with the method. The analyst should stay with the operator as long as is necessary to break him or her in on the new assignment. This period may be a matter of a few minutes or of several hours or even of days, depending on the complexity of the assignment.

Once the operator begins to get a feel for the method and works along systematically, the analyst can proceed with other work. However, he should not consider the installation phase complete until he has checked back several times during the first few days after installation to assure that the proposed method is working out as planned.

TEXT QUESTIONS

1. What are some of the communication skills that are important in selling a new method?
2. What are the principal elements of a well-written report?
3. What is meant by the "cash flow" appraisal technique?
4. What are the principal concerns of management with regard to a new method that is relatively costly to install?

GENERAL QUESTIONS

1. What is the relationship between return on capital investment and the risk associated with the anticipated sales of the product for which a new method will be used?
2. When do you feel that the oral report would be more important than the written report in getting approval for a new method?

PROBLEMS

1. How much capital could be invested in a new method if it is estimated that $5,000 would be saved the first year, $10,000 the second year, and $3,000 the third year? Management expects a 30 percent return on invested capital.
2. You have estimated the life of your design to be three years. You expect that a capital investment of $20,000 will be required to get it into

production. You also estimate, based upon sales forecasts, that the design will result in an after-tax profit of $12,000 the first year and $16,000 the second year, and a $5,000 loss the third year. Management has asked for an 18 percent return on capital investment. Should we go ahead with the investment to produce the new design? Explain.

SELECTED REFERENCES

Barish, Norman M. *Economic Analysis for Engineering and Manageriɑ! Decision-Making*. New York: McGraw-Hill Book Co., 1962.

Damerst, William A. *Clear Technical Reports*. New York: Harcourt-Brace-Jovanovich, Inc., 1973.

Emerson, C. Robert, and Taylor, William R. *An Introduction to Engineering Economy*. Davis, Calif.: Cardinal Publishers, 1973.

Fabrycky, W. J., and Thuesen, G. J. *Economic Decision Analysis*. Englewood Cliffs, N.J.: Prentice-Hall, Inc., 1974.

11

Making the job analysis and the job evaluation

Closely associated with installation of the proposed method is the job analysis of the work center and the resulting job evaluation. Every time a method is changed, the job description should be altered to reflect the conditions, duties, and responsibilities of the improved method. When a new method is introduced, it is important that a job analysis be made so that a qualified operator may be assigned to the work center and an appropriate base rate considered.

The factory cost of a product includes the cost of direct labor, direct material, and factory expense. Factory expense consists of such items as light, heat, rent, service supplies, and factory supervision. A close review of expense or overhead items will show that these costs are made up of but two components: material and labor. So, in effect, factory cost is made up of just labor and material. Therefore:

$$\text{Factory cost} = \text{Cost of labor} + \text{cost of material}$$

Material cost is easily determined, no matter what the product may be. In every instance, it is determined by multiplying the cost per unit of measure by the number of measures involved. Thus, some yardstick always governs its basis of cost. Steel is bought by the ton, castings and forgings by the piece, water and gas by the cubic foot, electricity by the kilowatt-hour, tubing by the foot, silver by the ounce, cloth by the square yard, oil by the gallon, and so forth. In every case, material cost can be prorated so that it is possible to determine the proportionate share each product item should absorb.

The only other item entering into factory cost is the cost of labor, either direct or indirect. Labor cost, in most commodities, represents the major portion of total costs. To have some conception of the true costs of specific products, standards must be established on labor elements. Hourly monetary rates mean nothing unless they are supplemented with performance standards. If an operator running a No. 2 Brown & Sharpe Universal Mill is paid a base rate of $4.20 per hour while milling a profile on a forging, there is no conception of the unit milling cost until a standard has been established on the job in terms of time. If a standard allows six minutes per piece, then a labor cost of $0.42 can be assigned to the op-

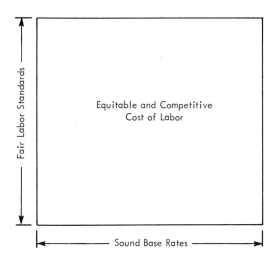

eration. However, time standards alone do not give the entire story with regard to labor costs. They form only one side of a rectangle whose area may be thought of as equitable and competitive costs of labor. The other side of the rectangle is sound base rates.

Base rates that are sound must assure money rates commensurate with the local rates for similar work; they must allow adequate differentials for jobs requiring higher skills and responsibilities; and they must be based upon techniques that can be explained and justified.

Job analysis

Appropriate base rates are a result of job evaluation, which was defined in Chapter 1 as a technique for equitably determining the relative worth of the different work assignments within an organization. The basis of job evaluation is job analysis, which is the procedure for making a careful appraisal of each job and then recording the details of the work so that it can be evaluated fairly by a trained analyst. Figure 11–1 illustrates an analysis of a clerical job for use in a point job evaluation plan. It should be clearly understood that before a job description is developed, all aspects of the opportunity should be carefully studied to assure that the best methods are being used and that the operator is thoroughly trained in the prescribed methods.

FIGURE 11–1
Job analysis for a shipping and receiving clerk

JOB TITLE ____Shipping and Receiving Clerk____ DEPT.____Shipping____
MALE _X_ FEMALE ____ DATE _____ TOTAL POINTS _280_ CLASS 5

JOB DESCRIPTION

Directs and assists in loading and unloading, counting, and receiving or rejecting purchased parts and supplies, and later delivers to proper departments.

Examines receivals out of line with purchase orders. Maintains file on all purchase
orders and/or shipping orders and keeps open orders up-to-date. Maintains daily and
weekly shipping reports and monthly inventory reports.

Assists in packing of all foreign and domestic shipments. Makes up request for inspection form on certain materials received and rejection form for all items rejected.

Job requires thorough knowledge of parking, shipping and receiving routine, plant
layout, shop supplies, and finished parts. Needs to have a knowledge of simple office
routine. Ability to work with other departments, as a service department, and to deal
effectively with vendors. Job requires considerable accuracy and dependability. The
effects of poor decisions include damaged receivables and shipments, inaccurate inventories, and extra material handling. Considerable lifting of weights up to 100 lbs. is
involved. Works in conjunction with two class 4 packers and shippers.

Job Evaluation	Degree	Points
Education	1	15
Experience and Training	2	50
Initiative and Ingenuity	3	50
Analytical Ability	3	50
Personality Requirements	2	30
Supervisory Responsibility	1	25
Responsibility for Loss	1	10
Physical Application	6	25
Mental or Visual Application	1	5
Working Conditions	5	20
		280

Typically, the various job responsibilities and authorities and the consequences resulting from poor decisions are items that would be included
in a job analysis. Also, the analysis should provide information regarding
the machines and tools used in connection with the job. The physical and
social conditions related to the job should be outlined.

Job evaluation

The main purpose of any job evaluation plan is to determine the proper compensation for the work performed on each job. A well-conceived job evaluation plan includes the following factors:

1. It provides a basis for explaining to employees why one job is worth more or less than another job.
2. It provides a reason to employees whose rate of pay is adjusted because of a change in method.
3. It provides a basis for assigning personnel with specific abilities to certain jobs.
4. It helps determine the criteria for a job when employing new personnel or making promotions.
5. It provides assistance in the training of supervisory personnel.
6. It provides a basis for determining where opportunities for methods improvement exist.

Four principal methods of job evaluation are being practiced in this country today. These are the classification method, point system, factor comparison method, and the ranking method.

The classification method, sometimes called the grade description plan, consists of a series of definitions designed to differentiate jobs into wage groups. Once the grade levels have been defined, each job is studied and assigned to the appropriate level on the basis of the complexity of its duties and responsibilities. This plan is used extensively in the United States Civil Service.

When using this method of job evaluation, the following steps must be taken:

1. Prepare a grade description scale for each type of job, for example, machine operations, manual operations, skilled (craft) operations, inspection.
2. Write the grade descriptions for each grade in each scale, using such factors as:
   a. Type of work and complexity of duties.
   b. Education necessary to perform job.
   c. Experience necessary to perform job.
   d. Responsibilities.
   e. Effort demanded.
3. Prepare job descriptions for each job. Classify each job by "slotting" (placing in a specific category) the job description into the proper grade description.

Both the point system and the factor comparison method are more objective and thorough in their evaluations of the various jobs involved, in that both of these plans make a study of the basic factors common to most jobs that influence their relative worth. Of the two plans, the point system is generally considered the more accurate method for occupational rating. In this method, all the different attributes of a job are compared directly with these attributes in other jobs.

When a point system is to be installed, the following procedure should be followed:

1. Establish and define the basic factors which are common to most jobs and which indicate the elements of value in all jobs.
2. Specifically define the degrees of each factor.
3. Establish the points to be accredited to each degree of each factor.
4. Prepare a job description of each job.
5. Evaluate each job by determining the degree of each factor contained in it.
6. Sum the points for each factor to get the total points for the job.
7. Convert the job points into a wage rate.

The factor comparison method of job evaluation usually has the following elements:

1. The factors that will establish the relative worth of all jobs are determined.
2. An evaluation scale is established which is usually similar to a point scale except that the units are in terms of money.
3. Job descriptions are prepared.
4. Key jobs are evaluated, factor by factor, by ranking each job from the lowest to the highest for each factor.
5. The wages paid on each key job are allocated to the various factors. The money allocation automatically fixes the relationships among jobs for each factor, and therefore establishes the ranks of jobs for each factor.
6. On the basis of the monetary values assigned to the various factors in the key jobs, other jobs are evaluated factor by factor.
7. A wage is determined by adding up the money value of the various factors.

The ranking method arranges jobs in their order of importance or according to their relative worth. This method became popular in the United States during World War II because of its simplicity and ease of installation. At this time, the National War Labor Board set up the requirement that all companies working on government contracts must have some type of wage classification system. The ranking method satisfied this requirement. Generally speaking, the ranking method is less objective than the other techniques; consequently, it necessitates greater knowledge of all jobs. For this reason, it has not been used extensively in recent years, but has been superseded by the other plans. The following steps are followed when installing the ranking method:

1. Prepare job descriptions.
2. Rank jobs (usually departmentally first) in the order of their relative importance.
3. Determine the class or grade for groups of jobs, using a bracketing process.
4. Establish the wage or wage range for each class or grade.

Selecting factors

It is generally considered preferable to use a small number of factors. Under the factor comparison method, most companies use five factors. In some point programs, 10 or more factors may be used. The objective is to use only as many factors as are necessary to provide a clear-cut difference among the jobs of the particular company. The elements of any job may be classified as to:

1. What the job demands that the employee bring in the form of physical and mental factors.
2. What the job takes from the employee in the form of physical and mental fatigue.
3. The responsibilities that the job demands.
4. The conditions under which the job is done.

The selection of factors is usually the first basic task that is undertaken when introducing job evaluation.

The National Electrical Manufacturers Association states that the relative value of a job is considered to depend on the following factors:

1. Education.
2. Experience.
3. Initiative and ingenuity.
4. Physical demand.
5. Mental and/or visual demand.
6. Responsibility for equipment or process.
7. Responsibility for material or product.
8. Responsibility for safety of others.
9. Responsibility for work of others.
10. Working conditions.
11. Hazards.

These factors are present in varying degrees in the various jobs, and any job under consideration will fall under some one of the several degrees of each factor. The various factors are not of equal importance. To give recognition to these differences in importance, weights or points are assigned to each degree of each factor, as shown in Table 11–1. Figure 11–2 illustrates a job rating and a substantiating data sheet based upon the plan of the National Electrical Manufacturers Association.

TABLE 11–1
Points assigned to factors and key to grades

Factors	1st degree	2d degree	3d degree	4th degree	5th degree
Skill					
1. Education	14	28	42	56	70
2. Experience	22	44	66	88	110
3. Initiative and ingenuity	14	28	42	56	70
Effort					
4. Physical demand	10	20	30	40	50
5. Mental and/or visual demand	5	10	15	20	25
Responsibility					
6. Equipment or process	5	10	15	20	25
7. Material or product	5	10	15	20	25
8. Safety of others	5	10	15	20	25
9. Work of others	5	10	15	20	25
Job conditions					
10. Working conditions.	10	20	30	40	50
11. Unavoidable hazards	5	10	15	20	25

Source: National Electrical Manufacturers Association.

Each degree of each factor is carefully defined so that it is evident what degree characterizes the work situation under study.

For example, education may be defined as appraising the requirements for the use of shop mathematics, drawings, measuring instruments, or trade knowledge. First-degree education may require only the ability to read and write, and to add and subtract whole numbers. Second-degree education could be defined as requiring the use of simple arithmetic, such as addition and subtraction of decimals and fractions, together with simple drawings and some measuring instruments, such as calipers and scale. It would be characteristic of two years of high school. Third-degree education may require the use of fairly complicated drawings, advanced shop mathematics, handbook formulas, and a variety of precision measuring instruments, plus some trade knowledge in a specialized field or process. It could be thought of as being equivalent to four years of high school plus short-term trades training. Fourth-degree education could require the use of complicated drawings and specifications, advanced shop mathematics, and a wide variety of precision measuring instruments, plus broad shop knowledge. It may be equivalent to four years of high school plus four years of formal trades training. Fifth-degree education may require a basic technical knowledge sufficient to deal with complicated and involved mechanical, electrical, or other engineering problems. Fifth-degree education could be thought of as being equivalent to four years of technical university training.

Experience appraises the length of time that an individual with the specified education usually requires to learn to perform the work satisfactorily from the standpoint of both quality and quantity. Here, first

JOB RATING - SUBSTANTIATING DATA
DORBEN MFG. CO.
UNIVERSITY PARK, PA.

JOB TITLE: Machinist (General)   CODE: 176   DATE: Nov. 12

FACTORS	DEG.	POINTS	BASIS OF RATING
Education	3	42	Requires the use of fairly complicated drawings, advanced shop mathematics, variety of precision instruments, shop trade knowledge. Equivalent to four years of high school or two years of high school plus two to three years of trades training.
Experience	4	88	Three to five years installing, repairing, and maintaining machine tools and other production equipment.
Initiative and Ingenuity	3	42	Rebuild, repair, and maintain a wide variety of medium-size standard automatic and hand-operated machine tools. Diagnose trouble, disassemble machine and fit new parts, such as antifriction and plain bearings, spindles, gears, cams, etc. Manufacture replacement parts as necessary. Involves skilled and accurate machining using a variety of machine tools. Judgment required to diagnose and remedy trouble quickly so as to maintain production.
Physical Demand	2	20	Intermittent physical effort required tearing down, assembling, installing, and maintaining machines.
Mental or Visual Demand	4	20	Concentrated mental and visual attention required. Laying out, setup, machining, checking, inspecting, fitting parts on machines.
Responsibility for Equipment or Process	3	15	Damage seldom over $300. Broken parts of machines. Carelessness in handling gears and intricate parts may cause damage.
Responsibility for Material or Product	2	10	Probable loss due to scrapping of materials or work, seldom over $150.
Responsibility for Safety of Others	3	15	Safety precautions are required to prevent injury to others; fastening work properly to face plates, handling fixtures, etc.
Responsibility for Work of Others	2	10	Responsible for directing one or more helpers a great part of time. Depends on type of work.
Working Conditions	3	30	Somewhat disagreeable conditions due to exposure to oil, grease, and dust.
Unavoidable Hazards	3	15	Exposure to accidents, such as crushed hand or foot, loss of fingers, eye injury from flying particles, possible electric shock, or burns.

REMARKS: Total 307 Points--assign to job class 4.

degree could involve up to three months; second degree, three months to one year; third degree, one to three years; fourth degree, three to five years; and fifth degree, over five years.

In a similar manner, each degree of each factor is identified with a clear definition and with specific examples when applicable.

Performing the evaluation

It is apparent that considerable judgment is needed to evaluate each job with respect to the degree required of each factor utilized in the plan. Consequently, it is usually desirable to have a committee perform the evaluation. A separate committee should be appointed for each department of the company or business. A typical committee would include a permanent chairman (usually from industrial relations or industrial engineering), a union representative, the department foreman, the department steward, and a management representative (usually from industrial relations).

When meeting, the committee should evaluate all jobs for the same factor before proceeding to the next factor. For example, all jobs in the department under study should be evaluated for degree of skill before proceeding to others factors, such as effort, responsibility, and job conditions. Using this pattern, the committee will measure the job rather than the individual filling the job.

Committee members should assign their degree evaluations independently of the other members. Then all members should discuss any differences that may exist until there is agreement on the level of the factor. It is wise not to end a meeting until the factor under study has been evaluated for all jobs being evaluated in the department.

Classifying the jobs

After all jobs have been evaluated, the points assigned to each job should be tabulated. The number of labor grades within the plant should now be decided. This number is a function of the range of points characteristic of the jobs within the plant. Typically, the number of grades runs from 8 (typical of smaller plants and lesser skilled industries) to 15 (typical of larger plants and higher skilled industries) (see Figure 11–3). For example, if the point range of all the jobs within a plant ranged from 110 to 365, the following grades could be established:

Grade	Score range (points)	Grade	Score range (points)
12	100–139	6	250–271
11	140–161	5	272–293
10	162–183	4	294–315
9	184–205	3	316–337
8	206–227	2	338–359
7	228–249	1	360 and above

It is not necessary to have like ranges for the various labor grades. It might be desirable to have increasing point ranges for more highly compensated jobs.

FIGURE 11–3

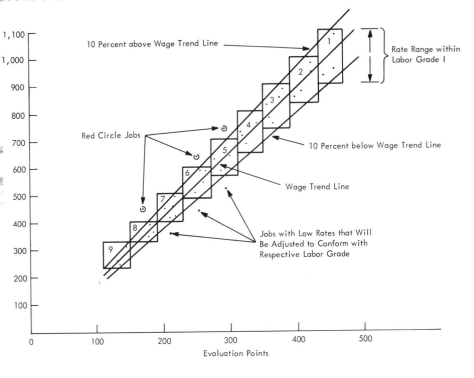

The jobs falling within the various labor grades should now be reviewed in relation to one another in order to assure fairness and consistency. For example, it would not be appropriate for a Class A machinist to be in the same grade level as a Class B machinist.

The next step is to assign hourly rates to each of the labor grades. These rates are based upon area rates for similar work, company policy, and the cost-of-living index. Frequently, a rate range is established for each labor grade. The total performance of each operator will determine his or her pay rate within the established range. Total performance refers to quality, quantity, safety, attendance, suggestions, and so on.

Installation of the job evaluation program

After area rates have been plotted against the point values of the various jobs, a rate versus point value trend line is developed. This trend line may or may not take the form of a straight line. Regression techniques are helpful in developing the trend line. After the trend line has been developed, it will be noted that several points will be both above and below the trend line. Points significantly above the trend line represent employees whose present rate is higher than that established by the job evaluation plan, and points significantly below the trend line represent employees whose present rate is less than that prescribed by the plan.

Employees whose rates are less than that called for by the plan should receive an immediate increase to the new rate. Employees whose rates are higher than that called for by the plan (such rates are referred to as "red circle rates") are not given a rate decrease. They are, however, not given an increase at the next contract review unless the adjustment results in a rate higher than that called for by the job evaluation plan. Of course, any new employee would be paid the new, lower rate.

Conclusion

It is important that employees understand the fairness of the job evaluation plan. It is also important that regular follow-up of the plan be done so that it is adequately maintained. Jobs do change, so it is necessary to review all jobs periodically and to make adjustments when necessary.

George Fry Associates undertook a comprehensive survey of job evaluation practices in over 500 companies. Some of the significant results from this survey are shown below.

	Percent
Firms using a standard job evaluation plan	66.0
Plan used:	
Ranking	3.5
Grade description.	1.0
Factor comparison	10.5
Point system	85.0
Average of hourly employees covered	65.0
Job evaluation function reports to:	
Industrial relations	69.0
Industrial engineering	16.0
Results used:	
In employment	87.0
In employee placement	88.0
In wage rate bargaining	68.0
Job evaluation program:	
Is recognized in union contract	83.0
Is an issue during negotiations	59.0
Has gone to arbitration (of these cases,	
management won 74 percent).	27.0

Understanding and acceptance:	*Good*	*Average*	*Poor*	*None*
By employees	7%	39%	49%	5%
By top-management	50	36	13	1
By middle-management	18	17	5	0
By first-line supervisors	35	52	13	0

It is especially unfortunate that over half of the companies responding to the survey found their employees' understanding of the plan to be poor or nonexistent. If a job evaluation plan is to succeed over the years, the vast majority of the employees should have at least an average understanding of how the plan works.

Job analysis and job evaluation are important steps after installation of the ideal method. Their principal purpose is to determine the relative worth of the jobs in a company. Job evaluation will provide the means for compensating all employees within an organization in proportion to their responsibilities and to the difficulty of their work. At the same time, it will lead to base pay rates in line with remuneration for similar work in the community. The benefits effected through job evaluation will improve personnel relations.

To predetermine factory costs, so necessary from the standpoint of submitting bids and quotations, it is essential that the cost of the various direct and indirect materials and of all labor be precisely predetermined.

Predetermined material costs can easily be calculated. Predetermined labor costs also can be readily computed if good labor standards developed by one of the work measurement techniques prevail. Standards determined by estimates and historical records usually will not be sufficiently accurate to meet competitive prices or will not allow manufacture at a necessary profit.

Although labor standards determine the "how long," it is necessary to have sound base rates to measure fairly the "how much" in terms of dollars and cents. Of the various methods of job evaluation used in establishing sound base rates, the point plans tend to give the most reliable results. Point job evaluation systems patterned after the techniques of the National Metal Trades and the National Electrical Manufacturers Association represent a logical approach toward developing the relative worth of the different work assignments within an organization.

TEXT QUESTIONS

1. What are the three basic components of cost?
2. Is time a common denominator of labor cost? Why or why not?
3. What is job analysis?
4. What four methods of job evaluation are being practiced in this country today?
5. Explain in detail how a "point" plan works.
6. What factors influence the relative worth of a job?
7. Why are estimates unsatisfactory for determining direct labor time standards?
8. What is the weakness of using historical records as a means of establishing standards of performance?

9. What work measurement techniques will give valid results when undertaken by competent trained analysts?
10. What are the principal benefits of a properly installed job evaluation plan?
11. Explain why a range of rates rather than just one rate should be established for every labor grade.
12. Explain what is meant by total operator performance.

GENERAL QUESTIONS

1. Should cost-of-living increases be given as a percentage of base rates or as a straight hourly increment? Why?
2. Why would a consulting firm such as George Fry and Associates undertake a comprehensive survey of job evaluation practices?

PROBLEMS

1. A job evaluation plan based upon the point system uses the following factors:
   a. Experience: maximum weight 200 points; five grades.
   b. Education: maximum weight 100 points; four grades.
   c. Effort: maximum weight 100 points; four grades.
   d. Responsibility: maximum weight 100 points; four grades.

   A floor sweeper is rated as 150 points, and this position carries an hourly rate of $3. A class 3 milling machine operator is rated as 320 points, which results in a money rate of $4.80 per hour. What grade of experience would be given to a drill press operator with a $3.80 per hour rate and point ratings of grade 2 education, grade 1 effort, and grade 2 responsibility?

2. A job evaluation plan in the Dorben Company provides for five labor grades, of which grade 5 has the highest base rates and grade 1 the lowest. The linear plan involves a range of 50 to 250 points for skill, 15 to 75 points for effort, 20 to 100 points for responsibility, and 15 to 75 points for job conditions. Five degrees are established for each of the four factors. Each labor grade has three money rates: a "low," a "mean," and a "high" rate.

   If the high money rate of labor grade 1 is $4 per hour and the high money rate of labor grade 5 is $10 per hour, what would be the mean money rate of labor grade 3? What degree of skill is required for a labor grade of 4 if second-degree effort, second-degree responsibility, and first-degree job conditions apply?

3. In the Dorben Company, the analyst has installed a point job evaluation plan covering all indirect employees in the operating divisions of the plant. Ten factors were used in this plan, and each factor was broken up into five degrees. In making the job analysis, the position shipping and receiving clerk was shown as having second-degree initiative and ingenuity, valued at 30 points. The total point value of this job was 250 points. The mini-

mum number of points attainable in the plan was 100, and the maximum was 500. If 10 job classes prevailed, what degree of initiative and ingenuity would be required to elevate the job shipping and receiving clerk from job class 4 to job class 5?

If job class 1 carries a rate of $4 per hour and job class 10 carries a rate of $4.60 per hour, what rate does job class 7 carry? (Note: Rates are based on the midpoint of job class point ranges.)

SELECTED REFERENCES

Otis, Jay, and Leukart, Richard H. *Job Evaluation: A Sound Basis for Wage Administration.* Englewood Cliffs, N.J.: Prentice-Hall, Inc., 1954.

Salvendy, Gabriel, and Seymour, Douglas W. *Prediction and Development of Industrial Work Performance.* New York: John Wiley & Sons, Inc., 1973.

Zollitsch, Herbert G., and Langsner, Adolph *Wage and Salary Administration.* 2d ed. Cincinnati: South-Western Publishing Co., 1970.

12

Time study requirements

The eighth step in the systematic procedure for developing the work center to produce the product is establishing time standards. Three techniques have been used to determine time standards: estimates, historical records, and work measurement procedures.

Estimates as a means of establishing standards were used to a greater extent in years past than is the case today. With increasing competition from foreign producers, there has been an increasing effort to establish standards based upon facts rather than judgment. Experience has shown that no individual can establish consistent and fair standards of production by the simple procedure of taking a look at a job and then judging the amount of time required to produce it. Where estimates are used, standards will be out of line on an average of about 25 percent. Compensating errors will sometimes diminish this figure, but experience has shown that over a period of time estimated values deviate substantially from measured standards. Both historical records and work measurement techniques will give much more accurate values than will the use of estimates based upon judgment alone.

Under the historical method, production standards are based upon the records of previously produced, similar jobs. In common practice, the worker "punches in" on a time clock every time he or she begins a new job, and then "punches out" on the job when he has completed it. This technique tells us how long it took to do a job, but never indicates how long it should have taken. Since operators wish to justify their entire working day, some jobs carry personal delay time, unavoidable delay time, and avoidable delay time to a much greater extent than they should, while other jobs do not carry their appropriate share of delay time. I have seen historical records that deviated consistently by as much as 50 percent on the same operation of the same job. Historical records as a basis of determining labor standards are better than no records at all. Such records will give more reliable results than will estimates based upon judgment alone, but they do not provide sufficiently valid results to assure equitable and competitive labor costs.

Any of the work measurement techniques—stopwatch time study, standard data, time formulas, or work sampling studies—represents a better way to establish fair production standards. All of these techniques are based upon facts. All consider each detail of the work and its relation to the normal time required to perform the entire cycle. Accurately established time standards make it possible to produce more within a given plant, thus increasing the efficiency of the equipment and the operating personnel. Poorly established standards, although better than no standards at all, will lead to high costs, labor dissension, and eventually the possible failure of the enterprise.

In order to have a successful installation of any of the work measurement techniques, there must be a wholehearted commitment by management. This commitment involves allocation of enthusiasm, time, and the necessary financial resources on a continuing basis.

A smoothly operating work measurement program requires considerable planning and effective communication to all members of the enterprise. Prior to the introduction of the program, objectives and policies should be clearly established, and properly trained and experienced analysts should be employed. Good communication is essential during installation and throughout the life of the program. All levels of management as well as the employees should be kept informed on the progress of installation and on the mechanics of the program.

As data from the work measurement system become available, they should be used. Sound standards have many applications that can mean the difference between the success and the failure of a business. They should be utilized for planning purposes, for the comparison of alternative methods, for effective plant layout, for determining capacities, for purchasing new equipment, for balancing the work force with the available work, for production control, for the installation of incentives, for standard cost and budgetary control, and so on.

In Chapter 1, time study was defined as a technique for establishing an allowed time standard to perform a given task. This technique is based upon measurement of the work content of the prescribed method, with due allowance for fatigue and for personal and unavoidable delays. Frequently, time study is defined by the layman as a method of determining a "fair day's work," and this concept will be discussed before the requirements and responsibilities of those associated with time study are explained. It is necessary to have a clear understanding of what is involved in a fair day's work.

A fair day's work

Practically everyone connected with industry in any way has often heard the expression "a fair day's work." Yet most of the people who have heard the expression would be perplexed if they were asked to define just what a fair day's work is. The intraplant wage rate inequities agreements of the basic steel industries[1] contain the provision that "the fundamental principle of the work and wage relationship is that the employee is entitled to a fair day's pay in return for which the company is entitled to a fair day's work." A fair day's work is defined in these agreements as the "amount of work that can be produced by a qualified employee when working at a normal pace and effectively utilizing his time where work is not restricted by process limitations." This definition does not make clear what is meant by *qualified employees, normal pace,* and *effective utilization.* Although all of these terms have been defined by the steel industries, a certain amount of flexibility prevails because firm benchmarks cannot be established on such broad terminology. For example, the term *qualified employee* is defined as "a representative average of those employees who are fully trained and able satisfactorily to perform any and all phases of the work involved, in accordance with the requirements of the job under consideration." This definition leaves a doubt as to what is meant by a "representative average employee."

Then the term *normal pace* is defined as "the effective rate of performance of a conscientious, self-paced, qualified employee when working neither fast nor slow and giving due consideration to the physical, mental, or visual requirements of the specific job." The intraplant wage rate inequities agreements specify as an example "a man walking without load, on smooth, level ground at a rate of three (3) miles per hour." Although the three miles an hour concept tends to tie down what is meant by normal pace, still a notable amount of latitude can prevail if we think of normal pace on the thousands of different jobs in American industry.

Again, a feeling of uncertainty arises when one considers the definition of *effective utilization.* This is explained in the agreements as "the maintenance of a normal pace while performing essential elements of the job during all portions of the day except that which is required for reasonable rest and personal needs, under circumstances in which the job is not subject to process, equipment or other operating limitations."

In general, a fair day's work is one that is fair to both the company and the employee. This means that the employee should give a full day's work for the time that he or she gets paid, with reasonable allowances for personal delays, unavoidable delays, and fatigue. He or she is expected to operate in the prescribed method at a pace that is neither fast nor slow, but one that may be considered representative of all-day performance by the experienced, cooperative employee.

Time study requirements

Certain fundamental requirements need to be realized before the time study is taken. If the standard is required on a new job, or if it is re-

---

[1] With United Steelworkers of America.

quired on an old job on which the method or part of the method has been altered, the operator should be thoroughly acquainted with the new technique before the operation is studied. It is also important that the method be standardized at all points where it is to be used before the study begins. Unless all details of the method and working conditions have been standardized, the time standards will have little value and will be a continual source of mistrust, grievances, and internal friction.

It is also important that the union steward, the departmental foreman, and the operator be cognizant of the fact that the job is to be studied. Each of these parties will then be able to make any specific plans in advance and to take the steps necessary to allow a smooth, coordinated study. The operator should verify that he is performing the correct method and should endeavor to acquaint himself with all details of the operation. The foreman should check the method to make sure that feeds, speeds, cutting tools, lubricants, and so forth, conform to standard practice as established by the methods department. Also, the foreman must investigate the amount of available material so that no shortage will take place during the study. If several operators are available for the study, the foreman should determine to the best of his ability which operator will be likely to give the most satisfactory results. The union steward should then make sure that only trained, competent operators are selected for time study observation. He should explain to the operator why the study is being taken and should answer pertinent questions raised by the operator from time to time.

Responsibilities of the time study man

All work involves varying degrees of skill and physical and mental effort. In addition to such variations in job content, there are differences in the aptitude, physical application, and dexterity of the workers. It is an easy matter for the analyst to observe an employee at work and to measure the actual time taken to perform a task. It is a considerably more difficult matter to evaluate all variables and determine the time required for the "normal" operator to perform the job.

Because of the many human interests and reactions associated with the time study technique, it is essential that there be full understanding on the part of the foreman, the employee, the union steward, and the time study analyst.

In general, the time study analyst is charged with these responsibilities:

1. To probe, question, and examine the present method to assure that it is correct in all respects before the standard is established.
2. To discuss the equipment, method, and operator's ability with the foreman before studying the operation.
3. To answer questions relating to time study practice or to a specific

time study that may be asked by the union steward, the operator, or the foreman.

4. To cooperate with the foreman and the operator at all times in order to obtain maximum help from both.
5. To refrain from any discussion with the operator under study or other operators that might be construed as criticism of the individual being studied.
6. To show on each time study complete and accurate information specifically identifying the method under study.
7. To record accurately the times taken to perform the individual elements of the operation being studied.
8. To evaluate the performance of the operator honestly and fairly.
9. To conduct himself at all times so that he obtains and holds the respect and confidence of the representatives of both labor and management.

The qualifications of a time study man that are necessary for him to meet successfully the responsibilities of his position are similar to those required for success in any field in which the major efforts are directed toward establishing ideal human relations.

First of all, a good time study man must have the mental ability to analyze diversified situations and make rapid, sound decisions. He should have an inquisitive, probing, and open mind that seeks to improve, and should always be cognizant of the "why" as well as the "how."

To supplement a keen mind, it is essential that the time study analyst have practical shop training in the areas in which he will be establishing standards. If he is to be associated with the metal trades, he should have a background as a journeyman machinist or the equivalent knowledge of the correct use and application of machines, hand tools, jigs, fixtures, and gages. This would include specific knowledge of cutting feeds, speeds, and depths of cuts to get the maximum results consistent with the desired quality of the product and ultimate tool life.

Since the time study man directly affects the pocketbook of the worker and the profit and loss statement of the company, it is essential that his work be completely dependable and accurate. Inaccuracy and poor judgment will not only affect the operator and the company financially, but may also result in complete loss of confidence by the operator and the union, which may undo harmonious labor relations that have taken management years to build up.

To achieve and maintain good human relations, the following personal requirements can be considered essential for the successful time study man:

1. Honesty.
2. Tact, human understanding.
3. Resourcefulness.
4. Self-confidence.

5. Good judgment, analytic ability.
6. Pleasing, persuasive personality, supplemented with optimism.
7. Patience, self-control.
8. Bountiful energy tempered with a cooperative attitude.
9. Well-groomed, neat appearance.
10. Enthusiasm for the job.

A review of all of the qualifications essential for a successful time study man may give the reader the opinion that he should be on a staff level with the president of the company. In fact, it is questionable whether all of these attributes would be required of the top executives of a concern. However, when one considers the magnitude of the labor relations problem today, it is essential that only the most competent people enter the field of time study. No other one individual within a company comes in contact with as many personnel from different levels of the organization as does the time study analyst. Therefore, it is imperative that he have the best qualifications.

In the time study man's approach to various employees, he should learn to recognize the human qualities in a worker and then be guided by a realization of the limitations of human nature. Thus, in order to receive cooperation, he must determine and follow through with the best method of approach to the worker. This calls for analysis of the employee's attitude toward his job, his fellow workers, the company, and the time study man himself.

The foreman's responsibility

Any and all foremen are management's representatives throughout the plant. Next to the operator, the foreman is closer to specific jobs than is any other man in the plant. In view of this, he must accept certain responsibilities in connection with the establishment of time standards.

To begin with, the foreman, in the interest of harmonious labor relations within the department he supervises, finds it mandatory that equitable time standards prevail. Both "tight" and "loose" standards are the direct cause of endless personnel problems, and the more of these that can be avoided, the easier and pleasanter will be his job. Of course, if all standards were loose, he would find his supervisory responsibilities relatively easy. However, this situation could not exist practically, since competition would not be met if all standards were loose.

The foreman should notify the operator in advance that his work assignment is to be studied. This clears the way for both the time study analyst and the operator. The operator has the assurance that his direct superior is cognizant of the fact that a rate is to be established on the job, and will have the opportunity to bring out specific difficulties that he feels should be corrected before a standard is set. The time study analyst will, of course, feel considerably more at ease if he knows that his presence is anticipated.

It should be the foreman's responsibility to see that the proper method established by the methods department is being utilized and that the operator selected is competent and has adequate experience on the job. Although the time study analyst is required to have a practical background in the area of work he is studying, he can hardly be expected to be infallible in specifications of all methods and processes. Thus, he should consider the foreman as his ally in verifying that the cutting tools are properly ground, that the correct lubricant is being used, and that a proper selection of feeds, speeds, and depths of cuts is being made.

If, for any reason, conditions are such that it is questionable whether a fair time study can be taken, the foreman should immediately make this fact known to the time study analyst.

In general, the foreman is responsible for assisting and cooperating with the time study man in any way that will aid in defining or clarifying an operation. He should carefully consider any suggestions for improvement brought out by the time study man, and should fully utilize his own background and influence to establish an ideal method in conjunction with the methods department, prior to stopwatch study.

The foreman is responsible for seeing that his operators use the prescribed method, and he should conscientiously assist and train all those employees coming under his jurisdiction in perfecting this method. He should freely answer any questions asked by the operator regarding the operation.

Any time a methods change takes place within his department, the foreman should notify the time study department immediately so that an appropriate adjustment of the standard can be made. This procedure should be followed regardless of the degree of the change. Methods changes would include such things as changes in material handling to and from the work station, in inspection procedure, in feeds and speeds, in work station layout, and in processes.

When a time study has been completed, the foreman should be required to sign the original study, thus indicating that he has complied with all his responsibilities on the study taken. Foremen who accept and carry out their responsibilities toward time study practice can be assured of operating harmonious departments that will be looked upon with favor by management, the union, and the employees themselves. Foremen who fall short of these responsibilities can contribute to the establishment of inequitable rates that will result in numerous labor grievances, pressure from management, and considerable dissatisfaction from the union.

The union's responsibility

Most unions are opposed to work measurement and would prefer to see all standards established by arbitration. However, the unions do recognize that standards are necessary for the profitable operation of a business and that management will continue to develop them by means of the principal work measurement techniques.

Furthermore, every union steward knows that poor time standards will cause him or her just as many problems as they will cause management. In the interest of operating a healthy union within a profitable business, the union should accept certain responsibilities toward time study.

Through training programs, the union should educate all of its members in the principles, theories, and economic necessity of time study practice. All of us tend to fear anything on which we are poorly informed. Operators can hardly be expected to be enthusiastic about time study if they know nothing about it. This is especially true in view of its background (see Chapter 2). Therefore, the union should accept the responsibility of helping to clarify and explain this important tool of management.

The union representative should make certain that the time study includes a complete record of the job conditions as to work method and work station layout. He should also ascertain that the current job description is accurate and complete.

Also, it is wise for the union representative to see that the elemental breakdown has been made with clearly defined end points, thus helping to assure the consistency of elemental times. He should assure that the study has been taken over a long enough period of time to accurately reflect all of the variations that normally take place in performing the operation, as well as the typical unavoidable delays. Time study is a sampling technique, and samples of insufficient size can lead to erroneous results.

The union should urge its members to cooperate with the time study analyst and to refrain from practices that would tend to place their performances at the low end of the rating scale. Encouraging operators to deceive the time study man will, in the final analysis, add up to but one thing: a poor standards structure that will include both loose and tight rates.

The union should accept the responsibility of seeing that updated standards are put into effect whenever a methods change is made. When methods are revised, the union should see that the time study department is notified through specified lines of authority.

Unions that train their members in the elements of time study, encourage cooperativeness, and stay abreast of management's program will benefit by more cooperation at the bargaining table, fewer work stoppages, and better satisfied members. Unions that encourage distrust of time study, and facilitate a program of keeping the operator in the dark," will be faced with a multitude of grievances from their members, a balky management negotiating team, and over a period of time, sufficient work stoppages to create hardships for all the parties concerned.

The operator's responsibility

Every employee should be sufficiently interested in the welfare of his company to give his wholehearted support to every practice and procedure inaugurated by management. Unfortunately, this situation is seldom realized, but it certainly can be approached if a company's management demonstrates its desire to operate with fair standards, fair base rates, good working conditions, and adequate employee benefits in the form of insurance and retirement programs. Once management has taken the initiative in these areas, every employee can be expected to cooperate in all operations and production control techniques.

The individual operators should be responsible for giving new methods a fair trial. They should wholeheartedly cooperate in helping to work out the "bugs" characteristic of practically every innovation. Suggestions for further improvement of the methods should be accepted as a part of each operator's responsibilities. The operator is closer to the job than anyone else, and he can make a real contribution to the company and to himself by doing his part in establishing ideal methods.

The operator should be responsible for assisting the time study analyst in breaking down the job into elements, thus assuring that all details of the job are specifically covered. He should also be responsible for working at a steady, normal pace while the study is being taken and should introduce as few foreign elements and extra movements as possible. He should be responsible for using the exact prescribed method and make no effort to deceive the time study analyst by introducing an artifical method with the thought of lengthening the cycle time and receiving a more liberal standard.

Conclusion

To many practicing time study men, the responsibilities we have assigned to the operator, the union, and the foreman may be considered a utopian goal which can never be realized. However, as stated, if management takes the initiative, these conditions can be approximated, and the result will be a competitive business that is profitable for all parties. For example, Howard M. Hinkel, personnel director of American Type Founders, Inc., reports that by establishing an "appreciation" course in time study and incentive techniques for the union stewards and committeemen in his company, the company has almost eliminated rate discussions that go beyond the operator's initial complaint.[2]

Time study procedures are the only known methods for supplying reasonably accurate information on the standard times that are so essential for the profitable, efficient operation of industry and business. To

---

[2] "Time Study Training for Stewards and Committeemen," *Factory Management and Maintenance,* August 1947.

assure the most valid results from time study practice, the time study analyst must receive complete cooperation from the foreman, the union steward, the operator, and company management.

Time study represents one of the most important and exacting forms of work in any industrial, commercial, or governmental enterprise. When intelligently used and fully understood by all parties, it offers marked benefits to workers, management, and the general public.

TEXT QUESTIONS

1. What type of reliability can we expect from estimates?
2. How is a fair day's pay determined by the intraplant wage rate inequities agreements of the basic steel industries?
3. What benchmark for normal pace is given under these agreements?
4. Why should the foreman sign the time study?
5. Why should the time study man have excellent personal qualities?
6. Explain how poor time standards increase the difficulties of the union steward.
7. How can management increase the cooperation of the union steward, the foreman, and the operator in their dealings with the time study analyst?

GENERAL QUESTIONS

1. Is it customary for the union to cooperate with the time study department to the extent recommended in this text? Is this cooperation likely to occur?
2. If the requirements for a time study man are so high, why is it that industry does not pay more for time study work?
3. How does the walking pace of three miles per hour agree with your concept of a normal performance?

PROBLEMS

1. It has been estimated that the typical plant that has not had the benefit of a methods, standards, and wage payment system is operating at about 50 percent of standard. A good methods program should increase productivity to 80 percent of standard. A standards program that is well conceived, operated, and maintained should increase the productivity to 95 percent of standard. And a well-designed and -implemented incentive system will further increase productivity to 120 percent of standard.

It has also been estimated that the cost of installing a complete methods, standards, and wage payment system will approximate $20,000 per year for every 100 people coming under the plan.

The XYZ Company employs 650 people. The total value added by production averages 80 percent of the payroll dollar. The average cost of labor (including fringe benefits) is $7.50 per hour.

What return on investment can be expected if the XYZ Company installs a complete methods, standards, and wage payment system? Show all your calculations.

SELECTED REFERENCES

Gomberg, William. *A Trade Union Analysis of Time Study.* 2d ed. Englewood Cliffs, N.J.: Prentice-Hall, Inc., 1955.

Griepentrog, Carl W., and Jewell, Gilbert. *Work Measurement: A Guide for Local Union Bargaining Committees and Stewards.* Milwaukee: International Union of Allied Industrial Workers of America, AFL-CIO, 1970.

Krick, Edward V. *Methods Engineering.* New York: John Wiley & Sons, Inc., 1962.

Presgrave, Ralph. *The Dynamics of Time Study.* 2d ed. New York: McGraw-Hill Book Co., 1945.

13

Time study equipment

The minimum equipment that is required to carry on a time study program includes a stopwatch, a time study board, time study forms, and a pocket calculator or slide rule.

In addition to the above, time-recording devices that are being used successfully and that offer some advantages over the stopwatch are time-recording machines, motion-picture cameras, and videotape equipment.

It will be noted that the equipment needed for time study or work measurement is not nearly as elaborate or costly as that required for micromotion study. In general, the criteria for success are the ability and personality of the time study analyst rather than the equipment he or she chooses to use.

The stopwatch

Several types of stopwatches are in use today, the majority of which are include in one of the following classifications:

1. Decimal minute watch (0.01 minute).
2. Decimal minute watch (0.001 minute).
3. Decimal hour watch (0.0001 hour).
4. Electronic stopwatch.

The decimal minute watch shown in Figure 13–1 has 100 divisions on its face, and each division is equal to 0.01 minute. Thus, a complete sweep of the long hand would require one minute. The small dial on the watch face has 30 divisions, and each of these divisions is equal to one minute. For every revolution of the sweep hand, the small hand will move one division, or one minute.

To start this watch, the side slide is moved toward the crown. To stop the watch and have the hands retain their respective positions, the side slide is moved away from the watch crown. To continue operation of the watch from the point where the hands stopped, the slide is moved toward the crown. To move both the sweep hand and the small

FIGURE 13–1
Decimal minute watch

Meylan Stopwatch Co.

hand back to zero, the crown is depressed. Releasing the crown will put the watch back into operation unless the side slide is moved away from the crown.

The decimal minute watch tends to be a favorite with time study men because of its ease of reading and recording. The sweep hand moves only 60 percent as fast as the long hand of the decimal hour watch, and thus terminal points are more discernible. In recording the time values, the analyst's task is simplified by the fact that the elemental readings are in hundredths of a minute, eliminating the need for the ciphers that are registered when using the decimal hour watch, which is read in ten thousandths of an hour.

The decimal minute watch (0.001 minute) is similar to the decimal minute watch (0.01 minute). In the former, each division of the larger hand is equal to one thousandth of a minute. Thus, it takes 0.10 minute for the sweep hand to circle the dial rather than one minute, as with the 0.01 decimal minute watch. This watch is used primarily for timing short elements for standard data purposes (see Chapter 18). In general, the 0.001 minute watch has no starting slide on its side, but is started, stopped, and returned to zero by successive depressions of the crown.

A special adaption of the decimal minute watch that many time study men find convenient to use is illustrated in Figure 13–2. The two long hands indicate decimal minutes and will complete one turn of the dial in one minute. The small dial is graduated in minutes, and a complete sweep of its hand represents 30 minutes.

To start this watch, the crown is depressed, and both sweep hands will

FIGURE 13–2
Decimal minute split (double-action)
watch

Meylan Stopwatch Co.

simultaneously start from zero. At the termination of the first element, the side pin is depressed, and this will stop the lower sweep hand only. The time study man can then observe the elapsed time for the element without the difficulty of reading a moving hand. He then depresses the side pin, and the lower hand will rejoin the upper hand, which has been moving uninterruptedly. At the termination of the second element, the side pin is depressed, and the procedure is repeated.

The decimal hour stop watch has 100 divisions on its face also, but each division on this watch represents 1 ten thousandth (0.0001) of an hour. Thus, a complete sweep of the large hand on this watch would represent one hundredth (0.01) hour, or 0.6 minute. The small hand registers each turn of the long hand, and a complete sweep of the small hand would take 18 minutes, or 0.30 hour (see Figure 13–3).

The decimal hour watch is started and stopped, and its hands are returned to zero, in the same manner as is the decimal minute (0.01) stopwatch.

Since the hour represents a universal unit of time in measuring output, the decimal hour watch is a practical timer and is widely used. Somewhat more skill is required to read this watch while timing short elements on account of the speed of the sweep hand. For this reason, some time study men prefer the decimal minute watch with its slower moving hand.

It is possible to mount four watches on a board, with linkage between them, so that during the course of a study the analyst can always read one watch which has stopped hands. Figure 13–4 illustrates such an arrangement. Here are shown four crown action watches which are actu-

FIGURE 13–3
Decimal hour watch

Meylan Stopwatch Co.

FIGURE 13–4
Time study board with four watches and form mounted, placed in proper position

FIGURE 13–5
Digital electronic time-study board

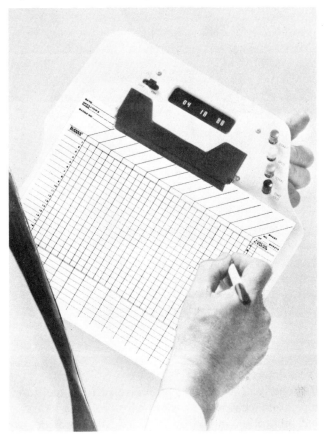

Meylan Stopwatch Corp.

ated by a lever shown at the right. First, pressure on the lever starts watch 1 (far left), cocks watch 2, stops watch 3, and starts watch 4. At the end of the first element, a clutch activating watch 4 is disengaged, and the lever is again pressed. This stops watch 1, starts watch 2, and resets watch 3 to zero, while watch 4 continues to run, as it will measure the overall time as a check. Watch 1 is now waiting to be read, while the next element is being timed by watch 2.

A more common practice is to use only one watch attached to the observation board, as illustrated in Figure 13–6.

Most stopwatches are produced so as to record times with accuracies of plus or minus 0.025 minutes over 60 minutes of operation. Government specifications of stopwatch equipment allow a deviation of 0.005 minutes per 30-second interval. All stopwatches should be checked periodically to verify that they are not providing "out-of-tolerance" readings. To help assure continued accuracy of reading, it is essential that stopwatches be

FIGURE 13–6
Eight-channel time study machine

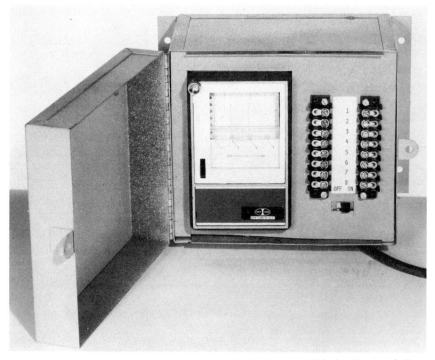

Meylan Stopwatch Corp.

maintained properly. They should be protected from moisture, dust, and abrupt temperature changes. Regular (about once a year is adequate) cleaning and lubrication should be provided. If the watches are not used regularly, they should be wound and allowed to run down periodically.

Today, totally electronic stopwatches are available at a cost of approximately $150. These watches provide resolution to one-hundredth second and accuracy to 0.003 percent. They weigh about 0.25 kilogram and are about 13 centimeters long by 5 centimeters wide and 5 centimeters deep. They permit timing any number of individual elements while also counting the total elapsed time. Thus, they can provide all of the advantages of snapback stopwatch study, with none of the disadvantages. Electronic stopwatches operate on rechargeable batteries. Typically, the batteries must be recharged after about 14 hours of continuous service.

Time-recording machines

Several versatile time study machines which facilitate the accurate measurement of time intervals are on the market today. These machines may be used in the absence of the time study analyst to measure the time that a facility is productive. For example, Figure 13–6 illustrates an eight-channel recorder where any two terminals may be connected to a normally open sensor which closes only when the machine or activity is productive. On the chart paper, a stylus records the state of the facility continuously. In the model illustrated, the activity of eight separate facilities can be recorded. Chart speeds ranging from 6 inches per hour to 480 inches per hour are available, depending upon the precision of measurement desired.

An adaptation of this equipment is its use with push button control, where each channel can be used in connection with a specific work element. This adaptation is especially useful in work sampling–type studies in which a professional wishes to self-evaluate the distribution of his time. For example, he may choose to assign the following to the eight channels:

Channel 1. Creative development.
Channel 2. Conference.
Channel 3. Dictation.
Channel 4. Incoming phone calls.
Channel 5. Outgoing phone calls.
Channel 6. Supervision and assigning work.
Channel 7. Reading mail.
Channel 8. Personal and interruptions.

By pushing a button related to the appropriate channel, he continuously accounts for the time expended during his workday.

Videotape and motion-picture

Videotape and motion-picture cameras are ideal for recording operator methods and elapsed time. Unfortunately, the cost of film and the delay necessitated by having to send the films out to be developed, prohibit the use of the motion-picture cameras in a great many instances. The high initial cost of videotape equipment may preclude its use.

Both of these picture-taking methods are especially useful in establishing standards by means of one of the synthetic motion time value techniques. By taking pictures of the operator and then studying them a frame at a time, the analyst can record the exact details of the method used and assign time values. It is also possible to establish standards by projecting the exposed films at the same speed that the pictures were taken and then performance rating the operator. All the facts necessary to accomplish a fair and accurate job of rating can be obtained by observing the exposed film. Then, too, potential methods improvements are revealed through the camera eye that would never be uncovered with the stopwatch or the time study machine.

These advantages, supplemented with the memomotion procedure, which allows longer cycle filming with minimum exposure of film, have increased the popularity of the camera as an effective tool of the time study analyst (see Chapter 8). Figure 8–9 illustrates a portion of a memomotion film taken at the rate of one frame per second.

The motion-picture camera described and illustrated in Chapter 8 will also do an excellent job when the analyst is filming for the purpose of establishing time standards.

The time study board

When the stopwatch is being used, it will be necessary to provide a suitable board to hold the time study form and the stopwatch. The board should be light, so as not to tire the arm, and yet strong and hard enough to provide a suitable backing for the time study form. One-quarter-inch plywood or a smooth plastic make suitable materials. The board should have arm and body contacts (see Figure 13–7) for comfortable fit and ease of writing while it is being held.

The watch is mounted in the upper right-hand corner of the board (for right-handed people), and to the left a spring clip holds the time study form. By standing in the proper position, the time study analyst can look over the top of his watch to the work station and follow the operator's movements while keeping both the watch and the time study form in his immediate field of vision.

FIGURE 13–7
Time study board for a right-handed observer (note the alignment chart on the board to assist the observer in determining the performance rating factor; see page 362 for an explanation of this alignment chart)

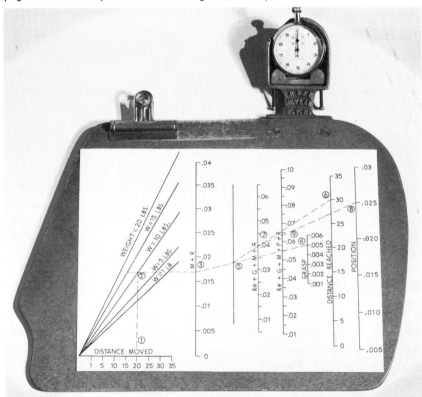

In order to assist the analyst in performance rating the operation under study, an alignment chart, developed by the author, may be attached to the board (see Figure 13–7). This chart permits the analyst to determine synthetically the allowed time for several of the effort elements comprised by the study. The ratio of the synthetic value for a particular element to the mean value actually taken by the operator serves as a guide for determining the performance factor. This chart establishes standards for the therbligs reach, grasp, move, position, and release when these are performed collectively.

Time study forms

On the time study form are recorded all details of the study. To date, there has been little standardization in the design of forms used by various industries. It is important that the form provide space to record all pertinent information concerning the method being studied. This is often done by constructing an operator process chart (see Chapter 6) on one side of the form. In addition to making a permanent record of the relative location of the tools and materials in the work area, the analyst should record such methods data as feeds, depths of cuts, speeds, and inspection specifications. Of course, it is also necessary to identify completely the operation being studied by including such information as the operator's name and number, operation description and number, machine name and number, special tools used and their respective numbers, department where the operation is performed, and prevailing working conditions. It is always better to provide too much than too little information concerning the job being studied. This will be discussed in a subsequent chapter.

The time study form should also include space for the signature of the foreman, indicating his approval of the method under observation. Likewise, the inspector should sign every study taken, acknowledging his or her acceptance of the quality of the parts being produced during the time study.

The form should be designed so that the analyst can conveniently record watch readings, foreign elements (see Chapter 14), and rating factors (see Chapter 15), and still use the sheet to calculate the allowed time. Figures 13–8 and 13–9 illustrate a time study form that has been developed by the author which allows sufficient flexibility to study practically any type of operation.

In this form, the various elements of the operation are recorded horizontally across the top of the sheet, and the cycles studied are recorded vertically row by row.

The "R" column is divided into two sections. The large area is for recording the watch reading, and in the small section marked "F" is shown the element performance factor if elemental rating is used. The "T" column is provided for elemental elapsed values (see Chapter 14).

FIGURE 13-8

Front of time study form designed for either elemental or overall rating

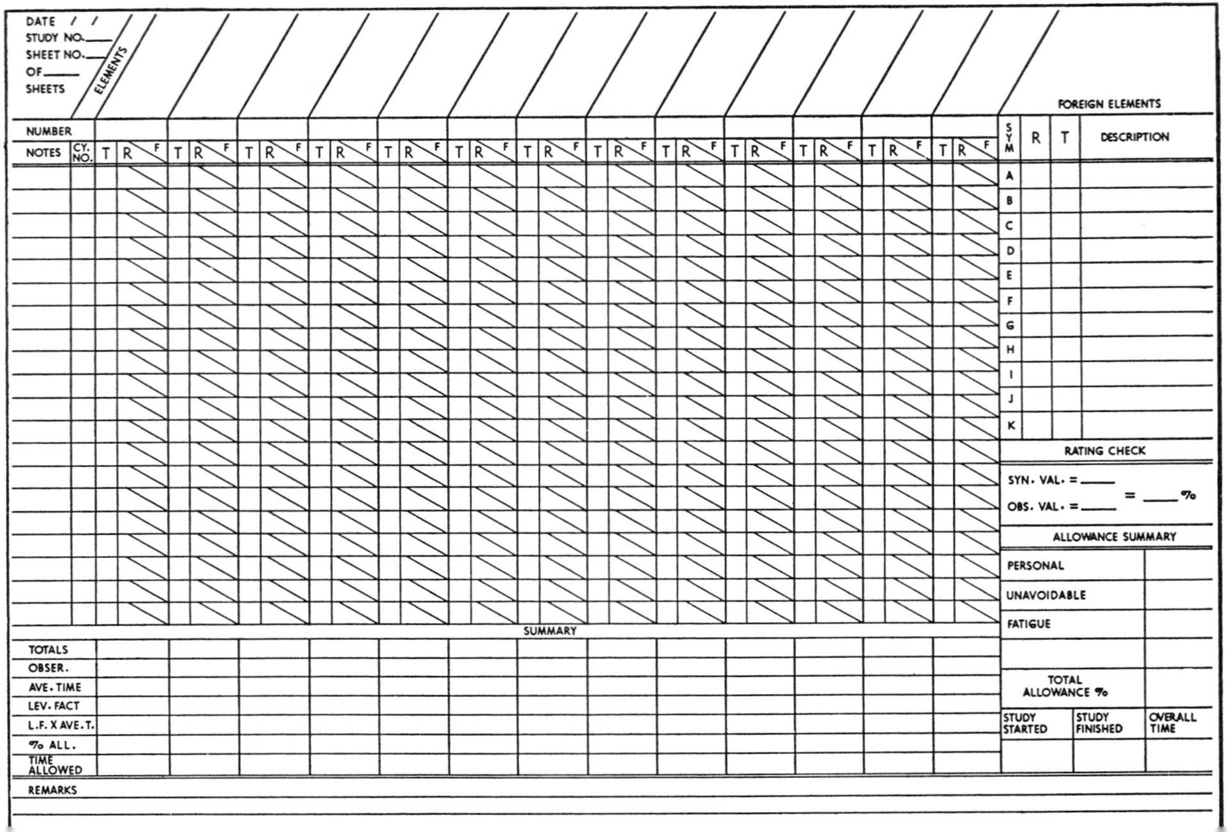

IGURE 13-9

ack of time study form—note operator chart (act breakdown) on which all details of the
ethod under study may be recorded

SKETCH

STUDY NO._____DATE_____

OPERATION_____

DEPT._____OPERATOR_____NO._____

EQUIPMENT_____

_____MCH. NO._____

SPECIAL TOOLS, JIGS, FIXTURES, GAGES_____ _____

CONDITIONS_____

MATERIAL_____

PART NO. _____DWG. NO. _____

PART DESCRIPTION_____

ACT BREAKDOWN		ELEM. NO.	SMALL TOOL NUMBERS, FEEDS, SPEEDS, DEPTH OF CUT, ETC.	ELEMENTAL TIME	OCC. PER CYCLE	TOTAL TIME ALLOWED
LEFT HAND	RIGHT HAND					

EACH PIECE_____ TOTAL

SET-UP HRS. PER C

FOREMAN INSPECTOR

OBSERVER APPROVED BY

Auxiliary equipment

The time study analyst will find it necessary to provide certain additional equipment to facilitate calculating studies rapidly and accurately. The foremost of these items is the electronic calculator, by means of which calculations involving multiplication, division, and proportion can be solved correctly and rapidly in a small fraction of the time required to work them out by longhand arithmetical procedures or slide rule computation.

Although, in most cases, modern machine tools, being self-powered, show their speeds in an obvious place, there will be occasions when the running speed will not be evident. Then, too, the speeds indicated by the manufacturer are based on pulley diameters which may have been altered during the setup or changed during the maintenance or overhaul of the machine.

In order to determine the speed being used, the time study analyst will find it convenient to use a speed indicator. This instrument has few parts, is simple in operation, and will give relatively accurate revolutions of shafting, wheels, and spindles in either direction.

Training equipment

Two inexpensive pieces of equipment should be available to assist in the training of time study men. The first of these, a random elapsed time describer, is illustrated in Figure 13–10. This device can be programmed (by means of a specially contoured cam) so that successive elements, in this case one through nine, are completed and so that each is accomplished in a known period. At the end of each element, a buzzer may sound and a bulb on the panel will light. The durations of the elements are recorded by the trainee as they take place. The trainee is signaled by the lighting of the bulb and by the buzzer at the end of each element. Either the buzzer or the light may be made inoperative if desired.

This exercise provides practice in reading the watch at terminal points and in recording the elapsed time. As an operator becomes proficient with data generated through the given cam, he is challenged by changing to a new cam capable of generating shorter elemental cycles. This device teaches the trainee the technique of using both sound and sight to identify terminal points. It also provides practice in handling two or more short elements that occur successively and are followed by a relatively long element.

A second helpful training tool is the metronome used by students of music. This device can be set to provide a predetermined number of beats per minute. For example, the metronome can be set to provide 104 beats per minute (the number of cards dealt per minute when dealing at a normal pace). By synchronizing the delivery of a card at a four-hand bridge table so that a card is delivered with each beat of the metronome,

FIGURE 13–10
Random elapsed time describer for training time study analysts

we are able to demonstrate normal pace. This speed of movement involving a series of reaches, grasps, moves, and releases can be easily identified with practice. To illustrate 80 percent performance, the instructor need only set his metronome to 83 beats per minute and then synchronize his card dealing accordingly. The instructor will find this device very helpful in demonstrating various levels of performance in card dealing. Once a new time study trainee becomes proficient in accurately rating hand movements involving the dealing of cards, he can make the transition to evaluating shop operations more easily.

TEXT QUESTIONS

1. What equipment is needed by the time study analyst to carry on a program of time study?
2. What features of the decimal minute watch make it attractive to time study analysts?
3. Where does the decimal minute watch (0.001 minute) have application?
4. How does the duration of one division on the decimal hour watch compare to one division on the decimal minute watch?
5. What is the principal advantage of the time-recording machine?
6. What is the memomotion procedure? What are its advantages and limitations?

7. Why is it desirable to provide space for an operator process chart on the time study form?

8. Describe the speed indicator and how it is used.

9. How many feet of film would be exposed on a five-minute cycle while using the memomotion technique?

10. How do government specifications of stopwatch equipment compare with the accuracy to which most stopwatches are produced?

11. Explain how a "random elapsed time describer" can be used to train new work measurement analysts.

12. How can the metronome be used as a training tool for performance rating?

GENERAL QUESTIONS

1. Explain the statement, "The criteria for success are the ability and personality of the time study analyst rather than the equipment he or she chooses to use."

2. Why hasn't industry adopted a standardized time study form?

3. Do you feel that the typical shopworker would favor the time study machine? Why or why not?

4. What would be the advantages of having a time study machine that could record and play back sound as well as record end points of progressive elements?

PROBLEMS

1. In order to demonstrate various levels of performance to a group of union stewards, the time study supervisor of the XYZ Company is using the metronome while dealing bridge hands. How many times per minute should the metronome beat to demonstrate the following levels of performance: 60 percent, 75 percent, 100 percent, 125 percent?

2. What would be the synthetic time value to reach 20 inches for a 15-pound casting, move it 15 inches, and place it in a fixture requiring 0.020 minutes position time and a grasp involving 0.004 minutes? See Chapter 19.

3. If government regulations permit an average deviation of 0.005 minutes per 30-second interval and a standard deviation of 0.001, what is the maximum deviation in seconds per minute that would be permitted at the 99 percent confidence level?

SELECTED REFERENCES

Barnes, Ralph M. *Motion and Time Study: Design and Measurement of Work.* 6th ed., New York: John Wiley & Sons, Inc., 1968.

Lowry, Stewart M.; Maynard, Harold B.; and Stegemerten, G. J. *Motion and Time Study,* 3d ed. New York: McGraw-Hill Book Co., 1940.

Mundel, M. E. *Motion and Time Study.* 4th ed. Englewood Cliffs, N.J.: Prentice-Hall, Inc., 1960.

Nadler, Gerald. *Work Design: A Systems Concept.* Rev. ed. Homewood, Ill.: Richard D. Irwin, Inc., 1970.

14

Elements of time study

The actual taking of a time study is both an art and a science. In order to be sure of success in this field, the analyst must have developed the art of being able to inspire confidence in, exercise judgment with, and develop a personable approach to everyone with whom he comes in contact. In addition, it is essential that his background and training have been such that he thoroughly understands and is able to perform the various functions related to each step in the taking of the study. These elements include selecting the operator, analyzing the job and breaking it down into its elements, recording the elapsed elemental values, performance rating the operator, assigning appropriate allowances, and working up the study.

Choosing the operator

The first approach to beginning a time study is made through the departmental foreman or the line supervisor. Upon review of the job in operation, both the foreman and the time study man should agree that the job is ready to be studied. If more than one operator is performing the work assignment on which the standard is to be established, several things should be taken into consideration in selecting just which operator to use in taking the study. In general, the operator who is average or somewhat above average in performance will give a more satisfactory study than will the low-skilled or highly superior operator. The average operator will usually perform the work consistently and systematically. His pace will tend to be in the approximate range of normal (see Chapter 15), thereby making it easier for the time study analyst to apply a correct performance factor.

Of course, the operator should be completely trained in the method being used and should possess a liking for his work and an interest in doing a good job. He should be familiar with time study procedures and practice, and have confidence in time study methods as well as in the time study analyst. It is desirable that the operator have a cooperative spirit, so that he will follow through willingly with suggestions made by both the foreman and the time study analyst.

At times, the analyst will have no choice as to whom he studies, in that the operation is performed by only one worker. In cases of this nature, the analyst must be very careful when he establishes his performance rating because the operator may be performing at either of the extreme ends of the rating scale. In one-worker jobs, it is also especially important that the method used be correct and that the analyst approach the operator tactfully.

Approach to the operator

The technique used by the time study man in approaching the selected operator will have much to do with the extent of the cooperation received. The operator should be approached in a friendly manner and be informed that the operation is to be studied. He should then be given the opportunity to ask any questions relative to such matters as the timing technique, method of rating, and application of allowances. In some instances, the operator may have never been studied before, and the time study man will find it worthwhile to answer all his questions frankly and patiently. He should encourage the operator to offer suggestions and, when the operator does so, should receive them willingly, thus recognizing that he respects the skill and knowledge of the operator.

The time study man should show interest in the worker's job, and at all times be fair and straightforward in his behavior toward the worker. This approach will establish the worker's confidence in the time study man's ability, and the analyst will find that the resulting respect and goodwill will not only help in establishing a fair standard, but will also make his future work assignments on the production floor pleasanter.

Analysis of materials and methods

Perhaps the most common mistake made by time study analysts is neglecting to make sufficient analysis and records of the method being studied. The time study form illustrated in Chapter 13 allows space for a sketch or photograph of the work area. If a sketch is made, it should be drawn to scale and should show all the details that affect the method. The sketch should clearly show the location of the raw material and finished parts bins with respect to the work area. Thus, distances that the operator must move or walk are clearly recorded. The location of all tools involved in the operation should be indicated, thus illustrating the motion pattern utilized in performing successive elements.

Immediately below the pictorical presentation of the method, space is frequently allotted for an operator process chart (see Chapter 6) of the method being studied. On high-activity work, it is recommended that this chart be completed before the actual timing of the operation begins. By completing this right- and left-hand chart, the analyst can entirely identify the method under study and can observe opportunities for methods improvement. Elemental breakdown of the study to be taken will be facilitated, and the analyst will be able to acquire a better conception of the skill being executed.

The value of completely identifying the method under study cannot be overemphasized. Since management usually guarantees a standard as long as the method studied is in effect, it is mandatory that the method studied be completely known. For example, after machining a casting, the operator may have been studied as placing the casting in a tote box on a pallet four feet from the work station. The element "lay aside finished casting" may have consumed 0.05 minute. Let us assume that some time after the standard was established the location of the skid was changed so that a chute and drop delivery could be used to remove the completed part. This could reduce the "lay aside finished casting" element to 0.01 minute. The savings of 0.04 minute per cycle would be substantial if the job could be completed in 0.40 minute or less. This small methods change could result in a loose rate that would be troublesome to the union, the company, and the worker's fellow employees. Unless a complete record of the method originally used to "lay aside finished casting" had been made, the time study department would have no authority to restudy the job.

More pronounced methods changes are often made without informing the time study department, such as changing the job to a different machine, increasing or decreasing feeds and speeds, or using different cutting tools. Changes of this nature, of course, seriously affect the validity of the original standard. Often the first time they are brought to the time study department's attention is when a grievance has been submitted that a rate is too tight, or the cost department complains about a loose standard. Investigation will frequently reveal that a change in method has been the cause of the inequitable standard. In order to know what part or parts of the job should be restudied, the analyst must have information as to the method used when the job was initially studied. If this information is not available and the rate appears too loose, the only recourse the analyst has is to let the loose rate prevail for the life of the job—a situation that will be resented by management—or to change the method again and then immediately restudy the job—a program that will be strongly criticized by the operator and his union.

Information on the type of material being processed should be recorded, as should information on the material used in cutting tools (ceramic, high-speed, carbide). A design change calling for a part to be made from "60–40 brass" rather than "70–30 brass" could have a pronounced effect

Motion and time study

on cycle time. Likewise, a change from high-speed tools to carbide tools could decrease machining time by more than 50 percent.

It has continually been emphasized that a job should not be time studied until it is *ready* to be studied, that is, until the method employed is correct. However, it has also been shown that methods must be continually improved in order to progress. A plant that does not continually emphasize methods improvement will fall into the doldrums and will eventually be unable to operate profitably. Since methods changes are continually taking place, it is necessary that a complete analysis of the existing materials and methods be made and recorded prior to beginning the actual watch readings.

Recording the significant information

Complete information should be recorded concerning the machines, hand tools, jigs or fixtures, working conditions, materials used, operation performed, name and clock number of the operator, department, date of the study, and time study observer's name. Perhaps all this detail will seem unimportant to the novice, but through experience he will realize that the more pertinent information he or she records, the more useful the study will be over the years to come. The time study will be a source of establishing standard data (see Chapter 18) and the development of formulas (see Chapter 20). It will also be useful in methods improvement, operator evaluation, tool evaluation, and machine performance.

When machine tools are used, the name, size, style, capacity, and serial or inventory number should be specified. For example, "No. 3 Warner & Swasey ram-type turret lathe, Serial No. 111408, equipped with 2-jaw S.P. air chuck" would identify the facility under use. Cutting tools should be described completely, such as "$\frac{1}{2}$-inch, two-flute, high-speed, 18–4–1 gun drill." Likewise, the descriptions of general facilities should be complete and well defined, such as "2-pound ball peen hammer," "1-inch micrometer," "combination square," "12-inch bastard file."

Dies, jigs, gages, and fixtures can be identified by their number and a short description. For example, "flange trimming die F–1156," "progressive trimming, piercing, forming die F–1202," "Go No-Go Thread Plug Gage F–1101."

The working conditions should be noted for several reasons. First, the prevailing conditions will have a definite relationship to the "allowance" added to the leveled or normal time. If conditions improve in the future, the allowance for personal time as well as fatigue may be diminished. Conversely, if for some reason it becomes necessary to alter the working conditions so that they become poorer than when the time study was first taken, it would follow that the allowance factor should be raised.

If the working conditions prevailing during the study are different from the normal conditions that exist on that job, then they would have an effect on the usual performance of the operator. For example, in a drop forge shop, if a study were taken on an extremely hot summer day, it can be readily understood that the working conditions would be poorer than they usually are, and operator performance would reflect the effect of the intense heat. The following examples will illustrate the description that should be included when recording the job conditions: "Normal for job, wet, hot (90° F), operator standing," or "poor, temperature 85° F, operator seated, clean, noisy."

It will be noted that it is customary to indicate first the working conditions as they compare with the average conditions for the job. This description is then followed by a short account of the exact conditions which were observed.

Raw materials are identified completely, giving such information as heat number, size and shape, weight, quality, and previous treatments.

The operation being performed is specifically described. For example, "broach ⅜ inch × ⅜ inch keyway in 1-inch bore" is considerably more explicit than the description "broach keyway." There could be several inside diameters on the part, each having a different keyway, and unless the hole that is being broached is specified and the keyway size is shown, misinterpretation may result.

The operator being studied should be identified by name and clock number. There could easily be two John Smiths in one company. Then, too, the clock number does not completely identify an employee, as labor turnover results in the assignment of the same clock number to more than one employee over a period of years.

The observer's position

Once the time study analyst has made the correct approach to the operator and has recorded all the significant information, he is ready to record the time consumed by each element.

The time study observer should take a position a few feet to the rear of the operator so as not to distract or interfere with him. It is important that the time study analyst stand while taking the study. A time study man who takes his studies while seated will be subject to criticism from the operators and will soon lose the respect of the production floor. Then, too, when standing, the observer is better able to move about and follow the movements of the operator's hands as the operator goes through his work cycle. Figure 14–1 illustrates the correct position of the observer with respect to the operator while studying a radial drill operation. It can be seen that the observer is in a position to have a simultaneous view of his watch, his time study form, and the hands of the operator.

During the course of the study, the observer should avoid any conversation with the operator, as this would tend to upset the routine of both the analyst and the worker.

FIGURE 14–1
Time study observer studying a radial drill press operation

Dividing the operation into elements

For ease of measurement, the operation is divided into groups of therbligs known as "elements." In order to divide the operation into its individual elements, the observer should watch the operator for several cycles. However, if the cycle time is relatively long (over 30 minutes), the
observer should write the description of the elements while taking the
study. If possible, the elements into which the operation is to be divided
should be determined before the start of the study. Elements should be
broken down into divisions that are as fine as possible and yet not so
small that accuracy of reading will be sacrificed. Elemental divisions of
about 0.04 minute are about as fine as can be read consistently by the experienced time study analyst. However, if the preceding and succeeding
elements are relatively long, an element as short as 0.02 minute can be
readily timed.

To identify end points completely and develop consistency in reading
the watch from one cycle to the next, elemental breakdown should consider sound as well as sight. Thus, the terminal points of elements can be
associated with the sounds made, such as when a finished piece hits the
container, when a facing tool bites into a casting, when a drill breaks
through the part being drilled, and when a pair of micrometers are laid
on a bench.

Each element should be recorded in its proper sequence and should include a basic division of work terminated by a distinctive sound or motion. Thus, the element "up part to manual chuck and tighten" would include the basic divisions reach for part, grasp part, move part, position part, reach for chuck wrench, grasp chuck wrench, move chuck wrench, position chuck wrench, turn chuck wrench, and release chuck wrench. The point of termination of this element would be evidenced by the sound of the chuck wrench being dropped on the head of the lathe. The element "start machine" could include reach for lever, grasp lever, move lever, and release lever. The rotation of the machine with the accompanying sound would identify the point of termination so that readings could be made at exactly the same point in each cycle.

Frequently, a standard elemental breakdown for given classes of facilities is adopted by the different time study analysts within a company to assure uniformity in establishing terminal points. For example, all single-spindle bench-type drill press work may be broken down into standard elements, and all lathe work may be composed of a series of pre-determined elements. Having standard elements as a basis for operation breakdown is especially important in the establishment of standard data (see Chapter 18).

The principal rules for accomplishing the elemental breakdown are:

1. Ascertain that all the elements being performed are necessary. If it develops that some are unnecessary, the time study should be discontinued and a methods study should be made to develop the proper method.
2. Keep manual and machine time separate.
3. Do not combine constants with variables.
4. Select elements so that terminal points can be identified by a characteristic sound.
5. Select elements so that they can be readily and accurately timed.

When dividing the job into elements, the analyst should keep machine or cutting time separate from effort or handling time. Likewise, constant elements (those elements for which the time will not deviate within a specified range of work) should be kept separate from variable elements (those elements for which the time will vary within a specified range of work).

Once the proper breakdown is made of all the elements constituting the operation, it is then necessary that each element be accurately described. The termination of one element is automatically the beginning of the succeeding element and is referred to as the "breaking point." The description of this terminal point or breaking point should be such that it will be readily recognized by the observer. This is especially important when the element does not include sound at its ending. On cutting elements, the feed, speed, depth, and length of the cut should be recorded immediately after the element description. Typical element descriptions are:

"Pk. up pt. from table & pos. in vise" and "Dr. ½″ D. 0.005″ F 1200 R.P.M." It will be noted that the analyst, in the interest of brevity, abbreviates and uses symbols to a great extent. This system of notation is acceptable only if the element is completely described in terminology and symbols meaningful to all who may come in contact with the study. Some companies use standard symbols throughout their plants, and everyone connected with a plant is familiar with the terminology.

When an element is repeated, it is not necessary to describe the element a second time, but only necessary to indicate in the space provided for a description of the element, the identifying number that was used when the element first occurred.

The time study form offers the flexibility required by diversified studies. For example, it will occasionally be impossible to record an element that successively repeats itself, due to the space limitations of the time study form. This can be handled by recording the watch readings of the repeating elements in the same column as the one in which the first occurrence of the element was recorded. Figure 14–2 illustrates this method of recording.

When there are more than 15 elements in the study, a second sheet should be used to record those elements in excess of 15. If more than 20 cycles are to be observed and the study comprises seven or fewer elements, then the right-hand half of the form can be utilized by repeating the elements and continuing the study in the open spaces. However, if more than seven elements occur and more than 20 cycles are to be recorded, a second sheet should be used. Figure 14–3 illustrates a 60-cycle study comprising four elements.

Taking the study

There are two techniques of recording the elemental time while taking the study. The continuous method, as the name implies, allows the stopwatch to run for the entire duration of the study. In this method, the watch is read at the breaking point of each element while the hands of the watch are moving. In the snapback technique, the watch is read at the termination point of each element, and then the hands are snapped back to zero. As the next element takes place, the hands move from zero. The elapsed time is read directly from the watch at the end of this element, and the hands are again returned to zero. This procedure is followed throughout the entire study.

When starting the study, the time study analyst should advise the operator that he is doing so, and should also let him know the exact time of day at which the study is being taken so that the operator will be able to verify the overall time. The time of day that the study is started is recorded on the time study form (see Figure 14–4), just before starting the stopwatch. Figures 14–4 and 14–5 illustrate a completed time study.

GURE 14–2

ethod of recording an element that successively repeats itself

ATE / /
UDY NO.—
EET NO.—
OF

EETS

CYCLE NO	1	2	3	4	5	6	7	8	9	10	11	12	13	14	15
1	5	12	16	26	35										
2			39	48	57										
3			62	71	80	101	9	19	48	57	62	68	201	11	32
4	37	44	49	60	71										
5			75	83	92										
6			97	307	17	36	43	52	81	90	93	400	35	44	63
7	69	77	82	91	501										
8			6	16	24										
9			27	38	47	66	78	89	608	17	23	29	59	69	88
10	93	701	5	14	24										
11			9	29	40										
12			44	52	64	84	93	805	35	46	52	58	85	94	920
13	25	33	37	47	56										
14			60	68	78										
15			83	96	1005	25	33	43	70	81	86	91	1125	35	55
16	60	68	73	82	91										
17			96	1206	15										
18			19	29	30	59	67	77	1302	11	17	23	54	64	84
19	90	97	1401	11	22										
20			26	37	47										

SUMMARY

OTALS															
BSER.															
VE. T.															
V. FACT															
F. C AV. T.															
% ALL.															
ME LLOWED															

FOREIGN ELEMENTS		GENERAL RATING CHECK	SYN. VAL. —— = %	REMARKS:
R	T		OBS. VAL.	
		STUDY STARTED	STUDY FINISHED / OVERALL TIME	
		A.M. —— P.M.	A.M. —— P.M. ——MIN.	
		F		
		G		
		H		

FIGURE 14–3
Time study form showing the recording of a four-element, 60-cycle study

DATE / /
STUDY NO.__
SHEET NO.__
OF __ SHEETS

NUMBER		1		2		3		4				1		2		3		4				1		2		3		4			
NOTES	CY. NO.	T	R F	T	R F	T	R F	T	R F	T	R F	T	R F	T	R F	T	R F	T	R F	T	R F	T	R F	T	R F	T	R F	T	R F	T	R
	1		4		14		20		23				64		73		80		83				53		65		72		75		
	2		27		36		42		46				87		97		503		6				80		91		97		1001		
	3		51		62		69		74				10		20		25		28				5		15		22		25		
	4		78		88		94		98				33		45		52		55				29		39		46		49		
	5		101		13		19		23				59		69		75		78				53		61		67		71		
	6		26		35		41		44				84		95		602		5				74		84		90		93		
	7		48		56		62		65				9		19		26		29				97		1106		13		16		
	8		69		79		85		88				34		43		49		52				20		30		37		40		
	9		91		200		5		9				55		63		68		72				44		54		60		63		
	10		13		22		28		31				76		91		98		702				67		76		82		85		
	11		35		45		51		54				5		15		21		24				89		99		1202		5		
	12		58		66		71		74				29		39		46		49				9		19		25		28		
	13		79		89		97		301				53		63		70		73				32		42		48		51		
	14		5		15		22		26				77		86		93		96				55		66		72		74		
	15		30		41		46		49				800		10		18		24				78		88		94		98		
	16		52		60		66		69				29		37		45		48				1302		12		18		21		
	17		73		83		89		92				52		62		69		72				25		35		41		44		
	18		96		406		12		15				76		85		91		94				49		61		68		71		
	19		19		29		35		38				98		908		15		18				75		85		91		95		
	20		42		51		57		60				23		35		45		48				1400		12		18		21		

SUMMARY

TOTALS														
OBSER.														
AVE. T.														
LEV. FACT														
L.F. & AV. T.														
% ALL.														
TIME ALLOWED														

FOREIGN ELEMENTS			GENERAL RATING CHECK	SYN. VAL. / OBS. VAL. = %			REMARKS:
	R	T					
A			STUDY STARTED	STUDY FINISHED	OVERALL TIME		
B			____ A.M. P.M.	____ A.M. P.M.	____ MIN.		
C			F				
D			G				
E			H				

FIGURE 14–4

Completed time study illustrating the continuous method of watch reading and overall performance rating (front of form)

		1	2	3	4	5	6	7	8	9	10	11	12	13	14	15

DATE 3/11
STUDY NO. 108
SHEET NO. 1 OF 1 SHEETS

SUMMARY

	TOTALS	OBSER.	AVE. T.	LEV. FACT	L.F.CAV.T.	%ALL.	TIME ALLOWED
1	2.80	20	.140	1.10	.154	15	.177
2	3.27	20	.163	1.10	.179	15	.206
3	.54	18	.047	1.10	.052	15	.060
4	12.62	18	.701	1.00	.701	10	.806
5	1.16	20	.058	1.10	.064	15	.711
6	3.61	19	.190	1.10	.208	15	.239
7	2.80	18	.147	1.10	.162	15	.186
8	1.99	20	.100	1.10	.110	15	.126
9	4.12	20	.206	1.10	.226	15	.260
10	.85	19	.045	1.10	.050	15	.058
11	1.84	19	.097	1.10	.107	15	.123
12	14.76	19	.777	1.00	.777	10	.855
13	1.81	19	.095	1.10	.104	15	.120
14	1.81	20	.090	1.10	.099	15	.114
15	3.67	20	.184	1.10	.202	15	.232

FOREIGN ELEMENTS

	R	T	
A	914 / 606	308	Get drink
B	1416 / 1234	182	Interrupted by foreman
C	2326 / 2275	51	Gage part
D	2914 / 2713	201	Chip in eye
E	3652 / 3041	611	Replace milling cutter

GENERAL RATING CHECK

SYN. VAL. #2 $\frac{.164}{.150} = 109\%$

OBS. VAL.

STUDY STARTED   10:32 A.M./P.M.
STUDY FINISHED 11:46 A.M./P.M.
OVERALL TIME   74 MIN.

F	
G	
H	

REMARKS:

FIGURE 14–5
Completed time study form (back of form)

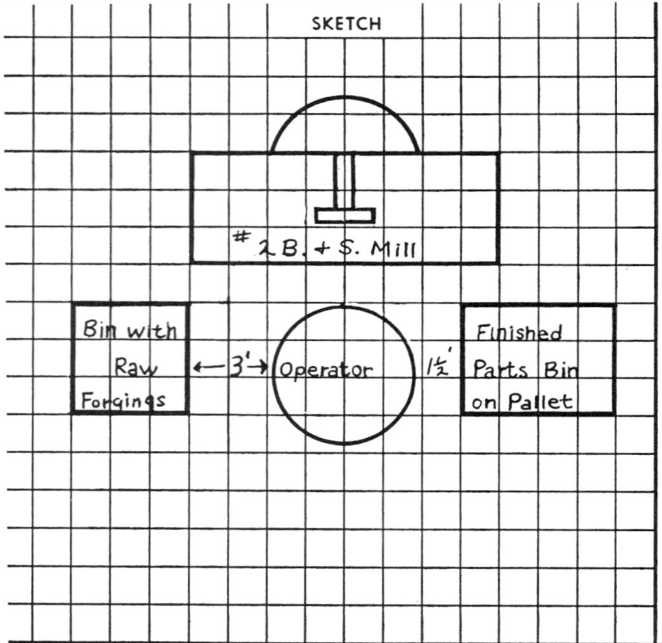

STUDY NO. 132   DATE 9-15

OPERATION Mill diagonal slots in regulator clamp

DEPT. 11-5   OPERATOR B.V. Harvey   NO. 14

EQUIPMENT #2 B. & S. Horizontal Mill

MCH. NO. 3-17694

SPECIAL TOOLS, JIGS, FIXTURES, GAGES Milling fixture
#E-14611, 18-4-1 high speed cutters

CONDITIONS Normal

MATERIAL X-4130 steel forging

PART NO. S-11694-1   DWG. NO. SA-11694

PART DESCRIPTION 2"clamp for series 411

ELEM. NO.	SMALL TOOL NUMBERS, FEEDS, SPEEDS, DEPTH OF CUT, ETC.	ELEMENTAL TIME	OCC. PER CYCLE	TOTAL TIME ALLOWED
1		.186	1	.186
2		.197	1	.197
3		.054	1	.054
4	95 R.P.M., 5⅝"/Min.	.763	1	.763
5		.076	1	.076
6		.252	1	.252
7		.189	1	.189

SKETCH

Bin with Raw Forgings   ←3'→ Operator

#2 B. & S. Mill

Finished 1½' Parts Bin on Pallet

ACT BREAKDOWN

LEFT HAND	RIGHT HAND
Reach 12", grasp (simple) forging on mill bed, move 12", hold.	Reach 8", grasp file, Move file to forging, file edge (3 strokes)
Position forging in vise	Move file 8", release. Reach 12" for vise handle, grasp handle, turn handle 90 to 180 degrees

Description	No.			95 R.P.M. 5⅛"/ Min.		
Move lever.	11	.117			1	.117
Get new piece from bin. Idle during mill cut	12	.846			1	.846
Reach for stop button, Reach 8" grasp control wheel, apply pressure. Move 45°, release wheel.	13	.124			1	.124
Reach 15" to part, grasp part, Reach 16" for vise handle, grasp Move part 8" and release. handle, turn handle 90-180°.	14	.117			1	.117
Reach 12" grasp brush, move brush to vise 12" and brush chips. Idle	15	.227			1	.227
Reach 8" to part, grasp, Move brush to table (12"), release Move 8" to vise, position brush reach for vise handle						
Grasp vise handle, turn handle 90 to 180 degrees.						
Idle: reach 8" grasp control wheel move 45 degrees and release						
Idle: Reach 14" for Start button, apply pressure.						
Reach 14" for control wheel, grasp wheel, turn wheel 15 degrees, release.						
Reach 18" grasp feed lever move lever Idle during mill cut Lay aside finished piece.						
Reach for control wheel Idle: 14", grasp wheel, turn 60 degrees.						
Reach 16" for stop button, Release control wheel, reach apply pressure. 12" and grasp vise handle, move handle 90°						
Reach for part 12", grasp, Reach 12", grasp brush, move Move part to table. brush to vise 12", brush release. chips, lay brush aside.						

EACH PIECE __3.593__ Min.

SET-UP _____

FOREMAN _W. H. Armstrong_

OBSERVER _George Thuning_

TOTAL __3.593__

INSPECTOR _C. A. Anderson_

APPROVED BY _B.W.M._

HRS. PER C __5.988__

The snapback method

The snapback method has certain advantages and disadvantages as compared with the continuous technique. These should be clearly understood before standardizing on one way of recording the values. In fact, some time study men find it desirable to use both methods, feeling that studies made up predominantly of long elements are adapted to snapback readings, while short-cycle studies are better studied with the continuous procedure.

Since elapsed elemental values are read directly in the snapback method, no clerical time is required for making successive subtractions, as is necessary with the continuous method. Then, too, elements that are performed out of order by the operator can be readily recorded without special notation. Proponents of the snapback method also bring out the fact that when using this procedure, it is not necessary to record delays, and that since elemental values can be compared from one cycle to the next, a decision can be made as to the number of cycles to study. Actually, it is erroneous to use observations of the past few cycles as a means of determining how many additional cycles should be studied. This practice can lead to studying entirely too small a sample.

W. O. Lichtner points out one recognized disadvantage of the snapback method, namely, that individual elements should not be removed from the operation and studied independently because elemental times are dependent upon the preceding and succeeding elements.[1] Consequently, by omitting such factors as delays, foreign elements, and transposed elements, erroneous values will prevail in the readings accepted.

One of the objections to the snapback method that has received considerable attention, particularly from labor groups, is the time that is lost while the hand is being snapped back to zero. Lowry, Maynard, and Stegemerten state: "It has been found that the hand of the watch remains stationary from 0.00003 to 0.000097 hour at the time of snapback, depending upon the speed with which the button of the watch is pressed and released."[2] This would mean an average loss of time of 0.0038 minute per element, which would be a 3.8 percent error in an element 0.10 of a minute in duration. Of course, the shorter the element, the greater the percent of error introduced, and the longer the element, the smaller the error. Although experienced time study men will tend in reading the watch to allow for the "snapback time" by reading to the next higher digit, it should be recognized that a sizable cumulative error can develop with the snapback method.

Another disadvantage of the snapback method is the tendency of the observer to become careless once he has established a value for the vari-

---

[1] W. O. Lichtner, *Time Study and Job Analysis* (New York: Ronald Press Co.).

[2] S. M. Lowry, H. B. Maynard, and G. J. Stegemerten, *Time and Motion Study and Formulas for Wage Incentives,* 3d ed. (New York: McGraw-Hill Book Co., 1940).

ous elements. He may anticipate what the reading will be and record the corresponding value without close attention to the true elapsed time.

In summary, the snapback procedure has the following disadvantages:

1. Time is lost in snapping back; therefore, a cumulative error is introduced into the study.
2. Short elements (0.06 minute and less) are difficult to time.
3. A record of the complete study is not always given in that delays and foreign elements may not be recorded.
4. Carelessness on the part of the time study man is promoted.
5. There is no verification of the overall time with the sum of the elemental watch readings.

The continuous method

The continuous method of recording elemental values is recommended for several reasons. Probably the most significant reason is that this type of study presents a complete record of the entire observation period and, as a result, appeals to the operator and his or her representatives. The operator is able to see that no time has been left out of the study and that all delays and foreign elements have been recorded. With all the facts clearly presented, it is easier to explain and sell this technique of recording times.

The continuous method is also better adapted to recording very short elements. Since no time is lost in snapping the hand back to zero, accurate values can be obtained on successive elements of 0.04 minute and on elements of 0.02 minute when followed by a relatively long element. With practice, a good time study man using the continuous method will be able to catch accurately three successive short elements (less than 0.04 minute) if they are followed by an element of about 0.15 minute or longer. He does this by remembering the watch readings of the terminal points of the three short elements and then recording their respective values while the fourth, longer element is taking place.

Of course, as previously pointed out, more clerical work is involved in calculating the study when the continuous method is used. Since the watch is read at the breaking point of each element while the hands of the watch continue their movements, it is necessary to make successive subtractions of the consecutive readings to determine elapsed elemental times. For example, the following readings might represent the terminal points of a 10-element study: 4, 14, 19, 121, 25, 52, 61, 76, 211, 16, and the elemental values of this cycle would be 4, 10, 5, 102, 4, 27, 9, 15, 35, and 5.

Recording the time consumed by each element

When recording the watch readings, the analyst notes only the necessary digits and omits the decimal point, thus giving as much time as possible to observe the performance of the operator. Thus, if a decimal minute watch is used, and the terminal point of the first element occurs at 0.08 minute, the analyst would record only the digit 8 in the "R" (reading) column of the time study form. If the decimal hour watch is used, and the end point of the first element is 0.0052, the recorded reading would be 52. Table 14–1 illustrates the procedure of using a decimal minute watch.

TABLE 14–1

Consecutive reading of watch in decimal minutes	Recorded reading
0.08	8
0.25	25
1.32	132
1.35	35
1.41	41
2.01	201
2.10	10
2.15	15
2.71	71
3.05	305
3.17	17
3.25	25

The small hand on the watch will indicate the number of elapsed minutes so that the observer can refer to it periodically to verify the correct first digit to record after the large hand sweeps past the zero. For example, after 22 minutes have passed in taking a given study, the observer may not recall whether the value to record after the termination of the element he is observing should be prefixed by "22" or "21." By glancing at the small hand of his watch, he will note that it has moved past the 22, thus letting him know that 22 is the correct prefix of the reading to be recorded.

All watch readings are recorded in consecutive order in the "R" column until the cycle is completed. Subsequent cycles are studied in a similar manner, and their elemental values are recorded.

Difficulties encountered

During the time study, the observer will encounter variations from the sequence of elements that he originally established, and occasionally he himself will miss specific terminal points. These difficulties tend to complicate the study, and the less often they occur, the easier it will be to calculate the study.

When the observer misses a reading, he should immediately indicate an "M" in the "R" column of the time study form. In no case should he make an approximation and endeavor to record the missed value, because this practice can destroy the validity of the standard established for the specific element. If the element were used as a source of standard data, appreciable discrepancies in future standards might result. Occasionally, the operator will omit an element, and this is handled by drawing a horizontal line through the space in the "R" column. This should happen infrequently, since it is a sign of an inexperienced operator or an indication of a lack of standardization of method. Of course, the operator can inadvertently omit an element—for example, by forgetting to "vent cope" in making a bench mold or by failing to "apply parting dust." If elements are omitted repeatedly, the analyst should stop the study and investigate the necessity of performing the omitted elements at all. This should be done in cooperation with the foreman and the operator so that the best method can be established. The observer is expected to be constantly on the alert for possibilities of better ways to perform the elements; as ideas come to his mind, he should jot them down in the "note" section of the time study form for future study and possible development.

Another variation with which the observer will be faced is the performance of elements out of sequence. This may happen fairly frequently when a new or inexperienced employee is studied on a long-cycle job made up of many elements. To avoid as many of these disturbances as possible is one of the prime reasons for studying a competent, fully trained employee. However, when elements are performed out of order, the observer should immediately go to the element that is being performed and draw a horizontal line through the middle of its "R" space; directly below this line is shown the time the operator began the element, and above it the completion time is recorded. This procedure is repeated for each element performed out of order and for the first element that is performed back in the normal sequence. Figure 14–4 illustrates a typical study, and elements 2, 3, and 4 on Cycle 6 show how the analyst handled the elements performed out of order.

During the time study the operator may encounter unavoidable delays, such as an interruption by a clerk or a foreman or tool breakage. Furthermore, the operator may intentionally cause a change in the order of work by going for a drink or stopping to rest. Interruptions of this nature occurring in the work cycle are referred to as "foreign elements."

Foreign elements can occur either at the breaking point or during the course of an element. The majority of foreign elements, particularly if they are controlled by the operator, occur at the termination of one of the elements constituting the study. When a foreign element occurs during an element, the observer will signify the event by an alphabetical designation in the "T" column of this element. If the foreign element occurs at the breaking point, the alphabetical designation will be recorded in the "T" column of the work element which follows the interruption. The letter *A* is used to signify the first foreign element, the letter *B* to signify the second, and so on.

As soon as the foreign element has been properly designated with an alphabetical symbol, a short description of it is made in the space provided immediately after the corresponding reference letter. The time that the foreign element begins is shown in the lower part of the "R" block of the foreign element section, and the watch reading at the termination of the foreign element is recorded in the upper part. These values can then be subtracted at the time the study is calculated, to determine the exact duration of the foreign element. This value is then shown in the "T" column of the foreign element section. Figure 14–4 illustrates the correct handling of several foreign elements.

Sometimes, investigation reveals that elements treated as foreign elements have a definite relation to the job being studied. In such cases, the elements should be considered irregular and the elapsed time should be leveled, the proper allowance added, and the result properly prorated to the cycle time so that a fair standard can be achieved.

Occasionally, a foreign element is of such short duration that it will be impossible to record it in the fashion outlined. Typical examples of this would be dropping a wrench on the floor and quickly picking it up, wiping one's brow with a handkerchief, or turning to speak briefly to the foreman. In cases similar to these examples, where the foreign element may be 0.06 minute or less, the most satisfactory method of handling the interruption is to allow it to accumulate in the element where it occurs and immediately circle the reading, indicating that a "wild" value has been encountered. A short comment should also be made in the "note" section of the time study form across from the element where the interruption occurred, thus justifying the circling procedure.

The number of cycles to study

The number of cycles that should be studied in order to arrive at an equitable standard is a subject that has caused considerable discussion among time study analysts as well as union representatives. Since the activity of the job, as well as its cycle time, directly influences the number of cycles that can be studied from an economic standpoint, one cannot be completely governed by the sound statistical practice that demands a certain size sample based on the dispersion of the individual element readings.

The General Electric Company has established Table 14–2 as a guide to the number of cycles to be observed.

The Westinghouse Electric Corporation has given consideration to activity as well as cycle time and has evolved the values shown in Table 14–3 as a guide for its time study men.

The mean of the sample of the observations should be reasonably close to the mean of the population. Thus, the analyst should take sufficient readings so that when their values are recorded a distribution of values

TABLE 14–2

Cycle time in minutes	Recommended number of cycles
0.10	200
0.25	100
0.50	60
0.75	40
1.00	30
2.00	20
2.00– 5.00	15
5.00–10.00	10
10.00–20.00	8
20.00–40.00	5
40.00–above	3

Source: Information taken from the Time Study Manual of the Erie Works of the General Electric Company, developed under the guidance of Albert E. Shaw, manager of wage administration.

TABLE 14–3

When the time per piece of cycle is over (hours)	Minimum number of cycles to study (activity)		
	Over 10,000 per year	1,000 to 10,000	Under 1,000
8.000	2	1	1
3.000	3	2	1
2.000	4	2	1
1.000	5	3	2
0.800	6	3	2
0.500	8	4	3
0.300	10	5	4
0.200	12	6	5
0.120	15	8	6
0.080	20	10	8
0.050	25	12	10
0.035	30	15	12
0.020	40	20	15
0.012	50	25	20
0.008	60	30	25
0.005	80	40	30
0.003	100	50	40
0.002	120	60	50
Under 0.002	140	80	60

with a dispersion characteristic of the population dispersion is obtained. Some concerns, in their training programs for time study analysts, have the observer take readings and plot the values to obtain a frequency distribution.[3] Although there is no assurance that the population of ele-

---

[3]For example, the Kurt Salmon Company.

mental times is normally distributed, still experience has shown that the variations in the performance of a given operator will approximate the normal bell-shaped curve (see Figure 14–6). The number of cycles that should be studied in order to assure a reli-

FIGURE 14–6
The normal frequency distribution

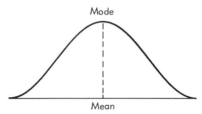

able sample may be determined mathematically, and this value, tempered with sound judgment, will give the analyst a helpful guide in deciding the length of observation.

Statistical methods may be utilized as an aid in determining the number of cycles to study. It is known that averages of samples ($\bar{x}$) drawn from a normal distribution of observations are distributed normally about the population mean μ. The variance of $\bar{x}$ about the population mean μ equals $\dfrac{\sigma^2}{n}$, where n equals the sample size and σ^2 equals the population variance.

Normal curve theory leads to the following confidence interval equation:

$$\bar{x} \pm z \frac{\sigma}{\sqrt{n}}$$

The above equation assumes that the population standard deviation is known. This, in general, is not true, but the population standard deviation may be estimated by the sample standard deviation s, where:

$$s = \sqrt{\frac{\displaystyle\sum_{i=1}^{i=n} (x_i - \bar{x})^2}{n - 1}}$$

or for computational purposes:

$$s = \sqrt{\frac{\Sigma x_i^2}{n - 1} - \frac{(\Sigma x_i)^2}{n(n - 1)}}$$

When estimating σ in this way, we are dealing with the quantity

$$\frac{\bar{x} - \mu}{\dfrac{s}{\sqrt{n}}}$$

which is not normally distributed except for large samples $(n > 30)$. Its distribution is the student's "t" distribution, which should be used in the formulas that follow. The confidence interval equation then is:

$$\bar{x} \pm t\,\frac{s}{\sqrt{n}} \tag{1}$$

For example, if 25 readings for a given element showed that $\bar{x} = 0.30$ and $s = 0.09$, there would be 95 percent confidence that μ would be contained in the interval 0.337 to 0.263 or that $\bar{x}$ is within ± 12.3 percent of μ. (See Table A3–3, Appendix 3, for values of t.)

$$\bar{x} \pm t\,\frac{s}{\sqrt{n}}$$

$$0.30 + 2.06\,\frac{0.09}{5} = 0.337$$

$$0.30 - 2.06\,\frac{0.09}{5} = 0.263$$

$$\frac{0.037}{30} = 12.3 \text{ percent}$$

If the accuracy is deemed unsatisfactory when equation (1) is utilized in the above manner, it is possible to solve for N, the required number of readings for a given accuracy, by equating $\dfrac{t\,s}{\sqrt{N}}$ to a percentage of $\bar{x}$:

$$\frac{ts}{\sqrt{N}} = k\bar{x}$$

$$N = \left(\frac{st}{k\bar{x}}\right)^2$$

where k = an acceptable percentage of $\bar{x}$.

In the above example, if $\bar{x}$ is required to be within ± 5 percent of μ with 95 percent confidence:

$$N = \left(\frac{st}{k\bar{x}}\right)^2$$

$$= \left(\frac{(0.09)(2.06)}{(0.05)(0.30)}\right)^2$$

$$= 152 \text{ observations}$$

If n readings have been taken on a time study, the selection of the most appropriate element for computing the desired number of readings may be a problem. It is recommended that the element having the greatest coefficient of variation $\frac{s}{\bar{x}}$ be selected for this purpose. It is also possible to solve for N before taking the time study by interpreting historical data of similar elements or by actually estimating $\bar{x}$ and s from several snapback readings of certain elements to be studied.

One concern has equated the number of observations in terms of the range of the cycle time and the average cycle time.[4] This has been expressed as:

$$N = 205 \frac{R}{\bar{x}} - 42$$

where:

N = Number of observations required
R = Range of cycle time
$\bar{x}$ = Mean cycle time

This has been derived as follows:

$$R = Ks$$

where:

K = average ratio between the range and the
standard deviation in a succession of samples
s = Standard deviation of the sample taken

To conform to the desired limits of 95 percent probability and 3 percent permissible error, the possible error, As, and the permissible error, $.03\bar{x}$, are assumed to be equal.
Therefore:

$$As = 0.03\bar{x}$$

$$s = \frac{0.03\bar{x}}{A}$$

where:

A = Probability factor at 95 percent

Substituting for s in the range formula:

$$R = \frac{(K)(0.03\bar{x})}{A}$$

The values of K and A vary with the number of observations (N) taken.

[4] A. C. Spark Plug Division, General Motors Corporation.

Substituting in the formula

$$R = \frac{(K)(0.03\bar{x})}{A}$$

the representative values in Table 14–4 are shown.

TABLE 14–4

N	K*	A†	$\bar{x} = 0.10$ minute	$\bar{x} = 0.20$ minute
40	4.3216	0.3222	0.0402	0.0805
50	4.4982	0.2859	0.0472	0.0944
75	4.8060	0.2310	0.0624	0.1248
100	5.0152	0.1990	0.0756	0.1512
125	5.1727	0.1774	0.0875	0.1749
150	5.2985	0.1617	0.0983	0.1967
175	5.4029	0.1494	0.1085	0.2169
200	5.4921	0.1396	0.1180	0.2360

* *K* values obtained from *Biometrika* by L. H. C. Tippett.
† *A* values obtained from *Statistical Methods for Research Workers* by R. A. Fisher.

On graph paper, plotting R (range) against $\bar{x}$ (average time) for various samples (N), we find a straight-line relationship from the formula $R = M\bar{x} + B$. Therefore, for sample sizes of 40, 150, 200, we get slope M:

$$N = 40 \qquad M = 0.402 \qquad B = 0$$
$$N = 150 \qquad M = 0.983 \qquad B = 0$$
$$N = 200 \qquad M = 1.180 \qquad B = 0$$

To introduce N into the above formula, N was plotted against M for various sample sizes, and the resulting straight-line formula is:

$$N = M_1 M + B$$
$$N = 205M - 42 \text{ (approximate)}$$

From the formula $R = M\bar{x} + B$, since $B = 0$ we get M in terms of R and $\bar{x}$ as:

$$M = \frac{R}{\bar{x}}$$

Substituting this expression from M in the formula $N = M_1 M + B$, we get

$$N = 205\frac{R}{\bar{x}} - 42$$

Given an equation involving N (number of observations), R (range of cycle time), and $\bar{x}$ (average cycle time), an alignment chart may be designed to facilitate the solution of N. The N alignment chart illustrated in Figure 14–7 is evolved as follows:

Rearranging the equation to conform to the general N chart equation:

$$\frac{205R}{\bar{x}} - N = 42$$

N varies from 0 to 240
R varies from 0 to 0.30
$\bar{x}$ varies from 0 to 0.75

Comparing with the general alignment chart equation

$$Af(u) + Bf(v) = Cf(w) + CK$$

Then:

$$R = f(u)$$
$$N = f(v)$$
$$O = f(w)$$
$$A = \frac{205}{\bar{x}}$$
$$B = -1$$
$$C = 1$$
$$K = 42$$

Select chart width 8 inches, chart height 12 inches.

FIGURE 14–7
N-type alignment chart for determining the number of time study observations required (above 40) for 95 percent probability and 3 percent error

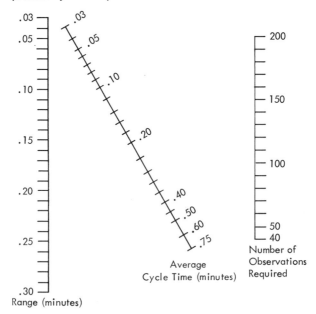

A. C. Spark Plug Division, General Motors Corp.

Develop scale modulus:

$$M_x = \frac{\text{Length of scale}}{\text{Range of } f(u)} = \frac{12}{0.30} = -40$$

$$\text{(when } B \text{ is negative, } M_x \text{ is negative)}$$

$$M_y = \frac{\text{Length of scale}}{\text{Range of } f(v)} = \frac{12}{240} = 0.05$$

Calculate A:

$$A = \frac{BDM_x}{AM_y + BM_x} = \frac{(-1)(8)(-40)}{\dfrac{205}{\bar{x}}(0.05) + (-1)(-40)} = \frac{0.320}{\dfrac{10.25}{\bar{x}} + 40}$$

Calculate E:

$$E = \frac{CM_xM_y(f(w) + K)}{AM_y + BM_x} = \frac{(1)(-40)(0.05)(42)}{\dfrac{205}{\bar{x}}(0.05) + (-1)(-40)}$$

$$E = \frac{-84}{\dfrac{10.25}{\bar{x}} + 40}$$

The scale equations:

$$X = M_xf(u) = -40R$$
$$Y = M_yf(v) = 0.05N$$

Values are then developed in tabular form as shown in Table 14–5.

TABLE 14–5
Sample

R	$x = -40R$	N	$y = 0.05N$	$\bar{x}$	$A = \dfrac{320}{\dfrac{10.25}{\bar{x}} + 40}$	$E = \dfrac{-84}{\dfrac{10.25}{\bar{x}} + 40}$
0.	0	0	0	0	—	—
0.05.	-2	40	2	.05	1.306	-0.343
0.20.	-8	150	7.5	.40	4.876	-1.280
0.30.	-12	240	12.0	.75	5.963	-1.565

Using these values, the N chart is laid out and plotted.

As has been pointed out, these mathematical derivations represent only a guide to the analyst and do not replace common sense and good judgment, which are still fundamental if the time study analyst is to do a good job in a practical manner.

The reader might rightfully ask, "When should the recommended number of cycles be observed?" If 30 cycles are to be taken, should they be taken as one successive group, as two groups of 15, or as three groups of 10? The reader should intuitively recognize that the mean of three groups of 10 taken at random times during the day will probably give a better estimate of the population mean than will one group of 30. The taking of a time study is a sampling procedure, and the average of several small samples usually provides more reliable estimates of parent parameters than does one sample of a size equivalent to the total of the several small samples. For example, if we had a barrel of 10,000 bolts, we would be more confident of the time quality of the entire 10,000 if we averaged the results from three independent samples. Thus, we might take one sample of 40 from one end, another sample of 40 from the middle, and a third sample of 40 from the other end. This would be a more reliable procedure than would taking a sample of 120 from one end.

The individual samples taken should be large enough so that their mean is somewhat stabilized. When stabilization occurs, the addition of one more cycle will not substantially alter the mean. For example, we might have the following averages given in Table 14–6.

TABLE 14–6

Average at the end of the	Average	Average at the end of the	Average
1st cycle	0.261	11th cycle	0.281
2d cycle	0.280	12th cycle	0.280
3d cycle	0.248	13th cycle	0.281
4th cycle	0.251	14th cycle	0.281
5th cycle	0.274	15th cycle	0.280
6th cycle	0.261	16th cycle	0.281
7th cycle	0.264	17th cycle	0.281
8th cycle	0.283	18th cycle	0.281
9th cycle	0.279	19th cycle	0.281
10th cycle	0.280	20th cycle	0.281

In Table 14–6 it is apparent that the average was stabilizing at the end of the 10th cycle and that nothing really practical was accomplished by the analyst in observing more than eleven cycles.

Rating the operator's performance

Before the observer leaves the work station, he should have given a fair performance rating to the study. On short-cycle, repetitive work, it is customary to apply one rating to the entire study. However, where the elements are of long duration and entail diversified manual movements, it is more practical to evaluate the performance of each element as it occurs during the study. The time study form illustrated in Figures 14–4 and 14–5 makes provision for both the overall rating and the individual element rating.

Since the actual time that was required to perform each element of the study was dependent to a high degree upon the skill and effort of the operator, it is necessary to adjust the time of the good operator up to normal and the time of the poor operator down to normal.

In the performance rating or leveling system, the observer evaluates the effectiveness of the operator in terms of his conception of a "normal" operator performing the same element. This effectiveness is given a value expressed as a decimal or percentage and assigned to the element observed. A "normal" operator is defined as a qualified, thoroughly experienced operator working under conditions as they customarily prevail at the work station, at a pace neither too fast nor too slow but representative of average. The normal operator exists only in the mind of the time study analyst, and the concept has been developed as a result of thorough and exacting training and experience in the technique of measuring wide varieties of work.

The basic principle of performance rating is to adjust the actual observed mean time for each acceptable element performed during the study to the time that would be required by the normal operator to perform the same work. To do a fair job of rating, the time study man must be able to disassociate himself from personalities and other varying factors, and consider only the amount of work being done per unit of time as compared to the amount of work that the normal operator would produce. Chapter 15 explains more fully the performance rating or leveling techniques in common use.

Apply allowances

It would be impossible for an operator to maintain an average pace for every minute of the working day, just as it would be impossible for a football game to entail 60 minutes of actual, continuous play time. Three classes of interruptions take place occasionally for which extra time must be provided. The first of these is personal interruptions, such as trips to the rest room and drinking fountain; the second is fatigue, which, as we all know, affects the strongest individual even on the lightest type of work. Lastly, unavoidable delays take place for which some allowances must be made, such as tool breakage, interruptions by the foreman, and slight tool trouble.

Since the time study is taken over a relatively short period, and since foreign elements have been removed in determining the normal time, an allowance must be added to the leveled or base time to arrive at a fair standard that can be achieved by the normal operator when exerting average effort. Chapter 16 outlines in detail the means for arriving at realistic allowance values.

Calculate the study

Once the analyst has properly recorded all the necessary information on the time study form, has observed an adequate number of cycles, and has properly performance rated the operator, he should thank the operator and proceed to the next step, which is computation of the study. In some plants, clerical help is used to compute the study, but in most cases the analyst himself does the work. The analyst usually prefers to analyze his own study because he is vitally interested in the resulting standard and does not care to let other individuals tamper with his calculations. Then, too, less chance of error will be risked if the analyst computes his own study, in that he will encounter no difficulty in interpreting his own notes and reading his own recorded values.

The initial step in computing the study is to verify the final stopwatch reading with the overall elapsed clock reading. These two values should check within plus or minus a half minute, and if a sizable discrepancy is noted, the analyst should check the study stopwatch readings for error.

When using the continuous method, a subtraction is made between the watch reading and the preceding reading, giving the elapsed time, and this is recorded in ink or red pencil. This procedure is followed throughout the study, each reading being subtracted from the succeeding one. It is important that the analyst be especially accurate in this phase. Carelessness at this point can completely destroy the validity of the study. If elemental performance rating has been used, a supplementary work sheet will be required to record the elapsed elemental time, and these values, after being multiplied by the leveling factor, will be recorded in ink or red pencil in the "T" spaces of the time study form.

Elements that have been missed by the observer are signified by placing an "M" in the "R" column, as previously explained. When computing the study, the analyst must disregard both the missed element and the succeeding one, since the subtracted value in the study will include the time for performing both elements.

Elements missed by the operator may be disregarded since they have no effect on the preceding or succeeding values. Thus, if the operator happened to omit element 7 of cycle 4 in a 30-cycle study, the analyst should have but 29 values of element 7 with which to calculate the mean observed time.

To determine the elapsed elemental time on out-of-order elements, it is merely necessary for the analyst to subtract the value appearing in the lower half of the "R" block from the value in the upper half.

For foreign elements, it is necessary to deduct the time required for the foreign element from the cycle time of the element in which it occurred. The time taken by the foreign element is obtained by subtracting the lower reading in the "R" space of the foreign element section of the time study form from the upper reading.

After all elapsed times have been calculated and recorded, they should be carefully studied for abnormality. There is no set rule for determining the degree of variation permitted to keep the value for calculation. If a broad variation on a certain element can be attributed to some influence that was too brief to be handled as a foreign element, yet long enough to affect substantially the time of the element, such as dropping a tool or blowing the nose, or if the variation may be attributed to errors in reading the stopwatch, then these values should be immediately circled and excluded from further consideration in working up the study. However, if wide variations are due to the nature of the work, then it would not be wise to discard any of these values.

Elements that are "paced" by the facility or process will have little variation from cycle to cycle, while considerably wider variation in "operator-controlled" elements would be expected. When unexplainable variations in time occur, the analyst should be quite careful before circling any such values. He must be aware that this is not a performance rating procedure, and by arbitrarily discarding high values or low values, he may end up with an incorrect standard. A good rule is, "When in doubt, do not discard the value."

If elemental rating is being used, then after the elemental elapsed time values have been computed, the normal elemental time is determined by multiplying each elemental value by its respective performance factor. This normal time is then recorded for each element in the "T" columns. The mean elemental normal value is determined next by dividing the number of observations into the total of the times in the "T" columns.

After the analyst has determined all elemental elapsed times, he should make a check to assure that no arithmetic errors have been made. This can best be done by adding the "total" column for each element and the foreign and abnormal element times. The total of these values must be equal to the last reading of the stopwatch. If this total does not equal the last watch reading, the study should be checked for arithmetic errors before proceeding with further computation. This check applies only when the continuous method is being employed.

If the overall study is leveled, then the means of the elapsed elemental times are computed, and the performance factor is applied to these values in the space provided to determine the various normal elemental time values.

After the normal elemental times have been evolved, the percentage allowance is added to each element to determine the allowed time. The nature of the job will determine the amount of allowance to be applied, as will be shown in Chapter 16. Suffice to say at this point that 15 percent is an average allowance for effort elements, and that 10 percent is representative of the allowance applied to process-controlled elements.

Upon determining the allowed time for each element, the analyst should then summarize these values in the space provided on the reverse side of the time study form to obtain the allowed time for the entire job. This is commonly referred to as the "standard" or the "rate" for the job.

Summary: Computing the study

To summarize, the steps taken in computing a typical study with continuous watch readings and overall performance rating are as follows:

1. Make subtractions of consecutive readings to obtain elemental elapsed times, and record in red pencil.
2. Circle and discard all abnormal or "wild" values where an assignable cause is evident.
3. Summarize the remaining elemental values.
4. Determine the mean of the observed values of each element.
5. Determine the elemental normal time by multiplying the performance factor by the mean elapsed time.
6. Add the appropriate allowance to the elemental normal values to obtain the elemental allowed times.
7. Summarize the elemental allowed times on the reverse side of the time study form to obtain the standard time. (Elements occurring more than once per cycle need be shown only once with the number of occurrences and the resulting product.)

If elemental leveling has been used on a continuous study, then the steps taken in computing the study would be:

1. Make subtractions of consecutive readings to obtain elemental elapsed times.
2. Determine the normal times of each individual element by multiplying the performance factor by the elapsed time value, and record them in red pencil in the "T" column.
3. Circle and discard all abnormal or "wild" values where an assignable cause is evident.
4. Determine the mean of the elemental normal times.
5. Add the appropriate allowance to the elemental normal values to obtain the elemental allowed times.
6. Summarize the elemental allowed times on the reverse side of the time study form to obtain the standard time. (Elements occurring more than once per cycle need be shown only once with the number of occurrences and the resulting product.)

TEXT QUESTIONS

1. Outline the characteristics that the operator should have if he or she is selected for time study.
2. What are the principal responsibilities of the time study observer?

3. What are some of the qualifications of the successful time study analyst?

4. How does the time study analyst "directly affect the pocketbook of the worker"?

5. Explain the statement that all foremen are "management's representatives throughout the plant."

6. Why does the foreman require fair time standards in his department?

7. What are the main responsibilities of the foreman in relation to time study work?

8. What is the significance of the foreman's signature on the time study?

9. What is a fair day's work?

10. Explain how having an appreciation course in the elements of time study made available to all employees will promote good labor relations.

11. What considerations should be given to the choice of the operator to be studied?

12. Of what use is the operator process chart on the time study form?

13. Why is it essential to record complete information on tools and the facility on the time study form?

14. Why are working conditions important in identifying the method being observed?

15. Why would a time study analyst who is hard of hearing have difficulty in recording his stopwatch readings of terminal points?

16. Differentiate between constant and variable elements. Why should they be kept separate when dividing the job into elements?

17. Explain how repeating elements may be recorded on the time study form.

18. What advantages does the continuous method of watch recording offer over the snapback method?

19. Why is the time of day recorded on the time study form?

20. What variations in sequence will the observer occasionally encounter during the course of the time study?

21. Explain what a foreign element is and how foreign elements are handled under the continuous method.

22. What factors enter into the determination of the number of cycles to observe?

23. Why is it necessary to performance rate the operator?

24. When should individual elements of each cycle be rated?

25. Define a "normal" operator.

26. For what reasons are allowances applied to the normal time?

27. What is the significance of a "circled" elapsed time?

28. What steps are taken in the computation of a time study conducted by the continuous overall performance rating procedure?

29. Based upon the Westinghouse guide sheet, how many observations should be taken on an operation whose annual activity is 750 pieces and whose cycle time is estimated at 15 minutes? What would be the number of observations needed according to the General Electric guide sheet?

GENERAL QUESTIONS

1. What do you feel the difference is between (*a*) a normal operator, (*b*) an average operator, and (*c*) a standard operator?
2. Why are time study procedures the only techniques for supplying reliable information on standard times?
3. To what positions of importance does time study work lead?
4. In what way can "loose" time standards result in poor labor relations?
5. Why are poor time standards a "headache" to union officials?
6. How would you approach a belligerent operator if you were the time study analyst?
7. You were using the General Electric guide sheet to determine the number of observations to study. It developed that 10 cycles were required, and after taking the study you used the standard error of the mean statistic to estimate the number of observations needed for a given confidence level. The resulting calculation indicated that 20 cycles should be studied. What would be your procedure? Why?

PROBLEMS

1. What would be the required number of readings if the analyst wanted to be 87 percent confident that the mean observed time was within ±5 percent of the true mean and he established the following values for an element after observing 20 cycles: 0.09, 0.08, 0.10, 0.12, 0.09, 0.08, 0.09, 0.12, 0.11, 0.12, 0.09, 0.10, 0.12, 0.10, 0.08, 0.09, 0.10, 0.12, 0.09?
2. The following data resulted from a time study taken on a horizontal milling machine:

   Pieces produced per cycle: 8.

   Mean measured cycle time: 8.36 minutes.

   Mean measured effort time per cycle: 4.62 minutes.

   Mean rapid traverse time: 0.08 minutes.

   Mean cutting time (power feed): 3.66 minutes.

   Performance factor: +15.

   Machine allowance (power feed): 10 percent.

   Effort allowance: 15 percent.

   An operator works on the job a full eight-hour day and produces 380 pieces.
   a. How many standard hours does the operator earn?
   b. What is his efficiency for the day?
   c. If his base rate is $5 per hour, compute his earnings for the day if he is paid in direct proportion to his output.
3. A work measurement analyst in the Dorben Company took 10 observations of a high-production job. He performance rated each cycle and then computed the mean normal time for each element. The element with the greatest dispersion had a mean of 0.30 minutes and a standard deviation

of 0.03 minutes. If it is desirable to have sampled data within ±5 percent of the true data, how many observations should this time study analyst take of this operation?

4. In the Dorben Company, the work measurement analyst took a detailed time study of the making of shell molds. The third element of this study had the greatest variation in time. After studying nine cycles, the analyst computed the mean and standard deviation of this element, with the following results:

$$\bar{x} = 0.42 \qquad S = 0.08$$

If the analyst wanted to be 90 percent confident that the mean time of his sample was within ±10 percent of the mean of the population, how many total observations should he have taken? Within what percent of the average of the total population is $\bar{x}$ at the 95 percent confidence level under the measured observations?

SELECTED REFERENCES

Barnes, Ralph M. *Motion and Time Study: Design and Measurement of Work.* 6th ed. New York: John Wiley & Sons, Inc., 1968.

Mundel, Marvin E. *Motion and Time Study: Principles and Practices,* 4th ed. Englewood Cliffs, N.J.: Prentice-Hall, Inc., 1960.

Nadler, Gerald. *Work Design: A Systems Concept.* Rev. ed. Homewood, Ill.: Richard D. Irwin, Inc., 1970.

15

Performance rating

While the time study observer is taking the study, he or she will carefully observe the performance of the operator during the entire course of the study. Seldom will the performance being executed conform to the exact definition of "normal," often referred to as "standard." Thus, it is essential that some adjustment be made to the mean observed time in order to derive the time required for the normal operator to do the job when working at an average pace. The actual time taken by the above-standard operator must be increased to that required by the normal worker, and the time taken by the below-standard operator must be reduced to the value representative of normal performance. Only in this manner can a true standard be established in terms of a normal operator.

Performance rating is probably the most important step in the entire work measurement procedure. Certainly, it is the step most subject to criticism, since it is based entirely upon the experience, training, and judgment of the work measurement analyst. Because of its importance, this chapter will bring out not only the most accepted techniques used today but will also review some of the more important historical methods, such as the Westinghouse system. An understanding of the historical philosophies and the resulting methods will be helpful to the student and practitioner in developing an appreciation for this very important step in the systematic procedure for methods and work measurement.

Definition

Performance rating is a technique for equitably determining the time required to perform a task by the normal operator after the observed values of the operation under study have been recorded. In Chapter 14, a "normal" operator was defined as a qualified, thoroughly experienced operator who is working under conditions as they customarily prevail at the work station, at a pace that is neither too fast nor too slow, but representative of average.

There is no one universally accepted method of performance rating,

340

although the majority of the techniques used are based primarily upon the judgment of the time study man. For this reason, the time study analyst must have the high personal characteristics discussed earlier in this text. No other component of the time study procedure is subject to as much controversy and criticism as the performance rating phase. Regardless of whether the rating factor is based upon the speed or tempo of the output or upon the performance of the operator when compared to that of the normal worker, judgment is still the criterion for the determination of the rating factor. Some attempts have been made to accumulate fundamental motion data and to compare these values to the mean observed values so as to arrive at a factor that can be used for the evaluation of the entire study. This technique, known as "synthetic rating," will be discussed later in this chapter.[1]

The concept of normal performance

Just as there is no universal method of performance rating, so there is no universal concept of normal performance. In general, a company engaged in the manufacture of low-cost, highly competitive products will have a "tighter" conception of normal performance than will a company producing a line of products protected by patents.

In defining what is meant by "normal" performance, it is helpful to enumerate benchmark examples that are familiar to all. When this is done, care must be taken that method and work requirements are carefully defined. Thus, if the benchmark of dealing 52 cards in 0.50 minute is established, a complete and specific description should be given of the distance of the four hands dealt, with respect to the dealer, and of the technique of grasping, moving, and disposing of the cards. Likewise, if the benchmark of 0.38 minute is established for walking 100 feet (3 miles per hour), it should be clearly explained whether this is walking on level ground or not, whether this is walking with or without load, and if with load, how heavy the load. Supplementing the benchmark examples should be a clear description of the characteristics of an employee carrying out a normal performance. A representative description of such an employee might be:

A workman who is adapted to the work, and has attained sufficient experience to enable him to perform his job in a workmanlike manner with little or no supervision. He possesses coordinated mental and physical qualities which enable him to proceed from one element to another without hesitation or delay, in accordance with the principles of motion economy. He maintains a good level of efficiency by his knowledge and proper use of all tools and equipment related to his job. He cooperates and performs at a pace best suited for continuous performance.

---

[1] Robert Lee Morrow, *Time Study and Motion Economy* (New York: Ronald Press Co., 1946).

The more clear-cut and specific the definition of normal, the better the definition will be. The definition should clearly describe the skill and effort involved in the performance, so that all the workers in a plant will have a comprehensive understanding of the concept of normal that prevails in the plant.

Uses and effects of the concept of normal performance

Even though personnel departments endeavor to provide only "normal" or "above normal" employees for each position within the company, still individual differences do exist. These differences among a given group of employees within the plant can become more pronounced as time goes on. Differences in inherent knowledge, physical capacity, health, trade knowledge, physical dexterity, and training will cause one operator to outperform another consistently and progressively. The degree of variation in the performance of different individuals has been estimated to approximate the ratio of 1 to 2.25.[2] Thus, in a random selection of 1,000 employees, the frequency distribution of the output would approximate the normal curve, with less than three cases on the average falling outside the three-sigma limits (99.73 percent of the time). If 100 percent should be taken as normal, then based upon a ratio of the slowest operator to the fastest as 2.25 to 1, $\bar{x} - 3_\sigma$ could equal 0.61, and $\bar{x} + 3_\sigma$ would then equal 1.39. This would mean that 68.26 percent of the people would be within a limit of plus or minus one sigma, or between performance rating values of 0.87 and 1.13. Graphically, the expected total distribution of the 1,000 people would appear as shown in Figure 15–1.

FIGURE 15–1
Normal distribution curve of the output of 1,000 people selected at random

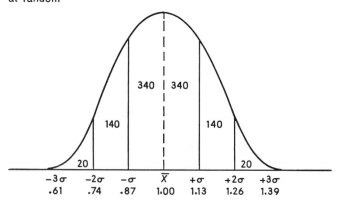

[2] Ralph W. Presgrave, *The Dynamics of Time Study,* 2d ed. (New York: McGraw-Hill Book Co., 1945).

Since employees are usually carefully screened by personnel departments before being assigned to specific jobs, the mean of any selected group may exceed the 100 percent figure representative of a sample taken arbitrarily from the population, and the dispersion of their output will probably be considerably less than the 2.25-to-1 ratio. In fact, some concerns feel that by using careful testing programs their selection of the right person for the job, supplemented by intensive training in the correct method of performance, will result in similar output within close limits by different operators assigned to the same job.

In the majority of cases, significant differences in output will prevail among those assigned to a given class of work, and it is necessary to adjust the performance of the operator being studied to a predetermined concept of normal.

The learning curve

Industrial engineers, human engineers, and other professionals interested in the study of human behavior have recognized that learning is time dependent. Even the simplest of operations may take hours to master. In complicated work, it may take days and even weeks before the operator can achieve coordinated mental and physical qualities enabling him or her to proceed from one element to another without hesitation or delay. A typical learning curve is illustrated by Figure 15–2.

Once the operator reaches the flattening section of the learning curve (as pictured in Figure 15–2), the problem of performance rating is simplified. However, it is not always convenient to wait this long in the

FIGURE 15–2
Typical productivity increase graph

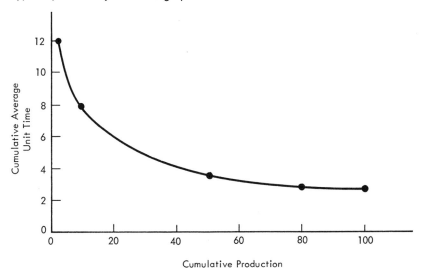

Cumulative Production

development of a standard. The analyst may be obliged to establish the standard at the point in the learning curve where the slope is greatest. It is in such cases that the analyst must have acute powers of observation and be able to execute mature judgment based upon thorough training so that an equitable normal time is computed.

It is helpful for the analyst to have learning curves available which are representative of the various classes of work being performed in the company. This information can be useful, not only in determining at what stage in the production it would be desirable to establish the standard, but also by providing a guide as to the expected level of productivity that the average operator with a known degree of familiarity of the operation will achieve after producing a fixed number of parts.

By plotting learning curve data on logarithmic paper, the analyst may be able to linearize the data and thus have it in a form that is easier to use. For example, by plotting both the dependent variable (cumulative average unit time) and the independent variable (cumulative production) shown in Figure 15–2 on log-log paper, a straight line is obtained, as shown in Figure 15–3.

FIGURE 15–3
Estimated unit times based on a 20 percent reduction each time the quantity doubles

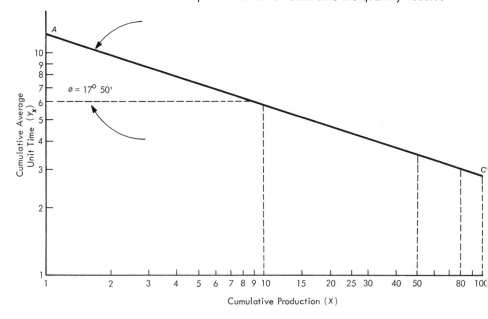

The analyst must recognize that a new learning curve situation does not necessarily result every time a new design is put into production. Former designs similar to new designs have a pronounced effect on the point at which the learning curve begins to flatten. Thus, if a company introduces a completely new design of a complex electronic panel, the assemply of this panel would involve a much different learning curve than would the introduction of a panel somewhat similar to one that had been in production for the past five years.

The theory of the learning curve proposes that as the total quantity of units produced doubles, the time per unit declines at some constant percentage. For example, if it is expected that a 90 percent rate of improvement will be experienced, then, as production doubles, the average time per unit will decline 10 percent.

Table 15–1 illustrates the decline in the cumulative average hours per

TABLE 15–1

Cumulative production	Cumulative average hours per unit	Ratio to previous cumulative average
1.	100.0	–
2.	90.0	90
4.	81.0	90
8.	72.9	90
16.	65.6	90
32.	59.0	90
64.	53.1	90
128.	47.8	90

unit of production with successive doubling of the production quantity where a 90 percent rate of improvement exists.

The reader should understand that the smaller the percent rate of improvement, the greater the progressive improvement with production output. Typical rates of learning that have been encountered are as follows: large or fine assembly work (such as aircraft), 70–80 percent; welding, 80 to 90 percent; machining, 90 to 95 percent.

Thus, the percent rate of improvement (learning) equals 100 times the cumulative average hours per unit at a given total production divided by the cumulative average hours per unit when the total production was 50 percent of the present quantity. For example, from Table 15–1:

$$\text{Percent learning} = \frac{81}{90} \times 100 = 90 \text{ percent or}$$

$$= \frac{72.9}{81} \times 100 = 90 \text{ percent}$$

When linear graph paper is used, the learning curve is a hyperbola of the form $Y_x = KX^N$. On log-log paper, the curve is represented by:

$$\log Y_x = \log K + N \log X$$

where

Y_x = Cumulative average value of X units
K = Value in time of the first unit
X = Number of units produced
N = Exponent representing the slope (tan ϕ in Figure 15–3)

By definition, the learning in percent is then equal to:

$$\frac{K(2X)^N}{K(X)^N} = 2^N$$

And taking the log of both sides:

$$N = \frac{\log \text{ of learning percent}}{\log 2}$$

For 80 percent learning:

$$N = \frac{\log \text{ of } 0.80}{\log \text{ of } 2} = \frac{9.9031 - 10}{0.3010}$$
$$= -0.322$$
$$\text{arc tan } 0.322 = 17°50'$$

In Table 15–2, the slopes of the common learning curve percentages are provided.

TABLE 15–2

Learning curve percentage	Slope
70	0.514
75	0.415
80	0.322
85	0.234
90	0.152
95	0.074

An example will help clarify the above relationships. Assume that 50 units are produced at a cumulative average of 20 man-hours per unit. It is desired to determine the learning curve percentage when 100 units are produced at a cumulative average of 15 man-hours per unit.

X_1 = Cumulative production to point one (50 units)
X_2 = Cumulative production to point two (100 units)
$\dfrac{X_2}{X_1}$ = 2 (when the production quantity is doubled)

and

$$\frac{\text{Tc.a. } 2}{\text{Tc.a. } 1} = \frac{K(X_2)^N}{K(X_1)^N} = 2^N$$

where:

Tc.a. = Cumulative average hours per unit for any number of units in logarithmic form

$$\log \text{Tc.a.} = \log K + N \log X$$
$$\log 20 = \log K + N \log 50$$
$$\log 15 = \log K + N \log 100$$
$$\log 20 - \log 15 = N (\log 50 - \log 100)$$
$$= \frac{\log 20 - \log 15}{\log 50 - \log 100}$$
$$= \frac{1.30103 - 1.17609}{1.69897 - 2.00000}$$
$$= \frac{0.12494}{0.30103}$$
$$= -0.415$$

and

$$2^{-0.415} = 75 \text{ percent (learning curve percentage)}$$

The above derivation is based on the "cumulative average learning curve" since the relationships refer to the average hours to produce a unit based on all units produced up to a particular point.

When working with the man-hours required to produce a specific unit, we are dealing with the "unit learning curve," which refers to the hours required to produce a specific unit. The log plot of the cumulative average is asymptotically parallel to the log plot of the unit curve. The cumulative average line is straight, while the individual line curves downward from unit 1 until it becomes parallel to the cumulative average line. For practical estimating purposes, the individual and cumulative curves become about parallel somewhere between the 10th and 20th unit.

To plot the unit time versus the quantity, two points where the quantity is above 20 may be calculated and the plotting made on log-log paper. To calculate the unit time value of the selected points, multiply the cumulative average time of these points by a conversion factor. The conversion factor used for making the unit time plot is $1 + N$. Thus, the conversion factor for an 85 percent learning curve would be:

$$1 + N = 1 + (-0.234) = 0.766$$

Multiplying this conversion factor by the cumulative average time for 20 units and the cumulative average time for 40 units would enable us to plot the unit time curve for the portion that is asymptotically parallel to the cumulative average plotting.

Characteristics of a sound rating system

The first and most important characteristic of any rating system is accuracy. Since the majority of rating techniques rely on the judgment of the time study observer, perfect consistency in rating cannot be expected.

However, rating procedures that readily permit the different time study analysts, within a given organization, to study different operators employing the same method to arrive at standards that are not more than 5 percent in deviation from the average of the standards established by the group are considered adequate. The rating plan that carries variations in standards greater than the plus or minus 5 percent tolerance should be improved or replaced.

Other factors being similar, the rating plan that gives the most consistent results will be the most useful. Inconsistency in rating will do more than any other one thing to destroy the operator's confidence in the time study procedure. For example, if a standard that allowed 7.88 hours to mill 100 castings were developed through stopwatch study, and at some later date a similar casting entailing a slightly longer cut were studied, and a standard of 7.22 hours established, considerable complaint would be registered by the operator, even though both studies were within the plus or minus 5 percent criterion of accuracy.

A rating plan that achieves consistency when used by the various time study analysts within a given plant, and yet is outside the accepted definition of normal accuracy, can be corrected. A rating plan that gives inconsistent results when used by the different time study men is sure to end in failure. Time study men who find difficulty in rating consistently after proper and thorough training will be better off if they seek some other means of livelihood. It is not difficult to correct the rating habits of an analyst who rates consistently high or consistently low, but it is most difficult to instill rating ability into the mind of an analyst who is rating considerably too high today and may be rating considerably too low tomorrow.

A rating system that is simple, concise, easily explained, and keyed to well-established benchmarks will be more successful than will a complex rating system requiring involved adjusting factors and computational techniques that confuse the average shop employee.

In view of the aforementioned accuracy limitations, every company will have, as time goes on, a number of standards which will be referred to as "tight" or "loose" by the production floor. If these inequities are within the 5 percent tolerance range, few grievances will develop. However, if it becomes possible for an operator to earn as much as 50 percent premium on one job, and he finds it difficult to even make standard on another, then general dissatisfaction can be anticipated for the whole rate program. At this point, it might be well to mention that inequities in different rates are not necessarily entirely due to poor performance rating. Loose rates frequently are due to methods improvements inaugurated over a period of time without restudy of the job from the time study point of view.

Rating at the work station

There is only one time when the performance rating should be done, and that is during the course of the observation of the elemental times. As the operator progresses from one element to the next, the analyst carefully evaluates his speed, dexterity, freedom from false moves, rhythm, coordination, effectiveness, and all the other factors influencing output by the prescribed method. It is at this time, and at this time only, that the performance of the operator in comparison to normal performance is most clearly evident to the observer. Once the performance has been judged and recorded, it should not be changed. This does not imply that faulty judgment on the part of the observer is not possible. In case the leveling is questioned, the job or operation should be restudied to prove or disprove the recorded evaluation.

Immediately after completing the study and recording the final performance factor, the observer should advise the operator as to his performance rating. Even if elemental rating is used, the analyst will be able to approximate the operator's performance and so advise him. This practice will give the operator an opportunity to express his opinion as to the fairness of the performance factor directly to the person responsible for its development. Thus, an understanding can be achieved prior to computation of the standard. Where this procedure is practiced, time study men will find that they receive greater respect from the operators and that they tend to be more conscientious in their rating of the operators' performance. Also, considerably fewer rate grievances will be submitted, because most rates will have been agreed upon, or at least satisfactorily explained and sold, before they are issued.

Rating elements versus overall study

The question of how frequently the performance should be evaluated during the course of a study often comes up. Although no set rule can be established as to the interval limit that permits concise rating, it can be said that in general the more frequently the study is rated, the more accurate the evaluation of the operator's demonstrated performance will be.

On short-cycle repetitive operations, little deviation in operator performance will be realized during the course of the average length study (15 to 30 minutes). In cases like this, it will be perfectly satisfactory to evaluate the performance of the entire study and to record the rating factor for each element in the space provided. Of course, power-fed or machine-controlled elements will be rated normal, or 1.00, as their speed cannot be changed or modified at will by the operator. In short-cycle studies, if the observer endeavors to performance rate each successive element of the entire study, he will be so busy recording values that he will be unable to do an effective job of observing, analyzing, and evaluating the operator's performance.

When the study to be taken is relatively long (over 30 minutes) or is made up of several long elements, then operator performance may be expected to vary during the course of the study. In such studies, it is important that performance be periodically evaluated and rated. Elements longer than 0.10 minute in duration can be consistently and accurately rated as they occur. However, if a study is made of a series of elements shorter than 0.10 minute, then no effort should be made to evaluate each element of each cycle of the study, as time will not permit this operation. It will be satisfactory to rate the overall time of each cycle or perhaps of each group of cycles.

METHODS OF RATING

The Westinghouse system

One of the oldest and at one time one of the most widely used systems of rating is the one developed by the Westinghouse Electric Corporation and outlined in detail by Lowry, Maynard, and Stegemerten.[3] In this method, four factors are considered in evaluating the performance of the operator. These are skill, effort, conditions, and consistency.

Skill is defined as "proficiency at following a given method," and can be further explained by relating it to craftsmanship, demonstrated by the proper coordination of mind and hands.

The skill of an operator is determined by his experience and his inherent aptitudes, such as natural coordination and rhythm. Practice will tend to develop skill, but it cannot entirely compensate for deficiencies in natural aptitude. All the practice in the world could not make major league baseball pitchers out of most male athletes.

A person's skill in a given operation increases over a period of time because increased familiarity with the work brings speed, smoothness of motions, and freedom from hesitations and false moves. A decrease in skill is usually caused by impairment of ability brought about by physical or psychological factors, such as failing eyesight, failing reflexes, and the loss of muscular strength or coordination. From this, it can readily be appreciated that a person's skill can vary from job to job and even from operation to operation on a given job.

According to the Westinghouse system of leveling or rating, there are six degrees or classes of skill that represent an acceptable proficiency for evaluation. These are: poor, fair, average, good, excellent, and super. The

---

[3] S. M. Lowry, H. B. Maynard, and G. J. Stegemerten, *Time and Motion Study and Formulas for Wage Incentives,* 3d ed. (New York: McGraw-Hill Book Co., 1940).

skill displayed by the operator is evaluated by the observer and rated as being in one of these six classes. Table 15–3 illustrates the characteristics of the various degrees of skill together with their equivalent numerical values. The skill rating is then translated into its equivalent percentage

TABLE 15–3
Skill

+0.15	A1	Superskill
+0.13	A2	Superskill
+0.11	B1	Excellent
+0.08	B2	Excellent
+0.06	C1	Good
+0.03	C2	Good
0.00	D	Average
–0.05	E1	Fair
–0.10	E2	Fair
–0.16	F1	Poor
–0.22	F2	Poor

Source: S. M. Lowry, H. B. Maynard, and G. J. Stegemerten, *Time and Motion Study and Formulas for Wage Incentives,* 3d ed. (New York: McGraw-Hill Book Co., 1940).

value, which ranges from plus 15 percent for superskill to minus 22 percent for poor skill. This percentage is then combined algebraically with the ratings for effort, conditions, and consistency to arrive at the final leveling, or performance rating factor.

Effort, according to this rating method, is defined as a "demonstration of the will to work effectively." It is representative of the speed with which skill is applied, and can be controlled to a high degree by the operator. When evaluating the effort given, the observer must exercise care to rate only the effective effort demonstrated. Many times an operator will apply misdirected effort with high tempo in order to increase the cycle time of the study and yet retain a liberal rating factor. As for skill, six classes of effort can be exhibited that are representative of acceptable speed for rating purposes: poor, fair, average, good, excellent, and excessive. Excessive effort has been assigned a value of plus 13 percent, and poor effort, a value of minus 17 percent. Table 15–4 gives the numerical values for the different degrees of effort and also outlines the characteristics of the various categories.

The conditions referred to in this performance rating procedure are those which affect the operator and not the operation. Conditions will be rated normal or average in more than a majority of instances, as conditions are evaluated in comparison with the way in which they are customarily found at the work station. Elements that would affect the working conditions include temperature, ventilation, light, and noise. Thus, if the temperature at a given work station was 60° F whereas it was customarily

TABLE 15–4
Effort

+0.13	A1	Excessive
+0.12	A2	Excessive
+0.10	B1	Excellent
+0.08	B2	Excellent
+0.05	C1	Good
+0.02	C2	Good
0.00	D	Average
–0.04	E1	Fair
–0.08	E2	Fair
–0.12	F1	Poor
–0.17	F2	Poor

Source: S. M. Lowry, H. B. Maynard, and G. J.
Stegemerten, *Time and Motion Study and Formulas
for Wage Incentives,* 3d ed. (New York: McGraw-Hill
Book Co., 1940).

maintained at 68° to 74° F, the conditions would be rated as lower than normal. Conditions which affect the operation, such as poor condition of tools or materials, would not be considered when applying the performance factor for working conditions. Six general classes of conditions have been enumerated, with values ranging from plus 6 percent to minus 7 percent. These "general state" conditions are listed as ideal, excellent, good, average, fair, and poor. Table 15–5 gives the respective values for these conditions.

TABLE 15–5
Conditions

+0.06	A	Ideal
+0.04	B	Excellent
+0.02	C	Good
0.00	D	Average
–0.03	E	Fair
–0.07	F	Poor

Source: S. M. Lowry, H. B. Maynard, and G. J.
Stegemerten, *Time and Motion Study and Formulas
for Wage Incentives,* 3d ed. (New York: McGraw-Hill
Book Co., 1940).

The last of the four factors that influence the performance rating is the consistency of the operator. Unless the snapback method is used, or unless the observer is able to make and record successive subtractions as he goes along, the consistency of the operator must be evaluated as the study is being worked up. Elemental time values that repeat constantly would, of course, have a perfect consistency. This situation occurs very infrequently, as there always tends to be dispersion due to the many variables, such as material hardness, the tool cutting edge, the lubricant, the skill and effort of the operator, erroneous watch readings, and the presence of foreign elements. Elements that are mechanically controlled would, of course, have values of near perfect consistency, but such elements are not rated. There are six classes of consistency: perfect, excellent, good, average, fair, and poor. Perfect consistency is rated plus 4 percent, and poor consistency is rated minus 4 percent, while the other classes fall in between these values. Table 15–6 summarizes these values.

TABLE 15–6
Consistency

+0.04	A	Perfect
+0.03	B	Excellent
+0.01	C	Good
0.00	D	Average
–0.02	E	Fair
–0.04	F	Poor

Source: S. M. Lowry, H. B. Maynard, and G. J. Stegemerten, *Time and Motion Study and Formulas for Wage Incentives,* 3d ed. (New York: McGraw-Hill Book Co., 1940).

No fixed rule can be cited as to the applicability of the consistency table. Some operations that are of short duration and tend to be free of delicate positioning manipulations will give relatively consistent results from one cycle to the next. Such operations would have a more exacting requirement of average consistency than would a job of long duration demanding high skill in its positioning, engaging, and aligning elements. The determination of the justified range of variation for a particular operation must be based to a large extent upon the time study analyst's knowledge of the work.

The time study analyst should be cautioned against the operator who consistently performs poorly in an effort to deceive the observer. This is easily accomplished by counting to oneself, thereby setting a pace that can be accurately followed. Operators who familiarize themselves with this performance rating procedure will sometimes perform at a pace that is consistent and yet is below the effort rating curve. In other words, they may be performing at a pace that is poorer than poor. In cases like this, the operator cannot be leveled. The study must be stopped and the situation brought to the attention of the operator or the foreman or both.

Once the skill, effort, conditions, and consistency of the operation have been assigned, and their equivalent numerical values established, the performance factor is determined by algebraically combining the four values and adding their sum to unity. For example, if a given job is rated C2 on skill, C1 on effort, D on conditions, and E on consistency, the performance factor would be evolved as follows:

Skill	C2	+.03
Effort.	C1	+.05
Conditions.	D	+.00
Consistency	E	−.02
Algebraic sum.		+.06
Performance factor.		1.06

The reader should again be cautioned that the performance factor is applied only to the effort, or manually performed, elements; all machine-controlled elements are rated 1.00.

The Westinghouse method of performance rating is adapted to the leveling of the entire study rather than to elemental evaluation. This method would prove cumbersome if it were used to level each element as soon as it took place. In fact, the time study form itself does not provide sufficient space to evaluate skill, effort, conditions, and consistency for each element of every cycle.

Many companies have modified the Westinghouse system so as to include only skill and effort factors entering into the determination of the rating factor. The contention is that consistency is very closely allied to skill, and that conditions are rated average in most instances. If conditions deviate substantially from normal, the study could be postponed, or the effect of the unusual conditions could be taken into consideration in the application of the allowance (see Chapter 16).

In 1949, the Westinghouse Electric Corporation itself developed a new rating method which was termed the "performance rating plan" to distinguish it from the leveling procedure just discussed. The performance rating plan is being used in most of the Westinghouse plants today.

In the performance rating plan, in addition to using the physical attributes displayed by the operator, the company made an attempt to evaluate the relationship between those physical attributes and the basic divisions of work. The characteristics and attributes that the Westinghouse performance rating technique considers are: (1) dexterity, (2) effectiveness, and (3) physical application.

These three major classifications do not in themselves carry any numerical weight, but have in turn been assigned attributes that do carry numerical weight, Table 15–7 gives the numerical values of the nine attributes that are evaluated under this system.

The first major classification, dexterity, has been divided into three attributes, the first of which is: *1. Displayed ability in use of equipment and tools, and in assembly of parts.*

When considering this attribute, the analyst is concerned primarily with the "do" portion of the work cycle after the "get" operations (reach, grasp, move) have taken place.

The second attribute under dexterity is: *2. Certainty of movement.*

TABLE 15–7

	+ Above		0 Expected	- Below	
DEXTERITY:					
1. Displayed ability in use of equipment and tools, and in assembly of parts.	6	3	0	2	4
2. Certainty of movement.	6	3	0	2	4
3. Coordination and rhythm.		2	0	2	
EFFECTIVENESS:					
1. Displayed ability to continually replace and retrieve tools and parts with automaticity and accuracy.	6	3	0	2	4
2. Displayed ability to facilitate, eliminate, combine, or shorten motions.	6	3	0	4	8
3. Displayed ability to use both hands with equal ease.	6	3	0	4	8
4. Displayed ability to confine efforts to necessary work.			0	4	8
PHYSICAL APPLICATION:					
1. Work pace.	6	3	0	4	8
2. Attentiveness.			0	2	4

In evaluating this attribute, the analyst is concerned with the number and the degree of hesitations, pauses, or roundabout moves. The basic divisions of accomplishment that will tend to give the operator a low rating for this attribute are change direction, plan, and avoidable delay. All of these affect certainty of movement.

The last attribute considered under dexterity is: *3. Coordination and rhythm.*

This attribute is evidenced by the degree of the displayed performance, by smoothness of motions, and by freedom from spasmodic spurts and lags.

The second major classification, effectiveness, has been defined as efficient, orderly procedure. This classification has been divided into four individual attributes. The first of these is: *1. Displayed ability to continually replace and retrieve tools and parts with automaticity and accuracy.*

Here the analyst evaluates the ability of the worker to repeatedly place tools, materials, and parts in specified locations and positions, and to retrieve them automatically and accurately by eliminating such ineffective basic divisions of work as searching and selecting.

The second of the individual attributes in effectiveness is: *2. Displayed ability to facilitate, eliminate, combine, or shorten motions.*

Here the analyst evaluates the proficiency of the basic divisions position, pre-position, release, and inspect. The transport therbligs are usually predetermined by the established method. However, a skilled worker will be able to eliminate or shorten the elements of pre-position, position, and inspect because of his manipulative ability.

The third attribute under effectiveness is: *3. Displayed ability to use both hands with equal ease.* Here the degree of effective utilization of both hands is rated.

The fourth and last attribute under effectiveness is: *4. Displayed ability to confine efforts to necessary work.*

This attribute is used to rate the presence of unnecessary work that could not be removed when taking the study. It carries only a negative weight, for no percentage is added when the work is confined to the necessary work because this condition is expected.

The third major classification, physical application, is defined as the demonstrated rate of performance and has two attributes. The first of these is: *1. Work pace.*

Work pace is rated by comparing the speed of movement to preconceived standards for the particular type of work under consideration.

The second attribute for physical application is: *2. Attentiveness.*

Attentiveness is rated as the degree of displayed concentration.

Both of the Westinghouse rating techniques demand considerable training in order that the time study analysts recognize the different levels of each of the attributes. Their training entails a 30-hour course of which approximately 25 hours are spent rating films and discussing the attributes and the degree to which each is displayed. The procedure generally followed is:

1.   A film is shown and the operation explained.
2.   The film is reshown and rated.
3.   The individual ratings are compared and discussed.
4.   The film is reshown, and the attributes pointed out and explained.
5.   Step 4 is repeated as often as is necessary to reach understanding and agreement.

Synthetic rating

In an effort to develop a method of rating that would not rely on the judgment of the time study observer and would give consistent results, R. L. Morrow established a procedure known as "synthetic leveling."[4]

The synthetic leveling procedure determines a performance factor for representative effort elements of the work cycle by comparing actual elemental observed times to times developed through the medium of fundamental motion data (see Chapter 19). Thus, the performance factor

---

[4] Morrow, *Time Study,* p. 241.

may be expressed algebraically as:

$$P = \frac{F_t}{O}$$

where:

P = Performance or leveling factor
F_t = Fundamental motion time
O = Observed mean elemental time for the
　　elements used in F_t

The factor thus determined would then be applied to the remainder of the manually controlled elements comprised by the study. Of course, machine-controlled elements are not rated, as is the case in all rating techniques.

A typical illustration of the application of synthetic rating is shown in Table 15–8.

TABLE 15–8

Element no.	Observed average time in minutes	Element type	Fundamental motion time in minutes	Performance factor
1	0.08	Manual	0.096	123
2	0.15	Manual	–	123
3	0.05	Manual	–	123
4	0.22	Manual	0.278	123
5	1.41	Power fed	–	100
6	0.07	Manual	–	123
7	0.11	Manual	–	123
8	0.38	Power fed	–	100
9	0.14	Manual	–	123
10	0.06	Manual	–	123
11	0.20	Manual	–	123
12	0.06	Manual	–	123

It will be noted that for element 1,

$$P = \frac{0.096}{0.08} = 120 \text{ percent}$$

and that for element 4,

$$P = \frac{0.278}{0.22} = 126 \text{ percent}$$

The mean of these is 123 percent, and this is the factor used for rating all of the effort elements. It can readily be seen that synthetic performance rating is a sampling technique.

It is essential that more than one element be used in establishing a synthetic rating factor because research has proven that operator performance will vary significantly from element to element, especially in complex work.

Actually, all experienced time study men unconsciously follow the synthetic rating procedure to some extent. From past experience, the time study man's mind is full of benchmarks that have been established on similar work. Consequently, he knows that the normal performance of advancing the drill of a 17-inch single-spindle Delta drill is 0.03 minute, and of indexing the hex turret of a No. 4 Warner & Swasey turret lathe is 0.06 minute, and of blowing out a vise or fixture with an air hose and laying the finished part aside is 0.08 minute. These benchmarks and many others, when compared to actual performance, certainly influence and even determine the rating factor given the operator.

A major objection to the application of the synthetic leveling procedure is the time required to construct a left- and right-hand chart of the elements selected for the establishment of basic motion times. The reader may well consider the desirability of establishing a standard for the entire job synthetically. This would eliminate the laborious task of recording elemental times, making subtractions, determining the mean elapsed time, determining the normal time synthetically for several elements so as to arrive at a performance factor, and applying the performance factor. In order to arrive at synthetic values rapidly and accurately, an alignment chart has been designed, which may be attached to the time study board (see Figure 13–6). Its application will be discussed later in this chapter. This technique permits the analyst to utilize synthetic values as a guide toward the establishment of a valid performance factor.

Speed rating

Speed rating is a performance evaluation method which gives consideration only to the rate of accomplishment of the work per unit time. In this method, the observer measures the effectiveness of the operator against the conception of a normal man doing the same work, and then assigns a percentage to indicate the ratio of the observed performance to normal performance. Particular emphasis is placed on the observer's having complete knowledge of the job before taking the study. It is evident that the pace of machine workers in a plant which produces aircraft engine parts would appear considerably slower to the novice than would the pace of machine workers who produce farm machinery components. The greater precision requirements of the aircraft work would require such care that the movements of the various operators would appear unduly slow to one who was not completely familiar with the work being performed.

In speed rating, 100 percent is usually considered normal. Thus, a rating of 110 percent would indicate that the operator was performing at a speed 10 percent greater than normal, and a rating of 90 percent would mean that he or she was performing at a speed 90 percent of normal. Some concerns using the speed rating technique have chosen to call 60 standard or normal. This is based on the standard hour approach, that is, producing 60 minutes of work every hour. On this basis, a rating of 80 would mean that the operator was working at a speed of 80/60, which equals 133 percent, or 33 percent above normal. A rating of 50 would indicate a speed of 50/60, or 83⅓ percent of normal or standard.

In the speed rating method, the analyst first makes an appraisal of the performance to determine whether it is above or below his conception of normal. He or she then executes a second judgment in an effort to place the performance in the precise position on the rating scale that will correctly evaluate the numerical difference between standard and the performance being demonstrated.

A form of speed rating referred to as "pace rating" has received considerable attention from the basic steel industry. In effect, pace rating is speed rating. However, in an effort to identify completely a normal pace on different jobs benchmarks have been provided on a broad range of work. Thus, in addition to card dealing, such effort operations as shoveling sand, coremaking, brick handling, and walking have been clearly identified as to method, and quantified as to normal rate of production. Once the time study analyst familiarizes himself with a series of benchmarks closely allied to the work he is going to study, he will be much better equipped to evaluate the speed being performed.

Objective rating

The rating procedure known as "objective rating," a method developed by M. E. Mundel, endeavors to eliminate the difficulty of establishing a normal speed criterion for every type of work.[5] In this procedure, a single work assignment is established to which all other jobs are compared as to pace. After the judgment of pace, a secondary factor is assigned to the job to allow for its relative difficulty. The factors that influence the difficulty adjustment are: (1) amount of body used, (2) foot pedals, (3) bimanualness, (4) eye-hand coordination, (5) handling or sensory requirements, and (6) weight handled or resistance encountered.

Numerical values, resulting from experiments, have been assigned for a range of degrees of each factor. The sum of the numerical values for each of the six factors comprises the secondary adjustment. By this method, the normal time can be expressed as follows:

$$T_n = (P_2)(S)(O)$$

where:

T_n = Computed established normal time
P_2 = Pace rating factor
S = Job difficulty adjustment factor
O = Observed mean elemental time

[5] M. E. Mundel, *Motion and Time Study*, 4th ed. (Englewood Cliffs, N.J.: Prentice-Hall, Inc., 1960).

This performance rating procedure will tend to give consistent results since comparing the pace of the operation under study to an operation that is completely familiar to the observer can be achieved more readily than can judging simultaneously all the attributes of an operation to a concept of normal for that specific job. The secondary factor will not effect inconsistency since this factor merely adjusts the rated time by the application of a percentage. This percentage value is taken from a table that gives values for the effects of various difficulties that are present in the operation being performed.

Operator selection

In an effort to eliminate the performance rating step entirely in the calculation of the standard, some concerns select the operators to be studied and then consider the average observed time as the normal time. When this method is utilized, more than one operator is usually studied, and enough cycles are observed so that a reliable average time (within plus or minus 5 percent of the population average) can be calculated. Of course, the success of this method is dependent upon the selection of the employees who are to be studied and their performance during the study. If the performances of the operators being observed are slower than normal, then too liberal a standard will result; conversely, if the observed operators produce at a pace more rapid than normal, then the standard will be unduly tight. There is always the possibility of having but one or two available operators and the chance that they may differ from normal. Then, in an effort to avoid delay in establishing a standard, the observer will make the study; the result will be a poor time standard.

Analysis of rating

As is true of all procedures requiring the exercise of judgment, the simpler and more concise the plan, the easier it will be to use and, in general, the more valid the results will be.

The performance rating plan that is easiest to apply, easiest to explain, and tends to give the most valid results is straight speed or pace rating augmented by synthetic benchmarks. As has been explained, 100 is considered normal in this procedure, and performance greater than normal is indicated by values directly proportional to 100. Thus, a rating of 120 would indicate that a performance of 20 percent higher than normal has been exhibited. A rating of 60 would indicate that the operator performed at a pace of only 0.60 of normal. The speed rating scale usually covers a range of from 0.50 to 1.50. Operators performing outside this productivity range of 3 to 1 may be studied, but this is not recommended. The closer the performance is to normal, the better will be the chance of achieving a fair normal time.

Four criteria will determine whether or not the time study analyst using speed rating can consistently establish values not more than 5 percent above or below the normal that would be representative of the average of a group of trained time study men. These are:

1. Experience in the class of work being performed.
2. Synthetic benchmarks on at least two of the elements being performed.
3. Selection of an operator who, from past experience, is known to give performances somewhere between 115 and 85 percent of normal.
4. Use of the mean value of three or more independent studies.

Certainly, the most important of these four criteria is experience in the class of work being performed. This does not necessarily mean that the analyst must at one time have been an actual operator in the work being studied, although this would be desirable. From his past personal experience, either by observation or operation, he should be sufficiently familiar with the work to understand every detail of the method being used. Thus, on a job being performed on a turret lathe, he should recognize the tooling, have knowledge of the correct speeds and feeds, the correct rake and clearance angles, the lubricant, horsepower requirements, method of holding the work, and so on. On an assembly job taking place in a fixture, the time study man should be familiar with the difficulty in positioning the components in the fixture, and should know the class of fit between all mating parts and have a clear understanding of the relationship between time and class of fit. He should know the proper sequence of events and the weights of all the parts being handled.

It has often been said that experience is the key to accurate performance rating, and this statement is certainly true. An analyst with 10 years' experience in the metal trades would find considerable difficulty in establishing standards in a woman's shoe factory, and of course the reverse would be true: a time study man with years of experience in the fabrication of shoes would be unable to establish equitable standards in a machine shop until he had acquired some experience in the line of work being performed.

Although it has been proven that operator performance varies from element to element, the average performance of two or more elements of the study will give a reasonably good estimate of the cycle performance. If the time study analyst has information that allows him to preestablish synthetically the normal time required to perform several of the elements involved in the study, he has an indication of the overall performance being executed.

The alignment chart shown on the time study board (see Figure 13–7) permits the analyst to establish synthetically the normal time for the elements "pick up pc. & position in fixture, jig, or die," "pick up pc. & bring to work station," "pick up pc. & lay aside."[6] Since very few work assignments can be performed without utilizing at least one of these elements, the time study analyst has at his disposal a useful guide for establishing a "correct" performance factor. For example, if a 10-pound casting is picked up and placed in a three-jaw turret lathe chuck, and the element entailed a reach of 30 inches, a move of 20 inches, a class 4 grasp (see Chapter 19 on MTM), and a P3SE position, the synthetic time could be determined graphically. By going from 20 inches on scale 1 vertically to 10 pounds on scale 2 and then moving horizontally to scale 3, a time of 0.017 minute for the move and release is obtained. This point connected to the G4 grasp of 0.005 minute (scale 4) gives a turning point on scale 5. This turning point on scale 5 connected with a reach of 30 inches on scale 6 gives a time of 0.04 minute for the reach, grasp, move, and release. This point connected with a P3SE position entailing 0.025 minute on scale 8 gives the time to perform the complete element, 0.065 minute on scale 9.

It is not intended to imply that the synthetic value will establish a rating for every element of each cycle of the study. The analyst through his experience and training in performance rating will be able to evaluate the operator quite precisely; the alignment chart and the resulting synthetic standards will provide an additional check and guide to the analyst, giving him more confidence in his ability to establish fair normal times.

Whenever more than one operator is available to be studied, the one who is thoroughly experienced on the job, who has a reputation of being receptive to time study practice, and who consistently performs at a pace near standard or slightly better than standard should be selected. The closer the operator performs to a normal pace, the easier it will be to level him. Sizable errors in judgment during rating are invariably a result of improper evaluation of an operator who is performing at either extremity of the rating scale. For example, if 0.50 minute is considered normal for dealing a deck of cards into four bridge hands, it will be found that performance within plus or minus 15 percent of this conception of normal will be fairly easy to identify. However, once the performance runs 50 percent faster or 50 percent slower than normal, considerably more difficulty is encountered in establishing an accurate rating factor for the performance being demonstrated.

It is good practice to take several (three or more) studies before arriving at the standard. The total error, due to rating and the determination of a mean elemental elapsed time that deviates from the population

---

[6] The author developed part of the data used in the design of this alignment chart after analyzing over 600 feet of film. The remainder of the data used was taken from Methods-Time Measurement tables.

mean, is minimized when the averages of several independent studies are used in computing the time standard. The independent studies can be made on the same operator at different times of the day, or on different operators. The point is that, as the number of studies increases, compensating errors will diminish the overall error. A recent example will clarify this point. The author, along with two other trained industrial engineers, reviewed performance rating training films involving 15 different operations. The results are tabulated in Table 15–9. The standard ratings were

TABLE 15–9
Performance ratings of three different engineers observing 15 different operations

Operation	Standard rating	Engineer A Rating	Engineer A Deviation	Engineer B Rating	Engineer B Deviation	Engineer C Rating	Engineer C Deviation	Average of Engineers A, B, C Rating	Average of Engineers A, B, C Deviation
1	110	110	0	115	+5	100	−10	108	−2
2	150	140	−10	130	−20	125	−25	132	−18
3	90	110	+20	100	+10	105	+15	105	+15
4	100	100	0	100	0	100	0	100	0
5	130	120	−10	130	0	115	−15	122	−8
6	120	140	+20	120	0	105	−15	122	+2
7	65	70	+5	70	+5	95	+30	78	+13
8	105	100	−5	110	+5	100	−5	103	−2
9	140	160	+20	145	+5	145	+5	150	+10
10	115	125	+10	125	+10	110	−5	120	+5
11	115	110	−5	120	+10	115	0	115	0
12	125	125	0	125	0	115	−10	122	−3
13	100	100	0	85	−15	110	+10	98	−2
14	65	55	−10	70	+5	90	+25	72	+7
15	150	160	+10	140	−10	140	−10	147	−3
Average of 15 operations	112	115.0		112.3		111.3		112.9	
Average deviation	0	+3.0		+0.3		−0.7		+0.9	

not disclosed until all 15 operations had been rated. The average deviation of all three engineers for the 15 different operations was only 0.87 of a point on a speed rating scale with a range of from 50 points to 150 points. Yet Engineer C was 30.0 high on operation 7, and Engineer B was 20.0 low on operation 2. In only one case (operation 3), when the known ratings were within the 70 to 130 range, did the average rating of the three analysts exceed plus or minus 5 points of the known rating.

The normal time for the operation should then be determined by averaging the normal times of the independent studies. This procedure will not only reduce the error inherent in the performance rating process, but will result in the determination of a mean measured time with less deviation from the population mean measured time.

Performance rating, like any other form of work involving judgment, must be accomplished by competent, well-trained individuals. If a person of this caliber is doing the work, and if he or she is equipped with a background of experience, has access to reliable synthetic time values, and uses sound judgment in operator selection, then reasonably accurate results are assured.

Training for rating

To be successful, the time study man must develop a record for setting standards correctly so that they will be accepted by both labor and management. Furthermore, his rates must be consistent so that he will maintain the respect of all parties. In general, the time study analyst is expected to regularly establish standards within plus or minus 5 percent of the true rate, when studying operators who are performing somewhere in the range of 0.70 to 1.30 of normal. Thus, if several operators are performing the same job, and different analysts, each studying a different operator, establish time standards on the job, then the resulting standard from each individual study should be within plus or minus 5 percent of the mean of the group of studies.

To assure consistency in rating, both with the rates which he himself has established and with the rates established by the other time study analysts in his plant, the time study man should continually participate in organized training programs. Of course, the training in performance rating should be more intense for the neophyte time study analyst.

One of the most widely used methods for training analysts in performance rating is the observation of motion-picture films illustrating diversified operations performed at different productivity levels. Each film has a known level of performance, and after it is shown on the screen, the correct rating is compared with the values established independently by the various trainees. If any of the time study analysts' values deviate substantially from the correct value, then specific information is given so as to justify the rating. For example, the observer may have been misled because there was high performance in the handling of the material to and from the work station, while poor performance prevailed during the cycle at the work station. Then, too, the analyst may have underrated the operator because of his apparently effortless sequence of motions, whereas the operator's smooth, rhythmic blending of movements is really an indication of high dexterity and manipulative ability.

The operations selected should be simple, yet should contain a number of fast motions. The observation of short elements cultivates both speed and concentration in the observer and trains the "second sight."

As successive films are shown, it is helpful for the analyst to plot his or her rating against the known values (see Figure 15–4). A straight line

FIGURE 15–4
Chart showing a record of seven studies with the analyst tending to rate
a little high on studies 1, 2, 4, and 6, and a little low on studies 3 and 7.
Only study 1 was rated outside the range of desired accuracy

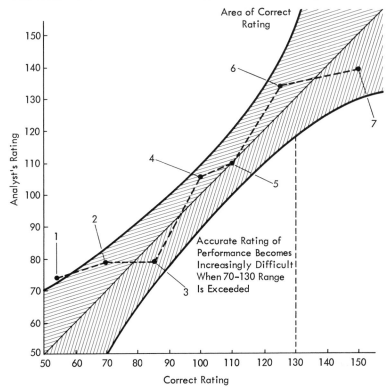

would indicate perfection, whereas high irregularities on both sides of the
line would indicate inconsistency as well as inability to evaluate performance.

In the example shown, the analyst rated the first film 75, whereas the
correct rating was 55. He rated the second 80, while the proper rating
was 70. In all but the first case, the analyst was within the company's established area of correct rating. It is interesting to note that this concern
indicates that only at the 100 percent, or normal, level of performance
is the plus or minus 5 percent criterion of accuracy valid. When performance is below 70 percent of normal or above 130 percent of normal,
a much larger than 5 percent error can be expected by the experienced
time study analyst.

It is also helpful to plot successive ratings on the X-axis and to indicate
the positive or negative magnitude of deviation from the known normal

FIGURE 15-5
Record of an analyst's rating factors on 15 studies

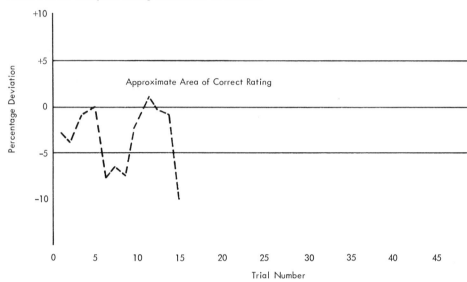

on the Y-axis (see Figure 15–5). The closer the time study man's rating comes to the X-axis, the more nearly correct he will be.

To determine quantitatively the ability of the analyst to performance rate, it is helpful to compute the percentage of the analyst's rating contained within specified limits of the known ratings. This can be done as follows:

1. Compute the mean difference ($\bar{x}_d$) between the rater's rating and the actual rating for n tests (n should be at least 15 observations).
2. Compute the standard deviation s_d of the differences in rating.
3. Compute the normal deviate z_1, where:

$$z_1 = \frac{+5 \text{ (or some other figure of accuracy)} - \bar{x}_d}{s_d}$$

4. Compute the normal deviate z_2, where:

$$z_2 = \frac{-5 \text{ (or some other figure of accuracy)} - \bar{x}_d}{s_d}$$

5. Compute the area under the normal distribution between ± 5 (or some other figure of accuracy) centered at $\bar{x}_d$ which is assumed equal to μ_d and s_d is assumed equal to σ_d.

For example, suppose that, after reviewing 15 rating films, a given analyst had the achievement shown in Table 15–10.

TABLE 15–10

Film no.	Correct rating	Analyst's rating	Difference (d)	Difference squared
1	115	105	−10	100
2	125	120	−5	25
3	85	95	+10	100
4	105	105	0	0
5	100	105	+5	25
6	95	110	+15	225
7	120	125	+5	25
8	140	150	+10	100
9	100	105	+5	25
10	60	75	+15	225
11	100	100	0	0
12	110	105	−5	25
13	90	90	0	0
14	130	125	−5	25
15	80	90	+10	100

The following computations would be made:

$$\bar{x}_d \text{ (mean difference)} = \frac{\Sigma d}{n} = \frac{50}{15} = 3.33$$

$$s_d \text{ (standard deviation)} = \sqrt{\frac{\Sigma d^2 - \dfrac{(\Sigma d)^2}{n}}{n-1}} = 7.7$$

$$z_1 = \frac{5.00 - 3.33}{7.7} = 0.217$$

and

$$P(z_1) = \int_0^{z_1} \frac{1}{\sqrt{2\pi}} e^{-\frac{z^2}{2}} \, dz = 0.0859$$

$$z_2 = \frac{-5 - 3.33}{7.7} = -1.08$$

and

$$P(z_2) = \int_0^{z_2} \frac{1}{\sqrt{2\pi}} e^{-\frac{z^2}{2}} \, dz = 0.3597$$

$$P(z_1) + P(z_2) = 0.4456$$

In this example, the analyst would be given a rating of 0.4456. This represents the portion of his ratings that would lie within plus or minus 5 rating points of the ideal ratings. This area of correct rating, together with the distribution of the difference between the analyst's rating and the correct rating, is shown in Figure 15–6. It would be advisable for the analyst to use the student's "t" distribution if the number of films observed is less than 15.

FIGURE 15–6
Graphic representation of the analyst's correct ratings (±5) and the distribution
of all his ratings

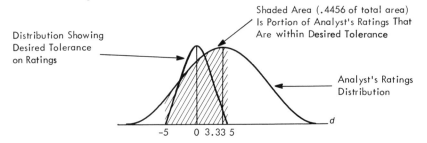

A recent statistical study involving the performance rating of 6,720 individual operations by a group of 19 analysts over a period of approximately two years confirmed several facts that had been accepted by most industrial engineers. This study concluded that the "level of performance" is a factor that affects significantly errors in rating performance. Time study men tend to overrate low levels of performance and to underrate high levels of performance.

It was also concluded that the operation studied has an effect on errors in rating performance. Thus, complex operations tend to be more difficult to performance rate, even by the experienced analyst, than do simpler operations. At low levels of performance, overrating is greater for difficult operations than for simple operations, while at high levels of performance, underrating is larger for the easy to perform operations.

Today, no known tests are available that can accurately evaluate the ability of a person to rate performance. Experience has shown, however, that unless an analyst shows a tendency toward consistency and proficiency after a brief training in rating, he may never be able to do an acceptable job in rating.

A survey made by one large manufacturer disclosed that the industrial engineering employee who was older in point of service did not level any more accurately than the newer man, that those in the higher work classifications did not level better than those in the lower classifications, and that those in machining areas did no better than those in assembly areas.

A group of undergraduate engineers were given 15 minutes' instruction in rating and then were requested to performance rate a film illustrating a series of heavy labor operations. The students were given no orientation in the class of work, and the operations observed differed substantially from the operations used in giving the students a concept of proper performance. It was interesting to note that the students were extremely successful in adjusting to heavy labor operations the concepts learned on light operations.

In another instance, 34 undergraduate engineering students were asked to performance rate five industrial films involving operations ranging from mold making to bench assembly. Four of these films were established as being representative of normal performance, and one was rated as 1.15, or 15 percent above normal. The results are shown in Table 15–11.

TABLE 15–11

Known rating	Number of students in each rating bracket					
	85-90	91-95	96-100	101-105	106-110	111-125
1.15 0		3	4	12	6	9
1.00 4		5	19	2	3	1
1.00 2		3	7	12	5	5
1.00 7		5	14	8	0	0
1.00 1		2	26	4	1	0

Note that the majority of the students did not deviate greatly from the known values.

Upon request, one division of the General Electric Company checked the validity of the performance rating technique on three different occasions. After brief training in the rating methods, the union steward and the foreman were requested to rate a given operation in conjunction with a time study analyst and the time study supervisor. At the completion of the study, the time study man rated the study without advising the other three men on the values he used, so that they would not be influenced by his opinion. These three men then went into their respective offices so that they would not be influenced by one another's opinions, and independently and secretly rated the operation. The results were as follows: time study man, 100; planning and wage payment supervisor, 99; foreman, 103; union steward, 103.

At a later date, a similar test case involving an entirely different class of work was undertaken. The results were: time study man, 100; planning and wage payment, supervisor 99; foreman, 103; union steward, 103.

On the third occasion, an entirely new set of conditions prevailed, and in this instance all four men rated identically—109!

The foregoing studies bring out (1) that the concept of normal performance can be taught quickly, and (2) that the concept is transferable to some degree to dissimilar operations.

By actual observation of different work assignments throughout a plant under the guidance of the time study supervisor, it is possible to achieve excellent training in rating. The supervisor will explain in detail the "why" of his values after the trainees have already written their ratings independently. The independent values should be recorded to determine the consistency of the group and the necessity for additional, and perhaps more intensive, training.

It has been proven that companies with extensive and well-designed programs of training in rating have been successful in largely eliminating the tendency to overrate and underrate.

TEXT QUESTIONS

1. Why has industry been unable to develop a universal conception of "normal performance"?
2. What factors enter into large variances in operator performance?
3. What are the characteristics of a sound rating system?
4. When should the performance rating procedure be accomplished? Why is this important?
5. What governs the frequency of performance rating during a given study?
6. Explain the Westinghouse system of leveling.
7. How does the Westinghouse performance rating method differ from the leveling method?
8. Under the Westinghouse rating system, why are "conditions" evaluated?
9. What is synthetic rating? What is its principal weakness?
10. What is the basis of speed rating, and how does this method differ from the Westinghouse system?
11. What is the purpose of the "secondary adjustment" in the objective rating technique? What factors are considered in the "secondary adjustment"?
12. What four criteria are fundamental for doing a good job in speed rating?
13. Why is training in performance rating a continuous process?
14. Using the data illustrated in Figure 15–4, determine the average percentage of correct rating within plus or minus 5 rating points.
15. Why is performance rating the most important step in the entire work measurement procedure?
16. Why should more than one element be used in the establishment of a synthetic rating factor?

GENERAL QUESTIONS

1. Would there be any objection to studying an operator who was performing at an excessive pace? Why or why not?
2. In what ways can an operator give the impression of high effort and yet produce at a mediocre or poor level of performance?
3. If an operator strongly objected to his performance rating factor upon completion of the study, what would your next step be if you were the time study analyst?
4. How would you go about maintaining a uniform concept of normal performance in a multiplant enterprise whose various plants are located in different sections of the country?

PROBLEMS

1. In the Dorben Company, 200 units of a new design were produced. The total time required to produce these 200 units was maintained by having

all the operators involved "punch in" and "punch out" on this line of work. The total recorded time was 25,412 hours. After 400 units of the new product were produced, it was noted that 42,808 hours had been utilized. From the data recorded, what is the learning curve associated with this new product?

2. The analyst is studying a complete assembly operation that takes place in his plant in conjunction with a new product line. He anticipates that an 85 percent rate of learning will take place. In order to develop helpful learning curve information for future planning, he wishes to compute the rate of learning that takes place on this assembly work.

   After 50 units were produced, the analyst noted a total assembly time charge of 1,000 man-hours. Only 500 more man-hours were needed to produce an additional 50 units. What learning curve percentage was taking place?

3. The work measurement analyst in charge of training time study analysts decided to have all trainees review 20 film loops, where the rate of each loop was known. Each trainee then computed his own record, which was based on the proportion of his ratings that fell within plus or minus five points of the known ratings.

   One analyst computed his average difference in rating as −4.08 points on the 20 films. The standard deviation was 6.4. What percentage of this analyst's ratings was contained within the desired rating?

   (Note: Assume that the sample values are the population values.)

4. The Dorben Company is using synthetic leveling on its low-skill highly repetitive operations. A time study analyst for the company finds that the mean time required by a given operator for element 2 averages 0.05 minutes. An MTM analysis of this element involves the following:

   One class A 20-inch reach; one class |C| grasp; one class C 24-inch move; one position involving a semisymmetrical assembly with light pressure and a relatively easy to handle part; one normal release.

   What performance factor will be assigned to the effort elements of this study? What would the allowed time for element 2 be if a P.D. & F. allowance of 20 percent were utilized?

5. In the Dorben foundry, an order of 20 large castings is being produced. For the first 10 castings, the average time per casting is 40 man-hours. What would the learning curve percentage be if the average time per casting were 35 hours upon completion of the order?

SELECTED REFERENCES

Barnes, Ralph M. *Motion and Time Study: Design and Measurement of Work.* 6th ed. New York: John Wiley & Sons, Inc., 1968.

Mundel, Marvin E. *Motion and Time Study: Principles and Practices.* 4th ed. Englewood Cliffs, N.J.: Prentice-Hall, Inc., 1960.

Nadler, Gerald. *Work Design: A Systems Concept.* Rev. ed. Homewood, Ill.: Richard D. Irwin, Inc., 1970.

16

Allowances

After the calculation of the normal time, sometimes referred to as the "rated" time, one additional step must be performed in order to arrive at a fair standard. This last step is the addition of allowance to take care of the many interruptions, delays, and slowdowns brought on by fatigue in every work assignment. For example, in planning a motorcar trip of 1,000 miles, we know that the trip cannot be made in 20 hours when driving at a speed of 50 miles per hour. An allowance must be added to take care of periodic stops for personal needs, for driving fatigue, for unavoidable stops brought on by traffic congestion and stoplights, for possible detours and the resulting rough roads, for car trouble, and so forth. Thus, we may estimate that we will make the trip in 25 hours, since we feel that 5 additional hours would be necessary to take care of all delays. Similarly, an allowance must be provided for the worker if the resulting standard is to be fair and readily maintainable by the average worker performing at a steady, normal pace.

It must be remembered that the watch readings of any time study are taken over a relatively short period of time, and that abnormal readings, unavoidable delays, and time for personal needs are removed from the study in the determination of the average or selected time. Therefore, the normal time has not provided for unavoidable delays and other legitimate lost time; consequently, some adjustment must be made to compensate for such losses.

In general, allowances are made to cover three broad areas. These are personal delays, fatigue, and unavoidable delays. The application of allowances is considerably broader in some concerns than in others. For example, a survey of 42 firms relative to what was ordinarily included in their allowances revealed the information shown in Table 16–1.

Allowances are frequently applied carelessly because they have not been established on sound time study data. This is especially true of allowance for fatigue, where it is difficult if not impossible to establish values based on rational theory. Many unions, well aware of this situation, have endeavored to bargain for additional fatigue allowance as if

TABLE 16–1

Allowance factor	No. of firms	Percent
1. Fatigue	39	93
A. General	19	45
B. Rest periods	13	31
Did not specify A or B	7	17
2. Time required to learn	3	7
3. Unavoidable delay	35	83
A. Man	1	2
B. Machine	7	17
C. Both, man and machine	21	50
Did not specify A, B, or C	6	14
4. Personal needs	32	76
5. Setup or preparation operations	24	57
6. Irregular or unusual operations	16	38

Source: J. O. P. Hummel, "Motion and Time Study in Leading American Industrial Establishments" (master's thesis, Pennsylvania State University).

it were a "fringe" issue. (Fringe benefits are company expenses which are not proportionate to employee output, such as insurance and pensions.) Allowance must be determined as accurately and correctly as possible; otherwise, all the care and precision that have been put into the study up to this point will be completely nullified.

Allowances are applied to three categories of the study. These are: (1) allowances applicable to the total cycle time, (2) allowances applicable to machine time only, and (3) allowances applicable to effort time only.

Allowances applicable to the total cycle time are usually expressed as a percentage of the cycle time and include such delays as personal needs, cleaning the work station, and oiling the machine. Machine time allowances include time for tool maintenance and power variance, while representative delays covered by effort allowances are fatigue and certain unavoidable delays.

There are two frequently used methods of developing standard allowance data. One is the production study, which requires an observer to study two or perhaps three operations over a long period of time. The observer records the duration of and reason for each idle interval (see Figure 16–1), and upon establishing a reasonably representative sample, he summarizes his findings to determine the percent allowance for each applicable characteristic. The data obtained in this fashion, like those for any time study, must be adjusted to the level of normal performance.

Since the observer must spend a long period of time in the direct observation of one or more operations, this method is exceptionally tedious, not only to the analyst but also to the operator or operators. Another disadvantage is that there is a tendency to take too small a sample, which may result in biased results.

The second technique of establishing the allowance percentage is

FIGURE 16–1
Lost time analysis chart

LOST TIME ANALYSIS OF TIME STUDY

Dwg._____ Part _____ Date _____

Operation _____

Symbol _____

A. Personal _____
B. Start work late _____
C. Stop work early _____
D. Talk with foreman or instructor _____
E. Talk with other persons _____
F. Search for tools _____
G. Search for drawings _____
H. Rework fault of operator _____
I. Rework fault of another operator _____
J. Rework fault of machine or fixtures _____
K. Idle-wait for crane (excess over allowed) _____
L. Idle-wait for inspector (excess over allowed) _____
M. Wait in line at tool crib (excess over allowed) _____
N. Wait in line at dispatch office (excess over allowed) _____
O. Wait in line at B/P station _____
P. Tool maintenance _____
Q. Oil machine _____
R. Clean work station _____
S. Circled readings (circled reading minus ave. for ele.) _____
T. Miscellaneous minor delays _____
U. Lost time developing methods during study _____
V. _____
W. _____
X. _____
Y. _____
Z. _____

                                                                      Total _____

1. Gross over-all _____ Mins. _____ Hrs.
2. Total lost _____   "   _____   "
3. % lost time compared with net actual (2 ÷ 4)   _____
4. Net actual or productive _____ Mins. _____ Hrs.
5. Allowed time   _____   "   _____   "

Note--Place lost time symbol alongside description of lost time on study and staple this card
to study.

through work sampling studies (see Chapter 21). This method involves the taking of a large number of random observations, thus requiring only part-time, or at least intermittent, services of the observer. In this method, no stopwatch is used, as the observer merely walks through the area under study at random times and notes briefly what each operator is doing.

The number of delays recorded, divided by the total number of observations during which the operator is engaged in productive work, will tend to equal the allowance required by the operator to accommodate the normal delays encountered.

In using work sampling studies for the determination of allowances, the observer must practice several precautionary measures. First, the observer must be careful that he does not anticipate his observations and that he records only the actual happenings. Then a given study should not cover dissimilar work, but should be confined to similar operations on the same general type of equipment. The larger the number of observations and the longer the period of time over which the data are taken, the more valid will be the results. Studies taken by R. L. Morrow indicate that fairly reliable results were obtained with 500 observations, while 3,000 observations gave very accurate results.[1] Daily observations should be taken over a span of at least two weeks.

Personal delays

Under the item of personal delays will come those cessations in work necessary for maintaining the general well-being of the employee. This will include trips to the drinking fountain and the rest room. The general working conditions and class of work will influence the time necessary for personal delays. Thus, working conditions involving heavy work performed at high temperatures, such as that done in the pressroom of a rubber-molding department or in a hot-forge shop, would require greater allowance for personal needs than would light work performed in comfortable temperature areas. Detailed production checks have demonstrated that a 5 percent allowance for personal time, or approximately 24 minutes in eight hours, is appropriate for typical shop working conditions. The amount of time needed for personal delays will, of course, vary to some extent with the person as well as the class of work. The 5 percent figure cited appears to be adequate for the majority of male and female workers.

Fatigue

Closely associated with the allowance for personal needs is the allowance for fatigue, although this allowance is usually applied only to the effort portions of the study. Fatigue allowances have not reached the state where their qualifications are based on sound, rational theories, and they probably never will. Consequently, next to performance rating, the fatigue allowance is the least defensible and the most open to argument of all the factors making up a time standard. However, fair fatigue allowances for different classes of work can be approximated by empirical means. Fatigue is not homogeneous in any respect; it ranges from strictly physical to purely psychological and includes combinations of both the physical and the psychological. It has a marked influence on some people, but apparently has little or no effect on others.

Whether the fatigue that sets in is physical or mental, the results are similar: there is a lessening in the will to work. The major factors that affect fatigue are well known and have been clearly established. Some of these are:

---

[1] Robert Lee Morrow, *Time Study and Motion Economy* (New York: Ronald Press Co., 1946).

1. Working conditions.
    a. Light.
    b. Temperature.
    c. Humidity.
    d. Air freshness.
    e. Color of room and environment.
    f. Noise.
2. Repetitiveness of work.
    a. Monotony of similar body movements.
    b. Muscular tiredness due to stressing some muscles.
3. General health of the worker, physical and mental.
    a. Physical stature.
    b. Diet.
    c. Rest.
    d. Emotional stability.
    e. Home conditions.

It is evident that fatigue can be reduced but never eliminated. In general, heavy work is diminishing in industry because of the marked progress in the mechanization of both material handling and processing elements. As industry becomes more automated, there will be less muscular tiredness due to the stressing of muscles. Thus, we have made real progress toward the decreasing of physical fatigue. The major problem of fatigue is not physical but psychological, and industry through its scientific selection programs is substantially reducing this factor by putting the right people on the right jobs. A person who has an unfavorable reaction to monotony is not placed on a monotonous job. Since it is not customary to provide fatigue allowance for the general health factors that influence the degree of fatigue, such conditions as emotional stability, rest, diet, and physical stature are usually considered in employee selection.

Because fatigue cannot be eliminated, proper allowance must be made for the working conditions and repetitiveness of work that influence the degree to which it sets in. Experiments have shown that fatigue must be plotted as a curve, not as a straight line. Figure 16–2 illustrates a typical work curve showing the relationship between the load in pounds and the time for handling each load.

Figure 16–3 shows the effect of fatigue in an assembly operation. After 80 blocks of work were completed (1 block involved 1,296 cycles), an experiment on exhaustion was performed in which the work load was increased by allowing only one five-minute break after the completion of a block of work. Many industrial studies have shown a drop in production toward the end of the working period which is attributable to fatigue alone. Usually the rate of production tends to increase during the early part of the day and then falls after the third hour. There is a short period of increased production after the lunch period, but this soon begins

FIGURE 16–2
Typical work curve

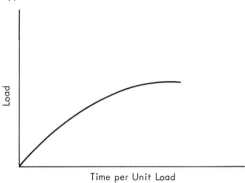

FIGURE 16–3
Learning curve for assembly work showing the impact of
financial incentive and the effect of fatigue after the
flattening of the learning curve

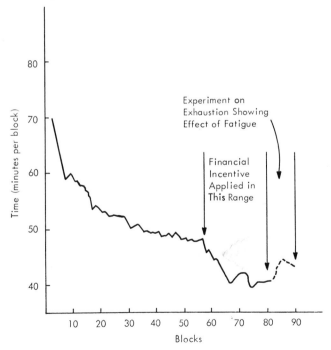

1 Block = 1,296 cycles.
Source: Redrawn from W. Rohmert and K. Schlaich, "Learning of Complex Manual Tasks," *International Journal of Production Research*, vol. 5, no. 2 (1966), p. 137.

to taper off, and output usually continues to decline for the balance of the working day.

Perhaps the most widely used method of determining the fatigue allowance is that of measuring the decline in production throughout the working period. Thus, the production rate for every quarter of an hour may be measured during the course of the working day. Any decline of production that cannot be attributed to methods changes or personal or unavoidable delays may be attributed to fatigue and expressed as a percentage. It must be recognized, however, that many outside factors, such as the state of health or outside interference, can influence the fatigue factor. Thus, many studies should be made in order to obtain a reasonable sample before deciding upon the final fatigue allowance for a given facility. Eugene Brey[2] has expressed the coefficient of fatigue as follows:

$$F = \frac{(T - t)\,100}{T}$$

where:

F = Coefficient of fatigue
T = Time required to perform the operation at the end of continuous work
t = Time required to perform the operation at the beginning of continuous work

Many attempts have been made to measure fatigue, none of which has been completely successful. The tests of fatigue may be classified into three groups: (1) physical, (2) chemical, and (3) physiological.

Physical tests include the various dynamometer tests of changes in the rate of working, such as the hand dynamometer, the mercury dynamometer, the water dynamometer, and the Martin spring balance for registering the force exerted by six different sets of large body muscles.

Chemical tests include the various techniques for analysis of the blood and the body secretions, such as saliva, so as to note changes resulting from fatigue.

The physiological tests of fatigue include the pulse count, blood pressure, the respiratory rate, oxygen consumption, and the production of carbon dioxide. Table 16–2 shows the rate of change in physiological reactions due to fatigue.

In recent years, considerable attention has been directed to the physiological requirements of various work assignments. This branch of the scientific study of the worker and his environment is being recognized as "occupational physiology." Here the effort per unit of time plus the physiological recovery time is measured for the different work assignments within a plant. As these studies progress, more quantitative data for esti-

---

[2] Eugene E. Brey, "Fatigue Research in Its Relation to Time Study Practice," *Proceedings, Time Study Conference, Society of Industrial Engineers.*

TABLE 16–2

Factor	Percent decrease	Percent increase
1. Pulse count	–	113.5
2. Pulse pressure	–	77.0
3. Respiratory rate	–	60.5
4. CO_2 combining power	41.0	–
5. Metabolism	–	43.0
6. White blood cells	–	57.0
7. Red blood cells	–	6.0
8. Blood pressure, diastolic	28.0	–
9. Blood pressure, systolic	–	20.5
10. Blood sugar	–	12.0

Source: Moss and Roe, *Physiological Reactions due to Fatigue.*

mating fatigue allowances will inevitably result. It is also expected that laws of physiological economy will be developed which will supplement and may even supersede some of the principles of motion economy.

In the development of equitable fatigue allowances, one of the foremost problems of time study is to determine at which portion of the fatigue curve a time study was made, and then to derive a fatigue allowance that can be used as a constant for work performed at the given work station in the future.

For most industrial operations, fatigue allowances have been arbitrarily broken into three elements, each of which has a spread of influence on the total fatigue allowance. These are: operations involving strenuous work, operations involving repetitive work, and operations performed under disagreeable working conditions. Of course, it is possible for more than one of these conditions to apply in any specific operation.

By taking controlled production studies from an adequate sample of work, it is possible to arrive at fatigue allowance values that will prove equitable for the various degrees of each of the above-enumerated factors. The apparent adequacy of fatigue allowances determined by measuring the decline in production through all-day production studies is due to the fact that the fatigue allowance for a given job is not a critical value, but may be safely established within a rather broad range.

The International Labour Office has tabulated the effect of working conditions in order to arrive at an allowance factor for personal delays and fatigue. This tabulation is shown in Table 16–3. The factors considered include: standing while working, abnormal positions demanded, use of force, illumination, atmospheric conditions, job attention required, noise level, mental strain, monotony, and tediousness.

In using this table, the analyst should determine an allowance factor for each element of the study. For example, element 3 of a given study may involve the application of a 40-pound force. Because of this use of

TABLE 16–3

A. Constant allowances:	
1. Personal allowance	5
2. Basic fatigue allowance	4
B. Variable allowances:	
1. Standing allowance	2
2. Abnormal position allowance:	
a. Slightly awkward	0
b. Awkward (bending)	2
c. Very awkward (lying, stretching)	7
3. Use of force, or muscular energy (lifting, pulling, or pushing):	
Weight lifted, pounds:	
5	0
10	1
15	2
20	3
25	4
30	5
35	7
40	9
45	11
50	13
60	17
70	22
4. Bad light:	
a. Slightly below recommended	0
b. Well below	2
c. Quite inadequate	5
5. Atmospheric conditions (heat and humidity)–variable	0–10
6. Close attention:	
a. Fairly fine work	0
b. Fine or exacting	2
c. Very fine or very exacting	5
7. Noise level:	
a. Continuous	0
b. Intermittent–loud	2
c. Intermittent–very loud	5
d. High-pitched–loud	5
8. Mental strain:	
a. Fairly complex process	1
b. Complex or wide span of attention	4
c. Very complex	8
9. Monotony:	
a. Low	0
b. Medium	1
c. High	4
10. Tediousness:	
a. Rather tedious	0
b. Tedious	2
c. Very tedious	5

force, an allowance of 9 percent would be used in the structure of the allowance computation for this element. It should be noted that this tabulation provides a basic 9 percent allowance to all effort elements, for personal delays, and fatigue. To this basic 9 percent allowance are added applicable variable allowances as enumerated in the table.

There are two ways of applying the fatigue allowance. It can be handled as a percentage which is added to the normal time, as has been explained. In this method, the allowance is based upon a percentage of the productive time only. As an alternative technique, the fatigue allowance can be handled through the establishment of periodic rest periods.

The former way is preferred because it allows the physically stronger employee to participate in greater earnings. If compulsory rest periods are used, the strong employee not as subject to fatigue as the average employee will be restricted in his or her output.

It should be recognized that rest periods will definitely reduce fatigue. If 10-minute rest periods are introduced into a plant, as they frequently are, the fatigue allowance that formerly prevailed should be proportionately modified. For example, if a bench assembly operation earned a fatigue allowance of 8 percent, and at some later date, through negotiation, a 10-minute rest period was provided in the morning and another 10-minute rest period in the afternoon, the fatigue allowance on this class of work would be diminished

$$\frac{20}{\text{Normal productive time}} = \underline{\qquad} \text{ percent}$$

The normal daily productive time on this class of work may be 400 minutes. The fatigue allowance accounted for by the 20-minute rest period would then be 20/400, or 5 percent. Thus, future standards in this area will carry a fatigue allowance of 8 percent minus 5 percent, or 3 percent.

Unavoidable delays

This class of delays applies to effort elements and includes such items as interruptions from the foreman, dispatcher, time study analyst, and others; material irregularities; difficulty in maintaining tolerances and specifications; and interference delays where multiple machine assignments are made.

As can be expected, every operator will have numerous interruptions during the course of the working day. These can be due to a wide range of reasons. The foreman or group leader may interrupt the operator to give him instructions or to clarify certain written information. Then the inspector may interrupt him to point out the reasons for some defective work that passed through the operator's work station. Interruptions frequently occur from planners, expediters, fellow workers, production personnel, time study men, dispatchers, and others.

Unavoidable delays are frequently a result of material irregularities. For example, the material may be in the wrong location, or it may be running slightly too soft or too hard. Again, it may be too short or too long, or may have excessive stock on it, as in forgings when the dies begin to wash out, or on castings due to incomplete removal of risers. When material deviates substantially from standard specifications, it may be necessary to restudy the job and establish allowed time for the extra elements introduced by the irregular material, as the customary unavoidable delay allowance may prove inadequate.

Machine interference

When more than one facility is assigned to an operator, there are times during the working day when one facility or more must wait until the operator completes his work on another facility he or she is servicing. As more facilities are assigned to the operator, the "interference" time delay is increased. In practice, machine interference has been found to "occur predominantly from 10 to 30 percent of the total working time, with extremes of from 0 to 50 percent."[3] The amount of machine interference is a function of the number of facilities assigned, the randomness of the required servicing time, the proportion of the service time to the running time, the length of the running time, and the mean length of the service time.

Although many expressions, tables, and charts have been developed to determine the magnitude of machine interference, the expression developed by Wright (*Mechanical Engineering,* vol. 58) is relatively simple and has proved to be satisfactory when the number of machines assigned is seven or more. When two, three, four, five, or six facilities are assigned, Wright recommends the use of empirical curves, as illustrated in Figure 16–4. For seven or more facilities, Wright developed:

$$I = 50[\sqrt{[(1 + X - N)^2 + 2N]} - (1 + X - N)]$$

where:

I = Interference expressed as a percentage of the mean attention time
X = Ratio of mean machine running time to mean machine attention time
N = Number of machine units assigned to one operator

Wright's formula for interference was developed from the solution of a problem in the congestion of telephone lines made by Thornton C. Fry of the Bell Telephone Laboratories.[4] The conditions applying to Dr. Fry's work included:

1. A telephone call requiring the use of a given trunk line will be delayed in the event that the trunk line is currently in use from a call on a different line. The second call will continue to be delayed until the first call is completed.
2. All telephone calls are of equal duration.

[3] H. B. Maynard, *Industrial Engineering Handbook* (New York: McGraw-Hill Book Co., 1956), pp. 3–78.

[4] Thornton C. Fry, *Probability and Its Engineering Uses* (Princeton, N.J.: D. Van Nostrand Co., Inc., 1928).

FIGURE 16–4
Interference in the percentage of attention time when the number of facilities assigned to one operator is six or less

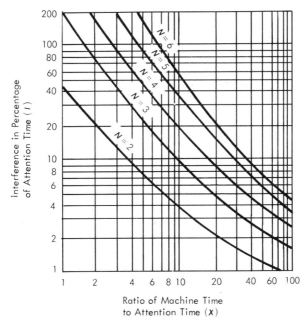

Ratio of Machine Time
to Attention Time (*x*)

3. Calls are assigned individually to groups of channels and collectively at random.

It can easily be seen that assumptions 2 and 3 do not always apply in multiple machine activity. However, as the number of machines increases, the random nature of servicing becomes more pronounced. Wright found his equation applicable in four entirely different industries where the number of machines assigned was greater than six.

For example, in the development of a standard for quilling production, an operator has been assigned 60 spindles. The mean running time per package (unit of output), determined by stopwatch study, is 150 minutes. The standard mean attention time per package, developed by time study, is 3 minutes.

The computation of the machine interference expressed as a percentage of the mean operator attention time would be:

$$I = 50[\sqrt{[(1 + X - N)^2 + 2N]} - (1 + X - N)]$$

$$I = 50\left[\sqrt{\left(1 + \frac{150}{3.00} - 60\right)^2 + 120} - \left(1 + \frac{150}{3.00} - 60\right)\right]$$

$$I = 50[\sqrt{(1 + 50 - 60)^2 + 120} - (1 + 50 - 60)]$$

$$I = 1,160 \text{ percent}$$

Thus, in this example we would have:

Machine running time . . . . 150.00 min.
Attention time, including the personal, fatigue, and
   unavoidable delay allowance . . . . 3.00 min.
Interference time (11.6) (3.00) . . . . <u>34.80 min.</u>
Standard time for 60 packages . . . . 187.80 min.
Standard time per package $\frac{187.80}{60}$ = 3.13 min.

The reader should recognize that the amount of interference that takes place is related to the performance of the operator. Thus, the operator demonstrating a low level of effort will experience more machine interference than would the operator who through higher effort reduces the time spent in attending the stopped machine. The analyst endeavors to determine the normal interference time which, when added to (1) the machine running time required to produce one unit and (2) the normal time spent by the operator in servicing the stopped machine, will equal the cycle time. The cycle time divided into the running time of each machine multiplied by the number of machines assigned to the operator will give the average machine running hours per hour. Thus, we have the expression:

$$O = \frac{NT_1}{C}$$

where:

O = Machine running hours per hour
N = Number of machines assigned to the operator
T_1 = Running time (hours) to produce one piece
C = Cycle time to produce one piece

and

$$C = T_1 + T_2 + T_3$$

where:

T_2 = Time (hours) spent by normal operator attending the stopped facility
T_3 = Time lost by normal operator working at normal pace because of interference

Using waiting line theory, analysts have developed tables where the interval between service times is exponential and where the service time is either constant or exponential. Table A3–14, Appendix 3, gives these values for various "k," which is the ratio of service time to facility running time $k = \frac{T_2}{T_1}$.

With reference to the previous example,

$$k = \frac{3.00}{150.00} = 0.02$$
$$N = 60$$

From Table A3–14, Appendix 3, with exponential service time and $k = 0.02$ and $N = 60$, we have a waiting time (interference delay) of 16.8 percent of the cycle time. Denoting the interference time by T_3, we have $T_3 = 0.168C$, where C is the cycle time to produce one unit per spindle. Then:

$$150 + 3.00 + 0.168C = C$$
$$0.832C = 153$$
$$C = 184 \text{ minutes}$$

and

$$T_3 = 0.168C = 30.9 \text{ minutes}$$

The reader can note the close agreement between the interference time computed by Wright's equation and that developed by the queuing model. However, as N (the number of machines assigned) becomes smaller, there is a greater proportion of difference between the two techniques.

Avoidable delays

It is not customary to provide any allowance for avoidable delays, which include visits with other operators for social reasons, uncalled-for work stoppages, and idleness other than rest to overcome fatigue. Of course, these delays may be taken by the operator at the expense of his output, but no allowance is provided for these cessations of work in the development of the standard.

Extra allowances

In typical metal trade and related operations, the allowance for personal, unavoidable, and fatigue delays usually approximates 15 percent. However, in certain cases, it may be necessary to provide an extra allowance to arrive at a fair standard. Thus, due to a substandard lot of raw material, it may be necessary to provide an extra allowance to take care of an unduly high generation of rejects caused by the poor material. Again, a situation may arise in which, because of the breakage of a jib crane, the operator is obliged to place a 50-pound casting in the chuck of his machine. Therefore, an extra allowance must be provided to take care of the additional fatigue brought on by the manual handling of the work.

Whenever practical, allowed time should be established for the additional work of any operation by breaking it down into elements and then including these times in the specific operation. If this is not practical, then an extra allowance must be provided.

A form of extra allowance frequently used, especially in the steel industry, is a percentage added to a portion or all of the cycle time when the operator must observe the process to maintain the efficient progress of the operation. This allowance is frequently referred to as "attention time" allowance, and is applied on such operations as:

1. The inspector in the electrolytic tinning operation while observing the tin plate coming off the line.
2. The first helper in an open hearth while observing the condition of the performance or molten bath.
3. The first helper in an open hearth while reporting to or receiving instructions from the melter.
4. The craneman in a shipping department while receiving directions or signals from the crane hooker.
5. The hi-mill roller in the seamless hot mill while required, during the rolling process, to watch for any change in the length of pipe to signal the operator to return the pipe if too short and adjust the rolls as required.[5]

The amount of allowance added to the actual required time for required attention by the U.S. Steel Company is 35 percent. Thus, an operator can earn 35 percent premium on a process-paced operation if he is able to keep the process running efficiently during the entire working day.

Using an extra allowance on operations where a large portion of the cycle time is based on the machine or process cycle is quite popular in plants which practice incentive wage payment. The allowance is added so that the operator may make earnings equivalent to those of operators assigned to operations that are not machine paced. A typical practice is to allow 30 percent extra allowance on the machine-controlled portion of the cycle time. This extra allowance provides the incentive for the operator to keep his facility productively employed during the entire working period.

Without this extra allowance, the operator would find it impossible to make the same earnings as his fellow employees. For example, if the machine-controlled portion of a cycle is two minutes and the operator-controlled portion is one normal minute, the operator would have to work at a pace 25 percent above normal in order to realize a 7 percent increase in productivity.

$$1.07 = \frac{3 \text{ minutes (normal cycle time)}}{1 \text{ minute (normal effort time)}}$$
$$1.25 \text{ (operator performance during effort time)}$$
$$+ 2 \text{ minutes (machine-controlled time)}$$

$1.07 - 1.00 = 0.07$ increase in productivity when the operator is working at 25 percent above normal during the effort part of the cycle

Now, if an extra allowance of 25 percent were added to the machine-controlled portion of the cycle, the operator would be able to achieve 25 percent incentive earnings if he worked at a pace 25 percent above

---

[5] From United States Steel Time Study Manual.

normal and did not utilize more allowance than was provided for personal delays, unavoidable delays, and fatigue.

Clean work station and oil machine

The time required to clean and lubricate the operator's machine may be classified as an unavoidable delay. However, this time, when performed by the operator, is usually included as a total cycle time allowance. The type and size of equipment, and the material being fabricated,

TABLE 16–4
Clean machine allowance chart

	Percent per machine		
Item	*Large*	*Medium*	*Small*
1. Clean machine when lubricant is used	1	¾	½
2. Clean machine when lubricant is not used	¾	½	¼
3. Clean and put away large amounts of tools or equipment	½	½	½
4. Clean and put away small amounts of tools or equipment	¼	¼	¼
5. Shut machine down for cleaning (this percentage is for machines equipped with chip pans, which are stopped at intervals to permit sweeper to clean away large chips)	1	¾	½

TABLE 16–5
Machine classification

Large machine	*Medium machine*
1. Turret lathe (20-in. chuck or over)	1. Turret lathe (10-in. to 20-in. chuck)
2. Boring mill (60 in. and over)	2. Boring mill (under 60 in.)
3. Punch press (100T and over)	3. Punch press (40T to 100T)
4. Planer (over 48 in.)	

TABLE 16–6
Oil machine allowance chart

	Percent per machine		
Item	*Large*	*Medium*	*Small*
1. Machine oiled or greased by hand	1½	1	½
2. Machine oiled automatically	½	½	½

will have considerable effect on the time required to clean the work station and lubricate the equipment. When these elements are included as part of the responsibilities of the operator, an applicable allowance must be provided. One concern has established the accompanying table of allowances to cover these items.

Frequently, the elements "clean work station" and "oil machine" are handled by giving the operator 10 or 15 minutes at the end of the day in which to perform this work. Of course, when this is done, the standards established would not include any allowance for cleaning and oiling the machine.

Power feed machine time allowance

The allowance required for power feed elements will frequently differ from that required for effort elements. Two factors are usually considered in applying allowances for power feed. These factors are power variance and tool maintenance.

Allowances are made for variance in power for reduced speeds brought on by belt slippage and for shutdowns because of minor repairs. In case a major repair to the facility is needed, then an extra allowance would be provided. This extra allowance would not be applied within the standard, but would be an independent standard covering the machine repair.

Tool maintenance allowance provides time for the operator to maintain his tools after the original setup. In the setup time, the operator is expected to provide first-class tools properly ground. Generally, little tool maintenance takes place during the course of the average production run. In long runs, of course, tools will have to be sharpened periodically. The percentage allowance for tool maintenance will vary directly with the number of perishable tools in the setup. For example, one manufacturer's tool maintenance allowance table is as follows:

Allowance for tool maintenance	*Percent*
1. One or more tools ground in tool crib	1
2. One tool sharpened by the operator	3
3. Two or more tools cutting at one time sharpened by the operator	6

Application of allowances

The fundamental purpose of allowances is to add enough time to the normal production time to enable the average worker to meet the standard when performing at a normal pace. It is customary to express the allowance as a multiplier so that the normal time, consisting of productive work elements, can be adjusted readily to the time allowed. Thus, if a 15 percent allowance were to be provided on a given operation, the multiplier would be 1.15.

Care must be exercised when including the allowance with the time study standard. It must be remembered that allowance is based on a percentage of the daily production time and not on the overall workday. For example, if a study revealed that in an eight-hour working day 50 minutes of delay time are to be allowed during 400 minutes of normal production time, then the percentage allowance applicable would be 50/400, or 12.5 percent.

Allowance is based on the normal production time, since it is this value that the percentage will be applied to on subsequent studies.

Typical allowance values

In an industrial survey comprising 42 different plants, the smallest average total allowance found to be in use was 10 percent. This was used in a plant producing household electrical appliances. The greatest average allowance was 35 percent, which was found to be in effect in two different steel plants. The average allowance of all the plants from which replies were received was 17.7 percent.

Typical allowances established in a certain plant for standard operations appear in Table 16–7. These values may or may not apply in other plants.

Summary

In establishing time study allowance values, the same care that prevails in taking individual studies should be exercised throughout. It would be folly to divide a job carefully into elements, precisely measure the duration of each element in hundredths of a minute, accurately evaluate the performance of the operator, and then arbitrarily assign an allowance picked at random. Practices of this nature will lead to inaccurate standards. If the allowances are too high, obviously manufacturing costs will be unduly inflated, and if the allowances are too low, tight standards will result, which will cause poor labor relations and the eventual failure of the system.

TEXT QUESTIONS

1. Compare the interference delay allowance using the queuing model and Wright's equation, where $N = 20$, mean facility running time is 120 minutes, and attention time is 3 minutes.
2. What three broad areas are allowances intended to cover?
3. To what three categories of the time study are allowances provided? Give several examples of each.
4. What are the two methods used in developing allowance standard data? Briefly explain the application of each technique.
5. Give several examples of personal delays. What percentage allowance seems adequate for personal delays under typical shop conditions?

TABLE 16–7
Time study allowances for standard operations

Symbol	Facility or operation	Method	Total applied to effort time	Total applied to machine time	Personal	Clean work station	Oil machine	Shutdown	Tool maintenance	Unavoidable delays and fatigue
21	Anneal	Oven	10	—	5	½	—	—	—	4½
22	Assembly	Bench	13	—	5	—	—	—	—	8
23	Assembly	Floor	14.5	—	5	—	—	—	—	9½
24	Blacksmith	Drop forge	21	—	7	1	—	—	—	13
25	Brake	Power	15	—	5	½	—	—	—	9½
26	Braze	Electric	15	—	5	½	½	½	—	9½
27	Drill	Hand feed	15	—	5	½	½	—	2	6½
28	Drill	Power feed	15	12	5	½	½	m-2	m-4	—
29	Engrave	Pantograph	18	—	5	½	½	e-½	e-2	e-6½
30	Lathe, English	Over 36 inches	15	15	5	2	1	m-2	m-5	e-7
31	Lathe, engine		15	15	5	2	1	m-2	m-5	e-7
32	Lathe, turret		17	15	5	2	1	m-2	m-5	e-9
33	Milling		16	15	5	2	1	m-2	m-5	e-8
34	Grinding	Blanchard	15	15	5	2	1	m-2	m-5	e-7
35	Grinding	Thread	17	15	5	2	1	m-2	m-5	e-9
36	Grinding	External and internal	16	15	5	2	1	m-2	m-5	e-8
37	Punch press	Up to 100 tons	14		5	½	1	½	—	7
38	Saw	Circular	14		5	½	½	½	1	6½
39	Saw	Do-All	15		5	½	½	1½	2	5½
40	Shear	Square	15		5	½	½	1		8
41	Welder	Spot	17		5	½	—	2	3	6½
42	Paint	Spray	17		5	2	—	1	1	8

m: applied to machine time only.
e: applied to effort time only.

6. What are some of the major factors that affect fatigue?
7. Under what groups have the tests of fatigue been classified?
8. What operator interruptions would be covered by the unavoidable delays allowance?
9. What percentage allowance is usually provided for avoidable delays?
10. When are "extra allowances" provided?
11. What fatigue allowance should be given to a job if it developed that it took 1.542 minutes to perform the operation at the end of continuous work and but 1.480 minutes at the beginning of continuous work?
12. Why are allowances based on a percentage of the productive time?
13. What is meant by "occupational physiology?
14. Define "attention time." Why is it necessary to apply an allowance to attention time?
15. Based upon the International Labour Office's tabulation, what would be the allowance factor on a work element involving a 42-pound pulling force if there was inadequate light and exacting work was required?
16. Explain what relation, if any, there is between financial incentives and the effect of fatigue.
17. What are the advantages of having operators oil and clean their own machines?

GENERAL QUESTIONS

1. Do you feel that the operation or the operator determines the extent to which personal delay time is utilized? Why or why not?
2. Why is fatigue allowance frequently applied only to the effort areas of the work cycle?
3. Should fatigue allowances vary with the different shifts for a given class of work? Why or why not?
4. What are the objections to determining fatigue allowances by measuring the decline of production not attributable to methods changes or personal or unavoidable delays?
5. Give several reasons for not applying an "extra" allowance to operations when the major part of the cycle is machine controlled and the internal time is small compared to the cycle time.
6. With the extension of automation and the tendency of many companies to go to the four-day week, do you feel that personal allowances should be increased? Why or why not?

PROBLEM

1. The work measurement analyst is planning to develop a table of allowances for a given class of work in his maintenance department, using the work sampling technique. The areas for which he wants to establish allowances and the variation within which he expects to find the actual allowance 95 percent of the time are as follows:

Personal allowance	3 to 7 percent
Crane wait	2 to 6 percent
Grind tools	5 to 9 percent
Avoidable delays	1 to 4 percent
Unavoidable delays	10 to 20 percent

How many random observations should he take? Over what period of time should he take these observations? Explain how he will determine when to take each day's observations?

SELECTED REFERENCES

International Labour Office. *Introduction to Work Study.* Geneva, Switzerland: Atar, 1964.

Barnes, Ralph M. *Motion and Time Study: Design and Measurement of Work.* 6th ed. New York: John Wiley & Sons, Inc., 1968.

Mundel, Marvin E. *Motion and Time Study: Principles and Practices.* 4th ed. Englewood Cliffs, N.J.: Prentice-Hall, Inc., 1960.

Nadler, Gerald. *Work Design: A Systems Concept.* Rev. ed. Homewood, Ill.; Richard D. Irwin, Inc., 1970.

17

The standard time

The standard time for a given operation is the time required for an average operator, fully qualified and trained and working at a normal pace, to perform the operation. It is determined by summarizing the allowed time for all of the individual elements comprised by the time study.

Elemental allowed times are determined by multiplying the mean elapsed elemental time by a conversion factor. Thus, we have the expression:

$$T_a = (M_t)(C)$$

where:

T_a = Allowed elemental time
M_t = Mean elapsed elemental time
C = Conversion factor found by multiplying the performance rating factor by one plus the applicable allowance

For example, if the mean elapsed elemental time of element 1 of a given time study was 0.14 minute and the performance factor was 0.90 and an allowance of 18 percent was applicable, the allowed elemental time would be:

$$T_a = (0.14)(0.90)(1.18) = (0.14)(1.06) = 0.148$$

Allowed elemental times are rounded off to three places to the right of the decimal point. Thus, in the preceding example, 0.1483 minute is recorded as 0.148 minute. If the result were 0.1485 minute, then the allowed time would be 0.149 minute.

Expressing the standard time

The sum of the elemental allowed times will give the standard in minutes per piece or hours per piece, depending on whether a decimal minute or decimal hour watch is used. The majority of industrial operations have relatively short cycles (less than five minutes); consequently, it is usually more convenient to express standards in terms of hours per hundred

Motion and time study

pieces. For example, the standard on a press operation might be 0.085 hour per hundred pieces. This is a more satisfactory method of expressing the standard than is 0.00085 hour per piece or 0.051 minute per piece. Thus, if the operator produced 10,000 pieces during the working day, he would have earned 8.5 hours of production, and would have performed at an efficiency of 106 percent. This is expressed as:

$$E = \left(\frac{He}{Hc}\right)\left(100\right)$$

where:

E = Percent efficiency
He = Standard hours earned
Hc = Clock hours on the job

In another instance, the standard time may have resulted in 11.46 minutes per piece. This would be converted into decimal hours per hundred pieces as follows:

$$S_h = 1.667 S_m$$

where:

S_h = Standard expressed in hours per hundred pieces
S_m = Standard expressed in minutes per piece
1.667 = Constant developed by converting minutes to decimal hours and multiplying by 100

Thus:

$$S_h = (1.667)(11.46)$$
$$= 19.104 \text{ hrs.}/C$$

If an operator produced 53 pieces in a given working day, the standard hours produced would be:

$$H_e = (0.01)(P_a)(S_h)$$

where:

H_e = Standard hours earned
P_a = Actual production in pieces
S_h = Standard expressed in hours per hundred

In this example:

$$H_e = (0.53)(19.104) = 10.125 \text{ hours}$$

Once the allowed time has been computed, the standard is released to the operator in the form of an operation card. The card can be run off on a Ditto machine, or some other duplicating process can be used. It will serve as the basis for routing, scheduling, instruction, payroll, opera-

FIGURE 17–1
Typical production operation card

PRODUCTION OPERATION CARD						
DESCRIPTION __Shower head face__ ____DWG. NO. __JB-1102__ ____PART NO. __J-1102-1__						
MADE FROM __2 1/2" diam. 70-30 extruded brass rod__						
Routing 9-11-12--14-12-18					DATE __9-15__	
OP. NO.	OPERATION	DEPT.	MACHINE AND SPECIAL TOOLS	SET-UP MINUTES	EACH PC. MINUTES	
1	Saw slug	9	J. & L. Air Saw	15 min		.077
2	Forge	11	150 Ton Maxi F-1102	70 min		.234
3	Blank	12	Bliss 72 F-1103	30 min		.061
4	Pickle	14	HCL. Tank	5 min		.007
5	Pierce 6 holes	12	Bliss 74 F-1104	30 min		.075
6	Rough ream and chamfer	12	Delta 17" D.P. F-1105	15 min		.334
7	Drill 13/64" holes	12	Avey D.P. F-1106	15 min		.152
8	Machine stem and face	12	#3 W.&S.	45 min		.648
9	Broach 6 holes	12	Bliss 74 1/2	30 min		.167
10	Inspect	18	F-1109, F-1110, F-1112		Daywork	

tor performance, cost, budgeting, and other necessary controls for the effective operation of a business. Figure 17–1 illustrates a typical production operation card.

Temporary standards

It is common knowledge that time is required to become proficient in any operation that is new or somewhat different.

Frequently, the time study man will be required to establish a standard on an operation that is relatively new and on which there is insufficient volume for the operator to reach his top efficiency. If the analyst bases his grading of the operator upon the usual conception of output, the resulting standard will seem unduly tight, and the operator will in all probability be unable to make any incentive earnings. On the other hand, if he takes into consideration the fact that the job is new and that the volume is low, and establishes a liberal standard, he may find himself in trouble if the size of the order is increased, or if a new order for the same job is received.

Perhaps the most satisfactory method of handling situations like this is through the issuance of temporary standards. By doing this, the time study analyst will establish the standard giving consideration to the difficulty of the work assignment and the number of pieces to be produced. Then, by using a learning curve for the type of work being studied and existing standard data, he can arrive at an equitable standard for the work at hand. The resulting standard will be considerably more liberal than if the job were being produced on a mass-production basis. The standard, when released to the production floor, will be clearly marked as a "temporary" standard and will show the maximum quantity for which it applies. It is good practice to issue temporary standards on vouchers of a different color than that of permanent standard vouchers so as to indicate clearly to all affected parties that the rate is temporary and is subject to restudy in the event of additional orders or an increase in the volume of the present order.

The analyst must be careful as to the number of temporary standards released, since too many such standards can result in a lowering of the approved conception of normal. Also, the operators may strongly object to the changing of temporary standards to permanent standards, since the tighter permanent standard appears to them as a rate or wage-cutting procedure. Only new work that is definitely foreign to the operator and that involves limited quantities of production should be considered as justifying the issuance of temporary standards. When temporary standards are released, they should be in effect for the duration of the contract or for 60 days, whichever period is shorter. Upon their expiration, they should be replaced by permanent standards. Several union contracts specifically state that temporary standards lasting longer than 60 days must become permanent.

Setup standards

The elements of work commonly included in setup standards involve all events that take place from the time the previous job was completed to the starting of the first piece of the present job. It is also customary to include in the setup standard the "teardown" or "put-away" elements: these include all items of work involved from the completion of the last piece to the setting up of the next job. Typical elements appearing in the setup standard would be:

1. Punching in on job.
2. Getting tools from tool crib.
3. Getting drawings from dispatcher.
4. Setting up machine.
5. Punching out on job.
6. Removing tools from machine.
7. Returning tools to crib.

In establishing setup times, the analyst uses the identical procedure followed in establishing standards for production. First, he should be assured that the best setup methods are in effect and that a standardized procedure has been adopted. Then the work is carefully broken down into elements, accurately timed, performance rated, and subjected to the appropriate allowance. The importance of valid setup times cannot be overemphasized especially in the case of job shops, where setup time represents a high proportion of the overall time.

The analyst must be especially alert when timing setup elements because he will not have the opportunity to get a series of elemental values for determining the mean times. Also, he will not be able to observe the operator perform the elements in advance and, consequently, will be obliged to divide the setup into elements while the study is taking place. Of course, setup elements for the most part are long in duration, and the analyst will find that he has a reasonable amount of time to break the job down, record the time, and evaluate the performance as the operator proceeds from one work element to the next.

There are two ways of handling setup times: first they can be distributed over a specific manufacturing quantity, such as 1,000 pieces or 10,000 pieces. This method is satisfactory only when the magnitude of the production order is standard. For example, industries that ship from stock and reorder on a basis of minimum-maximum inventories are able to control their production orders so as to conform to economical lot sizes. In cases like this, the setup time can be equitably prorated over the lot size. Suppose that the economical lot size of a given item was 1,000 pieces and that reordering was always done on the basis of 1,000 units. In the event that the standard setup time in a given operation was 1.50 hours, then the allowed operation time could be increased by 0.15 hour per 100 pieces in order to take care of the makeready and put-away elements.

This method would not be at all practical if the size of the order were not controlled. In a plant that requisitions on a job-order basis, that is, releases production orders specifying quantities in accordance with customer requirements, it would be impossible to standardize on the size of the work orders issued to the plant. Thus, this week an order for 100 units may be issued, and next month an order for 5,000 units of the same part. In the example cited above, the operator would be allowed but 0.15 hour to set up his machine for the 100-unit order, which would be inadequate. On the 5,000-unit order, he would be given 7.50 hours, which would be considerably too much time.

It is more practical to establish setup standards as separate allowed times. Then regardless of the quantity of parts to be produced, a fair standard prevails. In some concerns, the setup is performed by a person other than the operator who does the job. The advantages of having separate setup men are quite obvious. Lower skilled men can be utilized as operators when they do not have to set up their own facilities. Setups are more readily standardized and methods changes more easily introduced when the responsibility for setup rests with but one individual. Also, if sufficient facilities are available, continuous production performance can be achieved by having the next work assignment set up while the operator is working on his present job.

Partial setups

Frequently, it will not be necessary to set up a facility completely to perform a given operation, because some of the tools of the previous operation are required in the job that is being set up. For example, in hand screw machine or turret lathe setups, careful scheduling of similar work to the same machine will allow partial setups from one job to the next. Instead of having to change six tools in the hex turret, it may be necessary to change only two or three. This savings in setup time is one of the principal benefits of a well-formulated group technology program.

Since the sequence of work that is scheduled to a given machine seldom remains the same, it is difficult to establish partial setup times to cover all the possible variations. The standard setup for a given No. 4 Warner & Swasey turret lathe might be 0.80 hour. However, if this setup is performed after job X, it might take only 0.45 hour, and if it is performed after job Y, it may require 0.57 hour, while following job Z, 0.70 hour may be necessary, and so on. The possible variations in partial setup time are so broad that the only practical way to establish their values is on the basis of standard data (see Chapter 18) for each job in question.

In plants where setup times are relatively short (less than one hour) and where production runs are reasonably long, it is common practice to allow the operator the full setup time for each job he performs. This is advantageous for several reasons: first, the operators will be considerably more satisfied because of higher earnings, and they will tend to plan their work to the best possible advantage. This will result in more production per unit of time and lower total costs. Also, considerable time and paperwork are saved by avoiding the determination of a standard for the partial setup operations and its application in all pertinent cases. In fact, this saving tends to approach the extra amount paid to the operator resulting from the difference between the time required to make the complete setup and the time required to make the partial setup.

Maintenance of standard times

Considerable emphasis has been given to the necessity of establishing time standards that are fair. This means both fair to the worker and fair to management. Once fair standards have been introduced, it is equally important that they be maintained.

The standard time is directly dependent upon the method used during the course of the time study. Method, in the broad sense, refers not only to the tools and facilities being used, but also to such details as operator motion pattern, work station layout, material conditions, and working conditions. Since method controls the time standard, it is essential, if equitable standards are to be maintained, that methods changes and alterations be controlled. If methods changes are not controlled, inequities will soon develop in the standards established and much of the work spent on the development of consistent time standards will be undone.

Just as the financial records of a company are periodically subject to audit, so should all established time standards be checked at regular intervals to see whether they are in line with the method being used. The audit of time standards principally involves the investigation of the method currently being used by the operator. Frequently, minor changes will have been made by the operator, the foreman, or even the methods department, and no record of these changes will have been given to the time study analyst. It is not uncommon for workers to conceal methods changes for which they are responsible, so that they can increase their earnings or diminish their effort while achieving the same production. Of course, changes in method may develop which will increase the amount of time required to perform the task. These changes may be initiated by the foreman or inspector and may be of insufficient consequence, in his opinion, to adjust the standard.

As has been pointed out throughout this text, it is important that the operation being time-studied be analyzed for possible methods improvements prior to the establishment of the standard. Operation analysis, work simplification, motion study, and standardization of the method and conditions always precede work measurement. A standard does not get out of line if the method that was time-studied is maintained by the operator. If methods study has developed the ideal method, and if this method is standardized and followed by the operator, then there is less need to maintain time standards.

Frequently, however, methods changes will be introduced—both favorable and unfavorable. If these changes are extensive, they will be brought to the attention of management. Tight standards will be brought to management's attention by the operator. Standards that become very loose will be brought to management's attention through the payroll department where excessive earnings on the part of a given worker will be reported. However, it is the minor accruing methods changes that frequently take place unnoticed and weaken the entire standards structure. In order to maintain standards properly, the time study department should periodically verify the method being used with the method that was studied when the standard was established. This can readily be done by referring to the original time study where a complete description of the method employed was recorded. If this investigation reveals that the method has been changed adversely, then the reasons for the change should be investigated so that the better method can be employed. If the method has been improved, then the investigation should determine who was responsible for the innovation. If the operator developed the improvement, he or she should be justly rewarded through a "suggestion plan" or other means. Regardless of where the methods change came from, those elements affected should be restudied and the current standard introduced.

So as to first verify standards that may be out of line, the time study department will enlist the cooperation of the foreman. The foreman, being close to the operators coming under his jurisdiction, will at an early stage be aware of standards that may be either loose or tight. He will be able to advise the time study analyst as to the sequence in the audit of existing standards.

The audit frequency should be determined at the time the standard is developed. This is based upon an estimate of the number of hours of application of the standard in one year. For example, one large company uses the following data to determine the frequency of the audit of methods and standards:[1]

Hours of application per standard per year	Frequency of audit
0–10	Once per 3 years
10–50	Once per 2 years
50–600	Once per year
Over 600	Twice per year

The auditing of methods and standards, when properly done, takes time and, consequently, is expensive. However, it is important that this maintenance be carried out to help assure the success of the program. The audit should be carried out by a representative of the standards staff. Only completely qualified analysts should be used for auditing work. The auditor will observe the operation as performed to determine the correctness of the description, sequences, frequencies, conditions, and standard time allowances. It is usually satisfactory to check the accuracy of the standard time by measuring several cycles of the overall time, performance rating the data, and adding an appropriate allowance.

Conclusion

It is common practice to issue standards for setup operations, independent of "piece" standards, and to specify the allowed time in decimal hours or decimal minutes. Piece standards are expressed in hours per hundred pieces for ease in payroll computation, scheduling, and control. On extremely short-cycle operations, such as punch press work, die cast-

---

[1] By H. B. Brandt, former associate director, Industrial Engineering Division, Procter & Gamble Co.

ing, or forging, it may be more desirable to express standards in terms of hours per thousand pieces, since during the course of an eight-hour shift, several thousand pieces can be produced.

Work should be scheduled when feasible to take advantage of partial setups in order to improve delivery dates, decrease total cost, and allow greater remuneration for the operator. Where short setup times prevail and an extensive variety of small-volume orders are handled daily, no attempt should be made to evaluate partial setup times, but credit for full setup should be given the operators. Where setup times are relatively long, such as setting up a six-spindle automatic screw machine, and where production runs are substantial, then consideration should be given for evaluation of partial setup times. Credit for this time rather than for the complete setup time should be given to the operator.

Time standards must be maintained in order to assure a satisfactory rate structure. This calls for a continuing analysis of methods. All standards should be checked periodically to verify that the methods being employed are identical with those that were in use at the time the standards were established.

TEXT QUESTIONS

1. Define the term *standard time*.
2. How is the conversion factor determined?
3. Why is it usually more convenient to express standards in terms of time per hundred pieces rather than time per piece?
4. For what reason are temporary standards established?
5. Express the standard of 5.761 minutes in terms of hours per hundred pieces. What would the operator's efficiency be if 92 pieces were completed during a working day? What would his efficiency be if he set up his machine (standard for setup = 0.45 hours) and produced 80 pieces during the eight-hour workday?
6. What elements of work are included in the setup standard?
7. What is the preferred method of handling setup time standards?
8. How is the operator compensated on partial setups on hand screw machines?
9. Determine the conversion factor and the allowed time for a job that had an average time of 5.24 minutes and carried a performance factor of 1.15 and an allowance of 12 percent.
10. Explain why it is necessary that time standards be properly maintained.
11. Explain how you would use learning curves in the establishment of temporary standards.

GENERAL QUESTIONS

1. For what reasons might it be advantageous to express allowed times in minutes per piece?

2. How can the excessive use of temporary standards cause poor labor relations?
3. How does a company effectively maintain standards and avoid the reputation of being a "rate-cutter"?
4. If an audit of a standard revealed that the standard as originally established was 20 percent loose, explain in detail the methodology to be followed to rectify the rate.

PROBLEMS

1. Establish the money rate per hundred pieces from the following data:
   Cycle time (averaged measured time): 1.23 minutes
   Base rate: $5.40 per hour.
   Pieces per cycle: 4.
   Machine time (power feed): 0.52 minutes per cycle.
   Allowance: 17 percent on effort time; 12 percent on power feed time.
   Element 1 average time: 0.09 minutes.
   MTM time for element 1: 132 TMU. (One TMU = 0.00001 hr.)
   Plant uses synthetic performance rating.

2. The following data resulted from a time study taken on a horizontal milling machine:
   Pieces produced per cycle: 8.
   Average measured cycle time: 8.36 minutes.
   Average measured effort time per cycle: 4.62 minutes.
   Average rapid traverse time: 0.08 minutes.
   Average cutting time power feed: 3.66 minutes.
   Performance factor: +15 percent.
   Allowance (machine time): 10 percent.
   Allowance (effort time): 15 percent.

   The operator works on the job a full eight-hour day and produces 380 pieces. How many standard hours does the operator earn? What is his efficiency for the eight-hour day?

SELECTED REFERENCES

Graham, C. F. *Work Measurement and Cost Control*. London, England: Pergamon Press, 1965.

Rotroff, Virgil H. *Work Measurement*. New York: Reinhold Publishing Corp., 1959.

Sylvester, Arthur L. *The Handbook of Advanced Time-Motion Study*. New York: Funk & Wagnalls Co., 1950.

18

Standard data

Standard time data for the most part are elemental time standards taken from time studies that have been proved to be satisfactory. These elemental standards are classified and filed so that they can readily be abstracted when needed. Just as the housewife refers to her cookbook to determine how many minutes to cream butter and sugar, how long to beat the mixture, and thus how much time is required to bake the cake, so the analyst can refer to standard data and determine how long it should take the normal operator to pick up a small casting and place it in a jig, close the jig and lock the part with a quick-acting clamp, advance the spindle of the drill press, and perform the remainder of the elements required to produce the part.

The application of standard time data is fundamentally an extension of the same kind of process as that used to arrive at allowed times through the medium of stopwatch time study. The principle of standard data application certainly is not new; many years ago, Frederick W. Taylor proposed that each established elemental time be properly indexed so that it could easily be found and used in the establishment of time standards for future work. When we speak of standard data today, we refer to all the tabulated elemental standards, curves, alignment charts, and tables that are compiled to allow the measurement of a specific job without the necessity of a timing device, such as the stopwatch.

Work standards calculated from standard data will be relatively consistent in that the tabulated elements comprised by the data are a result of many proven stopwatch time studies. Since the values are tabulated, and it is only necessary to accumulate the required elements in establishing a standard, the various time study men within a given company will arrive at identical standards of performance for a given method. Therefore, consistency is assured for standards established by the different analysts within a plant as well as for the various standards computed by a given time study observer.

Standards on new work can usually be computed more rapidly through standard data than by means of stopwatch time study. The rapidity with which standards are established by means of standard data allows the establishment of standards on indirect labor operations, which is usually impractical if done by stopwatch methods. Typically, one work measurement analyst will establish 5 rates per day using stopwatch methods, while he will establish 25 rates a day with the standard data technique. The use of standard data permits the establishment of time standards over a wide range of work. Table 18–1 illustrates the coverage that is possible when standard data elements are determined.

TABLE 18–1

Operation	Number of time studies taken for development of standard data	Number of standards set in one year from standard data developed	Percent of standards set that would be covered by stopwatch studies
Coremaking	60	7,500	0.8
Snag grinding	40	656	6.1
Visual inspection	53	422	12.6
Turret lathe operation	100	600	16.7

Source: Phil Carroll, Jr., "Notes on Standard Elemental Data," *Modern Machine Shop*, April 1950, p. 176.

Development of standard time data

In the development of standard time data, it is necessary to distinguish constant elements from variable elements. A constant element is one for which the allowed time will remain approximately the same for any part within a specific range. A variable element is one for which the allowed time will vary within a specified range of work. Thus, the element "start machine" would be a constant, and the element "drill ⅜-inch diameter hole" would vary with the depth of the hole and the feed and speed of the drill.

As standard data are developed, they should be indexed and filed. Setup elements should be kept separate from those elements incorporated in the each piece time, and constant elements should of course be kept separate from the variable elements. Typical standard data would be tabulated as follows:

A. Machine or operation.
    1. Setup.
        a. Constants.
        b. Variables.
    2. Each piece.
        a. Constants.
        b. Variables.

Standard data are compiled from the different elements that have occurred on the time studies that have been taken on a given process over a period of time. Only those studies that have proven to be valid through use are employed. In tabulating standard data, the analyst should be careful to define end points clearly. Otherwise, there may be an overlapping of time in the recorded data. For example, in the element "out stock to stop" done on a bar feed No. 3 Warner & Swasey turret lathe, the element could include reaching for the feed lever, grasping the lever, feeding the bar stock through the collet to a stock stop located in the hex turret, closing the collet, and reaching for the turret handle. Then, again, this element may involve only the feeding of bar stock through the collet to a stock stop. Since standard data elements are compiled from a great number of studies taken by different time study men, care must be exercised in defining the limits or end points of each element. Figure 18–1 illustrates a form used to summarize data taken from an individual time study for the purpose of developing standard data on die-casting machines.

In order to fill a specific need in a standard data tabulation, the analyst may resort to work measurement of the particular element in question. This is handled quite accurately by using the "fast" watch (see Chapter 13), which records elapsed times in 0.001 of a minute. In this type of analysis, the snapback method is used to record the elapsed elemental time, as we are usually interested in determining the allowed time for only a few of the elements comprising the study. Upon completion of the observations, the elemental elapsed times are summarized and the mean determined, as in the case of a typical time study. The average values are then performance rated, and an allowance is added so as to arrive at fair allowed times.

Sometimes, because of the brevity of individual elements, it is impossible to measure their duration separately. It is possible to determine their individual values by timing groups collectively and using simultaneous equations to solve for the individual elements.

For example, element a might be "pick up small casting," element b might be "place in leaf jig," c might be "close cover of jig," d "position jig," e "advance spindle," and so on. These elements could be timed in groups as follows:

$$a + b + c \text{ equals element 1 equals 0.070 min.} = A \tag{1}$$
$$b + c + d \text{ equals element 3 equals 0.067 min.} = B \tag{2}$$
$$c + d + e \text{ equals element 5 equals 0.073 min.} = C \tag{3}$$
$$d + e + a \text{ equals element 2 equals 0.061 min.} = D \tag{4}$$
$$e + a + b \text{ equals element 4 equals 0.068 min.} = E \tag{5}$$

By adding these five equations:

$$3a + 3b + 3c + 3d + 3e = A + B + C + D + E$$

Then let

$$A + B + C + D + E = T$$
$$3a + 3b + 3c + 3d + 3e = T = 0.339 \text{ min.}$$

and

$$a + b + c + d + e = \frac{0.339}{3} = 0.113 \text{ min.}$$

Therefore:

$$A + d + e = 0.113 \text{ min.}$$

Then:

$$d + e = 0.113 \text{ min.} - 0.07 \text{ min.} = 0.043 \text{ min.}$$

since

$$c + d + e = 0.073 \text{ min.}$$
$$c = 0.073 \text{ min.} - 0.043 = 0.03 \text{ min.}$$

Likewise:

$$d + e + a = 0.061$$

and

$$a = 0.061 - 0.043 = 0.018 \text{ min.}$$

Substituting in equation (1)

$$b = 0.070 - (0.03 + 0.018) = 0.022$$

Substituting in equation (2)

$$d = 0.067 - (0.022 + 0.03) = 0.015 \text{ min.}$$

Substituting in equation (3)

$$e = 0.073 - (0.015 + 0.03) = 0.028 \text{ min.}$$

In determining standard data elements by means of simultaneous equations, extreme care must be exercised to be consistent when reading the watch at the terminal points of the established elements. Inconsistency in establishing terminal points will result in erroneous standard data elements.

Calculation of cutting times

Through the knowledge of feeds and speeds for different types of material, it is a relatively easy matter for the analyst to calculate and tabulate the cutting times for different machining operations. Table 18–2 gives recommended speeds and feeds for high-speed drills used on various kinds of material. This type of information is available in the various technical handbooks and can readily be obtained from the cutting tool manufacturers.

FIGURE 18–1

DIE-CASTING MACHINE

Mach. No.
Part No. _____ & Type _____ Operator _____ Date _____
                No. of Parts   Method of Placing   Total Wt. of Flsh,
Of _____ in Tote Pan _____ Parts in Tote Pan _____ Parts, Gate & Sprue _____
        No. of Parts   Liquid Metal _____
_____ per Shot _____ Plastic Metal _____ Chill ____ Skim ____ Drain ____
Capacity in Lbs.   Describe
Holding Pot _____ Greasing _____
_____

Describe
Loosening of Part _____

Describe
Location _____
_____

ELEMENTS	TIME	END POINTS
Get metal in holding pot	_____	All waiting time while metal is being poured in pot.
Chill metal	_____	From time operator starts adding cold metal to liquid metal in pot until operator stops adding cold metal to liquid metal in pot.
Skim metal	_____	From time operator starts skimming until all scum has been removed.
Get ladleful of metal	_____	From time ladle starts to dip down into metal until ladleful of metal reaches edge of machine or until ladle starts to tip for draining.
Drain metal	_____	From time ladle starts to tip for draining until ladleful reaches edge of machine.
Pour ladle of metal in machine	_____	From time ladleful of metal reaches edge of machine until foot starts to trip press.
Trip press	_____	From time foot starts moving toward pedal until press starts downward.
Press time	_____	Complete turnover of press.
Hold plunger down	_____	From time plunger stops downward motion until plunger starts moving upward.
Press button and raise slug	_____	From time plunger stops moving until slug is raised out of cavity.
Remove slug--drop slug--lift	_____	From time slug is raised out of cavity until slug is pushed into tote pan or pot.
Trip pedal to open dies	_____	From time foot starts moving to pedal until dies start to open.
Wait for dies to open	_____	From time dies start to open until die stops moving.
Remove part from die	_____	From time die stops moving until part is free of die cavity.
Place part in tote pan	_____	From time part is free of die cavity until part is placed in tote pan.

Drill press work

A drill is a fluted end-cutting tool used to originate or enlarge a hole in solid material. In drilling operations on a flat surface, the axis of drill is at 90° to the surface being drilled. When a hole is drilled completely

TABLE 18-2

A. Recommended drilling speeds for various materials with high-speed steel drills

Material	Recommended speed (surface feet per minute)
Aluminum and its alloys	200–300
Bakelite.	100–150
Brass and bronze, soft	200–300
Bronze, high tensile.	70–100
Cast iron, soft	100–150
Cast iron, hard.	70–100
Magnesium and its alloys.	250–400
Malleable iron	80–90
Nickel and monel metal	40–60
Slate, marble, and stone	15–25
Steel,.forgings	50–60
Steel, manganese (15% Mn)	15–25
Steel, soft	80–110
Steel, stainless	30–30
Steel, tool	50–60
Wrought iron	50–60
Wood	300–400

Note: Carbon steel drills should be run at speeds approximately 40 to 50 percent of those recommended for high-speed steel.

B. Recommended feeds for drills of various diameters

Diameter of drill (inches)	Feed (inches per revolution)
Under $\frac{1}{8}$	.001 to .002
$\frac{1}{8}$ to $\frac{1}{4}$	.002 to .004
$\frac{1}{4}$ to $\frac{1}{2}$	.004 to .007
$\frac{1}{2}$ to 1.	.007 to .015
1 and over.	.015 to .025

Note: It is best to start with a moderate speed and feed, increasing either or both after observing the action and condition of the drill.

Source: National Twist Drill & Tool Co.

through a part, the lead of the drill must be added to the length of the hole in order to determine the entire distance the drill must travel to make the hole. When a blind hole is drilled, the analyst does not add the lead of the drill to the hole depth, because the distance from the surface being drilled to the farthest penetration of the drill is the distance that the drill must travel in order to make the required depth of hole (see Figure 18–2).

Since the commercial standard for the included angle of drill points is 118°, the lead of the drill may readily be found through the expression:

$$l = \frac{r}{\tan A}$$

FIGURE 18–2
Distance L indicates the distance drill must travel when drilling through
(illustration at left) and when drilling blind holes (illustration at right)
(lead of drill is shown by distance 1)

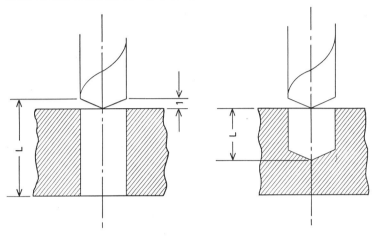

where:

$$l = \text{Lead of drill}$$
$$r = \text{Radius of drill}$$
$$\tan A = \text{Tangent of } \tfrac{1}{2} \text{ the included angle of the drill}$$

To illustrate, let us calculate the lead of a general purpose drill 1 inch
in diameter:

$$l = \frac{0.5}{\tan 59°}$$

$$l = \frac{0.5}{1.6643}$$

$$l = 0.3 \text{ inches lead}$$

Once the total length that the drill must move has been determined,
then the feed of the drill in inches per minute is divided into this distance
for the purpose of determining the drill cutting time in minutes.

Drill speed is usually expressed in feet per minute, and feed is expressed in thousandths of an inch per revolution. In order to change the
feed into inches per minute when the feed per revolution and the speed
in feet per minute are known, the student should substitute in the following equation:

$$F_m = \frac{3.82(f)(Sf)}{d}$$

where:

$$F_m = \text{Feed in inches per minute}$$
$$f = \text{Feed in inches per revolution}$$
$$Sf = \text{Surface feet per minute}$$
$$d = \text{Diameter of drill in inches}$$

For example, to determine the feed in inches per minute of a 1-inch drill running at a surface speed of 100 feet per minute and a feed of 0.013 inch per revolution, we should figure

$$F_m = \frac{(3.82)(0.013)(100)}{1} = 4.97 \text{ inches per minute}$$

If we wished to determine how long it would take for this 1-inch drill running at the same speed and feed to drill through 2 inches of a malleable iron casting, we should substitute in the equation:

$$T = \frac{L}{F_m}$$

where:

$$T = \text{Cutting time in minutes}$$
$$L = \text{Total length drill must move}$$
$$F_m = \text{Feed in inches per minute}$$

and we should have

$$T = \frac{2 \text{ (thickness of casting)} + 0.3 \text{ (lead of drill)}}{4.97}$$
$$= 0.464 \text{ minutes cutting time}$$

The cutting time so calculated does not include any allowance, which of course must be added to determine the allowed time. The allowance will include time for variations in material thickness and tolerance in the setting of stops, both of which affect the cycle cutting time to some extent. Also, personal and unavoidable delay allowance must be added in order to arive at an equitable allowed elemental time.

It must be remembered that all speeds are not available on the machine being used. For example, the recommended spindle speed for a given job might be 1,550 rpm; however, the fastest speed that the machine is capable of running may be 1,200 rpm. In that case, 1,200 rpm would be used and would be the basis of computing allowed times.

Lathe work

There are many variations of machine tools that can be classified in the lathe group. These would include the engine lathe, turret lathe, and automatic lathe (automatic screw machine). All of the lathe group machine tools are used primarily with stationary tools or with tools that translate over the surface to remove material from the revolving work, which may be in the form of forgings, castings, or bar stock. In some cases, the tool is revolved while the work is held stationary, as on certain stations of automatic screw machine work. Thus, a slot in a screwhead can be machined in the slotting attachment on the automatic screw machine.

Speeds and feeds are altered by many factors, such as the condition and design of the machine tool, type of material being cut, condition and design of the cutting tool, coolant used for cutting, method of holding the work, and method of mounting the cutting tool. Table 18–3 outlines the approximate cuts, feeds, and speeds for certain metallic and nonmetallic turning.

As in drill press work, feeds are usually expressed in terms of thousandths of an inch per revolution, and speeds in terms of surface feet per minute.

In order to determine the cutting time for L inches of cut, it is merely necessary to divide the length of cut in inches by the feed in inches per minute or, expressed algebraically:

$$T = \frac{L}{F_m}$$

where:

T = Cutting time in minutes
L = Total length of cut
F_m = Feed in inches per minute

and

$$F_m = \frac{3.82(S_f)(f)}{d}$$

where:

f = Feed in inches per revolution
S_f = Speed in surface feet per minute
d = Diameter of work in inches

Milling machine work

Milling refers to the removal of material with a rotating multiple-toothed cutter. While the cutter rotates, the work is fed past the cutter; thus, the milling machine differs from the drill press, where the work usually is stationary. In addition to machining plane and irregular surfaces, the milling machine is adapted for cutting threads, slotting, and cutting gears.

TABLE 18–3

Approximate cuts, feeds, and speeds for metalic, and nonmetallic turning (Tabular values are in fpm)

Class	Material, SAE No.	Cutting tool material	Cut, 0.005 to 0.015; feed, 0.002 to 0.005	Cut, 0.015 to 0.094; feed, 0.005 to 0.015	Cut, 0.094 to 0.187; feed, 0.015 to 0.030	Cut, 0.187 to 0.375; feed, 0.030 to 0.050	Cut, 0.375 to 0.750; feed, 0.030 to 0.090
Free-cutting steels	1112, X-1112 1120 1315, etc.	18-4-1 HSS	....	250– 350	175–250	80–150	55– 75
		Cast alloys	....	425– 550	315–400	215–300	100–210
		Sintered carbides	750–1,500	600– 750	450–600	350–450	175–350
Carbon and low-alloy steels	1010 1025	18-4-1 HSS	....	225– 300	150–200	75–125	45– 65
		Cast alloys	....	375– 500	275–350	180–250	100–175
		Sintered carbides	700–1,200	550– 700	400–550	300–400	150–300
Medium-alloy steels	1030 1050	18-4-1 HSS	....	200– 275	125–175	70–120	40– 60
		Cast alloys	....	325– 400	230–300	160–225	80–150
		Sintered carbides	600–1,100	450– 600	350–450	250–350	125–250
High-alloy steels	1060 1095 1350	18-4-1 HSS	....	175– 250	125–175	65–100	35– 55
		Cast alloys	....	250– 350	200–250	150–200	65–150
		Sintered carbides	500– 750	400– 500	300–400	200–300	100–300
Nickel steels	2330 2350	18-4-1 HSS	....	200– 275	130–180	70–110	45– 60
		Cast alloys	....	300– 375	225–300	145–200	85–140
		Sintered carbides	550– 800	425– 550	325–425	225–325	125–225
Chromium nickel-chrome steels	3120, 3450 5140, 52100	18-4-1 HSS	....	150– 200	100–125	50– 75	30– 50
		Cast alloys	....	230– 315	165–225	110–130	55–110
		Sintered carbides	425– 550	325– 425	250–325	175–250	75–175
Molybdenum steels	4130 4615	18-4-1 HSS	....	160– 210	110–140	60– 80	35– 55
		Cast alloys	....	250– 325	160–225	120–150	65–100
		Sintered carbides	475– 650	350– 475	275–350	200–275	100–200
Chrome, vanadium, and stainless steels	6120 6150 18 Cr-8Ni 6195	18-4-1 HSS	....	100– 150	80–100	50– 75	30– 50
		Cast alloys	....	210– 250	170–200	110–165	55–100
		Sintered carbides	375– 500	300– 375	250–300	175–250	75–175
Tungsten steels	7260, 18-4-1 annealed	18-4-1	....	120– 150	75–120	40– 75	25– 40
		Cast alloys	....	130– 175	110–130	80–100	35– 80
		Sintered carbides	325– 400	250– 325	200–250	150–200	50–150
Special steels	12-14 % Mn	18-4-1 HSS					
		Cast alloys					
		Sintered carbides	200– 250	125– 200	75–125	50– 75	
	Si elect. sheet ingot iron, etc.	18-4-1 HSS					
		Cast alloys	400– 500	300– 400	200–300	150–200	
		Sintered carbides	1,000–1,200	800–1,000	600–800	500–600	

TABLE 18-3 (continued)

Class	Material, SAE No.	Cutting tool material	Cut, 0.005 to 0.015; feed, 0.002 to 0.005	Cut, 0.015 to 0.094; feed, 0.005 to 0.015	Cut, 0.094 to 0.187; feed, 0.015 to 0.030	Cut, 0.187 to 0.375; feed, 0.030 to 0.050	Cut, 0.375 to 0.750; feed, 0.030 to 0.090
Cast iron	Soft Gray	18-4-1 HSS	....	120- 150	90-120	75- 90	35- 75
		Cast alloys	....	225- 300	160-220	125-160	70-125
		Sintered carbides	450- 600	350- 450	250-350	200-250	100-200
	Medium and malleable	18-4-1 HSS	....	120- 150	90-120	60- 90	30- 60
		Cast alloys	....	190- 225	150-190	120-150	60-120
		Sintered carbides	350- 450	250- 350	200-250	150-200	75-150
	Hard alloy	18-4-1 HSS	....	90- 125	60- 90	40- 60	20- 40
		Cast alloys	....	120- 170	80-120	55- 80	35- 55
		Sintered carbides	250- 300	150- 250	100-150	75-100	50- 75
	Chilled	18-4-1 HSS	10- 15	10- 30			
		Cast alloys					
		Sintered carbides	30- 50				
Copper-base alloys	Leaded, free-cutting, soft brass, and bronze	18-4-1 HSS	....	300- 400	225-300	150-255	100-150
		Cast alloys	....	500- 600	400-500	325-400	200-325
		Sintered carbides	1,000-1,250	800-1,000	650-800	500-650	300-500
	Normal brass, bronze low alloy	18-4-1 HSS	....	275- 350	225-275	150-225	75-150
		Cast alloys	....	375- 425	325-375	250-325	175-250
		Sintered carbides	700- 800	600- 700	500-600	400-500	200-400
	Tough copper, high tin, manganese, and aluminum bronzes, gilding metal	18-4-1 HSS	....	100- 150	75-100	50- 75	35- 50
		Cast alloys	....	225- 300	180-225	125-180	75-125
		Sintered carbides	500- 600	400- 500	300-400	200-300	100-200
Light alloys	Magnesium	18-4-1 HSS	500- 750	350- 500	275-350	200-275	125-200
		Cast alloys	700-1,000	500- 700	400-500	300-400	200-300
		Sintered carbides	1,250-2,000	800-1,250	600-800	500-600	300-500
	Aluminum	18-4-1 HSS	350- 500	225- 350	150-225	100-150	50-100
		Cast alloys	450- 650	300- 450	225-300	150-225	75-150
		Sintered carbides	700-1,000	450- 700	300-450	200-300	100-200
Plastics	Thermoplastic, thermosetting	18-4-1 HSS	650-1,000	400- 650	250-400	150-250	
		Cast alloys					
		Sintered carbides					
Abrasives	Glass, hard rubber, green ceramics, marble, slate	18-4-1 HSS	150- 250	75- 150			
		Cast alloys					
		Sintered carbides					

Source: Firth Sterling, Inc., Pittsburgh, Pa., and *Tool Engineers Handbook* (New York: McGraw-Hill Book Co. 1949)

In milling work, as in drill press and lathe work, the speed of the cutter is expressed in surface feet per minute. Feeds or table travel are usually represented in terms of thousandths of inch per tooth.

In order to determine the cutter speed in revolutions per minute from surface feet per minute and the diameter of the cutter, the following expression may be used:

$$N_r = \frac{3.82S_f}{d}$$

where:

N_r = Cutter speed in revolutions per minute

S_f = Cutter speed in feet per minute

d = Outside diameter of cutter in inches

To determine the feed of the work in inches per minute into the cutter, use the expression

$$F_m = fn_t N_r$$

where:

F_m = Feed of the work in inches per minute into the cutter

f = Feed in inches per tooth of cutter

n_t = Number of cutter teeth

N_r = Cutter speed in revolutions per minute

The number of cutter teeth suitable for a particular application may be expressed as:

$$n_t = \frac{F_m}{F_t \times N_r}$$

where:

F_t = Chip thickness

Table 18-4 gives suggested feeds and speeds for milling under average conditions.

In computing cutting time on milling operations, the analyst must take into consideration the lead of the milling cutter when determining the total length of cut under power feed. This can be determined by triangulation as illustrated in the example of slab-milling a pad shown in Figure 18-3.

In this case, the lead BC must be added to the length of the work (8 inches) in order to arrive at the total length that must be fed past the cutter. Clearance for removal of the work after the machining cut will, of course, be handled as a separate element because greater feed under rapid table traverse will be used. By knowing the diameter of the cutter, the analyst determines AC as being the cutter radius, and calculates the

TABLE 18–4
Feeds and speeds for milling (average conditions)

		WIDTH OF CUT IN INCHES															
		¼	½	¾	1	1¼	1½	2	2½	3	4	5	6	7	8	10	12
¼	Feed	4.5	4.5	4	4	4	3.5	3.5	3.5	3	3	3	2.5	2.5	2.5	2.25	2.25
	Speed	110	110	105	105	100	100	100	95	95	95	90	90	90	85	85	85
½	Feed	4.5	4	4	4	3.5	3.5	3.5	3	3	3	2.5	2.5	2.25	2.25	2	2
	Speed	105	105	100	100	95	95	95	90	90	90	85	85	85	80	80	80
¾	Feed	4	4	3.5	3.5	3.5	3	3	3	2.5	2.5	2.25	2.25	2	2	1.75	1.75
	Speed	100	100	95	95	90	90	90	85	85	85	80	80	80	75	75	75
1	Feed	4	3.5	3.5	3.5	3	3	3	2.5	2.5	2.25	2.25	2	2	1.75	1.5	
	Speed	95	95	90	90	85	85	85	80	80	80	75	75	75	70	70	
1¼	Feed	3.5	3.5	3.5	3	3	3	2.5	2.5	2.25	2	1.75	1.5	1.5	1.25		
	Speed	90	90	85	85	80	80	80	75	75	75	70	70	70	65		
1½	Feed	3.5	3.5	3	3	3	2.5	2.5	2.25	2	1.75	1.5	1.25	1			
	Speed	85	85	80	80	75	75	75	70	70	70	65	65	65			
2	Feed	3.5	3	3	3	2.5	2.5	2.25	2	1.75	1.5	1.25	1				
	Speed	80	80	75	75	70	70	70	65	65	65	60	60				
2½	Feed	3	3	2.5	2.5	2.25	2	1.75	1.5	1.25	1						
	Speed	80	75	75	70	70	65	65	65	60	60						
3	Feed	2.5	2.5	2.25	2	1.75	1.5	1.25	1								
	Speed	75	75	70	70	65	65	60	60								
4	Feed	2.5	2	1.75	1.5	1.25	1										
	Speed	75	70	70	65	65	60										
5	Feed	2	1.5	1.25	1												
	Speed	70	70	65	65												
6	Feed	1.75	1.5	1													
	Speed	70	65	65													

(Left axis label: DEPTH OF CUT IN INCHES)

Correction Factors

S.A.E. 1020	= 1.00
S.A.E.1035—Annld.	= .80
Tool Steel—Annld.	= .60
Alloy Steels:	
Free Machining	= .80
Medium	= .65
Tough	= .50
Cast Steel	= .40
Cast Irons:	
Hard	= .70
Medium	= .95
Soft	= 1.20
Malleable Iron	= .80
Brass and Bronze:	
Free Machining	= 2.50
Medium	= 1.60
Hard	= .80
Monel Metal	= .70
Magnesium Alloys	= 5.00
Aluminum	= 2.00

Feeds are given in Inches per Minute.
Speeds are given in Surface Feet per Minute.

To obtain the proper Feed and Speed for a given material, multiply both Feed and Speed figures given in table by the correction factor for that material.

Source: National Twist Drill & Tool Co.

height of the right triangle *ABC* by subtracting the depth of cut *BE* from the cutter radius *AE:*

$$BC = \sqrt{AC^2 - AB^2}$$

Let us assume, in the example above, that the cutter diameter is 4 inches, and that it has 22 teeth. The feed per tooth is 0.008 inch, and the cutting speed is 60 feet per minute. The cutting time is computed by using the equation

$$T = \frac{L}{F_m}$$

FIGURE 18–3
Slab-milling a casting 8 inches in length

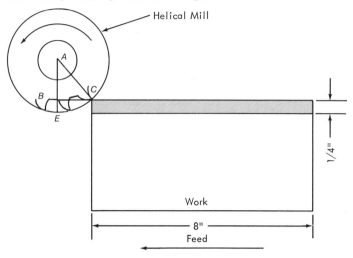

where:

$$T = \text{Cutting time in minutes}$$
$$L = \text{Total length of cut under power feed}$$
$$F_m = \text{Feed in inches per minute}$$

Then L would be equal to 8 inches $+ BC$ and

$$BC = \sqrt{4 - 3.06} = 0.975$$

Therefore:

$$L = 8.975$$
$$F_m = f n_t N_r$$
$$F_m = (0.008)(22)N_r$$

or

$$N_r = \frac{3.82 S_f}{d} = \frac{(3.82)(60)}{4} = 57.3 \text{ revolutions per minute}$$

Then

$$F_m = (0.008)(22)(57.3) = 10.1 \text{ inches per minute}$$

and

$$T = \frac{8.975}{10.1} = 0.888 \text{ minutes cutting time}$$

Through a knowledge of feeds and speeds, the time study analyst can determine the required cutting or processing time for various types of work that will be performed in his plant. The illustrations cited in drill press, lathe, and milling work are representative of the techniques used in establishing raw cutting times. To these values must be added the necessary applicable allowances so that fair elemental allowed values can be determined.

Determining horsepower requirements

When developing standard data times for machine elements, it is advisable to tabulate horsepower requirements for the various materials in relationship to depths of cut, cutting speeds, and feeds. The analyst will frequently use standard data for the planning of new work. So that he does not overload existing equipment, it is important that he have information as to the work load being assigned to each machine for the conditions under which the material is being removed. For example, in the machining of high-alloy steel forgings on a lathe capable of a developed horsepower of 10, it would not be feasible to take a ⅜-inch depth of cut while operating at a feed of 0.011 inch per revolution and a speed of 200 surface feet per minute. Table 18–5 indicates a horsepower requirement of 10.6 for these conditions. Consequently, the work would need to be planned for a feed of 0.009 inch at a speed of 200 surface feet, since in this case the required horsepower would be but 8.7. (See Table A3–12, Appendix 3, for horsepower requirements.)

Plotting curves

Because of space limitations, it is not always convenient to tabularize values for variable elements. By plotting a curve or a system of curves in the form of an alignment chart, the analyst can express considerable standard data graphically on one page.

Figure 18–4 illustrates a nomogram used for determining turning and facing time. For example, if the problem is to determine the production in pieces per hour to turn 5 inches of a 4-inch diameter shaft of medium carbon steel on a machine utilizing 0.015-inch feed per revolution and having a cutting time of 55 percent of the cycle time, the answer could be readily determined graphically. By connecting the recommended cutting speed of 150 feet per minute for medium carbon steel shown on scale 1 to the 4-inch diameter of the work shown on scale 2, a speed of 143 rpm is shown on scale 3. The 143-rpm point is connected with the 0.015-inch feed per revolution that is shown on scale 4. This line extended to scale 5 shows a feed of 2.15 inches per minute being used. This feed point connected with the length of cut shown on scale 6 (5 inches) gives the required cutting time on scale 7. This cutting time of 2.35 minutes, when connected with the percentage of cutting time shown on scale 8 (in this case, 55 percent), gives the production in pieces per hour on scale 9 (in this case, 16).

TABLE 18-5

Horsepower requirements for turning high-alloy steel forgings for cuts ⅜ inch and ½ inch deep at varying speeds and feeds

Surface feet	⅜-in. depth cut (feeds, in./rev.)						½-in. depth cut (feeds, in./rev.)					
	0.009	0.011	0.015	0.018	0.020	0.022	0.009	0.011	0.015	0.018	0.020	0.022
150	6.5	8.0	10.9	13.0	14.5	16.0	8.7	10.6	14.5	17.3	19.3	21.3
175	8.0	9.3	12.7	15.2	16.9	18.6	10.1	12.4	16.9	20.2	22.5	24.8
200	8.7	10.6	14.5	17.4	19.3	21.3	11.6	14.1	19.3	23.1	25.7	28.4
225	9.8	11.9	16.3	19.6	21.7	23.9	13.0	15.9	21.7	26.1	28.9	31.8
250	10.9	13.2	18.1	21.8	24.1	26.6	14.5	17.7	24.1	29.0	32.1	35.4
275	12.0	14.6	19.9	23.9	26.5	29.3	15.9	19.4	26.5	31.8	35.3	39.0
300	13.0	16.0	21.8	26.1	29.0	31.9	17.4	21.2	29.0	34.7	38.6	42.5
400	17.4	21.4	29.1	34.8	38.7	42.5	23.2	28.2	38.7	46.3	51.5	56.7

FIGURE 18–4
Nomogram for determining facing and turning time

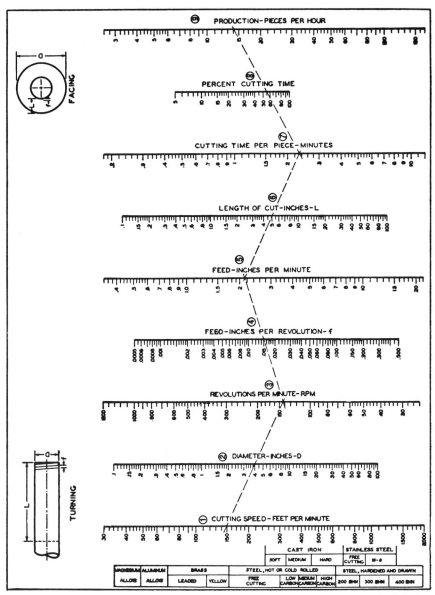

Crobalt, Inc.

There are, however, some distinct disadvantages to using curves. First, it is easy to introduce an error in reading from the curve because of a certain amount of interpolation that is usually required. Then, there is always the chance of outright error through incorrect reading or misalignment of intersections on the various scales.

When we refer to the plotting of curves to show the relationship between time and those variables that affect time, our solution may take the form of a single straight line, a curved line, a system of straight lines as in the ray chart, or a special arrangement of lines characteristic of an alignment chart or nomogram. In the plotting of simple, one-line curves, the analyst should observe certain standard procedures. First, it is standard practice to plot time on the ordinate of the charting paper, and the independent variable on the abscissa. If practical, all scales should begin at zero so that their true proportions can be seen. Lastly, the scale selected for the independent variable should have a sufficient range to utilize fully the paper on which the curve is being plotted. Figure 18–5

FIGURE 18–5

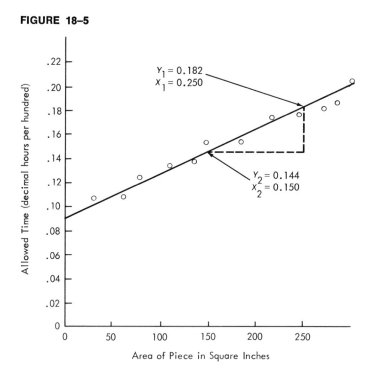

illustrates a chart expressing "forming" time in hours per hundred pieces for a certain gage of stock over a range of sizes expressed in square inches.

Each point in this chart represents a separate time study. Thus, 12 time studies were used in compiling data for this curve. Inspection of the plotted points revealed a straight-line relationship between the various studies. The equation for a straight line is:

$$Y = mx + b$$

(The equation for a straight line is also frequently expressed as: $Y = a + bx$.)

Y = Ordinate (hours per hundred pieces)

x = Abscissa (area of piece in square inches)

m = Slope of the straight line, or the proportionate change of time on the Y-axis for each unit change on the X-axis

b = Intercept of the straight line with the Y-axis when $x = 0$

The curve drawn by inspection shows an intercept value of 0.088 on the Y-axis, and the slope can be calculated by the equation:

$$m = \frac{y_1 - y_2}{x_1 - x_2}, \text{ where } x_1 y_1 \text{ and } x_2 y_2 \text{ are specific points on the curve}$$

$$= \frac{0.182 - 0.144}{250 - 150} = 0.00038$$

The least squares method

It is also possible to solve for m and b using the method of least squares. In this technique, the resulting slope and y-axis intercept will give a straight line where the sum of the squares of the vertical deviations of observations from this line is smaller than the corresponding sum of the squares of deviations from any other line. The two equations that are solved simultaneously are:[1]

$$\Sigma y = Nb + m\Sigma x \tag{1}$$
$$\Sigma xy = b\Sigma x + m\Sigma x^2 \tag{2}$$

The derivations of these equations are available in most mathematical texts.

In the example cited, we can solve for m and b using the least squares method as shown in Table 18–6. Substituting in equation (1) and (2):

$$12b = 1.834 - 2073m \tag{1}$$
$$2,073b = 348.82 - 453,801m \tag{2}$$

Multiplying equation (1) by 2,073 and equation (2) by 12:

$$24,876b = 3,801.882 - 4,297,329m$$
$$\underline{24,876b = 4,185.840 - 5,445,612m}$$
$$0 = -383.958 + 1,148,283m$$

[1] These equations are frequently expressed: $\Sigma y = na + b\Sigma x$ and $\Sigma xy = a\Sigma x + b\Sigma x^2$.

TABLE 18–6

Study	x or area	y or time	xy	x²
1........	25	0.104	2.60	625
2........	65	0.109	7.09	4,225
3........	77	0.126	9.70	5,929
4........	112	0.134	15.01	12,544
5........	135	0.138	18.63	18,225
6........	147	0.150	22.05	21,609
7........	185	0.153	28.31	34,225
8........	220	0.174	38.28	48,400
9........	245	0.176	43.12	60,025
10........	275	0.182	50.05	75,625
11........	287	0.186	53.38	82,369
12........	300	0.202	60.60	90,000
	2,073	1.834	348.82	453,801

or

$$m = \frac{383.958}{1,148,283} = 0.000334$$

and substituting in equation (1):

$$12b = 1.834 - 0.692$$
$$b = \frac{1.142}{12} = 0.095$$

Solving by regression line equations

As shown, the least squares method calls for solving simultaneous equations. These equations can get unwieldy. By solving these equations for the slope m and the y intercept b, direct substitution can be employed. Identical totals, as in the least squares solutions are used, which include:

$$\Sigma x, \; \Sigma x^2, \; \Sigma y, \; \Sigma xy$$

and N, the number of data. The regression line equation to solve for the constant b is:

$$b = \frac{(\Sigma x^2)(\Sigma y) - (\Sigma x)(\Sigma xy)}{(N)(\Sigma x^2) - (\Sigma x)^2}$$

and the slope m is computed as follows:

$$m = \frac{(N)(\Sigma xy) - (\Sigma x)(\Sigma y)}{(N)(\Sigma x^2) - (\Sigma x)^2}$$

Solving for b and m in the foregoing example, we get:

$$b = \frac{(453{,}801)(1.834) - (2{,}073)(348.82)}{(12)(453{,}801) - (2{,}073)^2}$$

$$= \frac{109{,}168}{1{,}148{,}283} = 0.095$$

and

$$m = \frac{(12)(348.82) - (2{,}073)(1.834)}{(12)(453{,}801) - (2{,}073)^2}$$

$$= \frac{383.96}{1{,}148{,}283} = 0.000334$$

Using standard data

For ease of reference, constant standard data elements are tabularized and filed under the machine or the process. Variable data can be tabularized or may be expressed in terms of a curve or an equation, and are also filed under the facility or operation class.

Where standard data are broken down so as to cover a given machine and class of operation, it may be possible to combine constants with variables and to tabularize the summary, thus allowing quick reference data that will express the allowed time to perform a given operation completely. Figure 18–6 illustrates welding data in which the constants "change electrode" and "arc" have been combined with the variables "weld cleaning" and "welding," and the result has been expressed in man-hours required to weld 1 inch for various sizes of welds.

Table 18–7 illustrates standard data for a given facility and operation class where again, elements have been combined so that it is necessary only to identify the job in question as to distance that the strip of sheet stock is moved per piece, in order to determine the allowed time for the complete operation.

TABLE 18–7
Standard data for blanking and piercing strip stock hand feed with piece automatically removed on Toledo 76 punch press

L (distance in inches)	T (time in hours per hundred hits)
1	0.075
2	0.082
3	0.088
4	0.095
5	0.103
6	0.110
7	0.117
8	0.123
9	0.130
10	0.137

FIGURE 18–6

CLASSIFICATION

KIND____Fillet_____

TYPE____Flat Position_____

ELECTRODE____E-6020 D.H._____

WELDING PROCEDURE GENERAL
PROCESS OF WELDING_____Shielded Metallic Arc_____POSITION ___Flat_____
MATERIAL____Mild Steel to Mild Steel_____P.D.S._1550, 1555___S.A.E.__1010____
ELECTRODE____E-6020_____D.H._____Convex_____Heavy_____
　　　　　　A.W.S. CLASS　　TYPE　　SHAPE OF WELD　　COATING
POWER SOURCE (A.C. OR D.C. – AND POLARITY IF D.C.)_____D.C. Straight_____
BACKING____None_____PEENING____None_____CHIPPING ____None_____
PREHEAT_____None_____
STRESS RELIEVING_____None_____

WELDING PROCEDURE–DETAILS

SIZE OF WELD	SIZE OF ELECTRODE	THICKNESS OF PLATE	NUMBER OF PASSES	WELDING CURRENT (AMPERES)	WELDING VOLTAGE (@ ARC)	*MAN HOURS PER INCH WELD	*WELDING SPEED FT./HR.
1/8	1/8	1/8	1	160-190	26-28	.0025	33.3
3/16	5/32	3/16	1	160-190	26-28	.0028	29.8
1/4	3/16	1/4	1	180-230	32-36	.0033	25.3
3/8	1/4	3/4	1	280-330	32-36	.0050	16.7
1/2	1/4	3/4	2	280-330	32-36	.0078	10.7
5/8	1/4	1"	2	280-330	32-36	.0123	6.8
3/4	1/4	1 1/2	4	280-330	32-36	.0196	4.3
1	1/4	1 1/2	6	280-330	32-36	.0318	2.6

*NOTE: INCLUDES CHANGE ELECTRODE TIME, ARC TIME, WELD CLEANING TIME AND WELDING TIME.

Setup elements also can frequently be combined or tabularized in combinations so as to diminish the time required for summarizing a series of elements. Table 18–8 illustrates standard setup data for No. 5 Warner & Swasey turret lathes applicable to a specific plant. In order to determine the setup time with this data, it is only necessary to visualize the tooling in the square and hex turret and refer to the table. For example, if a certain job required a chamfering tool, turning tool, and facing tool in the square turret, and needed two boring tools, one reamer, and a collapsible tap in the hex turret, the setup standard time would be determined as 69.70 minutes plus 25.89 minutes, or 95.59 minutes. The value is established by finding the relevant tooling under the "square turret" column (line 8) and the most time-consuming applicable tooling in the "hex turret" section, in this case, tapping. This gives a value of 69.7 minutes. Since three additional tools are located in the hex turret (1st bore, 2d bore, and

TABLE 18–8
Standard setup data for No. 5 turret lathes

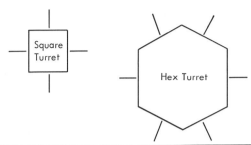

Basic tooling

			Hex turret				
No.	*Square turret*	*Partial*	*Cham-fer*	*Bore or turn*	*Drill*	*S. tap or ream*	*C. tap*	*C. die*
---	---	---	---	---	---	---	---	---
1.	Partial 31.5		39.6	44.5	48.0	47.6	50.5	58.5
2.	Chamfer 38.2		39.6	46.8	49.5	50.5	53.0	61.2
3.	Face or cut off 36.0		44.2	48.6	51.3	52.2	55.0	63.0
4.	Tn bo grv rad 40.5		49.5	50.5	53.0	54.0	55.8	63.9
5.	Face and chf. 37.8		45.9	51.3	54.0	54.5	56.6	64.8
6.	Fa and cut off 39.6		48.6	53.0	55.0	56.0	58.5	66.6
7.	Fa and tn or tn and cut off 45.0		53.1	55.0	56.7	57.6	60.5	68.4
8.	Fa, tn, and chf 47.7		55.7	57.6	59.5	60.5	69.7	78.4
9.	Fa, tn, and cut off 48.6		57.6	57.5	60.0	62.2	71.5	80.1
10.	Fa, tn, and grv. 49.5		58.0	59.5	61.5	64.0	73.5	81.6

11.	Circled basic tooling from above . . . . _____
12.	Each additional tool in square. . . . . 4.20*x*_____ = _____
13.	Each additional tool in hex . . . . 8.63*x*_____ = _____
14.	Remove and set-up three jaws . . . . 5.9 _____
15.	Set up subassembly or fixture . . . . 18.7 _____
16.	Set up between centers. . . . . 11.0 _____
17.	Change lead screw . . . . 6.6 _____

Total setup _____ min.

ream), we multiply 8.63 by 3 and get 25.89 minutes. Adding the 25.89 minutes to the 69.70 minutes gives us the required setup time.

More frequently, standard data are not combined but are left in their elemental form, thus giving greater flexibility in the development of time standards. Representative standard data applicable to a given plant would appear as shown in Table 18–9.

The data shown include the applicable personal delay and fatigue allowance. Let us see how this standard time data would be used to establish allowed setup and piece times to drill the two hold-down bolt holes in the cast-iron housing casting illustrated in Figure 18–7.

The allowed setup time would equal

$$B + C + D + F + G + H + I = 8.70 \text{ minutes}$$

TABLE 18–9
Standard data

Application: Allen 17-inch vertical single-spindle drill.
Work size: Small work—up to four pounds in weight and such that two or more parts can be handled in each hand.

Setup elements:	*Minutes*
A. Study drawing	1.25
B. Get material and tools and return and place ready for work	3.75
C. Adjust height of table	1.31
D. Start and stop machine	0.09
E. First-piece inspection (includes normal wait time for inspector)	5.25
F. Tally production and post on voucher	1.50
G. Clean off table and jig	1.75
H. Insert drill in spindle	0.16
I. Remove drill from spindle	0.14

Each piece elements:
1. Grind drill (prorate)	0.78
2. Insert drill in spindle	0.16
3. Insert drill in spindle (quick-change chuck)	0.05
4. Set spindle	0.42
5. Change spindle speed	0.72
6. Remove tool from spindle	0.14
7. Remove tool from spindle (quick-change chuck)	0.035
8. Pick up part and place in jig	
a. Quick-acting clamp	0.070
b. Thumbscrew	0.080
9. Remove part from jig	
a. Quick-acting clamp	0.050
b. Thumbscrew	0.060
10. Position part and advance drill	0.042
11. Advance drill	0.035
12. Clear drill	0.023
13. Clear drill, reposition part, and advance drill (same spindle)	0.048
14. Clear drill, reposition part, and advance drill (adjacent spindle)	0.090
15. Insert drill bushing	0.046
16. Remove drill bushing	0.035
17. Lay part aside	0.022
18. Blow out jig and part and lay part aside	0.081
19. Plug gage part	0.12 per hole

Element *A* is not included in the setup time because of the simplicity of the job. The drawing is not issued to the operator because the operation card provides him with necessary information relative to the drill jig number, drill size, plug gage number, and feed and spindle speed to be used. Element *E* is not included in the setup since first-piece inspection is not in this case performed by the inspector. According to the operation card issued, the operator will periodically inspect his own work with the go no-go gage provided. It is recommended that he check every 10th piece.

Since the casting in question would weigh less than four pounds, and two parts can easily be handled in one hand, the data outlined would be applicable.

After the analyst investigates the design of the drill jig to be used and makes an analytic motion study of the job in order to determine the

FIGURE 18–7
Cast-iron housing

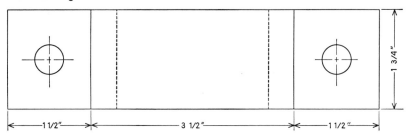

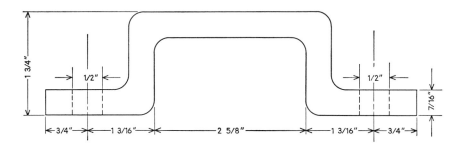

TABLE 18–10

Element number	Element	Allowed time (minutes)
8	Pk. pt. pl. jig (quick-acting clamp)	0.070
9	Rem. pt. jig (quick-acting clamp)	0.050
10	Pos. pt. & adv. dr.	0.042
13	Cl. dr. rep. pt. & adv. dr.	0.048
12	Cl. dr.	0.023
17	Lay pt. aside	0.022
19	Plug gauge pt. (10%)	0.012
1	Gr. dr. (once per 100 pcs.)	0.008
2	Insert dr. in sp. (once per 100 pcs.)	0.002
6	Rem. tool from sp. (once per 100 pcs.)	0.001
		0.278 min.

elements required to perform the operation, he will make his elemental summary. This would appear as shown in Table 18–10.

To this 0.278-minute time must be added the actual drilling time to drill the two holes. This can readily be determined as previously outlined. For a ½-inch diameter drill used for drilling cast iron, we should use a surface speed of 100 feet per minute and a feed of 0.008 inch per revolution.

One hundred surface feet per minute would equal 764 revolutions per minute.

$$\text{rpm} = \frac{12S_f}{\pi d}$$

where:

$$S_f = \text{Surface feet per minute}$$
$$\pi = 3.14$$
$$d = \text{Diameter of drill in inches}$$

However, investigation of the drill press for which this work has been routed reveals that 600 rpm or 900 rpm are the closest speeds that are available to the recommended 764 rpm. The analyst proposes to use the slower speed in view of his knowledge of the condition of the machine, and determines the drilling time as follows:

$$T = \frac{L}{F_m} \times 2$$

$$L = 0.437 + \frac{0.25}{\tan 59°} = 0.588 \text{ inches}$$

$$F_m = (600)(0.008)$$
$$= 4.8 \text{ inches per minute}$$

$$T = \left(\frac{0.588}{4.8}\right) 2 = 0.244 \text{ minute}$$

To this 0.244 minute cutting time must be added an appropriate allowance. If 10 percent allowance on the actual machining time is used, we should have a cutting time of 0.268 minute and a handling time of 0.278 minute, or a total time of 0.546 minute, to drill complete one casting on the single-spindle press available. This standard, supplemented with the setup time of 8.70 minutes, would be released as:

Setup: 0.145 hours
Each piece: 0.91 hours per hundred

Thus, if an operator set up this job, and ran 1,000 pieces in an eight-hour working day, he would be performing at an efficiency of 116 percent:

$$\frac{(0.91)(10) + 0.145}{8} = 116 \text{ percent}$$

Conclusion

When properly applied, standard data permit the establishment of accurate time standards in advance of the time that the job is to be performed. This feature makes their use especially attractive when estimating the cost of new work for quotation and subcontracting purposes.

Time standards can be achieved much more rapidly by using standard data, and the consistency of the standards established can be assured. Thus, this technique allows the economical development of indirect labor standards. For example, the standard data elements in Table 18–11 allow the rapid development of standards in the tabulating department of one

TABLE 18–11
Punch and verify standard elements (punch and verify section of tabulating department)

Code	Normal (min./occ.)	Description
P1001	0.377	*Get material from work table.* Operator gets up from workplace, walks (4 feet to 22 feet) to obtain material, returns to workplace.
P1002	0.356	*Put program card on machine.* Operator selects proper program card from small card file (4 × 7½ inch) at normal arm's length on worktable at left. Operator removes program cylinder from machine, attaches program card, and locks cylinder in machine. (Includes remove previous program card.)
P1003	0.016	*Remove cards from machine; stamp copy.* Operator removes cards from machine receptacle at left, stamps copy (invoices, order IBM list, etc.) with operator identification stamp.
P1004	0.071	*Tube code reference.* Operator refers to reference board to obtain tube-type tabulating code (6 digits). Reference board is indexed by RTMA tube type descriptions.
P1005	0.065	*Place cards and copy in work bin.* Operator places copy and cards in slots beneath worktable on left. Normal arm's reach.
P1006	0.377	*Give cards and copy to verifier.* Operator gets up, walks (4 to 22 feet), and places cards and copy on worktable of verifier.
P1007	0.300	*Punch group card.* Operator punches group card containing information common to the group of work (see individual description).
P1008	0.300	*Punch work card.* Operator punches work card containing information common to the particular unit (order, invoice, bill, etc.) (see individual description).
P1009	0.300	*Punch total card.* Operator punches card containing common or total information (see individual description).
P1010	Chart	*Punch detail card.* Operator punches detail card containing information partially duplicated from group. Works on total card and punches information relevant to the individual item (example: a particular tube type as an item on an order).
P1011	Chart	*Punch delivery date card.* Operator punches card bearing delivery dates of individual items.

company. Standards developed from standard data tend to be completely fair to both the worker and management in that they are the result of already proven standards. It can be pointed out that the elemental values used in arriving at the standards have proven satisfactory as components of established and acceptable standards in use throughout the plant.

The use of standard data simplifies many managerial and administrative problems in plants having a union which operates as a bargaining agent. Union contracts contain many clauses pertaining to such matters as the type of study to be taken (continuous or snapback), the number of cycles to be studied, who shall be studied, and who shall observe the study. These restrictions frequently make it difficult for the analyst to arrive at a standard that is equitable to both the company and the operator. By using the standard data technique, the analyst may avoid restrictive details. Thus, not only is the determination of a standard simplified, but the sources of tension between labor and management are alleviated.

In general, the more refined the elemental times, the greater the coverage possible for the data. Consequently, in job shop practice, it is practicable to have individual elemental values as well as grouped or combined values, so that the data for a given facility will have a flexibility that will allow the setting of rates for all types of work scheduled to the machine.

The use of fundamental motion data for establishing standards, in particular on short-cycle jobs, is becoming more widespread. This data is so basic that it alows the predetermination of standards on practically any class or type of manual elements. Of course, the "objective" or "do" basic divisions, such as "use," must be handled as variables, and tabularized data, curves, or algebraic expressions established. In Chapter 19, we will discuss the setting of rates for all types of work based on fundamental motion data techniques.

From the illustrative examples cited, it is evident that the application of standard data is an exacting technique. Careful and thorough training in methods and shop practice is fundamental before an analyst can accurately establish standards using standard data. It is essential that the analyst know and recognize the need for each element in the class of work for which he is setting rates. Supplementing this background, it is necessary that the man or woman working with standard data be analytic, accurate, thorough, conscentious, and completely dependable.

TEXT QUESTIONS

1. What do we mean by "standard data"?
2. What is the approximate ratio in the time required for setting standards by stopwatch methods and by standard data methods?
3. What advantages are there to establishing time standards by using standard data rather than by taking individual studies?
4. What would the time for the element "mill slot" depend upon?
5. Of what use is the "fast" watch in the compilation of standard data?
6. Compute the times for elements *a, b, c, d,* and *e* when elements $a + b + c$ were timed at 0.057 minute and $b + c + d$ were found to equal 0.078 minute; $c + d + e$, 0.097 minute, $d + e + a$, 0.095 minute, and $e + a + b$, 0.069 minute.
7. What would be the lead of a ¾-inch diameter drill with an included angle of 118°?
8. What would be the feed in inches per minute of a ¾-inch drill running at a surface speed of 80 feet per minute and a feed of 0.008 inch per revolution?

9. How long would it take the above drill to drill through a casting 2¼ inches thick?

10. How are feeds usually expressed in lathe work?

11. How long would it take to turn 6 inches of 1-inch bar stock on a No. 3 W. & S. turret lathe running at 120 feet per minute and feeding at the rate of 0.005 inch per revolution?

12. A plain milling cutter 3 inches in diameter with a width of face of 2 inches is being used to mill a piece of cold-rolled steel 1½ inches wide and 4 inches long. The depth of cut is ³⁄₁₆ inch. How long would it take to make the cut if the feed per tooth is 0.010 inch and a 16-tooth cutter running at a surface speed of 80 feet per minute is used?

13. What are some of the disadvantages of using curves to tabulate standard data?

14. What standard procedures should be followed in the plotting of simple curves?

15. What would be the horsepower requirements of turning a mild steel shaft 3 inches in diameter if a cut of ¼ inch with a feed of 0.022 inch per revolution at a spindle speed of 250 rpm were established?

GENERAL QUESTIONS

1. What do you think is the attitude of labor toward establishing standards through the use of standard data?

2. Using the drill press standard data shown in the text, preprice a drilling job with which you are familiar. How does this standard compare with the present rate on the job? (Be sure that identical methods are compared.)

3. Does complete standard data supplant the stopwatch? Explain.

4. If you were president of your local union, would you advocate that standards be established by standard data? Why or why not?

PROBLEMS

1. The analyst in the Dorben Company made 10 independent time studies in the hand paint spraying section of the finishing department. The product line under study revealed a direct relation between spraying time and product surface area. The following data were collected:

Study no.	Leveling factor	Product surface area	Standard time
1.	0.95	170	0.32
2.	1.00	12	0.11
3.	1.05	150	0.31
4.	0.80	41	0.14
5.	1.20	130	0.27
6.	1.00	50	0.18
7.	0.85	120	0.24
8.	0.90	70	0.23
9.	1.00	105	0.25
10.	1.10	95	0.22

Compute the slope and intercept constant, using regression line equations. How much spray time would you allow for a new part with a surface area of 250 square inches?

2. The work measurement analyst in the Dorben Company is desirous of developing an accurate equation for estimating the cutting of various configurations in sheet metal with a band saw. The data from eight time studies for the actual cutting element provided the following information:

No.	Lineal inches	Standard time
1.	10	0.40
2.	42	0.80
3.	13	0.54
4.	35	0.71
5.	20	0.55
6.	32	0.66
7.	22	0.60
8.	27	0.61

What would be the relation between the length of cut and the standard time, using the least squares technique?

3. The work measurement analyst in the XYZ Company wishes to develop standard data involving fast, repetitive manual motions for use in a light assembly department. Because of the shortness of the desired standard data elements, he is obliged to measure them in groups as they are performed on the factory floor. On a certain study, he is endeavoring to develop standard data for five elements which will be denoted as A, B, C, D, and E. Using a fast (0.001) decimal minute watch, the analyst studied a variety of assembly operations and arrived at the following data:

$$A + B + C = 0.131 \text{ minutes}$$
$$B + C + D = 0.114 \text{ minutes}$$
$$C + D + E = 0.074 \text{ minutes}$$
$$D + E + A = 0.085 \text{ minutes}$$
$$E + A + B = 0.118 \text{ minutes}$$

Compute the standard data values for each of the elements A, B, C, D, and E.

4. The work measurement analyst in the Dorben Company is developing standard data for prepricing work in the drill press department. Based upon the following recommended speeds and feeds, compute the power feed cutting time of ½-inch high-speed drills with an 118° included angle to drill through material that is 1-inch thick. Include a 10 percent allowance for P.D. & F.

Material	Recommended speed (ft./min.)	Feed (in./rev.)
Al (copper alloy)	300	0.006
Cast iron	125	0.005
Monel (R)	50	0.004
Steel (1112)	150	0.005

SELECTED REFERENCES

Derse, Joseph C. *Machine Operation Times for Estimators.* New York: Ronald Press Co., 1946.

Metcut Research Associates. *Machining Data Handbook.* Cincinnati: Metcut Research Associates, Inc., 1966.

Ostwald, Philip F. *Cost Estimating for Engineering and Management.* Englewood Cliffs, N.J.: Prentice-Hall, Inc., 1974.

Pappas, Frank G., and Dimberg, Robert A. *Practical Work Standards.* New York: McGraw-Hill Book Co., 1962.

Rotroff, Virgil H. *Work Measurement.* New York: Reinhold Publishing Corp., 1959.

19

Synthetic basic motion times

Since the time of Taylor, management has realized the desirability of having standard times assigned to the various basic divisions of accomplishment.

One of the earliest men to investigate this area was W. G. Holmes. He tabulated values based on the measurement of the various body members, including the finger, hand, foot, and arm, in a Time of Movements Chart.[1] Representative times established by Holmes are:

1. Hand with hinge movement at the wrist .0022 minute when moved 0 to 2 inches.
2. Arm with angular movement at the shoulder .0060 minute when moved 30 degrees.
3. Nerve reaction, eye to brain or reverse .0003 minute.

Actually, Holmes has taken his analysis to such a point that it is difficult and cumbersome to handle. He has provided times for nerve reactions from the eye to the brain, the knee to the brain, and the foot to the brain; also, times for mental processes and decisions, such as time to hear or smell, or realize contact.

R. M. Barnes has advocated a technique similar to Holmes's in the development of basic motion data. Barnes states: "There is a trend toward the use of standard time values for therbligs or combinations of therbligs and under certain conditions such time values may be more useful than the time values for longer elements obtained from stopwatch time studies."[2]

A related development was made by Harold Engstrom at the General Electric Company in Bridgeport, Connecticut. Engstrom described his technique as follows:

1. Analyze certain classes of work such as light bench assembly work or machine operation. The standards developed must be confined to par-

---

[1] W. G. Holmes, *Applied Time and Motion Study* (New York: Ronald Press Co., 1938), p. 244.

[2] Ralph M. Barnes, *Motion and Time Study,* 3d ed. (New York: John Wiley & Sons, Inc., 1949), p. 439.

ticular classes of work and no attempt made to embrace the entire gamut of industrial operations.

2. The jobs are then time studied to determine the basic elements of which they are comprised.
3. These time studied jobs are then photographed by means of a measurement camera for the purpose of micromotion analysis.
4. The evaluations of the effort of the worker are then made.
5. From the motion pictures previously taken, make combinations of therbligs into two major classifications, such as Get, which consists of Transport Empty and Grasp; Place, which consists of Transport Loaded, Preposition, Position, and Release Load. Dispose should be treated as Place.
6. These data are then tabulated.[3]

A. B. Segur developed time equations in 1924 as a result of analyzing micromotion films taken of highly efficient operators during World War I.

The films were taken originally in an effort to develop motion patterns to train handicapped workers. These early studies revealed that if the same motion was performed by various operators using the same motion pattern, the time involved in performing the motion was relatively constant.

It has never been a question of need, but a question of how the job of determining time values for the various basic divisions can be done practically. In recent years, considerable progress has been made in the assignment of time values to basic elements of work. These time values are referred to as synthetic basic motion times.

Definition of synthetic basic motion times

Synthetic basic motion times are a collection of valid time standards assigned to fundamental motions and groups of motions that cannot be precisely evaluated with ordinary stopwatch time study procedures. They are the result of studying a large sample of diversified operations with a timing device such as the motion-picture camera that is capable of measuring very short elements. The time values are synthetic in that they are often the result of logical combinations of therbligs. For example, a series of time values may be established for different categories of grasp. Within the grasp time may be included the therbligs search, select, and grasp. The time values are basic in that further refinement is not only difficult but impractical. Thus, we have the term *synthetic basic motion times.*

The necessity of synthetic basic motion times

Since 1945 there has been a growing interest in the use of synthetic basic motion times as a modern method of establishing rates quickly and

---

[3] *Proceedings, National Time and Motion Study Clinic, 1940,* pp. 45–46.

accurately without using the stopwatch or other time recording devices. In view of this, it is recommended that the student reread Chapter 7 after completing Chapter 18 on standard data. A by-product of synthesized time standards that has probably been as useful as, if not more useful than, the time standards themselves, is the development of methods consciousness to a refined degree in all parties associated with the establishment of standards using synthetic values. For this reason, the author believes that the entire subject of synthetic basic motion times should be meshed or integrated with the motion and micromotion techniques, even though it is also very closely allied to the work measurement phase.

Even with the laws of motion economy and the concept of basic division of accomplishment clearly established, the methods analyst without reliable time values for the basic divisions has only a portion of the facts necessary to engineer a method prior to beginning actual production. If he does not have time values of the basic divisions, how can he be sure that his proposed method is the best? With reliable motion times, he would be able to evaluate his proposed methods in terms of the average or normal worker, who would eventually be the recipient of his ideas. It is apparent that the method used determines the time to do a task, and it follows that if time values for all forms of activity were available, then the most favorable methods could be determined in advance.

Today, the practicing methods analyst has several sources of established synthetic values from which he can obtain information that may be helpful in his work.

Work-Factor[4]

The Work-Factor Company, Inc. is one of the pioneer concerns in establishing standards synthetically with motion-time values. Work-Factor data were made available in 1938, after four years of gathering values by the micromotion technique, stopwatch procedures, and the use of a "specially constructed photoelectric time machine."

The Work-Factor system has achieved flexibility by developing four different procedures of application, depending on the objectives of the analysis and the accuracy required. The procedures are known, respectively, as the detailed, simplified, abbreviated, and ready techniques.

Detailed Work-Factor is designed to provide precise time standards for either daywork measurement or incentive pay plans. It also provides a precise tool for methods analyses. It is primarily used for short-cycle operations and repetitive work. It is also commonly used for developing standard data.

---

[4] "Work-Factor" is the service mark (trademark) of the Work-Factor Company, Inc., identifying its services as consultants to industry and its system of predetermined fundamental motion-time standards for the motion times themselves and the techniques used to apply them to methods determination and work measurement.

Simplified Work-Factor is used when operations do not require as precise an analysis as is obtainable with detailed Work-Factor. It is generally used for medium quantity production. Simplified time standards can be obtained in about half the time as detailed, and the loss in accuracy normally does not exceed plus 5 percent.

Abbreviated Work-Factor is fast to apply. It is extremely useful for the short-order shop. Abbreviated varies from detailed about plus 10 to 12 percent.

Ready Work-Factor was developed to satisfy the need for a simple tool for evaluating manual work that could be taught readily to persons in functions other than time study, such as product design, estimating, manufacturing methods, and production and office supervision. When correctly applied, ready Work-Factor times may be expected to average within 0 to plus 5 percent of detailed Work-Factor.

Detailed Work-Factor system

Work-Factor recognizes the following variables that influence the time required to perform a task:

1. The body member making the motion, such as arm, forearm, finger-hand, foot.
2. The distance moved (measured in inches).
3. The weight carried (measured in pounds, converted into work factors).
4. The manual control required (care, directional control or steering to a target, changing direction, stopping at a definite location; measured in work factors).

Through analysis of films, Work-Factor concluded, as the Gilbreths did many years earlier, that finger motions can be performed more rapidly than arm motions, and that arm motions are made in less time than body motions. Work-Factor motion times have been compiled for the various body members. These include:

1. Finger-hand. Includes the movements of the fingers and thumb and movement of the hand about the wrist joint.
2. Arm. Includes the movements of the lower arm around the elbow when used as hinge joint and all movements of the whole arm hinged at the shoulder except for swivels. Movement of the hand, fingers, and lower arm may occur simultaneously.
3. Forearm swivel. Here the lower arm rotates about the axis of the forearm, such as when turning a screwdriver, or the full arm rotates about the entire axis and the swivel takes place at the shoulders.
4. Trunk. A forward, backward, sidewise, or swiveling motion of the trunk of the body.
5. Foot. Here are included motions of the foot when these are hinged at the ankle and the upper and lower leg remain in a fixed location.

All proponents of fundamental motion data techniques have recognized the element of distance in making reaches and moves and, in fact, all motions. The longer the distance, the more time is required, of course. Work-Factor has tabularized values for finger and hand movements from 1 inch up to 4 inches, and for arm movements from 1 inch to 40 inches. Distances are measured as a straight line between the starting and stopping points of the motion arc. The actual motion path is measured only when a change in direction is involved.

Following is a list of the points at which distance should be measured for the various body members:

Body member	*Point of measurement*
Finger or hand.	Fingertip
Arm	Knuckles (knuckle having greatest travel should be used)
Forearm swivel	Knuckles
Trunk.	Shoulder
Foot	Toe
Leg	Ankle
Head turn	Nose

Weight or resistance will influence time according to the magnitude of the part being moved, the body member being used, and the sex of the operator. It is measured in pounds for all body members except for "forearm swivel" motions where pound-inches of torque represent the unit of measurement.

Manual control is the most difficult variable to quantify. However, the Work-Factor system concluded that in the vast majority of instances work motions can be identified as involving one or more of the following four types of control:

1. Definite Stop Work-Factor. Here some manual control is required to stop a motion within a fixed interval. Definite stop does not exist when the motion is terminated by some physical barrier. The motion must be terminated by the muscular coordination of the operator.

2. Directional Control or Steer Work-Factor. Here manual control is necessary to bring or guide a part to a specific location or make a motion through an area with limited clearance.

3. Care or Precaution Work-Factor. Here manual control is exercised to prevent spilling or damage, such as moving a full vessel of acid or handling a thin piece of glass.

4. Change of Direction Work-Factor. Here manual control is exercised when the motion involves a direction change to reach a remote location, or to get around some obstacle. For example, to move a nut in back of a panel would require a change in direction once the moving hand reached the front of the panel.

A work factor has been defined as the index of the additional time required over and above the basic time. It is a unit for identifying the effect of the variables, manual control and weight. The other two variables that affect the time to perform manual motions—body member used and distance—do not employ work factors as a measure of magnitude. Here member used and inches represent the quantitative yardstick. The simplest or basic motions of a body member involve no work factors. As complexities are introduced into a manual motion through the addition of weight or control, work factors are added. And, of course, each work factor that is added represents an additional increment of time.

Table 19–1 illustrates the Work-Factor motion-time table, including all Work-Factor motion-time values. Note that there is a separate table for each body member and that each table contains values for distances. In the forearm swivel, motion distance is indicated in degrees while all other motions are shown in inches. At the bottom of each table are tabularized weight data giving weight or resistance limits for men and women. The forearm swivel motion resistance section is given in pounds-inches of torque, while all others are listed in pounds. The table gives the maximum resistance encountered that can be considered basic and also the top values for a specific number of work factors.

The Work-Factor System divides all tasks into eight "Standard Elements of Work." These are:

1. Transport (reach and move).
2. Grasp.
3. Pre-position.
4. Assemble.
5. Use.
6. Disassemble.
7. Mental process (inspect, react, and so on).
8. Release.

1. TRANSPORT. The element transport is the connecting link between the other standard elements. It is broken into two classifications:

a. Reach: When a body member is relocated to go to a destination, location, or object.
b. Move: When a body member is relocated to transport an object.

2. GRASP. The element grasp is the act of obtaining manual control of an object; it begins after the hand has moved directly to the object and ends when manual control has been obtained and a move can occur. The Work-Factor system recognizes three types of grasps:

a. Simple Grasp: Used for isolated easy-to-grasp objects and requires only one single motion.
b. Manipulative Grasp: Includes all grasps of isolated or orderly stacked objects that require more than one motion of the fingers to gain control. There may be arm motions, several finger motions, or combinations of both.

TABLE 19–1
Work-Factor motion-time table for detailed analysis time in Work-Factor* units

Distance moved, inches	Basic	Work-Factors 1	2	3	4	Distance moved, inches	Basic	Work-Factors 1	2	3	4
(A) Arm, measured at knuckles						(L) Leg, measured at ankle					
1	18	26	34	40	46	1	21	30	39	46	53
2	20	29	37	44	50	2	23	33	42	51	58
3	22	32	41	50	57	3	26	37	48	57	65
4	26	38	48	58	66	4	30	43	55	66	76
5	29	43	55	65	75	5	34	49	63	75	86
6	32	47	60	72	83	6	37	54	69	83	95
7	35	51	65	78	90	7	40	59	75	90	103
8	38	54	70	84	96	8	43	63	80	96	110
9	40	58	74	89	102	9	46	66	85	102	117
10	42	61	78	93	107	10	48	70	89	107	123
11	44	63	81	98	112	11	50	72	94	112	129
12	46	65	85	102	117	12	52	75	97	117	134
13	47	67	88	105	121	13	54	77	101	121	139
14	49	69	90	109	125	14	56	80	103	125	144
15	51	71	92	113	129	15	58	82	106	130	149
16	52	73	94	115	133	16	60	84	108	133	153
17	54	75	96	118	137	17	62	86	111	135	158
18	55	76	98	120	140	18	63	88	113	137	161
19	56	78	100	122	142	19	65	90	115	140	164
20	58	80	102	124	144	20	67	92	117	142	166
22	61	83	106	128	148	22	70	96	121	147	171
24	63	86	109	131	152	24	73	99	126	151	175
26	66	90	113	135	156	26	75	103	130	155	179
28	68	93	116	139	159	28	78	107	134	159	183
30	70	96	119	142	163	30	81	110	137	163	187
35	76	103	128	151	171	35	87	118	147	173	197
40	81	109	135	159	179	40	93	126	155	182	206

Weight, pounds:
Male	2	7	13	20	Up	Male	8	42	Up		
Female	1	3½	6½	10	Up	Female	4	21	Up		

Distance moved, inches	Basic	Work-Factors 1	2	3	4	Distance moved, inches	Basic	Work-Factors 1	2	3	4
(T) Trunk, measured at shoulder						(F, H) Finger-hand, measured at fingertip					
1	28	38	49	58	67	1	16	23	29	35	40
2	29	42	53	64	73	2	17	25	32	38	44
3	32	47	60	72	82	3	19	28	36	43	49
4	38	56	70	84	96	4	23	33	42	50	58
5	43	62	79	95	109	Weight, pounds:					
6	47	68	87	105	120	Male	⅔	2½	4	Up	
7	51	74	95	114	130	Female	⅓	1¼	2	Up	
8	54	79	101	121	139	(FT) Foot, measured at toe					
9	58	84	107	128	147	1	20	29	37	44	51
10	61	88	113	135	155	2	22	32	40	48	55
11	63	91	118	141	162	3	24	35	45	55	63
12	66	94	123	147	169	4	29	41	53	64	73
13	68	97	127	153	175	Weight, pounds:					
14	71	100	130	158	182	Male	5	22	Up		
15	73	103	133	163	188	Female	2½	11	Up		
16	75	105	136	167	193	(FS) Forearm swivel, measured at knuckles					
17	78	108	139	170	199	45 degrees	17	22	28	32	37
18	80	111	142	173	203	90 degrees	23	30	37	43	49
19	82	113	145	176	206	135 degrees	28	36	44	52	58
20	84	116	148	179	209	180 degrees	31	40	49	57	65

Weight, pounds:
				Torque, pound-inches:			
Male	11	58	Up	Male	3	13	Up
Female	5½	29	Up	Female	1½	6½	Up

Work-Factor* symbols:	Walking time				Visual inspection
W = weight or resistance		30-inch paces			Focus............20
S = directional control (steer)	Type	1	2	Over 2	Inspect.........30 per point
					React...........20
P = care (precaution)	General control	As single steps	260	120 + 80 per pace	Head turn:
U = change direction	Restricted	steps	300	120 + 100 per pace	45 degrees..............40
D = definite stop	Add 100 for 120- to 180-degree turn at start				90 degrees..............60
	Up steps (8 inches rise, 10 inches flat).......126				1 time unit = 0.006 second
	Down steps............................100				= 0.0001 minute
					= 0.00000167 hour

* Trademark.
By permission from H. B. Maynard, *Industrial Engineering Handbook* (New York: McGraw-Hill Book Co., 1956).

TABLE 19–2
Work-Factor® complex grasp table

COMPLEX GRASPS FROM RANDOM PILES

Column groups:
- **SOLIDS & BRACKETS** — [Thickness over 3/64″ (.0469″)]
- **THIN FLAT OBJECTS — THICKNESS**: (less than 1/64) 0-.0156″ ; (1/64 to 3/64) .0156″-.0469″
- **CYLINDERS AND REGULAR CROSS SECTIONED SOLIDS — DIAMETER**: 0-.0625″ (1/16) ; .0626″-.125″ (1/8) ; .1251″-.1875″ (3/16) ; .1876″-.5000″ (1/2) ; .5001″ & up (over 1/2)
- **Add for Entangled, Nested or Slippery Objects ***

SIZE (Major dimension or length)		S&B Blind	—Simo	S&B Visual	—Simo	Thin <1/64 Blind	—Simo	Visual	—Simo	Thin 1/64–3/64 Blind	—Simo	Visual	—Simo	Dia 0-.0625 Blind	—Simo	Dia .0626-.125 Blind	—Simo	Dia .1251-.1875 Blind	—Simo	Dia .1876-.5000 Blind	—Simo	Visual	—Simo	Dia .5001″& up Blind	—Simo	Visual	—Simo	Add	—Simo
.0000″-.0625″	1/16″ & less	120	172	B	B	—	—	—	—	131	189	B	B	S	S	S	S	S	S	S	S	S	S	S	S	S	S	17	26
.0626″-.1250″	over 1/16″ to 1/8″	79	111	B	B	108	154	B	B	85	120	B	B	85	120	S	S	S	S	S	S	S	S	S	S	S	S	12	18
.1251″-.1875″	over 1/8″ to 3/16″	64	88	B	B	102	145	B	B	74	103	B	B	79	111	74	103	S	64	S	S	S	S	S	S	S	S	12	18
.1876″-.2500″	over 3/16″ to 1/4″	48	64	B	B	72	100	B	B	56	76	B	B	79	111	68	94	64	88	S	S	S	S	S	S	S	S	8	12
.2501″-.5000″	over 1/4″ to 1/2″	40	52	B	B	64	88	B	60	48	64	B	44	62	85	56	76	56	76	44	58	B	58	S	40	S	32	8	12
.5001″-1.0000″	over 1/2″ to 1″	40	52	32	40	64	88	60	82	48	64	44	58	62	85	56	76	48	64	48	64	B	44	40	52	32	40	8	12
1.0001″-4.0000″	over 1″ to 4″	37	48	20	22	53	72	36	46	45	60	28	34	56	76	48	64	40	52	40	52	36	46	37	48	20	22	8	12
4.0001″ & up	over 4″	46	61	20	22	70	97	44	58	62	85	36	46	56	76	48	64	40	52	40	52	36	46	37	48	20	22	9	14

B = Use Blind column since visual grasp offers no advantage. S = Use Solid Table.

* Add the indicated allowances when objects: (a) are entangled (not requiring two hands to separate); (b) are nested together because of shape or film; (c) are slippery (as from oil or polished surface). When objects both entangle and are slippery, or both nest and are slippery, use double the value in the table.

Note: Special grasp conditions should be analyzed in detail.

c. Complex Grasp: Defined as the grasp of an object from a random (jumbled) pile. The system provides a complete table for complex grasps. Complex grasps involve more than one motion and sometimes include arm movements.

Objects to be grasped are classified as:

a. Cylinders and Regular Cross-sectioned Solids: Defined as all objects having cylindrical or regular (all sides and angles equal) cross sections, such as square, hexagons, and so on.

b. Thin, Flat Objects: Flat objects with an effective dimension of ³⁄₆₄ inch or less in thickness.

c. Solids and Brackets: Defined as objects over ³⁄₆₄ inch thick not otherwise classified as cylinders or thin flat objects.

3. PRE-POSITION. Pre-position occurs whenever it is necessary to turn and orient an object so that it will be in the correct position for a subsequent element. Pre-position frequently occurs on a percentage basis since the object will sometimes be in a usable position but must be oriented at other times. An example is a nail (0.100 in. $\times$ ³⁄₄ in.). In 50 percent of the cases, it will be grasped in a usable position; in the remaining 50 percent of the cases, it must be pre-positioned. Using the Work-Factor pre-position table (Table 19–3), the analysis would be: PP–0–50% = 24 units.

TABLE 19–3
Pre-position

			Type of pre-position (PP)		
Number of points satisfactory for use	Percent PP required	Optimum (³⁄₈ in. or over) 3F1	Very small (under ³⁄₈ in.) 3F1	Medium one hand (over 2 × 3 × ½ in.) 4F1	Medium two hands (over 3 × 3 × ½ in.) 2F1 + A4D	Large two hand (over 8 × 8 in 2F1W A8D
Two or more sides up:						
Four, three, or two opposite						
points	0	–	–	–	–	–
Two adjacent points	25	12	20	16	18	25
One point only	50	24	40	32	35	50
One specific side up:						
Four, three, or two opposite						
points	50	24	40	32	35	50
Two adjacent points	62½	30	50	40	44	63
One point only	75	36	60	48	53	75
From stock, etc.:	100	48	80	64	70	100

1. PP of more than one finger motion or one forearm swivel cannot be done simo with a move.
2. End for end PP in fingers, 3F1−50 percent, 24 units.
3. Other PP must be analyzed. Table is a guide.

4. ASSEMBLE. Assemble occurs whenever two or more objects are joined together, usually through mating or nesting. The system provides a complete assemble table. Assemble time depends on:

a. Size of Target: The target is the part of an assembly that accepts the plug.
b. Plug Dimensions: A plug is the part of an assembly that fits into the target.
c. Plug-Target Ratio: Assembly difficulty, and therefore assemble time, increases as the dimension of the plug approaches the dimension of the target. Therefore, assembly time is a function of the plug-target ratio.

$$\frac{\text{Plug dimension}}{\text{Target dimension}} = \text{Plug-target ratio}$$

d. Type (shape) of target: There are two types of targets in work factor terminology—closed and open. Closed targets are closed about the entire perimeter so that align motions are required along two axes. Open targets require align motions along one axis only.

When the above-listed facts are known, it is a simple matter to determine assemble time from the table.

Allowances are added for increased difficulty due to the distance between targets (two at a time), gripping distance (the distance from the hand to the end of the plug), and blind target (when the target is blind either before or during assembly).

5. USE. This element usually refers to machine time, special process time, and time that involves using tools. Use may involve manual motions, as in tightening a nut with a wrench or threading a pipe; in such cases, the motions are to be analyzed and evaluated according to all rules and time values obtained from the motion-time tables.

6. DISASSEMBLE. As the name implies, disassemble is the opposite of assemble. Usually it is a single motion. Time values are taken from the motion-time table.

7. MENTAL PROCESS. This applies to all mental activities and processes. It is the time interval in which reactions and nerve impulses take place.

Mental processes that can be measured are:

Eye motions	Inspections	Compute
Focus	Quality	Read
Shift	Quantity	Action
Reactions	Identity	Concept

8. RELEASE. Release, the opposite of grasp, is the act of relinquishing control of objects. There are three types:

a. Contact Release: Involves no motion. It is merely lifting the hand from an object.
b. Gravity Release: Occurs when objects drop from the grasp as contact is broken and before the releasing finger motions are complete.
c. Unwrap Release: Invloves unwrapping the fingers from around the grasped object and is not complete until the unwrapping motions are complete.

All time values on the Work-Factor motion-time table are in 0.0001 minute. These values are in terms of select time, which has been defined as "that time required for the average, experienced operator, working with good skill and good effort (commensurate with good health and physical and mental well-being) to perform an operation on one unit or piece." To determine the standard time, the analyst must add an allowance to the Work-Factor values, since the select time includes no allowance for personal needs, fatigue, unavoidable delays, or incentive.

FIGURE 19–1

OPERATION NAME Oldsmobile Case: 1st Draw Operation (2 Pin Die)   WORK FACTOR ANALYST
H. B. Amster

EQUIPMENT   Mach #1031: Bliss Double Action Draw Press 240 Tons   72.5 RPM

MATERIAL   CR Steel: Tmk = .050 = .003" Dia = 19.25"   Wt = 4.04 lbs.   SHEET 1 OF 1

NO.	LEFT HAND Elemental Description	Motion Analysis	Elem. Time	Cumu- lative Time	Elem. Time	Motion Analysis	RIGHT HAND Elemental Description	Mach.
1	Reach for blank	A20D	0080	0080	0080	A20D	Reach for blank	
2	Grasp blank	F1W	0023	0103	0103	0023	F1W	Grasp blank	
3	Carry blank to die	A40WSD	0159	0262	0262	0159	A40WSD	Carry blank to die	
4	Release & clear fingers	F3W	0028	0290	0290	0028	F3W	Release & clear fingers	
5	Place fingers on blank	F3D	0028	0318	0399	0109	A40D	Reach for trip lever	
6	Push blank against pins	A2P	0029	0347	0415	0016	F1	Grasp lever (No W Req'd)	
7	Withdraw hand (A10) & wait		0146	0493	0493	0078	A10WW	Pull lever to trip press (10 lbs.)	
8	Move hand to hold blank (turn Simo)	A30D	0096	0589	0589	0096	A30D	Reach for oil rag (turn 90°-Simo)	
9	Press down to hold blank	A1W	0026	0615	0606	0017	F2	Grasp rag	
10	Hold Blank at center	---	----	0691	0691	0085	A12UD	Carry rag & dip in oil pan	
11	" " " "	---	----	0723	0723	0032	A6	Raise from oil pan	
12	" " " "	---	----	0749	0749	0026	A4	Shake (squeeze Simo)	
13	" " " "	---	----	0804	0804	0055	A18	Carry rag to stack of blanks	
14	" " " "	---	----	0913	0913	0109	A40U	Apply oil (circular motion)	
15	Release blank	A1W	0026	0939	0955	0042	A10	Raise rag from blank	
16	Withdraw hand from blank	A16	0052	0991	0997	0042	A10	Strike to dislodge blank	
17	Approach blank	A3D	0032	1023	1048	0051	A15	Withdraw rag to side	
18	Grasp blank	F1W	0023	1046		---	---	Hold rag	
19	Turn blank over (release Simo)	2A14W	0138	1184		---	---	" "	
20	Move hand to blank center	A13D	0067	1251		---	---	" "	
21	Press down to hold blank	A1W	0026	1277	1277	0055	A18	Carry rag to blank	
22	Hold blank at center	---	----		1386	0109	A40U	Apply oil (circular motion)	
23	Release blank	A1W	0026	1412	1437	0051	A15	Toss rag near pan	
24	Hand idle (turn 90° Simo)	---	----		1517	0080	A20D	Reach for trip lever (turn Simo)	
25	" "	---	----		1533	0016	F1	Grasp handle (No W req'd)	
26	" "	---	----		1781	0248		Wait (.0493 + .1380 - .1533 = .0092 = .0248)	
27	Move hand to piece on punch	A20D	0080	1873	1873	0092	A15WW	Push lever to stop press (10 lbs.)	
28	Catch piece on palm	React	0020	1893	1896	0023	F1W	Release handle	
29	Carry piece to chute (balance)	A40WPD	0159	2052		---	---	Hand idle	
30	Toss to chute or stack	A5W	0043	2095		---	---	" "	
31	Turn to worktable	120°	0100	2195	2195	0100	120°	Turn to worktable	

A typical breakdown study of a draw operation on a Bliss double-action 240-ton press is shown in Figure 19–1. The symbols used in this analysis carry the following meanings:

W—Weight or resistance   A—Arm
S—Steering or directional control   L—Leg
P—Precaution or care   T—Trunk
U—Change direction   F—Finger
D—Definite stop   Ft—Foot
                         FS—Forearm swivel
                         H—Hand
                         HT—Head turn

In making a Work-Factor study, the analyst first lists all necessary motions made by both hands to perform the task; then he identifies each motion in terms of the distane moved, body member used, and work factors involved. In recording distances, fractional inches are not used. Thus, motions of 1 inch or less are recorded as exactly 1 inch, and motions greater than 1 inch in length are recorded as the nearest integral number. The analyst then selects from the table of values the appropriate figure for each of the basic motions and summarizes these to obtain the total time required by the normal operator to perform the task.

To this total time, percentage allowances for personal delays, fatigue, and unavoidable delays must be added to determine the allowed time.

Simplified Work-Factor system

Simplified Work-Factor is designed to measure work where cycle times are 15 minutes or longer and great precision is not required. Instead of recording individual motions and times, as in detailed analysis, simplified analysis employs larger units of time. This provides a tool with a considerable speed advantage in long-cycle jobs.

The simplified system affords a predetermined elemental time technique for nonrepetitive work and job shop-type operations.

Times in the simplified tables are selected averages and can be traced to the detailed tables. Simplified time values will generally be about 0–5 percent higher than the detailed values.

Essentially, the detailed rules apply to simplified with a few minor exceptions. The simplified Work-Factor table includes the following:

1. A table for transport motions, applied to all body members. Time values are classified in ranges, and the allowances for 2–3 and 4 work factors are averaged into single values.
2. A grasp table listing Work-Factor units for six grasp types, with variations for visual, simo, entangled, nested, and slippery conditions. These values cover the many values for simple, manipulative, and complex grasps in detailed Work-Factor. Three of the six values apply to complex grasps classed as medium, difficult, and very difficult, according to the type and size of the object being grasped. A parts classification table is provided as an objective guide for classifying grasps and pickups.
3. A pickup table providing simplified time values which combine the standard elements transport and grasp into values for picking up objects (that is, the motion sequence reach to object, grasp object, and move object to new location). The table is also used for place aside (that is, the motion sequence move object to new location, release object, and reach to next point in operation).
4. A pre-position table similar to the detailed table.
5. An assemble table in condensed form.
6. A visual inspection table.
7. A release table with values for six different release situations.
8. Simplified values for *walk and turn* and *apply pressure*.

TABLE 19–4
Work-Factor® assembly tables

AVERAGE NO. OF ALIGNMENTS (A1S Motions)

CLOSED TARGETS

Ratio of Plug Dia. ÷ Target Dia.

TARGET DIAMETER	To .224	.225 to .289	.290 to .414	.415 to .899	.900 to .934	.935 to 1.000
.875" & up	(D*) 18	(D*) 18	(D*) 18	(1/4) 25	(1/4**) 51	(1/4***) 59
.625" to .874"	(D*) 18	(D*) 18	(SD*) 18	(1/4) 25	(1/4**) 51	(1/4***) 59
.375" to .624"	(SD*) 18	(SD*) 18	(1/4) 25	(1/2) 31	(1/2**) 57	(1/2***) 65
.225" to .374"	(1/2) 31	(1) 44	(1) 44	(1 1/2) 57	(1 1/2**) 83	(1 1/2***) 91
.175" to .224"	(1) 44	(1) 44	(1) 44	(1 1/2) 57	(1 1/2**) 83	(1 1/2***) 91
.125" to .174"	(1) 44	(1 1/4) 51	(1 1/2) 57	(1 1/2) 57	(1 1/2**) 83	(1 1/2***) 91
.075" to .124"	(2 1/2) 83	(2 1/2) 83	(2 1/2) 83	(2 1/2) 83	(2 1/2**) 109	(2 1/2***) 117
.025" to .074"	(3) 96	(3) 96	(3) 96	(3) 96	(3**) 122	(3***) 130

OPEN TARGETS

Ratio of Plug Dia. ÷ Target Dia.

TARGET DIAMETER	To .224	.225 to .289	.290 to .414	.415 to .899	.900 to .934	.935 to 1.000
.875" & up	(D*) 18	(D*) 18	(D*) 18	(D*) 18	(1/4**) 51	(1/4***) 59
.625" to .874"	(D*) 18	(D*) 18	(D*) 18	(SD*) 18	(1/4**) 51	(1/4***) 59
.375" to .624"	(SD*) 18	(SD*) 18	(SD*) 18	(1/2) 31	(1/2**) 57	(1/2***) 65
.225" to .374"	(1/4) 25	(1/2) 31	(1/2) 31	(3/4) 38	(3/4**) 64	(3/4***) 72
.175" to .224"	(1/2) 31	(1/2) 31	(1/2) 31	(3/4) 38	(3/4**) 64	(3/4***) 72
.125" to .174"	(3/4) 38	(1) 44	(1) 44	(1) 44	(1**) 70	(1***) 78
.075" to .124"	(1 1/4) 51	(1 1/4) 51	(1 1/4) 51	(1 1/4) 51	(1 1/4**) 77	(1 1/4***) 85
.025" to .074"	(1 1/2) 57	(1 1/2) 57	(1 1/2) 57	(1 1/2) 57	(1 1/2**) 83	(1 1/2***) 91

* Letters indicate Work Factors in move preceding Assembly.

** Requires A(X)S Upright for all ratios of .900 and greater. (Table value includes A1S Upright.)

*** Requires A(Y)S Upright and A(Z)P Insert for all ratios of .935 and greater. (Table value includes A1S Upright and A1P Insert.)

Abbreviated Work-Factor system

Abbreviated Work-Factor is a rapidly applied technique for determining the approximate time required to perform the manual portion of any job. It is equally applicable to jobs done in the home, field, office, or factory. It is particularly advantageous for estimating labor costs in advance of actual production on all types of work. It can be used for establishing incentive rates on small quantity nonrepetitive or long-cycle operations.

Abbreviated Work-Factor is convenient for studying operations of many minutes or hours in length. In some circumstances the motion descriptions and time values are applied as rapidly as the operator performs his task. Hence, when one cycle of the operation is completed, the time standard will be essentially completed. The speed with which abbreviated applications are made depends on the required accuracy, the analyst's skill, and the nature of the work studied.

Abbreviated values are so compiled that if in some situations accuracy is sacrificed, time is overallowed rather than underallowed. For this reason, abbreviated standards allow greater overall time than do those established by detailed or simplified Work-Factor. They average about 12 percent more time than do detailed analyses.

Ready Work-Factor

Although the ready technique is less precise than detailed Work-Factor, experience to date indicates that the deviations from detailed are not large and that the resulting time values are suitable for a wide range of applications.

The ready techniques employs the ready unit 0.0010 minutes (10 Work-Factor units) as its unit of time.

All ready units in the Ready Work-Factor Time Tables are simple numbers arranged in easily remembered sequences, such as 1, 2, 3, and 4, or 3, 5, 7, 9, and 11.

Only a few simple rules of application are employed. These are explained in a Ready Work-Factor manual.

Simple illustrations of correct analyses of each standard element are established as guideposts. Objects familiar and available to all, such as pencils, bricks, and washers, are used. These examples form a framework of reference for correct application. It is relatively easy for an analyst to decide that the specific element he is evaluating is easier than one guidepost and more difficult than another.

Through simpler, more graphic terminology and a minimum number of times values, Ready Work-Factor creates a work-time relationship in the minds of its users. Time values become handy thoughts, easily applied mentally and conversationally. The practical person who uses Ready

Work-Factor learns to visualize time with the same ease as he does tools and materials.

Methods-Time Measurement (MTM–1)

In 1948 the text *Methods-Time Measurement* was published, giving time values for the fundamental motions reach, move, turn, grasp, position, disengage, and release.[5] The authors defined MTM as "a procedure which analyzes any manual operation or method into the basic motions required to perform it, and assigns to each motion a pre-determined time standard which is determined by the nature of the motion and the conditions under which it is made."[6]

The MTM data, like those of Work-Factor, are the result of frame-by-frame analysis of motion-picture films involving diversified areas of work. The data taken from the various films were "leveled" (adjusted to the time required by the normal operator) by the Westinghouse technique. This method of performance rating is described in Chapter 15. The data were then tabulated and analyzed to determine the degree of difficulty caused by variable characteristics. For example, it was found that not only distance but also the type of reach affected reach time. Further analysis indicated that there were five distinct cases of reach, each requiring different time allotments to perform for a given distance. These are:

1. Reach to object in fixed location, or to object in other hand, or on which other hand rests.
2. Reach to single object in location which may vary slightly from cycle to cycle.
3. Reach to object jumbled with other objects so that search and select occur.
4. Reach to a very small object or where accurate grasp is required.
5. Reach to indefinite location to get hand in position for body balance, or next motion, or out of way.[7]

Also, move time was influenced not only by distance and the weight of the object being moved, but by the specific type of move. Three cases of move were found. These are:

1. Move object to other hand or against stop.
2. Move object to approximate or indefinite location.
3. Move object to exact location.[8]

Table 19–5 summarizes MTM values that have been developed to date. It will be noted that the time values of the therblig grasp vary from 2.0 TMUs to 12.9 TMUs (1 TMU equals 0.00001 hour), depending on the

---

[5] H. B. Maynard, G. J. Stegemerten, and J. L. Schwab, *Methods-Time Measurement* (New York: McGraw-Hill Book Co., 1948).

[6] Ibid.

[7] Ibid.

[8] Ibid.

TABLE 19–5

TABLE I—REACH—R

Distance Moved Inches	Time TMU				Hand In Motion		CASE AND DESCRIPTION
	A	B	C or D	E	A	B	
¾ or less	2.0	2.0	2.0	2.0	1.6	1.6	A Reach to object in fixed location, or to object in other hand or on which other hand rests.
1	2.5	2.5	3.6	2.4	2.3	2.3	
2	4.0	4.0	5.9	3.8	3.5	2.7	
3	5.3	5.3	7.3	5.3	4.5	3.6	B Reach to single object in location which may vary slightly from cycle to cycle.
4	6.1	6.4	8.4	6.8	4.9	4.3	
5	6.5	7.8	9.4	7.4	5.3	5.0	
6	7.0	8.6	10.1	8.0	5.7	5.7	
7	7.4	9.3	10.8	8.7	6.1	6.5	
8	7.9	10.1	11.5	9.3	6.5	7.2	C Reach to object jumbled with other objects in a group so that search and select occur.
9	8.3	10.8	12.2	9.9	6.9	7.9	
10	8.7	11.5	12.9	10.5	7.3	8.6	
12	9.6	12.9	14.2	11.8	8.1	10.1	
14	10.5	14.4	15.6	13.0	8.9	11.5	D Reach to a very small object or where accurate grasp is required.
16	11.4	15.8	17.0	14.2	9.7	12.9	
18	12.3	17.2	18.4	15.5	10.5	14.4	
20	13.1	18.6	19.8	16.7	11.3	15.8	
22	14.0	20.1	21.2	18.0	12.1	17.3	E Reach to indefinite location to get hand in position for body balance or next motion or out of way.
24	14.9	21.5	22.5	19.2	12.9	18.8	
26	15.8	22.9	23.9	20.4	13.7	20.2	
28	16.7	24.4	25.3	21.7	14.5	21.7	
30	17.5	25.8	26.7	22.9	15.3	23.2	

TABLE II—MOVE—M

Distance Moved Inches	Time TMU				Wt. Allowance			CASE AND DESCRIPTION
	A	B	C	Hand in Motion B	Wt. (lb.) Up to	Factor	Constant TMU	
¾ or less	2.0	2.0	2.0	1.7	2.5	0	0	A Move object to other hand or against stop.
1	2.5	2.9	3.4	2.3				
2	3.6	4.6	5.2	2.9	7.5	1.06	2.2	
3	4.9	5.7	6.7	3.6				
4	6.1	6.9	8.0	4.3	12.5	1.11	3.9	
5	7.3	8.0	9.2	5.0				
6	8.1	8.9	10.3	5.7	17.5	1.17	5.6	
7	8.9	9.7	11.1	6.5				
8	9.7	10.6	11.8	7.2				
9	10.5	11.5	12.7	7.9	22.5	1.22	7.4	B Move object to approximate or indefinite location.
10	11.3	12.2	13.5	8.6				
12	12.9	13.4	15.2	10.0	27.5	1.28	9.1	
14	14.4	14.6	16.9	11.4				
16	16.0	15.8	18.7	12.8	32.5	1.33	10.8	
18	17.6	17.0	20.4	14.2				
20	19.2	18.2	22.1	15.6				
22	20.8	19.4	23.8	17.0	37.5	1.39	12.5	
24	22.4	20.6	25.5	18.4				C Move object to exact location.
26	24.0	21.8	27.3	19.8	42.5	1.44	14.3	
28	25.5	23.1	29.0	21.2				
30	27.1	24.3	30.7	22.7	47.5	1.50	16.0	

TABLE III—TURN AND APPLY PRESSURE—T AND AP

Weight	Time TMU for Degrees Turned											
	30°	45°	60°	75°	90°	105°	120°	135°	150°	165°	180°
Small— 0 to 2 Pounds	2.8	3.5	4.1	4.8	5.4	6.1	6.8	7.4	8.1	8.7	9.4
Medium—2.1 to 10 Pounds	4.4	5.5	6.5	7.5	8.5	9.6	10.6	11.6	12.7	13.7	14.8
Large— 10.1 to 35 Pounds	8.4	10.5	12.3	14.4	16.2	18.3	20.4	22.2	24.3	26.1	28.2

APPLY PRESSURE CASE 1—16.2 TMU. APPLY PRESSURE CASE 2—10.6 TMU

M.T.M. Association

TABLE 19-5 *(continued)*

TABLE IV—GRASP—G

Case	Time TMU	DESCRIPTION
1A	2.0	Pick Up Grasp—Small, medium or large object by itself, easily grasped.
1B	3.5	Very small object or object lying close against a flat surface.
1C1	7.3	Interference with grasp on bottom and one side of nearly cylindrical object. Diameter larger than $\frac{1}{2}''$.
1C2	8.7	Interference with grasp on bottom and one side of nearly cylindrical object. Diameter $\frac{1}{4}''$ to $\frac{1}{2}''$.
1C3	10.8	Interference with grasp on bottom and one side of nearly cylindrical object. Diameter less than $\frac{1}{4}''$.
2	5.6	Regrasp.
3	5.6	Transfer Grasp.
4A	7.3	Object jumbled with other objects so search and select occur. Larger than $1'' \times 1'' \times 1''$.
4B	9.1	Object jumbled with other objects so search and select occur. $\frac{1}{4}'' \times \frac{1}{4}'' \times \frac{1}{8}''$ to $1'' \times 1'' \times 1''$.
4C	12.9	Object jumbled with other objects so search and select occur. Smaller than $\frac{1}{4}'' \times \frac{1}{4}'' \times \frac{1}{8}''$.
5	0	Contact, sliding or hook grasp.

TABLE V—POSITION*—P

CLASS OF FIT		Symmetry	Easy To Handle	Difficult To Handle
1—Loose	No pressure required	S	5.6	11.2
		SS	9.1	14.7
		NS	10.4	16.0
2—Close	Light pressure required	S	16.2	21.8
		SS	19.7	25.3
		NS	21.0	26.6
3—Exact	Heavy pressure required.	S	43.0	48.6
		SS	46.5	52.1
		NS	47.8	53.4

*Distance moved to engage—1″ or less.

TABLE VI—RELEASE—RL

Case	Time TMU	DESCRIPTION
1	2.0	Normal release performed by opening fingers as independent motion.
2	0	Contact Release.

TABLE VII—DISENGAGE—D

CLASS OF FIT	Easy to Handle	Difficult to Handle
1—Loose—Very slight effort, blends with subsequent move.	4.0	5.7
2—Close — Normal effort, slight recoil.	7.5	11.8
3—Tight — Considerable effort, hand recoils markedly.	22.9	34.7

TABLE VIII—EYE TRAVEL TIME AND EYE FOCUS—ET AND EF

Eye Travel Time $=15.2 \times \frac{T}{D}$ TMU, with a maximum value of 20 TMU.

where T = the distance between points from and to which the eye travels.
D = the perpendicular distance from the eye to the line of travel T.

Eye Focus Time = 7.3 TMU.

TABLE 19–5 (concluded)

TABLE IX—BODY, LEG, AND FOOT MOTIONS

DESCRIPTION	SYMBOL	DISTANCE	TIME TMU
Foot Motion—Hinged at Ankle.	FM	Up to 4″	8.5
With heavy pressure.	FMP		19.1
Leg or Foreleg Motion.	LM —	Up to 6″	7.1
		Each add'l. inch	1.2
Sidestep—Case 1—Complete when leading leg contacts floor.	SS-C1	Less than 12″	Use REACH or MOVE Time
		12″	17.0
		Each add'l. inch	.6
Case 2—Lagging leg must contact floor before next motion can be made.	SS-C2	12″	34.1
		Each add'l. inch	1.1
Bend, Stoop, or Kneel on One Knee.	B,S,KOK		29.0
Arise.	AB,AS,AKOK		31.9
Kneel on Floor—Both Knees.	KBK		69.4
Arise.	AKBK		76.7
Sit.	SIT		34.7
Stand from Sitting Position.	STD		43.4
Turn Body 45 to 90 degrees—			
Case 1—Complete when leading leg contacts floor.	TBC1		18.6
Case 2—Lagging leg must contact floor before next motion can be made.	TBC2		37.2
Walk.	W-FT.	Per Foot	5.3
Walk.	W-P	Per Pace	15.0

TABLE X—SIMULTANEOUS MOTIONS

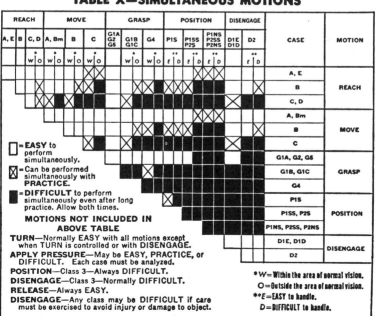

classification of the grasp, Two cases of release and 18 cases of position also influence time.

The steps followed in applying MTM technique are similar to those followed for Work-Factor. First, the analyst summarizes all left-hand and right-hand motions required to perform the job properly. Then he determines from the methods-time data tables the leveled time in TMUs for each motion. The nonlimiting motion values should be circled or deleted, as only the limiting motions will be summarized, provided that it is "easy" to perform the two motions simultaneously (see section X of Table 19–5), to determine the time required for a normal performance of the task. For example, if the right hand reached 20 inches to pick up a nut, the classification would be R20C and the time value would be 19.8 TMUs. If, at the same time, the left hand reached 10 inches to pick up a cap screw, a designation of R10C with a TMU value of 12.9 would be in effect. The right hand would be limiting, and the 12.9 value of the left hand would not be used in calculating the normal time.

The tabulated values do not carry any allowance for personal delays, fatigue, or unavoidable delays, and when these values are used for establishing time standards, an appropriate allowance must be added to the summary of the synthetic basic motion times.

MTM—General Purpose Data

MTM—GPD has been defined by the M.T.M. Association as a "dynamic system of universally applicable motion patterns built with M.T.M. which is used in place of the more detailed M.T.M. procedures in building standard data for specific work applications wherever motion patterns are used."

MTM—General Purpose Data is sponsored by the M.T.M. Association and is available to anyone who wishes to use it. The data are leased by the association.

The General Purpose Data system involves two levels of data. The first, basic level provides data for the combination of the more common sequences of the detailed MTM motions. The second, multipurpose level combines the more common sequences of the first level. For example, the first-level "GET" table includes codes that represent the time for a "reach," a "grasp," and a "release." The first-level "PLACE" table includes codes that represent the time for a "move" and a "position." A typical second-level code would be the combination of the first-level "get" and the first-level "place" into one code called "multi-get." This multi-get would represent the time for a reach, a grasp, a move, a position, and a release.

The coding used for the identification of MTM—General Purpose Data is "alphamnemonic." The code has seven characters, the first five being letters and the last two, numbers. For example, the code BGT-EV-06 would identify the data as being:

B	denotes basic (first) level
GT	signifies get category
E	implies an easy get
V	indicates that the object is in a variable location
06	denotes a 6-inch reach

The first-level or basic data cover 13 categories with 353 codes. The six most used categories within this basic data are:

GET
PLACE
ELEMENTAL
BODY
READ
WRITE

The remaining seven categories of the data are:

ACTUATE
THREADED FASTENER
LUBRICATE
DIP
CLEAN
TOOL USE
INSPECT & TEST

The second-level or multipurpose data were published in 1964 and contain approximately 200 codes involving these five categories:

GET
PLACE
BODY MOTIONS
VISING
CLAMPING

Additional categories currently being reviewed for inclusion as supplements to the second-level listings include:

FASTEN—THREADED AND NONTHREADED
OBJECT HANDLING
PACKAGING
TOOL USE
ACTUATE

The M.T.M. Association has established the following requirements for obtaining MTM—G.P.D.:

1. The company must be a member in good standing with the Association.
2. At least one of the employees of the company must be a qualified appli-
cator of MTM, having taken the prescribed 105-hour training course and
passed the Association's test for certification.
3. The company must certify employment of or access to qualified guidance
in the use of M.T.M. and M.T.M.—G.P.D. This guidance can be obtained
in one of two ways:
    a. Have one of their M.T.M. applicators certified by the Association Test-
    ing as an in-plant instructor.
    b. Employ the services of a licensed instructor from one of the 17 man-
    agement council members of the association.

The M.T.M. Association has stated that MTM—General Purpose
Data will give an accuracy within 15 percent (when compared with
MTM) for cycle times over 0.120 minutes in 95 percent of the cycles
analyzed. As the cycle time increases, the accuracy of the General Purpose Data will improve.

MTM–2

In an effort to further the application of MTM to work areas where
the detail of MTM–1 would economically preclude its use, the International Directorate of the M.T.M. Association initiated a research project
to develop less refined data suitable for the majority of motion sequences.
The result of this effort was MTM–2, which has been defined by the
M.T.M. Association of the United Kingdom as

a system of synthesized M.T.M. data and is the second general level of M.T.M.
data. It is based exclusively on M.T.M. and consists of:
1. Single basic M.T.M. motions.
2. Combinations of basic M.T.M. motions.
The data is adapted to the operator and is independent of the work place
or equipment used. It is not possible to replace any element in M.T.M.–2 by
means of other elements in M.T.M.–2.

In general, MTM–2 should find application in work assignments
where:

1. The effort portion of the work cycle is more than one minute in length.
2. The cycle is not highly repetitive.
3. The manual portion of the work cycle does not involve a large number of either complex or simultaneous hand motions.

It has been observed that the variability between MTM–1 (MTM)
and MTM–2 is dependent to a large extent upon the length of the cycle.
This is reflected in Figure 19–2, where the range of percentage deviation
of MTM–2 from MTM is shown. This range of "error" is considered to
be the expected range 95 percent of the time.

IGURE 19-2
ercentage variation of MTM–1 when compared with MTM–2 at increasing cycle lengths

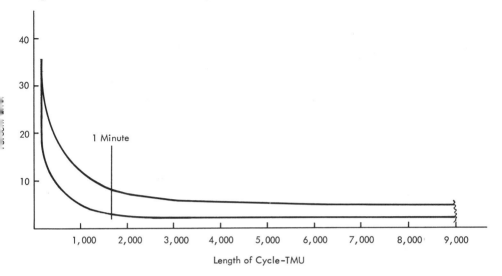

MTM–2 recognizes 10 classes of actions, which are referred to as "categories." These 10 categories and their symbols are:

GET	G
PUT	P
GET WEIGHT	GW
PUT WEIGHT	PW
REGRASP	R
APPLY PRESSURE	A
EYE ACTION	E
FOOT ACTION	F
STEP	S
BEND & ARISE	B
CRANK	C

In using MTM–2, distances are estimated by classes and affect the times of the GET and PUT categories. As in MTM–1, the distance moved is based on the length of the path traveled by the knuckle at the base of the index finger in the case of hand motions and is measured at the fingertips if only the fingers move.

The codes for the five tabulated distance classes are:

Inches	Code
0–2	5
Over 2–6	15
Over 6–12	30
Over 12–18	45
Over 18	80

The categories GET and PUT are usually considered simultaneously. Three variables affect the time required to perform both of these variables. These variables are the case involved, the distance traveled, and the weight handled. The reader should recognize that GET can be considered a composite of the therbligs reach, grasp and release, while PUT is a combination of the therbligs move and position.

Three cases of GET have been identified as A, B, and C. Case A implies a simple contact grasp, such as when the fingers push an ashtray across the desk. If an object such as a pencil is picked up by simply enclosing the fingers with a single movement, a case B grasp is employed. If the type of grasp is neither an A nor a B, then a case C GET is being employed.

The analyst can resort to the decision diagram (Figure 19–3) to assist in the determination of the correct case of GET.

FIGURE 19-3

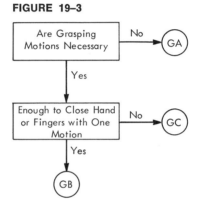

Tabular values in TMUs of the three cases of GET applied to each of the five coded distances are shown in Table 19–6.

TABLE 19-6
Tabular values for GET

		Distance code				
Case	Description	5	15	30	45	80
GA	No grasp necessary	3	6	9	13	17
GB	Simple closing of fingers or hand	7	10	14	18	23
GC	Any other grasp	14	19	23	27	32

Just as there are three cases of GET, so there are three cases of PUT. The decision diagram (Figure 19–4) may be utilized to determine which case of PUT is applicable. An explanation of the three cases of PUT, as

FIGURE 19-4

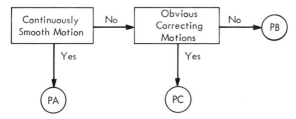

well as tabular values for each class applied to the five coded distances, is given in Table 19–7.

TABLE 19-7
Tabular values for PUT

		Distance code				
Case	Description	5	15	30	45	80
PA	Smooth continuous motion	3	6	11	15	20
PB	Some irregularity in motion pattern	10	15	19	24	30
PC	Correcting motions obvious	21	26	30	36	41

PUT is accomplished in one of two ways: insertion and alignment. An insertion involves placing one object into another, such as a shaft into a sleeve, while an alignment involves orienting a part on a surface, such as bringing a rule up to a line.

Tolerance limits have been established to provide guidelines as to the most appropriate case of PUT. When one part is inserted into another and more than 0.4-inch clearance exists, the parts can usually be brought together by utilizing a smooth, continuous motion, and consequently a case PA would be involved. If the clearance were reduced to less than 0.4-inch so that some irregularity in motion pattern occurred, yet no pressure was required to mate the objects, the case would be PB. And if a close fit were involved where slight pressure was required to mate the objects or if a loose fit and a difficult to handle part were encountered, thus necessitating corrective mating, the case would be cataloged as PC.

For alignment-type PUT the following dimensional relations can be used as guidelines. If an object is placed adjacent to a line to a tolerance of less than $\frac{1}{16}$ inch, the case is classified as PC. However, if the tolerance is more than $\frac{1}{16}$ inch but less than $\frac{1}{4}$ inch, the case is classified as PB, and if the tolerance is more than $\frac{1}{4}$ inch, the case is classified as PA.

Weight is considered in MTM–2 in a fashion similar to MTM–1.

The time value addition for "Get Weight" (GW) has been estimated as 1 TMU per effective kilogram. Thus, if a load of 6 kilograms is handled by both hands, the time addition due to weight would be 3 TMUs since the effective weight per hand is 3 kilograms.

For "Put Weight" (PW), additions have been estimated at 1 TMU per 5 kilograms of effective weight up to a maximum of 20 kilograms.

The category regrasp (R) has been defined as under MTM–1. Here, however, a time of 6 TMUs has been assigned. The authors of MTM–2 point out that for a regrasp to be in effect, control must be retained by the hand.

Apply pressure has been assigned a time of 14 TMUs. The authors point out that this category can be applied by any member of the body and that the maximum permissible movement for an apply pressure is ¼ inch.

Eye action (E) is allowed under either of the following cases:

1. When it is necessary for the eye to move in order to see the various aspects of the operation involving more than one specific section of the work area.
2. When the eye must concentrate on an object so as to recognize a distinguishable characteristic.

The estimated value of (E) is 7 TMUs. The value is only allowed when E must be performed independently of hand or body motions.

Crank (C) occurs when the hands or fingers are used to move an object in a circular path of more than ½ revolution. A PUT is indicated in cranks of less than ½ revolution. Only two variables are associated with the category crank under MTM–2. These are the number of revolutions and the weight or resistance involved. A time of 15 TMUs is allotted for each complete revolution. Where weight or resistance is significant, PW is applied to each revolution.

Foot movements are allowed 9 TMUs, and step movements 18 TMUs. The time for step movement is based upon a 34-inch pace. The decision

FIGURE 19–5

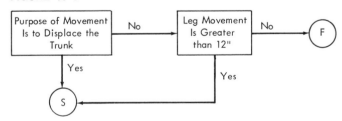

diagram (Figure 19–5) can be helpful in ascertaining whether a given movement should be classified as a step or a foot movement.

The category BEND & ARISE (B) occurs when the body changes its vertical position. Typical movements characteristic of B include sitting down, standing up, and kneeling. A time value of 61 TMUs has been assigned to B. The authors indicate that when an operator kneels on both knees the movement should be classed as 2 B.

A summary of MTM–2 values is shown in Table 19–8. The reader

TABLE 19–8
A summary of MTM–2 data (all time values are in TMUs)

Code	GA	GB	GC	PA	PB	PC
– 5.	3	7	14	3	10	21
–15.	6	10	19	6	15	26
–30.	9	14	23	11	19	30
–45.	13	18	27	15	24	36
–80.	17	23	32	20	30	41

GW – 1/KG　　PW – 1/5 KG

A	R	E	C	S	F	B
14	6	7	15	18	9	61

should recognize, as in the case of MTM–1, that motions performed simultaneously with both hands cannot always be performed in the same

FIGURE 19–6
MTM–2 simultaneous hand
motion allowances

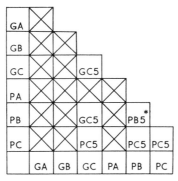

* If PB _____ is performed simultaneously with PB _____, an addition of PB5 is made only if the actions are outside the normal area of vision.

time as motions performed by one hand only. Figure 19–6 reflects motion patterns where the time required for simultaneous motions is the same as that required for motions performed by one hand. In these instances, an *X* appears. When additional time is required to perform simultaneous motions, the magnitude of this extra amount is given with the appropriate coding.

As with all fundamental motion data systems, the novice should not try to apply the data until he has been properly trained in their use and application.

MTM–3

The last level of Methods-Time Measurement is known as MTM–3. This level was not developed to replace MTM or MTM–2 but as a supplement to them. MTM–3 is intended to be used in work situations where, in the interest of saving time at the expense of some accuracy, it is a better alternative than either MTM or MTM–2.

MTM–3 can be effectively used to study and improve methods, evaluate alternative methods, develop standard data and formulas, and establish standards of performance. MTM–3 should not be used in connection with operations that require either eye focus or eye travel time since the data do not consider these motions. H. B. Maynard and Company has also advised against using MTM–3 in work situations where the frequency is high. In fact, the company has stated: "Manual motions with a frequency higher than 10 are not permitted in an MTM–3 analysis."[7] It has also stated that MTM–3 "should not be used on cycles that are less than 4 minutes in duration."

Table 19–9 presents MTM–3 data. Ten time standards ranging from 7

TABLE 19–9
Maynard Research Council, Inc.: MTM–3

		Handle		*Transport*	
Inches	*Code*	*HA*	*HB*	*TA*	*TB*
6	6	18	34	7	21
6	32	34	48	16	29
		SF 18		B 61	

Do not attempt to use these data in any way unless you understand their proper application. This statement is included as a word of caution to prevent difficulties resulting from misapplication of the data.

to 61 TMU form the basis for the development of any standard subject to the limitations heretofore noted.

Once the analyst is properly trained in utilizing MTM–3, standards can be developed quite rapidly by means of this technique.

---

[7] William Antis, John M. Honeycutt, Jr., and Edward N. Koch, *The Basic Motions of MTM* (Pittsburgh: The Maynard Foundation, 1971).

Applying synthetic basic motion times

As a means of improving methods, a general knowledge of the fundamentals of synthetic motion-time standards will prove invaluable. For example, if the methods analyst has a grounding in MTM, he will design work stations to utilize the "G1A" grasps requiring but 2 TMUs rather than the more difficult grasps which require as much as 12.9 (G4C) TMUs. Likewise, he will endeavor to design for utilizing contact releases rather than normal releases and for symmetrical positioning rather than semisymmetrical or nonsymmetrical positioning. One of the important uses for any of the synthetic basic motion-time techniques lies in the area of methods. Once an analyst has an appreciation of these techniques, he will find himself looking more critically at each and every work station, thinking about how improvements may be made. Reaches and moves of 20 inches will appear unduly long to him, and he will immediately consider the savings possible through shortened motion patterns. Positioning elements that involve heavy pressure are automatically signs of a need for improvement. Operations that require eye travel and eye focus time can usually be improved.

Of course, standards of performance may be established through synthetic basic motion times. If the data are to be used for this purpose, a much greater knowledge of the techniques of application is required. In no case should an analyst endeavor to establish time standards to be used for rate purposes until he has had thorough training and has demonstrated his ability to use the data precisely. Thus, the analyst must know whether the distance moved is the linear distance taken by the hand or the circumferential distance taken by the arc that the hand makes. He must know whether the distance is measured from the center of the hand, the knuckles, or the fingertips. He must know when application of pressure prevails and when it does not. He must clearly understand the elements of alignment and orientation as they affect positioning time. These and many other considerations must be mastered before an analyst can expect to establish consistent and accurate time standards with this tool.

It is only after thorough training that a group of analysts will arrive at consistent time standards when using any of the synthetic motion-time techniques. In no case should any analyst endeavor to establish time standards with only a superficial knowledge of these techniques.

The development of standard data

Today, one of the broadest uses of predetermined times is in the development of standard data elements. With standard data, operations can be prepriced much faster than by the laborious procedure of summarizing long columns of fundamental motion times. In addition to saving time, applying standard data reduces clerical errors, since less arithmetic is involved.

With sound standard data, it is economically feasible to establish standards on maintenance work, material handling, clerical and office work, inspection work, and other indirect and expense operations. Thus, elements involving long cycles and consisting of many short duration elements may be prepriced economically with standard data. For example, one company developed standard data applicable to radial drill operations in its toolroom. Standard data were developed for the elements required to move the tool from one hole to the next and to present and back off the drill. These standard data elements were then combined in one multivariable chart so that they could be summarized rapidly (see Table 19–10).

Figure 19–7 illustrates another application of synthetic basic motion times in the development of standard data. The work sheet shown gives all the standard elements for trimming blanks on punch presses. It is a matter of only minutes to predetermine a standard for trimming blanks with weights of up to 40 pounds.

Another example illustrating the flexibility of synthetic basic motion times is the development of a formula for prepricing a clerical operation. This formula for sorting time slips included the following elements:

1. Pick up pack of departmental time slips and remove rubber band.
2. Sort time slips into direct labor (incentive), indirect labor, and daywork.
3. Record the total number of time slips.
4. Get pile of time slips, put rubber band around pack, and put aside.
5. Get pile of time slips and bunch.
6. Sort incentive time slips into "parts" time slips.
7. Count piles of incentive "parts" time slips.
8. Record the number of "parts" time slips and the number of incentive time slips.
9. Sort "parts" time slips into numerical sequence.
10. Bunch piles of numerical time slips and place in one pile on desk.

Each element was analyzed from the fundamental motion standpoint. Synthetic values were assigned and variables determined. Resulting from this procedure was an algebraic equation that allowed the rapid prepricing of the clerical operation.

Frequently, the stopwatch can be a helpful complement in developing standard data elements. Some portions of an element may be more readily determined by synthetic basic motion times, and other portions may be better adapted to stopwatch measurement.

It is wise to verify standard data elements with a decimal minute (0.001) watch as an added check on their validity. When predetermined times are converted to standard data elements that have been verified,

TABLE 19–10

Multivariable chart with two points of entry and with reference included for element: up spindle, swing and traverse head, and down spindle (Western radial drill)

Traverse of Head in Inches — Maximum Swing of Head in Inches

Traverse of Head (in.)	Maximum Swing of Head (in.)
2	0–1 / 2 3 / 4 8 10
3	0–1 / 2 3 6 / 8 10
4	0–1 2 3 / 4 6 10
5	0–1 / 2 3 6 / 10
6	0–2 3 / 4 8 10
7	0–1 / 2 4 6 / 10
8	0–1 2 3 / 6 10
9	0–1 / 3 6 10
10	0–1 / 3 6 10

Depth of Previous Hole in Inches

Reference Line	0–1	2	3	4	5	6	7
	.010	.013	.015	.017	.019	.022	.024
.038	.048	.051	.053	.055	.057	.060	.062
.040	.050	.053	.055	.057	.059	.062	.064
.042	.052	.055	.057	.059	.061	.064	.066
.044	.054	.057	.059	.061	.063	.066	.068
.046	.056	.059	.061	.063	.065	.068	.070
.048	.058	.061	.063	.065	.067	.070	.072
.050	.060	.063	.065	.067	.069	.072	.074
.052	.062	.065	.067	.069	.071	.074	.076
.055	.065	.068	.070	.072	.074	.077	.079
.058	.068	.071	.073	.075	.077	.080	.082
.060	.070	.073	.075	.077	.079	.082	.084
.063	.073	.076	.078	.080	.082	.085	.087
.066	.076	.079	.081	.083	.085	.088	.090
.069	.079	.082	.084	.086	.088	.091	.093
.072	.082	.085	.087	.089	.091	.094	.096
.076	.086	.089	.091	.093	.095	.098	.100
.079	.089	.092	.094	.096	.098	.101	.103
.083	.093	.096	.098	.100	.102	.105	.107
.088	.098	.101	.103	.105	.107	.110	.112

FIGURE 19–7
Standard data for prepricing press operations

1. Length of blank_____	4. No. of cuts_____	7. _____
2. Width of blank_____	5. Machine_____	8. _____
3. Weight of blank_____	6. _____	9. _____

A. P/U Blank-Asy. To Stop-Day.--Piece Aside

Weight	Toss Pc. Aside	Place Pc. Aside	Stack Piece		
4 lbs.	.054	.062	.064	X____occ.=	_____
4-14 lbs.	.071	.081	.084	X____occ.=	_____
14-26 lbs.	.098	.109	.113	X____occ.=	_____
26-40 lbs.	.103	.115	.119	X____occ.=	_____
40 lbs.	.108	.121	.126	X____occ.=	_____

B. Invert Part 180°

Weight	Major Dimension Inverted					
	5"-9.9"	10"-19.9"	20"-29.9"	30"-39.9"		
4 lbs.	simo	simo	.030	.034	X____occ.=	_____
4-14 lbs.	simo	.031	.037	.041	X____occ.=	_____
14-26 lbs.	.028	.035	.042	.046	X____occ.=	_____
26-40 lbs.	-	.039	.047	.052	X____occ.=	_____
40 lbs.	-	-	.050	.057	X____occ.=	_____

C. Dsy. Pc.-Rotate 90°-Piece

Weight	Major Dimension Rotated			
	5"-9.9"	10"-19.9"	20"-29.9"	30"-39.9"
4 lbs.	.030	.036	.040	.043
4-14 lbs.	.038	.043	.050	.053
14-26 lbs.	-	.056	.062	.065
26-40 lbs.	-	.063	.070	.073
40 lbs.	-	.070	.077	.081

Dsy. Pc.-Rotate 180°Asy. Piece

Weight	Major Dimension Rotated			
	5"-9.9"	10"-19.9"	20"-29.9"	30"-39.9'
4 lbs.	.044	.053	.062	.068
4-14 lbs.	.055	.066	.075	.084
14-26 lbs.	-	.080	.090	.101
26-40 lbs.	-	.089	.100	.113
40 lbs.	-	.098	.110	.125

D. Trip Machine
1. When foot has to reach to pedal   .009 X____occ. per blank=_____(includes trip)
2. When foot stays on pedal   .003 X____occ. per blank=_____(includes trip)
3. Look for pedal   .011 X____occ. per blank=_____(does not include 1 & 2)

E. Machine Times

Machine	Value	
Small Niagara	.008 X____occ. per blank=_____	
All Others	.010 X____occ. per blank=_____	

F. Material Handling
1. Push material truck to work area at beginning of operation ____no value
   (this value is usually included in the preceding operations)
2. Load pieces - fr.table or skid to machine bed .076 ÷ ____no. of pcs. =____
   fr.mach. bed to table or skid   .076 ÷ ____no. of pcs. =____
3. Exchange tote pans .327 ÷ ____no. of pcs. in pan= ____
4. Walk to rear of machine and return.
   (1). Small Niagara - .038÷ ____pcs. per sheet = ____value per blank.
   (2). All others   - .069÷ ____pcs. per sheet = ____value per blank.
5. Remove scrap at rear of machine.
   (1). Empty tote pan of scrap .327÷ ____no. of blanks cut per pan = _____.
   (2). Remove scrap by hand   .078÷ ____no. of blanks cut per 20 lbs. load = ____.
6. Cover finished blank with separator - .023÷ ____no. of pcs. covered = _____

Total Select Time_____

they will establish fair standards that are more consistent than those established through stopwatch procedure.

Summary

Basic Motion Timestudy, Work-Factor, Methods-Time Measurement, Dimensional Motion Times, and Motion-Time Analysis, as well as several leading industries, have made substantial contributions to a fund of knowledge available for the analysis of fundamental motions. Many years ago, Frederick W. Taylor visualized the development of standards for basic divisions of work similar to those currently in use. In his paper on "Scientific Management," he brought out that the time would come when a sufficient volume of basic standards would be developed so as to make further stopwatch studies unnecessary.

Synthetic basic motion-time values are becoming more accurate as additional studies are being made. There is still a need for further research, testing, and refinement in this area. For example, there is a question as to the validity of adding basic motion times for the purpose of determining elemental times in that therblig times may vary once the sequence is changed. Thus, the time for the basic element "reach 20 inches" may be affected by the preceding and succeeding elements and is not dependent entirely on the class of reach and distance. In general, it might be said that all fundamental motion techniques that have proved successful have made an effort to take care of the additivity of the elements of the motion in one way or another. Methods-Time Measurement, for instance, has provided three types of motion for the consideration of the movement of the hand, and several cases according to the nature of a move or reach. Work-Factor recognizes five elements of difficulty. Similar observations might be made about other fundamental motion techniques.

What has been done already has proved applicable in most instances. However, there are still some gaps to be filled. For example, more research needs to be undertaken in the development of data for combined motions. Reaches that involve a move during the first part of the motion are a typical illustration of the areas in which more basic study is needed. Likewise, moves that involve palming a component while en route are typical of combination motions that require more research.

In the analysis of motion patterns with existing data, it is important that the analyst consider not only the main purpose of the motion pattern, but also its complexity and characteristics. For example, if the hand is empty while moving toward an object, the motion will be classified as a reach; but if the hand holds an object while moving toward another object, not only the main purpose of the motion but also what the hand does with the object during the motion should be considered. If the hand is palming an object while reaching for something else, the motion cannot be classified as a basic reach. The time necessary to perform the combined motion depends on factors other than distance. The physical characteristics of the motion should not be neglected. When an object is being palmed while the hand is moving, in addition to the move, a simultaneous operation takes place. It is likely that the result might be a reduction in the average speed. This would allow the hand to establish control of the object during the distance moved. The longer the distance, the more time the hand has to palm the object in that particular motion.

Thus, the longer the combined motion, the more the motion approaches the time required for a simple reach of the same distance.

Additional study should be given to operations that involve simultaneous motions. Intuitively, there should be a relationship between the degree of bimanualness that the normal operator is able to perform and the motion pattern employed.

Further study will continue to increase the accuracy and the resulting application of synthetic basic motion times. Today, next to the stopwatch, this technique is the most important tool of the work measurement analyst.

One of the paramount values of basic motion times is that the analyst who endeavors to apply the synthetic procedure must carefully examine all factors affecting the motion patterns of a job. This minute scrutiny inevitably reveals opportunities for the refinement of work methods.

In one company, $40,000 was allocated in an effort to provide advanced tooling to increase the rate of production on a brazing operation. Prior to retooling, a work measurement study of the existing method was made, using fundamental motion data, and it developed that by providing a simple fixture and rearranging the loading and unloading area, production could be increased from 750 to 1,000 pieces per hour. The total cost of the synthetic basic motion-time study was $40, and as a result of the study it was not necessary to embark on the $40,000 retooling program.

A manager of one of our leading electrical appliance companies has emphasized that basic motion times permit predetermining whether or not expenditures for new facilities and tools are warranted, and can accurately forecast the cost of the reductions that may be achieved by such expenditures.

The combining of synthetic basic motion times, in the form of standard data, permits the application of basic motion times to a wide variety of work. Standard data thus developed are finding application on indirect and expense operations as well as on direct work. It is in this form that synthetic basic motion times have their broadest application.

It can be concluded that predetermined time systems have a definite place in the field of methods and work measurement and that they are no better than the people using them; that is, they should not be installed without professional help or a complete understanding of their application.

TEXT QUESTIONS

1. About when did synthetic time values begin to be used in industrial work?
2. What are the advantages of using synthetic basic motion times?
3. What variables are considered by the Work-Factor technique?
4. How did Work-Factor develop its values?
5. What is the time value of one TMU?

6. Who pioneered the MTM system?

7. How did the development of the basic motion times values differ from the development of both MTM and Work-Factor?

8. Who was originally responsible for thinking in terms of developing standards for basic work divisions? What was his contribution?

9. Calculate the equivalent in TMUs of 0.0075 hours per piece, of 0.248 minutes per piece, of 0.0622 hours per hundred, of 0.421 seconds per piece, of 10 pieces per minute.

10. How is MTM related to the analysis of method?

11. For what reasons was MTM–2 developed? Where does MTM–2 have special application?

12. What classes of action are recognized by MTM–2?

13. Explain the relationship of fundamental motion times to standard data.

14. If you have finished drilling a hole 3 inches deep on a Western radial drill, how long would it take to present the drill and drill a second hole in a steel forging ½ inch in diameter and 3 inches deep? Traverse: 6 inches; swing of head: 8 inches; 0.007-inch feed; 50 feet per minute surface speed.

15. How has Work-Factor endeavored to take care of the additivity of the elements of a motion pattern?

GENERAL QUESTIONS

1. What does the future hold for synthetic basic time values?

2. Which of the two techniques outlined is easier to apply?

3. Which method, in your opinion, will give the most reliable results? Why?

4. Describe as vividly as you can how you would explain to a worker in your forge shop who knows nothing about MTM what it is and how it is applied.

5. Give several objections to the application of MTM which you might receive from a worker, and explain how you would overcome them.

6. Some companies have been experiencing a tendency for their time study men to become more liberal in their performance rating over a period of years. How do fundamental motion data offset this tendency toward creeping loose standards?

7. Is there consistency between MTM–1 and MTM–2 in the handling of simultaneous motions?

PROBLEMS

1. Determine the time for the dynamic component of M20 B20.

2. A 30-pound bucket of sand having a coefficient of friction of 0.40 is pushed 15 inches away from the operator with both hands. What would be the normal time for the move?

3. A ¾-inch diameter coin is placed within a 1-inch diameter circle. What would be the normal time for the position element?

4. Give the MTM breakdown required to grasp the cotter pin with the pliers shown in the following sketch.

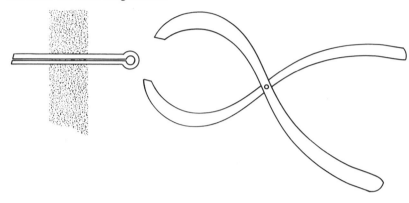

SELECTED REFERENCES

Antis, William, Honeycutt, John M., Jr., and Koch, Edward N. *The Basic Motions of MTM*. 3d ed. Pittsburgh: The Maynard Foundation, 1971.

Bailey, Gerald B., and Presgrave, Ralph *Basic Motion Time-Study*. New York: McGraw-Hill Book Co., 1958.

Birn, Serge A., Crossan, Richard M., and Eastwood, Ralph W. *Measurement and Control of Office Costs*. New York: McGraw-Hill Book Co., 1961.

Geppinger, H. C. *Dimensional Motion Times*. New York: John Wiley & Sons, Inc., 1955.

Karger, Delmar W., and Bayha, Franklin H. *Engineered Work Measurement*. New York: Industrial Press, 1957.

Maynard, Harold B., Stegemorten, G. J., and Schwab, John L. *Methods Time Measurement*. New York: McGraw-Hill Book Co., 1948.

Quick, Joseph H., Duncan, James H., and Malcolm, James A. *Work-Factor Time Standards*. New York: McGraw-Hill Book Co., 1962.

20

Formula construction

Formula construction as applied to time study involves the design of an algebraic expression or a system of curves that allows the establishment of a time standard in advance of production by enabling known values peculiar to the job to be substituted for the variable elements. A time study formula represents a simplification of standard data and has particular application in nonrepetitive work where it is impractical to establish standards on the basis of an individual time study for each job.

Application of formulas

Time formulas are applicable to practically all types of work. They have been successfully used in office operations, foundry work, maintenance work, painting, machine work, forging, coil winding, grass cutting, window washing, floor sweeping, welding, and many other areas. If sufficient time studies are collected to give a reliable sampling of data, it is possible to design a formula for a given range of work in any type of job.

It is important to recognize that the formula should be applied only to jobs that fall within the limits of the data used in developing the formula. If the boundaries of the formula are extended without the supporting proof of individual time studies, erroneous standards, with all the dangers brought about by inequitable rates, may result.

Once a formula has been built to cover a given operation, it should be immediately applied to all pertinent jobs within the range for which it was designed that do not already carry time standards.

Advantages and disadvantages of formulas

The advantages of using formulas rather than individual time studies for setting standards parallel those of using standard data. They may be summarized as follows:

1. More consistent time standards are established.

2. The duplication of time study effort on similar operations is eliminated.
3. Standards may be established much more rapidly.
4. A less experienced, less trained man may be used for establishing time standards.
5. Accurate, rapid estimates for labor costs may be made before production is begun.

Probably the most significant advantage of the formula over the standard data method is that a less valuable man is needed to work with formulas than with standard data. Any high school graduate who is proficient in algebra will be able to work out a time study formula and solve for the allowed time required to perform the task. Then, too, standards are established more rapidly by using formulas than by accumulating standard data elements. Since columns of figures must be added in the standard data method, there is a greater chance for omission or arithmetic error in setting a standard than in using a formula.

In the development of formulas, caution must be exercised in the treatment of constants. There is a natural tendency to treat more elements as constants than should be. Thus, errors can creep into the formula design. It is true that a formula will give consistent results, and so it is also true that if it is not accurate, it will give standards that are consistently wrong.

Another disadvantage of the formula lies in its application. Sometimes, in order to arrive at a standard at the earliest possible time, industry will use formulas in instances where the variables are beyond the range of the data used in developing the formula. Thus, the formula may be used where it has no application, and the resulting value will be far from valid.

Characteristics of the usable formula

A time study formula must be both completely reliable and practical if it is to be used with confidence. For the expression to be reliable, it must always give accurate results. If the developed formula is accurate, it should verify the individual standards used in its development within plus or minus 5 percent. The greater the number of studies that can be used in the development of the formula, the better the chances for ending up with a reliable formula. At least 10 independent studies should be available for a given class of work before the design of a formula is attempted. If tabularized standard data are available, these standards can be used for derivation of the formula. Frequently, when a formula is built, standard data, existing individual time studies, and new individual time studies will comprise the basic data used in its construction. Of course, it is understood that all mathematical calculation must be free from error before complete confidence can be vested in a formula.

The formula will give as accurate results as the data that entered into its derivation. The studies being used and the elements they comprise must show consistency in their end points and also in the method used, if the formula is to give consistently valid results.

For a formula to be practical, it should be as clear, concise, and simple as possible. The simplest formulas are best understood and most easily applied. Cumbersome expressions involving the taking of terms to powers should be avoided. Symbols of unknowns should not be repeated throughout the formula but should appear in only one place, together with their applicable suffixes, prefixes, and coefficients. The area of work that each symbol represents should be specifically identified. Liberal substantiating data should be included in the formula report so that any qualified, interested party can clearly identify the derivation of the formula. It is important that the limitations of the formula be noted by describing in detail its applicable range. Formulas so constructed will allow their users to apply them rapidly and accurately, with little difficulty in obtaining the required information.

Steps to follow in formula construction

The first step in formula construction is to determine what class of work is involved and what range of work is to be measured. For example, a formula might be developed for curing bonded rubber parts between two and eight ounces in weight. The class of work covered by the formula would be "curing molded parts," and the range of work involved would be two to eight ounces. After this overall analysis has been made, and a clear understanding of what is required has been obtained, the next step will be collecting the formula data. This step involves gathering former studies and standard data elements that have proven to be satisfactory, as well as taking new studies, in order to obtain a sufficiently large sample to cover the range of work for which the formula is needed. It is important that like elements in the different studies be consistent in their end points. This is essential in determining the variables that influence time, as well as in arriving at an accurate value for constant elements.

The elemental time study data are then posted on a work sheet for analysis of the constants and variables. The constants are combined and the variables analyzed so that the factors influencing time can be expressed either algebraically or graphically.

Once the constant values have been selected and the variable elements equated, then the expression is simplified by combining constants and unknowns where possible. The next procedure is to develop the synthesis in which the derivation of the formula is fully explained, so that the person using it, and any other interested party, will be fully cognizant of its application and development.

Before the formula is put into use, it should be thoroughly checked for accuracy, consistency, and ease of application. Once this has been done and the formula report describing the method used, working conditions, limitations of application, and so forth, has been completely written up, then the formula is ready for installation.

In collecting the individual time studies to be used for constructing the formula, it is perfectly acceptable for the observer to use existing time studies if they have been proven to be satisfactory, if the constant and variable elements in the studies have been properly separated, and if the studies were taken under prevailing conditions and methods. When taking new studies for formula work, the analyst should use the same exacting care, principles, and procedures as when taking a study for an individual standard. However, when taking studies for a time formula, the observer should break the various studies into like elements, with end points terminating at identical places in the work cycle. Also, the studies should be made of different operators to get as large a cross section as possible, and the jobs selected should cover the entire range of the proposed formula.

The number of studies that are needed to construct a formula will be influenced by the range of work for which the formula is to be used, the relative consistency of like constant elements in the various studies, and the number of factors that influence the time required to perform the variable elements.

FIGURE 20–1

			S-1	S-2	S-3	S-4
SHEET NO. __1__ OF 1 SHEETS		**MASTER TABLE (**				
FORMULA __73__	STUDY					
DATE __June 10,__	OPERATOR		Petrecca	Winters	Ekey	Kumpf
	Performance Factor		110	110	90	110
	Diameter of Core		1 7/8	2 1/4	1 7/8	2 3/
PART __Cylindrical Core__	Length of Core		7 7/8	6 1/2	9 1/4	4 1/
OPERATION __Made core from oil sand mix__	Number of Clamps		1	1	2	1
PERFORMED ON __Bench__	L/D Ratio		4.2	2.89	4.93	1.55
	Area		2.76	3.97	2.78	5.93
	Volume		21.7	25.8	25.6	25.2
COMPILED BY __J. Bodesky__	C1 + D1 + F1		1.242	1.244	1.499	.890

SYMBOL	OPERATION DESCRIPTION	NORMAL TIME MINUTES	REFERENCE	Operation Class				
A-1	Close core box	.046	Average	C	.049	.041	.050	.053
B-1	Clamp core box (C- clamps)	.112 N	Time vs.No.Cl.	V	.110	.111	.243	.120
C-1	Fill partly full of sand	Y x CT	Time v.L/D v.Vol.	V	.085	.094	.093	.119
D-1	Ram	Y x CT	Time v.L/D v.Vol.	V	.225	.255	.242	.248
E-1	Place rod and wire	.0153L+.03	Time vs.Length	V	.150	.103	.168	.088
F-1	Fill and ram	Y x CT	Time v.L/D v.Vol.	V	.35	.668	.900	.280
G-1	Strike off with slick	.0157A+.07	Time vs. Area	V	.115	.135	.129	.152
H-1	Remove vent wire	.047	Average	C	.048	.050	.048	.056
J-1	Rap box	.043	Average	C	.039	.042	.045	.050
K-1	Remove clamps	.061N	Time vs.No.Cl.	V	.057	.062	.116	.048
M-1	Open box	.046	Average	C	.046	.040	.038	.052
N-1	Roll out core	.0057L+.045	Time vs.Length	V	.098	.082	.102	.075
P-1	Clean box	$\sqrt{.0067}+.000016V^2$	Time vs.Volume	V	.107	.112	.120	.110

As has been stated, at least 10 studies should be available before a formula is constructed. If less than 10 are used, the accuracy of the formula may be impaired through incorrect curve construction and data that are not representative of typical performance. Of course, the more studies that are used, the more data will be available and the more the normal conditions that prevail will be reflected.

Analyzing the elements

After a sufficient number of time studies have been gathered, it is helpful to summarize the data on one work sheet for analysis purposes. Figure 20–1 illustrates a "Master Table of Detail Time Studies" form which has been designed for this purpose. In addition to the information called for on the form, any specific information, affecting the variable elements, such as surface area, volume, length, diameter, hardness, radius, and weight, should be posted under its corresponding study number in the "Job Characteristics" section of the form.

Here the name of the operator studied, his rating factor, and the part number of the job are also shown. A separate column has been assigned for recording all of this pertinent information for each time study.

In the left-hand column headed "Symbol," an identifying term is placed for each element. The letters of the alphabet followed by a suffix

TAIL TIME STUDIES

	S-6	S-7	S-8	S-9	S-10	S-11	S-12	S-13	S-14	S-15	S-16	S-17
n	Markley	Noyes	Kinachan	Winters	Kumpf	Petrecca	Judd	Judd	Geiger	Plasan	Ekey	Noyes
	105	105	98	120	100	108	95	102	117	85	.95	100
4	2	1 7/8	1 1/8	1	1 3/4	1 1/4	1 1/4	1 1/4	7/8	1 1/8	2	1 7/8
2	8	4 1/4	10	13 1/8	5 1/8	6 3/4	7 5/8	4 1/8	8 1/4	6 1/4	10	12 5/8
	1'	1	2	1	1	1	1	1	1	1	2	2
	4	2.27	8.89	13.13	2.73	5.40	6.10	3.30	9.43	5.56	5.0	6.31
	3.14	2.78	.99	.78	2.40	1.23	1.23	1.23	.60	.99	3.14	2.78
	25.1	11.8	9.90	10.3	12.3	8.27	9.33	5.05	4.96	6.19	31.4	35.1
7	1.316	.733	1.489	2.138	.804	1.159	1.181	.797	1.277	1.032	1.740	2.299
	.042	.047	.050	.041	.043	.042	.047	.048	.049	.046	.043	.048
	109	.087	.232	.218	.098	.114	.097	.128	.091	.099	.229	.240
	.141	.082	.129	.206	.073	.102	.079	.062	.133	.100	.107	.195
	.245	.160	.344	.436	.194	.281	.252	.090	.342	.162	.285	.480
	.142	.093	.180	.222	.110	.125	.140	.090	.142	.120	.182	.210
	.725	.258	.889	1.290	.392	.665	.850	.432	.802	.626	1.089	1.390
	.120	.104	.078	.077	.198	.089	.084	.086	.074	.085	.119	.106
	.042	.041	.052	.051	.043	.042	.042	.044	.040	.051	.052	.045
	.041	.045	.057	.048	.040	.039	.042	.041	.050	.038	.047	.041
	.039	.062	.116	.120	.071	.059	.056	.070	.058	.058	.140	.145
	.041	.045	.054	.046	.066	.038	.065	.045	.040	.035	.038	.052
	.091	.068	.108	.120	.080	.082	.090	.065	.090	.071	.110	.118
	.118	.090	.089	.088	.090	.087	.085	.085	.085	.088	.138	.150

number are frequently used. Thus, we should have the symbols A–1, B–1, C–1, and so on. When more than 26 elements are involved, then the alphabet is repeated, and the suffix 2 is used. These symbols are employed for identification and reference from the Master Table to elements grouped in the synthesis.

Under the "Operation Description," every element that has occurred on the individual time studies is recorded. The element description should be expressed clearly, so that any interested person in the future will be able to visualize exactly the work content of the element. A "C" or a "V" in the column headed "Operation Class" indicates whether an element is a constant or a variable.

Next, the normal elemental times from the individual studies are recorded in their appropriate spaces. After these data have been posted, the elemental time values for each element are compared and the reasons for variance are determined. As can be expected, a certain amount of variation will prevail, even in the constant elements, because of inconsistency in performance rating. In general, the constant elements should not deviate substantially, and the allowed time for each constant may be determined by averaging the values of the different studies. This average time should then be posted in the "Normal Time," column, and in the "Reference" column the word *average* should be shown.

The variable elements will tend to vary in proportion to some characteristic or characteristics of the work, such as size, shape, or hardness. These elements will have to be studied carefully to determine which factors influence the time, and to what extent. By plotting a curve of time versus the independent variable, it is frequently possible for the analyst to deduce an algebraic expression in terms of the time representing the element. This procedure is explained in a later section of this chapter. If an equation can be computed, then it should be shown in the "Normal Time" column, and the curve or curves used in its deduction should be referred to in the adjacent "Reference" column.

Compute expressions for variables

If the analysis of the elemental data reveals that one variable characteristic governs the elemental time, then the analyst should construct a graphic relationship following the procedure outlined in Chapter 18 (see Figure 18–4). Thus, time is the dependent variable in the plotting, since it is the value we want to predict. In those cases where time is to be estimated in terms of a single independent variable, y will be referred to as the random variable, time, whose distribution depends on the independent variable, x. In most relations that we are estimating, x is not random; it is fixed for all practicality, and we are concerned with the mean of the corresponding distribution of time y (the successive elements observed through stopwatch analysis) for given x (see Figure 20–6).

Plotted data may take a number of forms—the straight line, parabola, hyperbola, and ellipse, exponential forms, or no regular geometric form whatsoever. Graphic procedures may be quite useful for obtaining predicting equations.

If the data plots as a straight line on uniform graph paper, then we can equate time as a function of the slope times the variable v and the y intercept, as discussed on page 421.

Another frequent linear relationship that the reader should be cognizant of is the reciprocal function expressed by the equation:

$$y = \frac{1}{b + mv}$$

It can be seen that the reciprocal function involves a linear relationship for values of $\frac{1}{y}$ and v since, by taking the reciprocal of both sides of the equation, we get:

$$\frac{1}{y} = b + mv$$

The hyperbolic curve takes the form illustrated in Figure 20–2 and is expressed by the equation:

$$\frac{y^2}{a^2} - \frac{x^2}{k^2} = 1$$

If the plotted data take the form of a segment of the hyperbola, k may be computed graphically by drawing a tangent to the curve through the origin and using the resulting line as the diagonal of the rectangle

FIGURE 20–2
Segment of hyperbolic curve

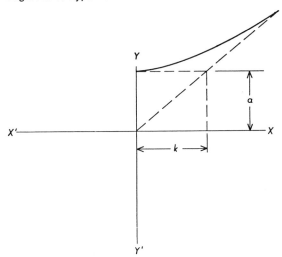

whose height is a, the distance from the origin to the y intercept. Substituting time and the variable characteristic for x and y would give:

$$T = \sqrt{a^2 + \frac{(v^2)(a^2)}{k^2}}$$

FIGURE 20–3
Segment of ellipse

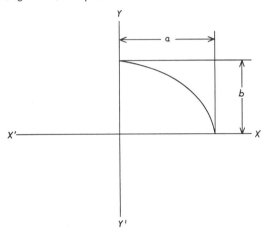

The ellipse segment shown in Figure 20–3 is expressed by the equation:

$$\frac{x^2}{a^2} + \frac{y^2}{b^2} = 1$$

When the plotted time values take the form of an ellipse, a and b may be computed graphically and their values substituted in the expression:

$$T = \sqrt{b^2 - \frac{b^2 v^2}{a^2}}$$

In other instances, the data may take the form of a parabola, as illustrated in Figure 20–4. This curve is expressed by the equation:

$$x^2 = \frac{a^2 y}{b}$$

and after substituting graphic solutions for a and b, we have the equation:

$$T = \frac{b v^2}{a^2}$$

As in previous illustrations, T is equal to time and v is equal to the variable characteristic governing the elemental time.

One additional relationship that has much application in formula development is the power function. This is of the form:

$$y = b m^v$$

FIGURE 20–4
Segment of parabola

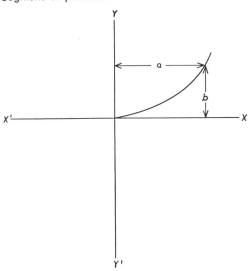

When the data plotted on uniform graph paper do not follow a straight line, parabola, hyperbola, or circle, the analyst should try using semilogarithmic paper so as to determine whether the points will fall close to a straight line on a transformed scale. If the paired data give a straight line when plotted on semilog paper, then the curve of y on x is exponential. By this we mean that for any given x, the mean of the distribution of the y's (times in our work) is given by bm^x.

To illustrate, in the element "strike arc and weld," the following data were obtained from 10 detailed studies:

Study no.	Size of weld	Minutes per inch of weld
1	$1/8$	0.12
2	$3/16$	0.13
3	$1/4$	0.15
4	$3/8$	0.24
5	$1/2$	0.37
6	$5/8$	0.59
7	$11/16$	0.80
8	$3/4$	0.93
9	$7/8$	1.14
10	1	1.52

When the data were plotted on rectangular coordinate paper, a smooth curve resulted (see Figure 20–5). The data were then plotted on semi-

FIGURE 20–5
Plotted curve on regular coordinate paper takes exponential form

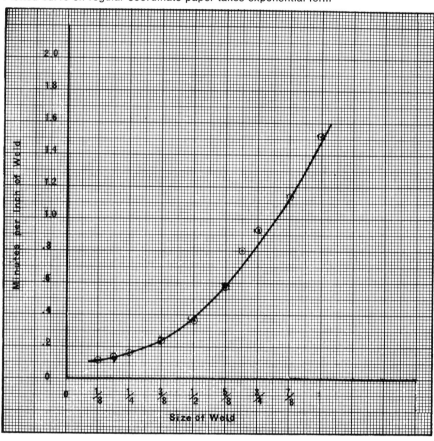

logarithmic paper, and the straight line as shown in Figure 20–6 was obtained.

This plotting can now be stated in equation form by the following derivation:

Select any two points on the straight line:

$$\text{Point 1} \quad X = 1, \qquad Y = \log 1.52$$
$$\text{Point 2} \quad X = \tfrac{3}{8}, \qquad Y = \log 0.24$$

Solving for the slope m:

$$m = \frac{Y_1 - Y_2}{X_1 - X_2}$$
$$= \frac{\log 1.52 - \log 0.24}{1 - \tfrac{3}{8}}$$
$$= 1.28$$

FIGURE 20–6
Data plotted on semilogarithmic paper take the form of a straight line with the equation $Y = AB^x$

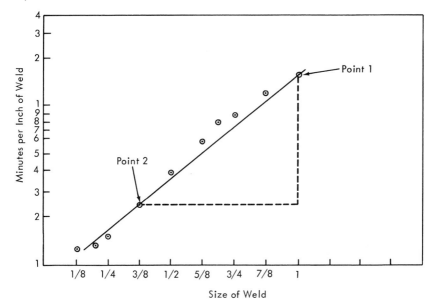

To determine the equation of the line from the equation $Y - Y_1 = m(X - X^1)$, we get:

$$\log Y - \log 1.52 = 1.28(X - 1)$$
$$\log Y - 0.18184 = 1.28X - 1.28$$
$$\log Y = 1.28X - 1.10$$

From the exponential form

$$Y = AB^x$$
$$\log Y = \log A + X \log B$$

and

$$\log A = -1.10$$
$$A = 0.07944$$
$$\log B = 1.28$$
$$B = 19.05$$

and

$$Y = (0.07944)(19.05)^x$$

Rounding off:

$$Y = (0.08)(19)^x$$

This equation would then be a component of the formula expression and would be less cumbersome than referring to the curve shown in Figure 20–6. It can be checked as follows:

With ½-inch weld:

$$\text{Time} = (0.08)(19)^{0.5}$$
$$= 0.35 \text{ minutes}$$

This checks quite closely with the time study value of 0.37 minutes.

Sometimes, plotting the data on logarithmic paper will result in a straight line. For example, the equation $Y = AX^m$ written in logarithmic form becomes:

$$\log Y = \log A + m \log X$$

It is evident here that $\log Y$ is linear with $\log X$, and when the data are plotted on logarithmic paper, a straight line will be obtained.

Graphic solutions for more than one variable

When more than one variable affects time, it is often practical to use graphic techniques even if the relationships are nonlinear.

For example, it is possible to solve for time when we have two or more variables, one or more of which are nonlinear, by constructing a chart for each variable. The first chart will plot the relationship between time and one of the variables in selected studies in which the values of the other variables tend to remain constant. The second chart will show the relationship between the second variable and time that has been adjusted so as to remove the influence of the first variable. The procedure is extended with a third chart if there is an additional variable affecting the time required to perform the element.

Let us look at an example to illustrate the application of this procedure. It is decided to construct a formula for rolling various widths and lengths of 0.072-inch thick, cold-rolled sheet metal on a bench roll. Element 1, shown on the master table of detail time studies, is "pk. up pc. & pos.," and element 3 is "lay aside pc. & pos." Since both of these elements involve the handling time of the same amount of stock, their values are combined so as to simplify the final algebraic expression. The values shown in Table 20–1 were obtained.

First inspection seemed to suggest that the time for element 1 plus element 3 would vary with the area of the part handled. However, upon plotting, it was apparent that something else was influencing the time required to perform these elements. Further analysis revealed that the long and narrow pieces required considerably more time than the nearly square parts, even though their respective areas were about the same. It was decided that two variables were affecting time: these were the area of the part and the increased difficulty of handling the longer pieces. The latter may be expressed as a ratio of length to width.

TABLE 20–1

Study no.	Width	Length	Element 1 + element 3
1.	2	16	0.18
2.	3	18	0.13
3.	8	1	0.09
4.	10½	32	0.14
5.	7	84	0.55
6.	8½	69	0.38
7.	20	30	0.22
8.	11	55	0.26
9.	10	75	0.68
10.	20½	41	0.66
11.	10½	90	1.80
12.	16	64	1.44

It was noted that four of the 12 studies appearing on the master work sheet (studies 5, 6, 7, and 8) were taken on stock having areas of about 600 square inches. It was also apparent that these four studies had a relatively wide spread of "length divided by width" ratios which was apparently responsible for the range in allowed elemental times.

Since these four studies had about the same areas, the effect of L/W could be shown by plotting a simple curve. This was done as shown in Figure 20–7. Now, in order to see what the effect of area is on the

FIGURE 20–7
Time plotted against the ratio of length to width for constant areas of about 600 square inches

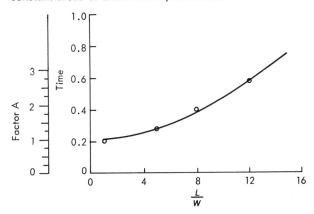

Area = C = 600 square inches.

elemental time, it will be necessary to adjust all the tabularized time study values so as to take into account the influence of the variable L/W. To get a factor that can be used to divide the time study values so as to arrive

at an adjusted time that varies with area, we construct a scale parallel
to the Y axis and call it "Factor A" (adjustment factor). By extending
the lowest point on our L/W curve horizontally to the adjustment axis,
we get a distance from the origin that can be evaluated as unity on the
adjustment axis. Proportional values can then be constructed so as to get
a scale on the adjustment axis.

Now by taking the L/W value of all the studies appearing on the
master table, we can determine adjustment factors graphically for each
study by moving horizontally to the Factor A axis from the specific point
on the L/W versus Time curve. Once the adjustment values have been
determined for each study, then the adjusted time can be computed by
dividing the allowed elemental time study values by the adjustment factor.
The resulting adjusted times can then be plotted against area (see
Figure 20–8).

FIGURE 20–8
Adjusted time plotted against area

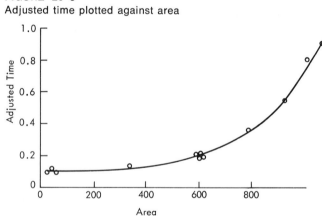

The system of curves so developed represents the graphic solution and
can be used for establishing standards within their range in the following
manner. First, the area and the length over width ratio of the sheet is calculated. Using the L/W ratio, we can refer to the first curve to obtain
an adjustment factor. Referring to the second curve, we can determine
an adjusted time. The product of the adjusted time and the adjustment
factor will give the allowed time to perform elements 1 and 3.

By extending this procedure, it is possible for the analyst to solve
graphically elemental time values when more than two variable characteristics govern the elemental time.

Sometimes variables will remain relatively constant within a specific
group. However, once the limits of the specific group are extended, the
variable will tend to show a pronounced change in value. For example,
the handling of boards of 1-inch white pine to a planer may be classified:

Group	Allowed time
Small (up to 300 square inches)	0.070 min.
Medium (300 to 750 square inches)	0.095 min.
Large (750 to 1,800 square inches)	0.144 min.

This method of grouping will tend to give erroneous values at the extremities of each group. Thus, in the above example, a board of 295 square inches would be allowed 0.070 minute handling time while one of 305 square inches would be given 0.095 minute. The chances are that in the first case, 0.070 minute would represent a tight standard, and that in the latter case, 0.095 minute would be somewhat loose.

When elements are grouped in the manner shown, it is important that the grouping be clearly and specifically defined so that there will be no questions as to the category in which an element belongs.

Least squares and regression techniques

Although graphical procedures may be quite useful in establishing a predicting equation for one dependent variable as a function of one independent variable, one may want to resort to more sophisticated methods of curve fitting, such as least squares and regression techniques. This is

FIGURE 20–9
Four distributions of time for four values of x and the resulting true curve passing through the means of these distributions

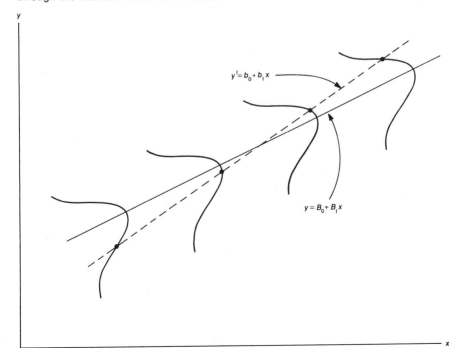

indeed even more true when one dependent variable y is viewed as a function of more than one independent variable. However, least squares and regression techniques are often mistakenly employed by the novice, with resultant blunders.

As has already been pointed out, we are interested in determining the mean of the distribution of times for a given variable x. The reader should recognize that there is a distribution of the random variable time for a given value of the independent variable x. Figure 20–9 illustrates such a relationship with the true straight line of the equation $y = B_0 + B_1 x$ passing through the mean of each of the four distributions of y. This relationship is referred to as the regression curve of y on x.

When the analyst is gathering his data, he is only able to approximate the mean of the time distributions. Any observed time may differ from the true mean by some finite amount. In order to determine the value of the estimates b_0 and b_1 of the parameters B_1 and B_0 for a given set of data (n paired observations $x_i y_i$), where it is assumed that the regression of y on x is linear, we want to determine the equation of the line that best fits the data. By best fit, we mean the equation of the line $y' = b_0 + b_1 x$, where the sum of the squared vertical distances $(y - y')$ for the points comprised by the data is a minimum.

The normal equations involving the parameters b_1 and b_0 whose values for the line of best fit of the data are:

$$\sum_{i=1}^{n} y_i = b_0 n + b_1 \sum_{i=1}^{n} x_i \tag{1}$$

$$\sum_{i=1}^{n} x_i y_i = b_0 \sum_{i=1}^{n} x_i + b_1 \sum_{i=1}^{n} x_i^2 \tag{2}$$

These equations can readily be solved simultaneously to provide the numerical values of the slope b_1 and the y intercept b_0. For example, referring to page 422, we can solve for b_1 and b_0 using the above equations as follows:

TABLE 20–2

Study	x or area	y or time	xy	x^2
1.	25	0.104	2.60	625
2.	65	0.109	7.09	4,225
3.	77	0.126	9.70	5,929
4.	112	0.134	15.01	12,544
5.	135	0.138	18.63	18,225
6.	147	0.150	22.05	21,609
7.	185	0.153	28.31	34,225
8.	220	0.174	38.28	48,400
9.	245	0.176	43.12	60,025
10.	275	0.182	50.05	75,625
11.	287	0.186	53.38	82,369
12.	300	0.202	60.60	90,000
	2,073	1.834	348.82	453,801

Substituting in equation (1) and (2):

$$12b_0 = 1.834 - 2,073b_1 \tag{1}$$
$$2,073b_0 = 348.82 - 453,801b_1 \tag{2}$$

Multiplying equation (1) by 2,073 and equation (2) by 12:

$$24,876b_0 = 3,801.882 - 4,297,329b_1$$
$$\underline{24,876b_0 = 4,185.840 - 5,445,612b_1}$$
$$0 = -383.958 + 1,148,283b_1$$

or

$$b_1 = \frac{383.958}{1,148,283} = 0.000334$$

Substituting in equation (1):

$$12b_0 = 1.834 - 0.692$$
$$b_0 = \frac{1.142}{12} = 0.095$$

The above equations can be transformed so that we can solve directly for the slope b_1 and the y intercept b_0 by substituting in the following relationships:

$$b_0 = \frac{(\Sigma x^2)(\Sigma y) - (\Sigma x)(\Sigma xy)}{(n)(\Sigma x^2) - (\Sigma x)^2}$$

$$b_1 = \frac{(n)(\Sigma xy) - (\Sigma x)(\Sigma y)}{(n)(\Sigma x^2) - (\Sigma x)^2}$$

The least squares technique is also applicable in plotting curves and exponential relationships as straight lines on semilog and log-log paper.

For example, in the case of the exponential relationship involving semilog paper where:

$$\log y = \log b_0 + x \log b_1$$

we can solve for the y intercept b_0 from the equation:

$$\log b_0 = \frac{(\Sigma x^2)(\Sigma \log y) - (\Sigma x)(\Sigma x \log y)}{(n)(\Sigma x^2) - (\Sigma x)^2}$$

and for the slope b_1 from the equation:

$$\log b_1 = \frac{(n)(\Sigma x \log y) - (\Sigma x)(\Sigma \log y)}{(n)(\Sigma x^2) - (\Sigma x)^2}$$

In those cases where the dependent variable y is a nonlinear function of one independent variable x, and a simple graphic procedure is not indicated, we may endeavor to fit our data by the polynomial equation:

$$y' = b_0 + b_1x + b_2x^2 + b^{\mathfrak{s}}x^3 \ldots , b_qx^q$$

For example, let us assume that we have a set of n points of data $(y_i x_i)$. In order to estimate the coefficients b_0, b_1, b_2, b_3 . . . , b_q we minimize:

$$\sum_{i=1}^{n} \left[y_i - (b_0 + b_1 x_i + b_2 x_i{}^2 + b_3 x_i{}^3 \ldots , b_q x_i{}^q) \right]^2$$

Here we are merely applying the least squares technique by minimizing the sum of the squares of the vertical distances from the points to the curve. By differentiating partially with respect to b_0, b_1, b_2 . . . , b_q, equating these partial derivatives to zero, and rearranging some of the terms, we obtain $q + 1$ normal equations. These are:

$$\Sigma y = n b_0 + b_1 \Sigma x + \ldots , b_q \Sigma x^q$$
$$\Sigma xy = b_0 \Sigma x + b_1 \Sigma x^2 + \ldots , b_q \Sigma x^{q+1}$$
$$\Sigma x^q y = b_0 \Sigma x^q + b_1 \Sigma x^{q+1} \ldots , b_q \Sigma x^{q+q}$$

Multiple regression

At times the analyst will recognize that more than one independent variable is influencing the dependent variable (time). If two independent variables are involved in a linear relationship, then we are fitting a plane to a set of n points so as to minimize the sum of the squares of the vertical distances from the points to the plane. Thus, we are minimizing

$$\sum_{i=1}^{n} \left[y_i - (b_0 + b_1 x_i + c_1 z_i) \right]^2$$

(See Figure 20–10.)

FIGURE 20–10
Two independent variables characterize the
best fit of a plane in a geometric cube

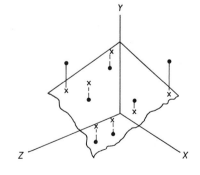

The resulting normal equations are:

$$\Sigma y = n b_0 + b_1 \Sigma x + c_1 \Sigma z$$

$$\Sigma xy = b_0 \Sigma x_1 + b_1 \Sigma x^2 + c_1 \Sigma xz$$
$$\Sigma zy = b_0 \Sigma z + b_1 \Sigma xz + c_1 \Sigma z^2$$

Let us look at an example utilizing the multiple regression technique. The analyst decides to construct a formula for bagging hardware in burlap bags of various sizes. He finds that time is not only dependent upon the number of components placed in each bag but also upon the total weight of the components placed in each bag. From a master table of detailed time studies see page 472), the information shown in Table 20–3

TABLE 20–3

Study no.	y Time	x Total weight of components	z Number of components
1.	0.264	6.62	11
2.	0.130	1.15	6
3.	0.202	5.61	8
4.	0.126	4.01	4
5.	0.220	6.14	9
6.	0.332	7.25	14
7.	0.222	7.02	6
8.	0.155	1.75	8
9.	0.345	5.45	17
10.	0.256	6.91	10
	2.252	51.91	93

was gathered for the second element, which is "place hardware components in bag."
Substituting

$$\Sigma x = 51.91, \ \Sigma z = 93, \ \Sigma x^2 = 312.75$$
$$\Sigma xz = 515.27, \ \Sigma z^2 = 1{,}003, \ \Sigma y = 2.252$$
$$\Sigma xy = 12.772, \ \Sigma zy = 23.429$$

into the normal equations gives us:

$$2.252 = 10b_0 + 51.91b_1 + 93c_1$$
$$12.772 = 51.91b_0 + 312.75b_1 + 515.27c_1$$
$$23.429 = 93b_0 + 515.27b_1 + 1{,}003c_1$$

The solution of this system of equations provides the following results: $b_0 = 0.016$, $b_1 = 0.0139$, $c_1 = 0.0147$. It is now possible to estimate the time y to bag parts having the total weight x and the number of components z by direct substitution in the equation:

$$y^1 = 0.016 + 0.0139x + 0.0147z$$

Multiple regression techniques are also applicable for nonlinear cases. The interested reader can explore these areas in several of the texts on regression analysis.

Some precautions

As previously stated, regression methods may present difficulties of interpretation and analysis to the uninformed. These difficulties are primarily concerned with the statistical and engineering analyses used in the selection of the number of independent variables and with the significance that each variable plays in the predicting equation. The form of the equation selected (that is, linear or nonlinear) and the inclusion of cross products of the variables in the equation when interaction effects are present may further complicate the analysis. Under these circumstances, it may be advantageous to consult with an analyst competent in regression techniques. However, in the interest of ease of application, it is a good general practice to make predicting equations as simple as possible. From a statistical point of view, it is also true that complicated predicting equations may result in time estimates that are little, if any, more reliable than equations of a simpler form.

Develop synthesis

The purpose of the synthesis in the formula report is to give a complete explanation of the derivation of the various components entering into the formula so as to facilitate its use. Furthermore, a clearly developed synthesis will help in explaining and selling the formula in the event that questions as to its suitability come up at some later date.

In order that the final expression may be in its simplest form, all constants and symbols should be combined wherever possible, with due respect to the accuracy and flexibility of the formula. Elemental operations may be classified under specific headings, such as preparations, handling by hand, handling by jib crane, and operation. Constant values will appear under each elemental class. These values can be combined to simplify the final expression of the specific formulas. Details of the combinations are explained in the synthesis. A typical illustration of a synthesis for a constant element would be:

$$A1 + C1 + \text{constant from equation } 4 + H1 = 0.11 + 0.17$$
$$+ 0.09 + 0.05 = 0.42 \text{ minute}$$

In the treatment of variable elements, a more detailed explanation is usually required due to the added complexities associated with this class of elements. For example, we may wish to combine elements $B1$ and $D1$, in that they have been influenced by the ratio of L/D. The two equations may be:

$$B1 = 0.1\frac{L}{D} + 0.08$$

$$D1 = 0.05\frac{L}{D}$$

and the combination of these would give

$$B1 + D1 = 0.15\,\frac{L}{D} + 0.08$$

Thus, the synthesis would clearly show where the 0.15 L/D came from.

Compute expression

The final expression may not be entirely in algebraic form. It may have been more convenient to express some of the variables in terms of systems of curves, nomographs, or single curves. Then, too, some of the variable data may have been made up into tables, and these will be referred to by a single symbol in the formula. A typical example of a formula would be:

$$\text{Normal time} = 0.07 + \text{Chart 1} + \text{Curve 1} + \frac{0.555N}{C}$$

where:

$$N = \text{Number of cavities per mold}$$
$$C = \text{Cure time in minutes}$$

In applying this formula, the analyst would refer to Chart 1 in order to pick out the applicable tabularized value. He would also refer to Curve 1 to obtain the numerical value of another variable entering into the formula. Knowing the number of cavities in the mold and the cure time in minutes, he would substitute in the latter part of the expression and summarize the values of the three variables with the constant time of 0.07 minute to get the resulting normal time.

Check for accuracy

Upon completion of the formula, the analyst should verify it before releasing it for use. The easiest and fastest way to check the formula is to use it to check existing time studies. This can best be done by tabularizing the results under the following headings: "Part Number," "Time Study Value," "New Formula Value," "Difference," and "Percentage Difference."

Any marked differences between the formula value and the time study value should be investigated and the cause determined. The formula would be expected to show an average difference of less than 5 percent from the time study values of those studies used in its derivation. If, at this point, the formula does not appear to have the expected validity, then the analyst should accumulate additional data by taking more stopwatch studies. Once the formula has been satisfactorily tested over a range of work, it will be acceptable to those who are affected directly and indirectly by its application. At this point, it would be time well spent for the analyst to go over the formula with the foreman of the department where it will be used. After the foreman understands its derivation and application and has verified its reliability, the analyst can be assured of his backing on standards resulting from its use.

Write formula report

All data, calculations, derivations, and applications of the formula should be consolidated in the form of a complete report prior to putting the formula into use. Then all the facts relative to the process employed, operating conditions, and scope of formula will be available for future reference.

The Westinghouse Electric Corporation includes 14 separate sections in its formula reports. These are:

1. Formula number, division, data, and sheet number.
2. Part.
3. Operation.
4. Work station.
5. Normal time.
6. Application.
7. Analysis.
8. Procedure.
9. Time studies.
10. Table of detail elements.
11. Synthesis.
12. Inspection.
13. Wage payment.
14. Signature of constructor and approver.

Formula number

For ease of reference, all paperwork entering into the development of the formula should be identified by the assignment of a number to the formula. The formula number is made up of a prefix which identifies the department or division in which the formula will be used, and a number which gives the chronological arrangement of the formula with reference to other formulas being used in the department. Thus, the formula 11N–56 would be the 56th formula designed for use in department 11N. The date that the formula was put into use is also included, so as to connect working conditions in effect at the time the formula was designed with standards established through its application.

Part

The part number or numbers together with drawing numbers should be given, with a concise description of the work, so that the products for which the formula has application are clearly defined. For example,

"Parts J–1101, J–1146, J–1172, J–1496, side plates ranging from 12×24 inches to 48×96 inches" would identify the parts for which the formula may be used.

Operation and work station

The operation covered by the formula should be clearly stated, such as "roll radius" or "fabricate inner and outer members" or "broach keyways" or "assemble farm tank." In addition, the work station should be completely described, with information as to equipment, jigs, fixtures, and gages, and their size, condition, and serial number. A photograph will often help to define the method in effect at the time of compilation of the formula.

Normal time

When expressing the formula for the normal time, separate equations should be used for the "setup" time and the "each piece" time. Immediately following the formula should be a key which outlines the meaning of the symbols used in each equation. All tables, nomograms, and system of curves should be included at this point, as well as a sample solution, so that the person using the formula clearly understands it.

Application

After the expressions for the allowed time have been concisely stated, a clear explanation of the application of the formula should be given. This should describe in detail the nature of the work on which the formula may be used, and specifically state the limits within which it may be applied. For example, the application of a formula for curing bonded rubber parts may be written as follows: "This formula applies to all curing operations done in 24×28-inch platen presses when the number of cavities range between 8 and 100 and the cubic inches of rubber per cavity range between 0.25 and 3."

Analysis

Under the analysis is given a detailed account of the entire method employed, which will include the tools, fixtures, jigs, and gages and their application; workplace layout; methods of material handling; method of obtaining supply materials; and nature of setup, with details as to how the operator is assigned the job, the distance to the tool crib, and other pertinent data.

In the analysis section of the report is included information as to the breakdown of the allowances used in the formula. The reason for any special or extra allowance should be clearly stated. The allowance for personal time, fatigue, and unavoidable delays should be shown independently, so that if any question comes up in the future as to the inclusion of an allowance in the formula, it can be clearly shown what allowances were included and why.

Procedure

After the analysis of the job has been completed, the procedure used by the operator in performing the work should be written up in detail. The best way of doing this is to include all the elements appearing on the master work sheet in their correct chronological sequence. While abbreviations were liberally used on the time studies and on the master table, care should be exercised to avoid their use in writing up the operator's procedure. Individual elements should be written in exact detail. Then all parties will have no doubt as to the work elements that fall within the scope of the formula.

Time studies

The formula report need not include the actual time studies used in the compilation of the formula, as they are usually available in their own independent files. However, reference should be made to the time studies used. This can be done in tabular form as follows:

Time study no.	Part	Drawing	Plant	Date taken	Taken by
S-112	J-1102	JB-1102	A	9-15-	J. B. Smith
S-147	J-1476	JA-1476	B	10-24-	A. B. Jones
S-92	J-1105	JB-1105	B	6-11-	J. B. Smith

Table of detail elements

This table is used as a reference source for information as to the normal element time and its derivation. The information for composing this table is taken from the master table of detailed elements and is recorded in chart form as follows:

Symbol	Element description	Normal time	Reference
Al	Close core box	0.046 min.	Average
Cl	Fill partly full of sand	A × C.T.	All studies
Jl.	Rap box	0.043 min.	Average

Synthesis

After recording the table of detailed elements, the analyst includes the synthesis of the report. The synthesis, as previously explained, tells the manner in which the allowed time was derived.

Inspection, payment, and signatures

Upon completion of the synthesis the constructor records the following information in order: the inspection requirements appearing on the drawings of the jobs covered by the formula; the type of wage payment plans where the formula will be used, such as daywork, piecework, and group incentive; and finally his signature and that of his supervisor.

Representative formula

The following formula report will illustrate and help clarify the procedures discussed in this chapter:

Formula No.: M–11–No. 15
Date: September 15, 19–
Sheet: 1 of 15

Part: Cylindrical oil sand mix cores ⅞″ in diameter to 2¾″ diameter and 4″ long to 13″ long.
Operation: Make core complete in wood core box.
Work Station: 30″ × 52″ Bench 36″ high.
Normal Time: Piece time in decimal minutes equals:

$$0.173N + 0.0210L + \sqrt{0.0067 + 0.000016V^2} + Y \times C.T. + 0.0157A + 0.327$$

Where

$$N = \text{Number of ``C'' clamps}$$
$$L = \text{Length of core in inches}$$
$$V = \text{Volume of core in cubic inches}$$
$$Y = \text{Adjustment factor (Curve 1)}$$
$$CT = \text{Adjusted time (Curve 2)}$$
$$A = \text{Cross-sectional area of core}$$

Example:

Calculate normal time to make a green sand core 2″ in diameter and 8″ long. Volume would be 25.1 cubic inches and with 8″ length, only one clamp would be required. L/D would be equal to 8/2 or 4.

Then:

$$\text{Normal time} = (0.173)(1) + (0.0210)(8) + \sqrt{0.0067 + (0.000016)(25.1^2)}$$
$$+ (1.5)(.76) + (0.0157)(3.14) + 0.327 = 1.987 \text{ minutes.}$$

APPLICATION

This formula applies to all cylindrical cores made of oil sand mix between diameters ranging from ⅞ inch to 2¾ inches and lengths ranging from 4 inches to 13 inches. This work is performed manually in wooden split core boxes.

ANALYSIS

The following hand tools are provided the operator for making cores covered by this formula: C-clamps, slick, and rammer. The work is done at a bench with the operator standing. Oil sand mix for producing the cores is piled on the floor by the move man approximately 4 feet from the operator. Periodically, the operator replenishes a supply of sand on his bench from the inventory on the floor, through the use of a short-handled shovel. Rods and reinforcing wires are requisitioned from the storeroom by the operator. He usually acquires a full day's supply with each requisition.

The work voucher placed above the operator's work station by the departmental foreman indicates the sequence of jobs to be performed and is the authority for the operator to begin a specific job. Operation cards and drawings must be obtained by the operator from the tool crib.

The standard allowance which must be added when applying this formula is 17 percent. This involves 5 percent for personal delays, 6 percent for unavoidable delays, and 6 percent for fatigue.

PROCEDURE

The working procedure followed in making oil sand mix cores covered by this formula includes 14 elements exclusive of setup and put-away elements. These are:

1. Pick up two sections of core box and close together.
2. Clamp core box shut using one clamp for cores 9 inches or less and two clamps for cores greater than 9 inches in length.
3. Fill core box partly full of sand (about one third depending on core).
4. Ram the sand solid in the core box.
5. Place the vent rod and reinforcing wire.
6. Fill the remainder of the core box with sand and ram solid.
7. With slick, strike off sand on both ends of core box so sand core is flush with box.
8. Remove the vent wire from the core and lay aside.
9. Rap box lightly.
10. Remove clamps and lay aside.
11. Open core box.
12. Roll out core on rack beside work station.
13. Clean core box with kerosene rag.

TIME STUDIES

The following summary includes the time studies used in developing this formula:

Time study no.	Part number		Plant	Date taken	Taken by
S–13	P–1472	PB–1472	A	6–15	Black
S–111	P–1106	PB–1106	A	12–11	Black
S–45	P–1901	PB–1901	A	7–13	Hirsch
S–46	P–1907	PB–1907	B	7–14	Black
S–47	P–1908	PA–1908	B	7–14	Black
S–32	P–1219	PA–1219	A	8–10	Black
S–76	P–1711	PA–1711	A	11–12	Obe
S–70	P–1701	PB–1701	B	11–9	Obe
S–17	P–1311	PB–1311	B	6–16	Black
S–18	P–1312	PB–1312	B	6–16	Black
S–59	P–1506	PB–1506	A	7–26	Hirsch
S–60	P–1507	PB–1507	A	7–26	Hirsch
S–50	P–1497	PB–1497	B	7–19	Obe
S–51	P–1498	PA–1498	B	7–20	Obe
S–52	P–1499	PA–1499	B	7–21	Obe
S–53	P–1500	PA–1500	B	7–21	Obe
S–54	P–1501	PA–1501	B	7–22	Obe

TABLE OF DETAIL ELEMENTS

Symbol	Element description	Decimal minute normal time	Reference
A–1	Close core box	0.046	Average
B–1	Clamp core box	$0.112N$	Time vs. no. clamps
C–1	Fill partly full of sand	$Y \times CT$	Time vs. L/D & vol. Time vs. vol.
D–1	Ram	$Y \times CT$	Time vs. L/D & vol. Time vs. vol.
E–1	Place rod and wire	$0.0153L + 0.03$	Time vs. length
F–1	Fill and ram	$Y \times CT$	Time vs. L/D & vol. Vs. vol.
G–1	Strike off	$0.0157A + 0.07$	Time vs. area
H–1	Remove vent wire	0.047	Average
J–1	Rap box	0.043	Average
K–1	Remove clamps	$0.061N$	Time vs. no. clamps
M–1	Open box	0.046	Average
N–1	Roll out core	$0.0057L + 0.045$	Time vs. length
P–1	Clean box	$\sqrt{0.0067 + 0.000016V^2}$	Time vs. volume

SYNTHESIS

Standard time = Sum of allowed elemental time
$$= \text{A–1} + \text{B–1} + \text{C–1} + \text{D–1} + \text{E–1} + \text{F–1} + \text{G–1}$$
$$+ \text{H–1} + \text{J–1} + \text{K–1} + \text{M–1} + \text{N–1} + \text{P–1}$$

Since Elements C–1, D–1, and F–1 are all dependent on the variables volume and length/diameter, their values may be combined before plotting on coordinate paper. The combined values for these three elements for the 17 time studies are:

Study no.	C-1 + D-1 + F-1
S-13	1.045
S-111.	1.017
S-45	1.235
S-46	0.647
S-47	1.325
S-32	1.110
S-76	0.500
S-70	1.362
S-17	1.932
S-18	0.659
S-59	0.988
S-60	1.181
S-50	0.584
S-51	1.277
S-52	0.888
S-53	1.481
S-54	2.065

For reference purposes, the combination of C–1 + D–1 + F–1 will be designated R–1 [see Figures 20–11 and 20–12]. Curves 1 and 2 are used to solve for the allowed time for these three elements by first getting the adjustment factor from curve 1 and then getting the adjusted time from curve 2. The

FIGURE 20–11

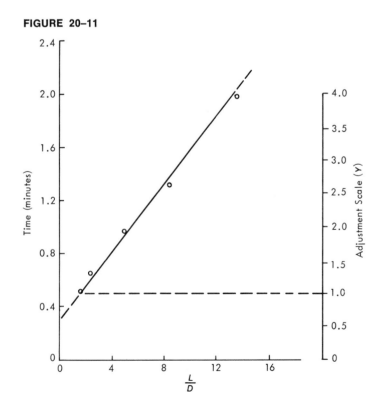

FIGURE 20–12

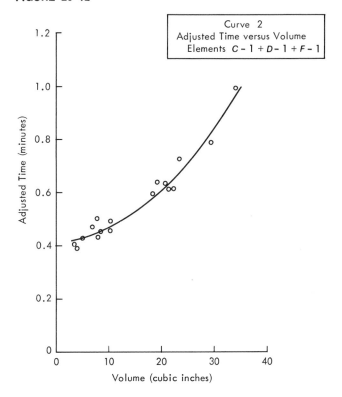

product of these two values will give the allowed time for C–1 + D–1 + F–1.

The time for element E–1 (place rod and wire) has been plotted on curve 3 [see Figure 20–13], where its relationship to core length is shown.

This curve can be expressed algebraically by the equation:

$$T = fv + C$$

or

$$\text{Time} = (\text{Slope})(\text{length}) + y \text{ intercept}$$

Solving graphically:

$$f = \frac{0.210 - 0.100}{12 - 4.8} = \frac{0.110}{7.20}$$
$$= 0.0153$$
$$C = 0.03 \text{ (from graph)}$$

Thus:

$$\text{Time} = 0.0153L + 0.03$$

Element G–1 (strike off) has also been solved as a straight-line relation-

FIGURE 20–13

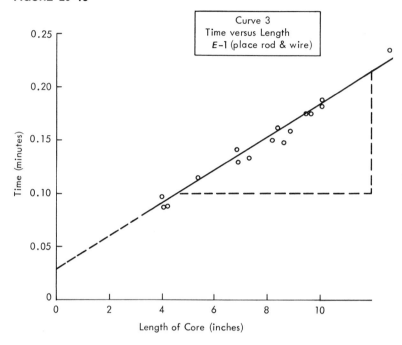

ship. Here a cross-sectional area of the core has been plotted against time (see curve 4) [Figure 20–14]. In this case

$$\text{Time} = (f)(\text{area}) + 0.07$$

and

$$f = \frac{0.132 - 0.08}{4 - 0.7} = \frac{0.052}{3.3}$$
$$f = 0.0157$$

Then:

$$\text{Time} = 0.0157A + 0.07$$

Element N–1 (roll out core) is shown on curve 5 [see Figure 20–15]. Here

$$\text{Time} = f(\text{length}) + 0.045$$

and

$$f = \frac{0.125 - 0.068}{14 - 4}$$
$$= 0.0057$$

FIGURE 20–14

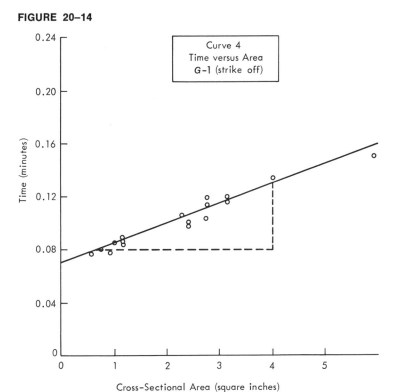

Cross-Sectional Area (square inches)

Then:

$$\text{Time} = 0.0057L + 0.045$$

The time for element P–1 (clean box) when plotted against volume of core, gives a hyperbolic relationship as shown in curve 6 [Figure 20–16]. This relationship with time can be expressed algebraically as follows:

$$\text{Time} = \sqrt{a^2 + \frac{v^2 a^2}{k^2}}$$

$$= \sqrt{0.082^2 + \frac{(v^2)(0.082)^2}{400}}$$

$$= \sqrt{0.0067 + 0.000016v^2}$$

Elements A–1, H–1, J–1, M–1 were classified as constants with the following respective allowed times determined by taking their average values: 0.046, 0.047, 0.043, 0.046. When these constants are added to the sum of the y intercept values of equations (3), (4), and (5), a total constant time of 0.327 minute is determined:

```
A-1 . . . . . . . . . . . . . . . . . . . . . 0.046
Curve 3 intercept . . . . . . . . . . . 0.030
H-1 . . . . . . . . . . . . . . . . . . . . . 0.047
Curve 4 intercept . . . . . . . . . . . 0.070
J-1 . . . . . . . . . . . . . . . . . . . . 0.043
M-1 . . . . . . . . . . . . . . . . . . . . 0.046
Curve 5 intercept . . . . . . . . . . . 0.045
```
$$C_T = 0.327 \text{ minute}$$

Both elements B–1 and K–1 are proportional to the number of clamps *N*. Thus, these two elements may be combined for simplicity as follows:

$$B\text{--}1 = 0.112N$$
$$K\text{--}1 = 0.061N$$
$$S\text{--}1 = 0.173N$$

Since elements E–1 (place rod and wire) and N–1 (roll out core) are proportional to length of core, we may combine their equated relationships with length and designate the combined total by the symbol T–1.

FIGURE 20–15

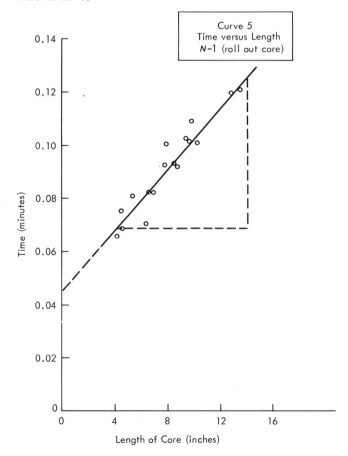

FIGURE 20–16

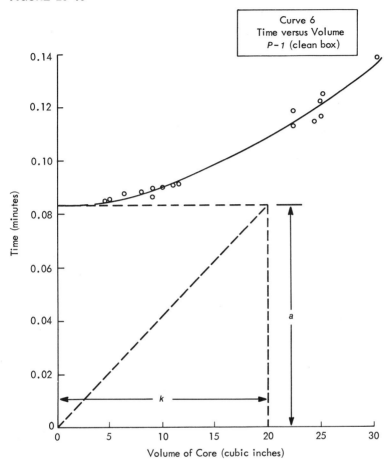

Thus:

$$E\text{–}1 = 0.0153L$$
$$N\text{–}1 = 0.0057L$$
$$\overline{T\text{–}1 = 0.0210L}$$

Simplifying our initial equation:

Normal time = S–1 + T–1 + P–1 + R–1 + G–1 + C$_T$
$$= 0.173N + 0.0210L + \sqrt{0.0067 + 0.000016V^2}$$
$$+ Y \times CT + 0.0157A + 0.327$$

INSPECTION

The only inspection requirement on the cores covered by this formula is visual inspection done by the operator at the time the core is rolled free from the core box.

PAYMENT

This formula is to be used in development of allowed times that will be applied to the individual standards in a one-for-one incentive wage payment plan.

APPROVED: *John Purdy*   *Scott Dickinson*
Supervisor   Formula Constructor

Use of the digital computer

Once a formula has been designed, it can be used for establishing standards on all work fitting the parameters of the formula. The resulting time standards can then be tabulated for rapid reference.

FIGURE 20–17
Program flowchart symbols

General Operation Symbol: Used for any operation which creates, alters, transfers or erases data, or any other operation for which no specific symbol has been defined in the Standard.

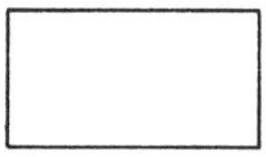

Subroutine (Predefined Process) Symbol: Used when a section of program is considered as a single operation for the purpose of this flowchart.

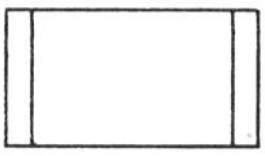

Generalized Input/Output Symbol: Used to represent the input/output (I/O) function of making information available.

Branch Symbol: Has one entry line and more than one exit. The symbol contains a description of the test on which the selection of an exit is based. The various possible results of this test are shown against the corresponding exits.

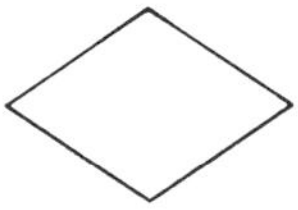

Offpage Connector Symbol: Used as a linkage between two blocks of logic that are to be found on separate pages of the flowchart. The symbol is only used on the "exit" page, on the "entry" page an onpage symbol is used.

Onpage Connector Symbol: Used as a linkage between two blocks on the same page, when it is not desirable to connect them using a linkage line. The label of the block to which the connection is being made is written inside the symbols.

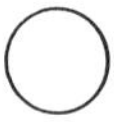

Terminal Symbol: Used as the beginning or end of a flowline (e.g., start or end of a program).

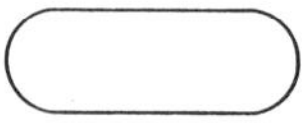

Annotation Symbol: Used to add additional information to a symbol or block of program.

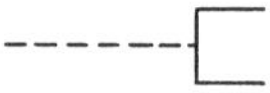

Flowlines (Linkage Lines): Used to show the flow between blocks of a flowchart. The normal flow is from top to bottom and left to right of the page. The programmer may dispense with the use of the direction arrows when the chart follows the normal flow. They must be used, however, for any portion of the diagram which does not follow the normal flow.

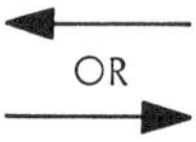

Rather than solve the formula innumerable times with a slide rule or desk calculator, the analyst should consider the electronic computer to compute in the least possible time *all* standards for which the formula is applicable.

A simple example of an equation to determine knurling times will suffice to illustrate both flow charting and programming. Let the equation for knurling times be given by

$$T = 0.1A + 1.5$$

where:

$$T = \text{Normal time}$$
$$A = \text{Surface area of knurled shaft}$$

Suppose it is desired to generate a table of times (T) for areas (A) where A varies from 3 to 80 square inches by increments of $\frac{1}{4}$ inch. The flowchart for this very simple example is given in Figure 20–18. The parenthetical numbers relate to program statement numbers as they appear in the program.

FIGURE 20–18
Knurling times flowchart

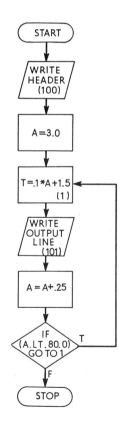

FIGURE 20–19

KNURLING TIMES
T=.1A+1.5

AREA	TIME
3.00	1.800
3.25	1.825
3.50	1.850
3.75	1.875
4.00	1.900
4.25	1.925
4.50	1.950
4.75	1.975
5.00	2.000
5.25	2.025
5.50	2.050
5.75	2.075
6.00	2.100
6.25	2.125
6.50	2.150
6.75	2.175
7.00	2.200
7.25	2.225
7.50	2.250
7.75	2.275
8.00	2.300
8.25	2.325
8.50	2.350
8.75	2.375
9.00	2.400
9.25	2.425
9.50	2.450
9.75	2.475
10.00	2.500
10.25	2.525
10.50	2.550
10.75	2.575
11.00	2.600
11.25	2.625
11.50	2.650
11.75	2.675
12.00	2.700
12.25	2.725
12.50	2.750
12.75	2.775
13.00	2.800
13.25	2.825

Programming

Let's continue with the knurling times example which has been flow charted. It is easy to translate it into FORTRAN. For the achievement of repetitive formula computation, the branching IF statement is used. Every READ and WRITE statement requires a FORMAT statement, the number of which is given by the second of the couple in the parenthetic expression after each; the first number represents the device: six being the printer. The source program is as follows:

```
C     PRØGRAM TØ PRØDUCE TABLE ØF KNURLING
      TIMES
C     EQUATIØN—TX.1A+1.5
      WRITE  (6,100)
      A = 3.0
1   T = .1*A + 1.5
```

```
      WRITE  (6,101)  A,T
      A = A + .25
      IF  (A.LT.80.0)  GO  TO  1
      STØP
  100 FØRMAT (1H1,10X,14HKNURLING TIMES/1H ,12X,9HT =
      .1A + 1.5/1HO,8X,4HAREA1,10X,4HTIME/)
  101 FØRMAT (1H ,F12.2,F15.3)
      END
```

The job control language cards are not included in the source program, as often these are unique to the computer installation. Generally, this requires less than 10 seconds compile and execute time on a System/360 model 50 or above. The result of this program is shown in Figure 20–19.

Conclusion

Through the design of a time study formula, it is possible to establish standards in a fraction of the time required by taking individual studies. However, before a formula is released for use, the mathematics used in its development should be carefully checked to assure that the expression is correct. Then, too, several test cases should be tried to be certain that the formula will establish true, consistent standards. The formula should be clearly identified as to its range of application, and in no case should standards be established with data that are beyond the scope of the formula.

In summary, the following steps represent the chronological procedure in time study formula design:

1. Collect data.
   a. Using time studies already available.
   b. Using standard data already available.
   c. Taking new time studies.
2. Compile master work sheet and identify formula.
3. Analyze and classify elements.
   a. Constants.
   b. Variables.
4. Develop synthesis.
5. Compute the final expression.
6. Check mathematics of developed formula.
7. Test formula.
8. Write formula report.
9. Use formula.

By systematically following these nine steps, the analyst will have little difficulty in designing reliable time study formulas.

Once a formula has been designed, it can be programmed on the digital computer and solved for all possible variations within the scope of the formula. Thus, tabulated standards can be made available for broad variations in the work assignment.

TEXT QUESTIONS

1. What advantages does the formula offer over standard data in establishing time standards?
2. Is the use of time study formulas restricted to machine shop operations where feeds and speeds influence allowed times? Explain.
3. What are the characteristics of a sound time study formula?
4. What is the danger of using too few studies in the derivation of a formula?
5. What is the function of the synthesis in the formula report?
6. Write the equation of the ellipse with its center at the origin and axes along the coordinate axes and passing through $(2, 3)$ and $(-1, 4)$.
7. Find the equation of the hyperbola with its center at $(0, 0)$ and $a = 4$, $b = 5$, foci on the y axis.
8. What 14 sections make up the formula report?
9. What nine steps represent the chronological procedure in the design of time study formulas?
10. Explain in detail how it is possible to solve graphically for time when two variables are influential and they cannot be combined.
11. Develop an algebraic expression for the relationship between time and area from the following data:

Study no.	1	2	3	4	5
Time	4	7	11	15	21
Area	28.6	79.4	182	318	589

12. If the data shown in Figure 20–1 were plotted on logarithmic paper, would the plotting be a straight line? Why or Why not?
13. What types of problems lend themselves to solution on the digital computer?

GENERAL QUESTIONS

1. Would the company union prefer standards to be set by formulas or standard data? Why or why not?
2. Would it be necessary for the "chart and formula" designer to have a background in time study work? Why or why not?
3. If an operator objected strongly to a rate established through a formula, explain in detail how you would endeavor to prove to him that the rate was fair.

PROBLEMS

1. In the Dorben Company, the work measurement analyst planned to develop standard data on a new milling machine that was recently installed. The material being cut in one group of studies involved a 1½-inch width of cut, with lengths varying from 4 inches to 30 inches. For this work, a plain carbide-tipped milling cutter 3 inches in diameter and a width of face of 2 inches was being used. Depths of cut ranged from ³⁄₁₆ inch to ⁷⁄₁₆ inch.

   Give the equation that can be programmed to provide cutting time in terms of d (depth of cut) and l length of casting being milled). In all cases, the feed per tooth is 0.010 inch and the 16-tooth cutter is running at a surface speed of 80 feet per minute.

2. The work measurement analyst in the Dorben Company was studying the hand filing and polishing of external radii. Six studies provided the following information:

Study	Size of radii	Minutes per inch
1	³⁄₈	0.24
2	½	0.37
3	⁵⁄₈	0.59
4	¹¹⁄₁₆	0.80
5	¾	0.93
6	1	1.52

   These data plotted as a straight line on semilog paper where time (the dependent variable) was the logarithmetic scale. Develop an algebraic equation for estimating the time of filing and polishing various radii.

3. In the assembly department of the Dorben Company, various hardware is bagged for shipment. The time study analyst desires to design a formula or a system of curves to establish standards on this work. From the master table of detailed time studies, the following information was gathered for the second element, which was "place hardware components in bag":

Study no.	Time (min.)	Weight of components (lb.)	Bag size	No. of components
1	0.264	6.62	4	11
2	0.130	1.15	2	6
3	0.186	5.61	3	8
4	0.169	2.91	2	6
5	0.126	4.01	2	4
6	0.220	6.14	3	9
7	0.200	5.50	3	6
8	0.332	7.25	4	14
9	0.222	7.02	4	6
10	0.155	1.75	2	8
11	0.345	5.45	4	17
12	0.256	6.91	4	10

   From these data, design a formula or a system of curves for establishing the standard for this element.

4. In the blanking of various leather components from animal skins, the analyst noted a relationship between standard time and the area of the

component. After taking five independent time studies, he observed the following:

Study no.	Area of leather component (sq. in.)	Standard time (min.)
1	5.0	0.07
2	7.5	0.10
3	15.5	0.13
4	25.0	0.20
5	34.0	0.24

Derive an algebraic expression to preprice the blanking of the various leather components.

5. The analyst in the Dorben Company decides to develop a formula for prepricing a particular assembly operation involving different sizes of work. The assembly operation involves three constant elements and one variable element. The constant elements are determined from MTM data. The classifications of the fundamental motions on the constant elements are as follows:

Element 1: One R10C, one G1B, one P2SSD, one AF (max), one M20C with 2# weight, and one RL (case 1)

Element 2: One eye travel with T = 20 and D = 10, one R12B, one G1A, one M10C, one P1SSE, ten T 30° 2#

Element 3: One R10A, one G1B, one M20B, one RL2

The variable element is based on the following data:

Study number	Standard time (min.)	Surface area (sq. in.)
1	0.282	7
2	0.163	5
3	0.022	2
4	0.120	4.5
5	0.227	6

Develop the algebraic expression for establishing standards for this operation for parts with surface areas of up to seven square inches.

6. The industrial engineer in the Dorben Company is in the process of building a formula to preprice the manufacture of a line of specialty forgings. The data gathered indicated a nonlinear relationship between forging volume and standard time for the variable elements related to positioning in the die and the actual forging. The data taken were as follows:

Study number	Time (min.)	Forging volume (cu. in.)
1	0.130	30
2	0.110	24
3	0.103	20
4	0.088	10
5	0.083	5
6	0.120	27

Develop an algebraic expression for the computation of standard time for any forging having a volume of between 5 and 30 cubic inches.

SELECTED REFERENCES

Allendoerfer, C. B., and Oakley, C. O. *Principles of Mathematics.* New York: McGraw-Hill Book Co., 1955.

Denbow, Carl H., and Goedicke, Victor. *Foundations of Mathematics.* New York: Harper & Row, 1959.

Lowry, Stewart M., Maynard Harold B., and Stegemerten, B. J. *Motion and Time Study.* 3d ed. New York: McGraw-Hill Book Co., 1940.

Rotroff, Virgil H. *Work Measurement.* New York: Reinhold Publishing Corp., 1959.

21

Work sampling studies

Work sampling is a technique used to investigate the proportions of total time devoted to the various activities that are comprised by a job or work situation. The results of work sampling are effective for determining allowances applicable to the job, for determining machine utilization, and for establishing standards of production. This same information can be obtained by time study procedures. Work sampling is a method that will frequently provide the information faster and at considerably less cost than by stopwatch techniques.

In conducting a work sampling study, the analyst takes a comparatively large number of observations at random intervals. The ratio of the number of observations of a given state of activity to the total number of observations taken will approximate the percentage of time that the process is in that given state of activity. For example, if 10,000 observations at random intervals over a period of several weeks showed that a given automatic screw machine was turning out work in 7,000 instances, and that in 3,000 instances it was idle for miscellaneous reasons, then it would be reasonably certain that the downtime of the machine would be 30 percent of the working day, or 2.4 hours, and that the effective output of the machine would be 5.6 hours per day. The application of work sampling was first made by L. H. C. Tippett in the British textile industry. Later, under the name "ratio-delay" study, it received considerable attention is this country.[1] The accuracy of the data determined by work sampling depends on the number of observations; unless the sample size is of sufficient quantity, inaccurate results will occur.

The work sampling method has several advantages over that of acquiring data by the conventional time study procedure. These are:

1. It does not require continuous observation by an analyst over a long period of time.
2. Clerical time is diminished.

---

[1] Robert Lee Morrow, *Time Study and Motion Economy* (New York: Ronald Press Co., 1946).

3. The total man-hours expended by the analyst are usually much fewer.[2]
4. The operator is not subjected to long-period stopwatch observations.
5. Crew operations can be readily studied by a single analyst.

The theory of work sampling is based on the fundamental laws of probability. If at a given instant an event can either be present or absent, statisticians have derived the following expression, which shows the probability of x occurrences of an event in n observations:

$$(p + q)^n = 1$$
$$p = \text{Probability of a single occurrence}$$
$$q = (1 - p) \text{ the probability of an absence of occurrence}$$
$$n = \text{Number of observations}$$

If the above expression, $(p + q)^n = 1$, is expanded according to the binomial theorem, the first term of the expansion will give the probability that $x = 0$, the second term $x = 1$, and so on. The distribution of these probabilities is known as the binomial distribution. Statisticians have also shown that the mean of this distribution is equal to np and that the variance is equal to npq. The standard deviation is, of course, equal to the square root of the variance.

One may now logically ask, of what value is a distribution that allows only one event to either occur or not occur? For the answer to this, consider the possibility of taking one condition of the work sampling study at a time. All the other conditions can then be considered as nonoccurrences of this one event. Using this approach, we can now proceed with the discussion of binomial theory.

Elementary statistics tells us that as n becomes large, the binomial distribution approaches the normal distribution. Since work sampling studies involve quite large sample sizes, the normal distribution is a satisfactory approximation of the binomial distribution. Rather than use the distribution of the number of occurrences a with mean np and variance npq, it is more convenient to use the distribution of a proportion with a mean of p, i.e., $\frac{np}{n}$, and a standard deviation of

$$\sqrt{\frac{pq}{n}} \left(\text{i.e., } \frac{\sqrt{npq}}{n} \right)$$

as the approximately normally distributed random variable.

In work sampling studies, we take a sample of size n in an attempt to estimate p. We know from elementary sampling theory that we cannot expect the $\hat{p}$ ($\hat{p} = $ the proportion based on a sample) of each sample to be the true value of p. We do, however, expect the $\hat{p}$ of any sample to fall within the range of $p \pm 2$ sigma approximately 95 percent of the time. In other words, if p is the true percentage of a given condition,

---

[2] *Ibid.,* p. 334.

we can expect the $\hat{p}$ of any sample to fall outside the limits $p \pm 2$ sigma only about 5 times in 100 due to chance alone. This theory will be used to derive the total sample size required to give a certain degree of accuracy. It will also be used for subsample sizes a little later.

Illustrative example

To clarify the fundamental theory of work sampling, it would be helpful to interpret the results of an experiment. Let us assume the following circumstances: we have observed one machine that has random breakdowns for a 100-day period. During this period, we have taken eight random observations per day.

Let:

n = Number of observations per day
N = Total number of random observations
x_i = Number of breakdown observations observed in n random observations on day i
x = 0, 1, 2, 3, . . . , n
i = 1, 2, 3, . . . , k
N_x = Number of days that the experiment showed x number of breakdowns in n random observations
$P(x)$ = Probability of x observations down in n observations by the binomial distribution

$$P(x) = \frac{n!}{x!(n-x)!} p^x q^{n-x}$$

$$p + q = 1$$

where:

p = Probability of machine being down
q = Probability of machine running

$$\tilde{P} = \frac{x_i}{n} = \text{Observed proportion of downtime per day}$$

$$\hat{P} = \frac{\sum\limits_{i=1}^{k} x_i}{N}$$

An all-day time study for several days revealed that $p = 0.5$. Table 21–1 shows the number of days in which x breakdowns were observed ($x = 0, 1, 2, 3, . . . , n$) and the expected number of breakdowns given by our binomial model.

The reader can observe the close agreement between the observed number of days that a specified number of breakdowns were encountered (Nx) and the expected number computed theoretically $100P(x)$ when $p = 0.5$. A hypothesis that the theoretical information shows close

TABLE 21-1

x	Nx	P(X)	100P(X)
0	0	0.0039	0.39
1	4	0.0312	3.12
2	11	0.1050	10.5
3	23	0.2190	21.9
4	27	0.2730	27.3
5	22	0.2190	21.9
6	10	0.1050	10.5
7	3	0.0312	3.12
8	0	0.0039	0.39
	100	1.00*	100*

* Approximately.

enough agreement to the observed information for the theoretical binomial to be accepted, may be tested by the chi-square (χ^2) distribution. The χ^2 distribution tests whether the observed frequencies in a distribution differ significantly from the expected frequencies.

In the example, where the observed frequency is N_x and the expected frequency is $100\ P(x)$:

$$\chi^2 = \sum_{x=0}^{k} \frac{[N_x - 100P(x)]^2}{100P(x)}$$

That is, the quantity under the summation is distributed approximately as χ^2 for k degrees of freedom. In this example we have $\chi^2 = 0.206$.

The analyst must determine whether the calculated value of χ^2 is sufficiently large to refute a null hypothesis that the difference between the observed frequencies and the computed frequencies is due to chance alone. This experimental value of χ^2 is so small that it could easily have occurred through chance causes alone. Therefore, we accept the hypothesis that the experimental data "fits" the theoretical binomial distribution.

In the typical industrial situation, p (which was known to have a value of 0.5) is unknown to the analyst. The best estimate of p is $\hat{p}$, which may be computed as $\frac{x}{n}$. As the number of observations taken at random per day (n) increases, $\hat{p}$ will approach p. However, with a limited number of random observations, the analyst is concerned with the accuracy of $\hat{p}$.

If a plot of $P(x)$ versus x were made from the above example, it would appear as shown in Figure 21-1.

When n is sufficiently large, regardless of the actual value of p, the binomial distribution will very closely approximate the normal distribution. This tendency can be seen in the above example when p is approximately 0.5. When p is near 0.5, n may be small and the normal can be a good approximation to the binomial.

FiGURE 21-1

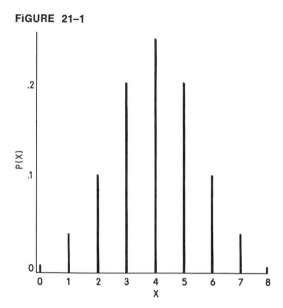

When using the normal approximation, we set

$$u = p$$

and

$$\sigma p = \sqrt{\frac{pq}{n}}$$

To approximate the binomial distribution the variable z used for entry in the normal distribution (see Table A3–2, Appendix 3) may take the following form:

$$z = \frac{\hat{p} - p}{\sqrt{\frac{pq}{n}}}$$

Although p is unknown in the practical case, we can estimate p from $\hat{p}$, and can determine the interval within which p lies, using confidence limits. For example, we can imagine, that the interval defined

$$\hat{p} - 2\sqrt{\frac{\hat{p}\hat{q}}{n}}$$

and

$$\hat{p} + 2\sqrt{\frac{\hat{p}\hat{q}}{n}}$$

contains p 95 percent of the time.

Graphically, this may be represented as:

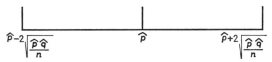

We can derive the expression for finding a confidence interval for p as follows: let us suppose that we want an interval which will contain p 95 percent of the time; that is, a 95 percent confidence interval. For n sufficiently large, the expression,

$$z = \frac{\hat{p} - p}{\sqrt{\frac{\hat{p}\hat{q}}{n}}}$$

is approximately a standard normal variable. Therefore, we can set the probability

$$P\left(z_{.025} < \frac{\hat{p} - p}{\sqrt{\frac{\hat{p}\hat{q}}{n}}} < z_{.975} \right) = 0.95$$

Rearranging the inequalities then gives:

$$P\left(\hat{p} - z_{.975} \sqrt{\frac{\hat{p}\hat{q}}{n}} < p < \hat{p} + z_{.975} \sqrt{\frac{\hat{p}\hat{q}}{n}} \right) = 0.95$$

remembering that $- z_{.025} = z_{.975} = 1.96$, or approximately 2. The interval with approximately a 95 percent chance of containing p is then

$$\hat{p} - 2\sqrt{\frac{\hat{p}\hat{q}}{n}} < p < \hat{p} + 2\sqrt{\frac{\hat{p}\hat{q}}{n}}$$

These limits imply that the interval defined contains p with 95 percent confidence since z has been selected as having a value of 2.

The underlying assumptions of the binomial are that p, the probability of a success (the occurrence of downtime), is constant each random instant that we observe the process. Therefore, it is always necessary to take random observations when taking a work sampling study.

Selling work sampling

Before beginning a program of work sampling, it is a good idea for the analyst to sell the use and the reliability of the tool to all members of the organization who will be affected by the results. If the program is to be used for establishing allowances, it should be sold to the union and the foreman, as well as company management. This can be done by having several short sessions with representatives of the various interested parties and explaining examples of the law of probability, thus illustrating why ratio delay procedures will work. Unions as well as workers look with favor upon work sampling techniques once the procedure is fully explained, since work sampling is completely impersonal, does not utilize the stopwatch, and is based upon accepted mathematical and statistical methods.

In the initial session at which work sampling is explained, the simple study in tossing unbiased coins may be used. All participants will, of course, readily recognize that a single coin toss stands a 50–50 chance of being heads. When asked how they would determine the probability of heads versus tails, they will undoubtedly propose tossing a coin a few times to find out. When asked whether two times is adequate, they will say no. Ten times may be suggested, and the response will be: "That may not be adequate." When 100 times is suggested, the group will agree: "That should do it with some degree of assurance." This example will firmly implant the principal requisite of work sampling: adequate sample size to ensure statistical significance.

The instructor would next discuss the probable results of tossing four unbiased coins. It can be explained that there is only one arrangement in which the coins can fall showing no heads, only one arrangement that permits all heads. However, three heads or one head can result from four possible arrangements. There are six possible arrangements that will give two heads.

With all 16 possibilities thus accounted for, the group will recognize that if four unbiased coins are tossed continually, they will distribute themselves as shown in Figure 21–2.

After the above explanation and the demonstration of this distribution by making several tosses and recording the results, it will readily be ac-

FIGURE 21–2
Distribution of number of heads with
infinite number of tosses using four
unbiased coins

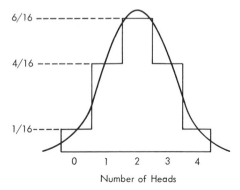

Number of Heads

cepted that 100 tosses could demonstrate a normal distribution. Furthermore, it will be understood that a thousand tosses would probably approach a normal distribution more closely and that 100,000 would give a nearly perfect distribution, but a distribution not sufficiently more accurate than the 1,000-toss distribution to be economically worth the extra effort. Here the idea of approaching significant accuracy rapidly at first, and then at a diminishing rate, can be established.

Now it can be pointed out that a machine or operator could be in a heads or tails state. For example, a machine could be running (heads) or idle (tails). An example showing 360 observations that have been made on a turret lathe could be demonstrated. These observations might give data like those shown in Figure 21–3.

FIGURE 21–3

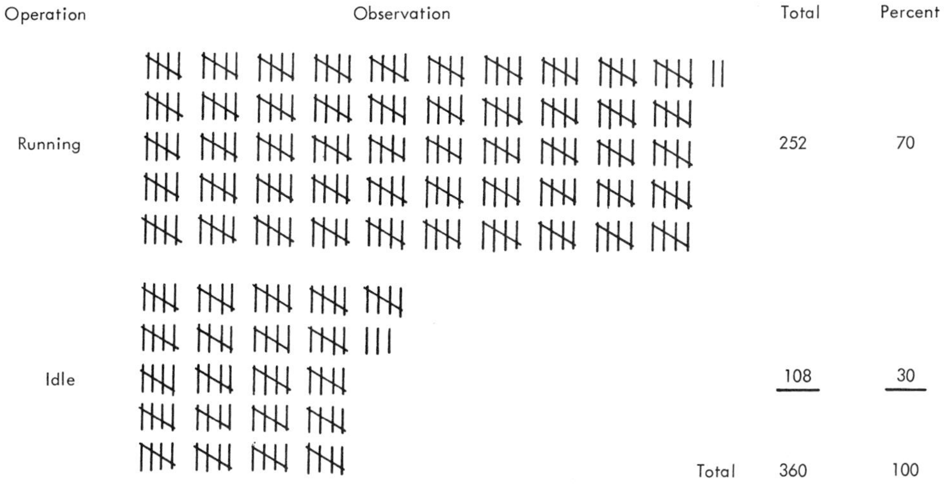

Operation	Observation	Total	Percent
Running	＼＼＼ ＼＼＼ ＼＼＼ ＼＼＼ ＼＼＼ ＼＼＼ ＼＼＼ ＼＼＼ ＼＼＼ ＼＼＼ ‖ ＼＼＼ ＼＼＼ ＼＼＼ ＼＼＼ ＼＼＼ ＼＼＼ ＼＼＼ ＼＼＼ ＼＼＼ ＼＼＼ ＼＼＼ ＼＼＼ ＼＼＼ ＼＼＼ ＼＼＼ ＼＼＼ ＼＼＼ ＼＼＼ ＼＼＼ ＼＼＼ ＼＼＼ ＼＼＼ ＼＼＼ ＼＼＼ ＼＼＼ ＼＼＼ ＼＼＼ ＼＼＼ ＼＼＼ ＼＼＼ ＼＼＼ ＼＼＼ ＼＼＼ ＼＼＼ ＼＼＼ ＼＼＼ ＼＼＼ ＼＼＼ ＼＼＼ ＼＼＼	252	70
Idle	＼＼＼ ＼＼＼ ＼＼＼ ＼＼＼ ＼＼＼ ＼＼＼ ＼＼＼ ＼＼＼ ＼＼＼ ⦀ ＼＼＼ ＼＼＼ ＼＼＼ ＼＼＼ ＼＼＼ ＼＼＼ ＼＼＼ ＼＼＼ ＼＼＼ ＼＼＼ ＼＼＼ ＼＼＼	108	30
		Total 360	100

A cumulative plot of "running" would level off, giving an indication of when it would be safe to stop taking readings (see Figure 21–4). All those in attendance would understand that "idle" machine time could be broken down into the various types of interruptions and delays and could be accounted for and evaluated.

Once the validity of work sampling has been sold up and down the line, the analyst should clearly define his problem. If he wishes to establish allowance data, he should make a summary of all elements customarily included in the various allowance categories. To do this, he would be wise to make a preliminary survey of the class of work for which the allowances are to be determined. It may be necessary to spend several hours on the production floor observing all the types of delays encountered, so that the element listing can be complete.

FIGURE 21–4
Cumulative percentage of running time

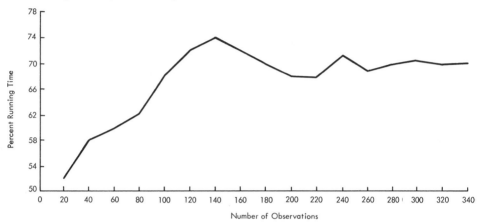

Number of Observations

Planning the work sampling study

After the analyst has explained the method and obtained the approval of the supervisor involved, he is ready to perform the detailed planning that is essential before beginning actual observations.

His first step is to make a preliminary estimate of the activities on which he is seeking information. These would be the activities referred to earlier. This estimate may involve one or more activities. Frequently the estimate can be made from historical data. If the analyst does not feel that he can make a reasonable estimate, he should work sample the area or areas involved for a short period (two or three days) and use the information obtained as the basis of his estimates (see Figure 21–3).

Once the preliminary estimates have been made, he should determine the accuracy that is desired of the results. This can best be expressed as a tolerance within a stated confidence level.

Having made a preliminary estimate of the percentage occurrences of the elements being studied, and having determined an accuracy desired at a given confidence level, the analyst will now make an estimate of the number of observations to be made. Knowing how many observations need to be taken and the time that is available to conduct the study, he can determine the frequency of observations.

His next step will be to design the work sampling form or card on which the data will be tabulated and the control charts that will be used in conjunction with the study.

Determining the observations needed

To determine the number of observations needed, the analyst has to know how accurate his results must be. The larger the number of observations, the more valid the final answer will be. Three thousand observations will give considerably more reliable results than will 300. However, if the accuracy of the result is not the prime consideration, 300 observations may be ample.

In random sampling procedures, there is always the chance that the final result of the observations will be beyond the acceptable tolerance. However, sampling errors will diminish as the size of the sample increases. The standard error σ_p of a sample proportion or percentage as shown in most textbooks on statistics may be expressed by the equation:

$$\sigma_p = \sqrt{\frac{pq}{n}} = \sqrt{\frac{p(1-p)}{n}}$$

where:

σ_p = Standard deviation of a percentage

p = True percentage occurrence of the element being sought, expressed as a decimal

n = Total number of random observations upon which p is based

By approximating the true percentage occurrence of the element being sought (which we can designate as $\hat{p}$), and by knowing the allowable standard error, it is possible to substitute in the above expression and compute n:

$$n = \frac{\hat{p}(1-\hat{p})}{\sigma_p{}^2}$$

For example, it is desired to determine the number of observations required with 95 percent confidence so that the true proportion of personal and unavoidable delay time is within the interval 6–10 percent. It is expected that the unavoidable and personal delay time encountered in the section of the plant under study is 8 percent. These assumptions are expressed graphically in Figure 21–5.

FIGURE 21–5
The tolerance range of the percentage of unavoidaþle delay allowance required within a given section of a plant

In this case, $\hat{p}$ would equal 0.08, and $\sigma_{\hat{p}}$ would equal 1 percent, or 0.01. Using these values, we can solve for n as follows:

$$n = \frac{0.08(1-0.08)}{(0.01)^2}$$

$$= 736 \text{ observations}$$

Frequently it is desirable to express the interval of the element being sought as a percentage element tolerance. In the example given, we use a ± 25 percent element tolerance (0.02/0.08). To simplify the calculation of *n* for varying tolerances of the element being sought for a given confidence level, it is possible to construct an alignment chart. Figure 21–6

FIGURE 21–6
Nomogram to determine the number of random observations for various confidence levels

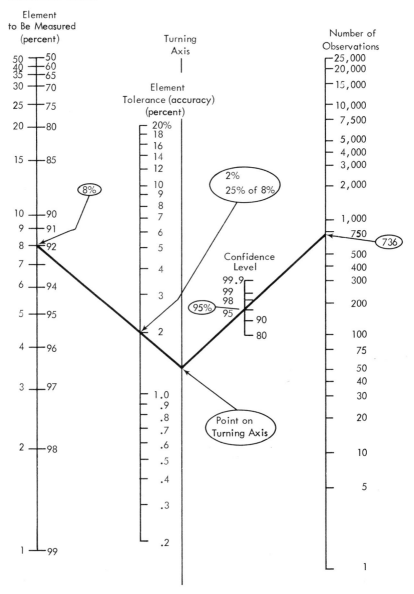

illustrates such an alignment chart for 95 percent confidence limits (± 2 standard deviations).

After the initial estimate of the number of observations has been obtained, a more accurate estimation of p may be computed. By using the above equation,

$$n = \frac{\hat{p}(1 - \hat{p})}{\sigma_p{}^2}$$

the achieved accuracy is again computed. If the achieved accuracy is as good as, or better than, the desired accuracy, we accept this $\hat{p}$ as the true value of the population value p. If the desired accuracy has not been obtained, more observations are taken, and the above process is repeated.

Determining the frequency of the observations

The frequency of the observations depends for the most part on the number of observations required and the time limit placed on the development of the data. For example, if 3,600 observations were needed and the study was to be completed in 30 calendar days, we should need to obtain approximately:

$$\frac{3{,}600 \text{ observations}}{20 \text{ working days}} = 180 \text{ observations per working day}$$

Of course, the number of analysts available and the nature of the work being studied will also influence the frequency of the observations. For example, if only one analyst is available, and if he is accumulating allowance data on a limited battery of facilities, it may be impractical for him to take 180 observations during one working day. Once the number of observations per day has been determined, then the actual time that the analyst records his observations must be selected. To obtain a representative sample, it is important that observations be taken at all times of the working day.

In the example cited, we may assume that one analyst is available and that he is studying a battery of 20 turret lathes to determine the personal and unavoidable delay allowance. He has calculated that 180 observations per day will be required, and since he has 20 machines to observe, he must make nine random trips to the machine floor every eight-hour working day for a period of 20 days. The time of day selected for these nine observations should be chosen at random daily. Thus, no set pattern should be established from day to day for the time when the analyst appears on the production floor.

One method that may be used is to select nine numbers daily from a statistical table of random numbers, ranging from 1 to 48 (see Appendix 3, Table A3–4). Let each number carry a value in minutes equivalent to 10 times its size. The numbers selected can then set the time in minutes from the beginning of the working day to the time for taking the observations. For example, the random number 20 would mean that the analyst should make a series of observations 200 minutes after the beginning of the shift. If the working day began at 8:00 A.M., then at 20 minutes after 11:00 A.M. an inspection of the 20 turret lathe operators (in the example cited above) would be made.

Designing the work sampling form

The analyst will need to design an observation form to best record the data that will be gathered during the course of the work sampling study. No standard form should be used, since each work sampling study is unique from the standpoint of the total number of observations needed, the random times that observations will be made, and the information being sought. The best form is the one that is tailored to the objectives of the study.

A work sampling study form is illustrated in Figure 21–7. This form was designed to determine the time being utilized for various productive and nonproductive states in a maintenance repair shop. It can be seen that this form was designed to accommodate 20 random observations during the working day.

The use of control charts

The control chart techniques used so extensively in statistical quality control work can be applied readily to work sampling studies. Since work sampling studies deal exclusively with percentages or proportions, the "p" chart is used most frequently.

To understand how the "p" chart can be of value in a work sampling study, the reader should understand the theory behind control charting. While a complete discussion of control chart theory is impracticable in a text of this nature, a brief discussion will now be presented to enable the reader to see the logic in using control charts.

The first problem encountered in setting up a control chart is the choice of limits. In general, a balance is sought between the cost of looking for an assignable cause when none is present and not looking for an assignable cause when one is present. As an arbitrary choice, the three-sigma limits will be used throughout the remaining discussion for establishing control limits on the "p" chart.

Suppose that p for a given condition is 0.10 and samples of size 180 are taken each day. By substituting in the equation for n (page 519), control limits of ± 0.07 are obtained. The nomogram in Figure 21–8 is designed to give the correct three-sigma limits for various sample sizes and various values of p. A control chart similar to Figure 21–9 could then be constructed. The p' values for each day would be plotted on the chart. (Note that $\dfrac{\Sigma p'}{N} = \hat{p}$,

FIGURE 21-7

WORK SAMPLING STUDY

Main Repair Shop _____

Number Working This Study _____ Date _____ By _____

Remarks _____

Obs. No.	Random Time	Productive Occurrences							Nonproductive Occurrences							Total Observations	Percentage Productive	Percentage Nonproductive
		Mch	Weld	Pipe Fit	Gen. Labor	Elect.	Carpen.	Janitor	Get Tools	Grind Tools	Wait Job	Wait Crane	Confer Foreman	Personal	Idle			
1																			
2																			
3																			
4																			
5																			
6																			
7																			
8																			
9																			
10																			
11																			
12																			
13																			
14																			
15																			
16																			
17																			
18																			
19																			
20																			
Totals																			

FIGURE 21–8
Nomogram for finding control limits on the results of work
sampling

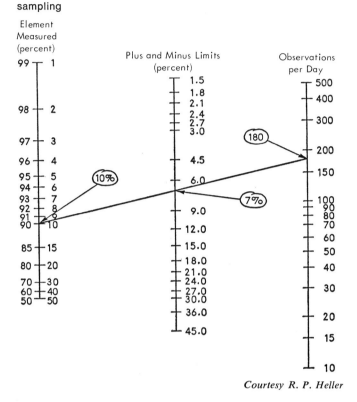

Courtesy R. P. Heller

FIGURE 21–9

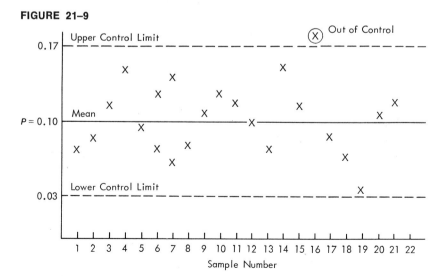

Sample Number

where N equals the number of days the work sampling study was made, and approximately the same number of observations were made each day of the study.)

What does a control chart indicate? In quality control work, we say that the control chart indicates whether or not the process is in control. In a similar manner, the analyst in work sampling considers points beyond three-sigma limits of p as out of control. Thus, a certain sample that yields a value of p' is assumed to have been drawn from a population with an expected value of p if p' falls within the plus or minus three-sigma limits of p. Expressed another way, if a sample has a value p' that falls outside the three-sigma limits, it is assumed that the sample is from some different population or that the original population has been changed.

$$3\sigma_p = 3 \sqrt{\frac{p(1 - p)}{n}} \tag{1}$$

As in quality control work, points other than those out of control may be of some statistical significance. For example, it is more likely that a point will fall outside the three-sigma limits than that two successive points will fall between the 2- and 3-sigma limits. Hence, two successive points between the two- and three-sigma limits would indicate that the population had changed. Series of significant sets of points have been derived. This idea is discussed in most statistical quality control texts under the heading "Theory of Runs."

A hypothetical example will show how control charts can facilitate a work sampling study. Company XYZ wishes to measure the percentage of machine downtime in the lathe department. An original estimate shows downtime to be approximately 0.20. The desired results are to be within ± 5 percent of p with a level of significance of 0.95. The sample size is computed to be 6,400. It was decided to take the 6,400 readings over a period of 16 days at the rate of 400 readings per day. A p' value was computed for each daily sample of 400. A p chart was set up for $p = 0.20$ and subsample size $N = 400$ (see Figure 21–10). Each day, readings were taken and p' was plotted. On the third day, the point for p' went above the upper control limit. An investigation revealed that there had been an accident in the plant and that several of the men had left their machines to assist the injured employee to the plant hospital. Since an assignable cause of error was discovered, this point was discarded from the study. If a control chart had not been used, these observations would have been included in the final estimate of p.

On the fourth day the point for p' fell below the lower control limit. No assignable cause could be found for this occurrance. The industrial engineer in charge of the project also noted that the p' values for the first two days were below the mean p. He decided to compute a new value for p using the values from days 1, 2, and 4. The new estimate of p turned

FIGURE 21–10

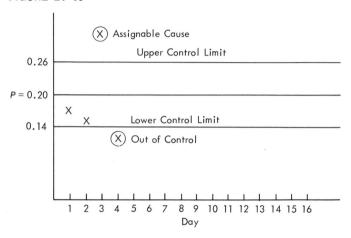

out to be 0.15. To obtain the desired accuracy, *n* is now 8,830 observations. The control limits also change, as shown in Figure 21–11. Observations were taken for 12 more days, and the individual *p′* values were plotted on the new chart. As can be seen, all the points fell within the control limits. A more accurate value of *p* was then calculated, using all 6,000 observations. The new estimate of *p* was determined to be 0.14. A

FIGURE 21–11

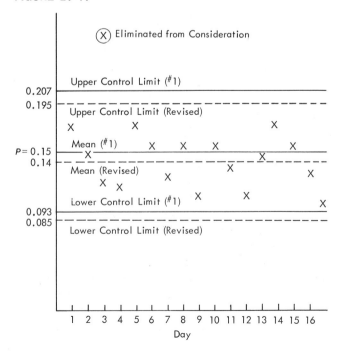

recalculation of achieved accuracy showed it to be slightly better than the desired accuracy. As a final check, new control limits were computed using p equal to 0.14. The dashed lines superimposed on Figure 21–11 showed that all points were still in control using the new limits. If a point had fallen out of control, it would have been eliminated, and a new value of p would have been computed. This process would then have been repeated until the desired accuracy was achieved and all p' values were in control.

One may ask, "Is there any reason to assume that since the percentage downtime is 0.14 today that it will be 0.14 a year from now?" Improvement should be a continuing process, and percentage downtime should diminish. One purpose of work sampling is to determine areas of work that might be improved. After discovering such areas, an attempt is made to improve the situation. Control charts can be used to show the progressive improvement of work areas. This idea is especially important if work sampling studies are used to establish standard times, for such standards must be changed whenever conditions change if they are to remain realistic.

Observing and recording the data

A representative sample form for a shift study is shown in Figure 21–12. Here six random observations of each facility were made per shift. A digit designating the particular observation was indicated in the space provided for the state of each facility under study. Since 14 facilities were being studied, a total of 84 observations were made per shift.

As the analyst approaches the work area, he should not anticipate the recording he expects to make. He should walk to a point a given distance from the facility, make his observation, and record the facts. It might be helpful to make an actual mark on the floor to show where the analyst must stand before making his observations. If the operator or machine being studied is idle, the analyst should determine the reason for idleness, confirming the reason with the line foreman before making the proper entry. The analyst should learn to take visual observations, making his written entries after leaving the scene of the work area. This will minimize a feeling of being watched on the part of the shopworkers, and they will perform in their accustomed manner.

To help assure that the operators perform in their usual fashion, it is advisable to inform them of the purpose of the study. The fact that no watch is used tends to relieve the operators of a certain mental tension; little difficulty is experienced in getting their full cooperation.

Using a random activity analysis camera

Even if the requisites of work sampling are observed, the data will tend to be biased when the technique is used for studying people only.

FIGURE 21-12

Observation form for recording events on 14 facilities during a work sampling study

WORK SAMPLING OBSERVATION RECORD

STUDY NO. 46

PLANT DuBois STUDY AREA Section C of Heavy Machine Shop

REMARKS Were Producing Three Cylinder Turbines

DATE 5/13

OBSERVER R. Guild

MACHINE	DWG.	CUTTING	SETUP	MACHINE IDLE	CRANE WAIT	WAIT-INSPECTION	AID INSPECTION	WAIT-TOOLS NOT AVAILABLE	WAIT-TOOL TROUBLE	CONFER WITH OTHER SHIFT	TOOL HANDLING	GET OR GRIND TOOLS	CONFER WITH FOREMAN, INSP.	WAIT FOR JOB	REMOVE CHIPS CLEAN TABLE	MISCELLANEOUS	NO OPERATOR
20' VBM	933967	3 5	4						1	6		2					
16' VBM Ⓢ	983469 / 90J384	2 5	1	3									4 6 1				
28' VBM	34E9	2 6	4	3 5					1								
12' VBM	96J961	2 5	4						1	3 6							
16' PLANER	98J184	3 4 5 6	1							2							
8' IMM	97J686	3 4								2		5					
16' VBM	96J939	1 2 3 4 5 6															
14' PLANER	92J518	1 2 3 4 5 6															
72" E. LATHE	97J43	2 4 5 6	1							3							
96" E. LATHE E-1810	30F307	2 3 4	1 6							5							
96" E. LATHE	36F463	2 3 5 6	1 4														
160" E. LATHE	93J771	6				5			1	4							
11-1/2' PLNR Ⓢ	90J158 / 96J798	2 4 5 6	3						1	1							
32' VBM	96J739	1 2 3 4 5 6										2					
		40	13	10	1	2	1		3	9		2	3				= 84

This is because the arrival of the observer at the work center will immediately influence the activity of the operator. The operator will have a tendency to become productively engaged as soon as he sees the analyst approaching the work center. Then, too, there is a natural tendency for the observer to record what *has just* happened or what *will be* happening rather than what *is* actually happening at the exact moment of the observation.

The use of a random activity analysis camera similar to that described in Chapter 8 will permit unbiased work sampling studies of problems involving people. In a work sampling study recently made by the author to determine the attentiveness of each worker, the value of the random activity analysis camera was significantly demonstrated. Work sampling studies of the same activity were made both by conventional, manual means and by means of the random activity analysis camera. The results of the two methods were then compared. The work center was a data processing installation, where the workers involved were required to keypunch and verify various quantities of data processing machine cards. The study was concerned only with the elements "working" and "not working." Working time included the elements adjusting cards, removing cards, punching cards, and so on, while not working involved absence from the work station and idleness.

The data collected with the random activity analysis camera revealed a 12.3 percent greater "not working" average than was indicated by the personal observation method. Table 21–2 summarizes the daily "Absent and Not Working" percentages for the two work sampling methods.

TABLE 21–2
Percentage of absent and not working time for women engaged in data processing activities

Day	1	2	3	4	5	6	7	8	9
Camera study	42.9	47.1	48.9	47.9	45.7	53.2	50.7	49.3	43.6
Personal observation study	36.8	40.7	37.1	32.1	32.1	31.9	35.2	34.5	42.0

During a nine-day period, 2,520 observations were made with the random activity analysis camera, and during the next nine-day period, 2,170 observations were made by manual observation. The daily observations made by means of the two techniques are shown in Table 21–3.

A "*t*"-test was made comparing the data gathered by both techniques to determine whether the difference in the data was significant. Accordingly, the null hypothesis was established, stating that work sampling results obtained by personal observations and those obtained by memomotion camera studies are the same.

Table 21–4 represents the difference in the percentage of "Absent and

TABLE 21–3

Day	Number of observations with camera	Number of personal observations
1...............	280	280
2...............	280	140
3...............	280	294
4...............	280	280
5...............	280	280
6...............	280	210
7...............	280	210
8...............	280	252
9...............	280	224
Total..........	2,520	2,170

TABLE 21–4
Difference between the results of a camera study and of a personal observation study in determining the percentage of absent and not working times

Day...........	1	2	3	4	5	6	7	8	9
Difference	6.1	6.4	11.8	15.8	13.6	21.3	15.5	14.8	1.6

Not Working" times observed during the nine-day study by the two methods.

The mean difference is $\dfrac{\Sigma X}{n} = \dfrac{106.9}{9} = 11.9$

If there were no bias between the two methods, we should expect that the mean difference would not differ significantly from 0.

With the assumed value of the population mean discrepancy of 0, we can compute a "t" value from the expression:

$$t = \frac{(\bar{X} - \bar{x})\,\sqrt{n - 1}}{s}$$

where:

$s = $ The standard deviation of the differences

$$s^2 = \frac{\Sigma d^2}{n} - \bar{d}^2$$

$$= 33.04$$

And

$$s = 5.75$$

Thus

$$t = \frac{(11.9 - 0)\ \sqrt{8}}{5.75}$$

$$= 5.86$$

By comparing the computed value of "*t*" (5.86) with the tabularized value at the 0.001 probability level for eight degrees of freedom (see Appendix A3–3), 5.041, we can conclude that the difference in the data obtained from the two methods is highly significant.

Work sampling for the establishment of allowances

One of the most extensive uses of work sampling has been in establishing allowances to be used in conjunction with normal times to determine allowed times. However, the technique is also being used for establishing standards of production, determining machine utilization, allocating work assignments, and improving methods.

As explained in Chapter 16, the determination of time allowances must be correct if fair standards are to be realized. Prior to the introduction of work sampling, allowances for personal reasons and unavoidable delays were frequently determined by taking a series of all-day studies on several operations, and then averaging the results. Thus, the number of trips to the rest room, the number of trips to the drinking fountain, the number of interruptions, and so forth could be recorded, timed, and analyzed, and a fair allowance determined. Although this method gave the answer, it was a costly, time-consuming operation that was fatiguing to both the analyst and the operator. Through ratio delay study, a great number of observations (usually over 2,000) are taken at different times of the day, and of different operators. Then, if the total number of legitimate occurrences other than work that involve the operators is divided by the total number of working observations, the result will tend to equal the percentage allowance that should be given the operator for the class of work being studied. The different elements that enter into personal and unavoidable delays can be kept separate, and an equitable allowance determined for each class or category. Figure 21–13 illustrates a summary of a ratio delay study for determining unavoidable delay allowances on bench, bench machine, machine, and spray operations. It will be noted that in 26 cases out of 2,895 observations made on bench operations, there were interferences. This indicated an unavoidable delay allowance of 0.95 percent on this class of work.

Work sampling for the determination of machine utilization

Machine utilization is readily determined by the work sampling technique in the same manner that is used in establishing allowances. To

FIGURE 21–13

Summary of interruptions on various classes of work taken from a work sampling study for determining unavoidable delay allowance

XYZ ELECTRIC PRODUCTS, INC.

PLANT _26_ DEPARTMENT _4_

SUMMARY SHEET (RATIO DELAY) DATE _4/22/54_

OPERATION	Engineer	Supply	Quality	Mechanic	Supervision	Turning on Light	Miscellaneous	Working		Number of Interferences	TOTAL OBSER.	PER CENT ALLOWANCE
Bench	1	//	1	0	12	1	119	2,750		26	2,895	.95
Bench Machine	0	2	0	0	5	0	69	984		7	1,060	.71
Machine	0	1	6	//	9	0	29	1,172		27	1,228	2.30
Spray	0	0	0	7	36	0	262	1,407		43	1,712	3.06

REMARKS: Summary Sheet for Interference Allowances. See observation sheet for details of Miscellaneous. Miscellaneous was included in working category because in all cases downtime was credited to the operators.

NOTE: ALL OBSERVATIONS ARE TO BE TAKEN AT RANDOM.

give the reader a clearer understanding of the steps involved, a case history will be related.

It was desired to gain information on machine utilization in a section of a heavy machine shop. Management had estimated that the actual cutting time in this section should be 60 percent of the working day in order to comply with the quotations being submitted. There were 14 facilities involved in this section of the shop, and it was estimated that approximately 3,000 observations should be taken to get the accuracy desired.

A work sampling form was then designed (see Figure 21–12) to accommodate the 16 possible states that each of the facilities under study might be in at the time of an observation.

To assure random observations, a random pattern of visitation to the shop area was established. Six observtions of the 14 facilities were made during each shift. To get the required total number of observations, 36 separate shifts were observed—12 studies on each of the 3 shifts.

Since there were 14 machines and 6 observations of each per shift, a quick check for 84 separate readings per sheet assured complete coverage of the shift (see Figure 21–12). Each trip took the analyst about 15 to 20 minutes. Thus, he was occupied on this work only about two hours per shift, leaving him free to perform his other work during the remaining six hours.

FIGURE 21–14
Cumulative percentage machine cutting

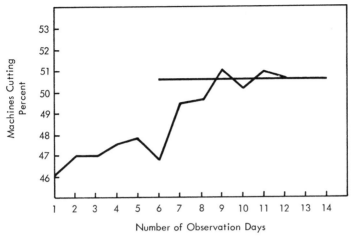

Number of Observation Days

Since it was principally desired to learn the status of the actual cutting time in this section, a cumulative percentage machine cutting chart was kept (see Figure 21–14). At the beginning of each day's study, all previous cutting observations were taken cumulatively as a ratio to the total observations to date. By the end of the 10th day of study, the percentage of machine cutting time began to level off at 50.5 percent.

After 36 shifts had been studied, the sums of all observations in each category divided by the total number of observations resulted in percentages that represented the distribution of the cutting time, setup time, and the various delay times listed. Figure 21–15 illustrates the summary sheet of this study. It will be noted that cutting time amounted to 50.7 percent. The percentage of time required by the various delays indicated areas for methods improvement that would help increase the cutting time.

Work sampling in establishing indirect and direct labor standards

Some companies are finding work sampling applicable for establishing incentive standards on indirect as well as direct labor operations.[3] The technique is the same as that used for allowance determination. A large number of random observations are taken; then the percentage of the total number of observations that the facility or operation is working will approximate the percentage of the total time that it is truly in that state.

One company's technique for taking work sampling studies on clerical operations is to take observations on each operator at one-minute intervals. The various elements of work in which the operator may be engaged

---
[3] Aluminum Company of America and Douglas Aircraft Co.

FIGURE 21–15
Work sampling summary sheet

DATE ____7-15____

OBSERVER ____R. Guild____

MACHINE	DWG.	CUTTING	SETUP	MACHINE IDLE	CRANE WAIT	WAIT-INSPECTION	AID INSPECTION	WAIT-TOOLS NOT AVAILABLE	WAIT-TOOL TROUBLE	CONFER WITH OTHER SHIFT	TOOL HANDLING	GET OR GRIND TOOLS	CONFER WITH FOREMAN/INSP.	WAIT FOR JOB	REMOVE CHIPS CLEAN TABLE	MISCELLANEOUS	NO OPERATOR	
20' VBM		101	7	14	2	3		1		2	37	5	3			6	35	216
16' VBM		102	34	14	15	3	1	1		1	28	5	7	4		1		216
28' VBM		119	34	10	5	5	2				18	2	1		2		18	216
12' VBM		109	24	12	13	6	1			3	26	6	2	3		5	6	216
16' PLANER		127	17	6	9	2					22			15		6	12	216
8' IMM		64	18	17	16	3				2	30	7	3			28	28	216
16' VBM		147	19	10	14	3	1				15	2			2		3	216
14' PLANER		140	8	5	7	2				2	17	3		11		3	18	216
72" E.LATHE		99	13	12	7	3				1	32	8	3		2		36	216
96" E.LATHE		89	9	29	18	11	1			2	29	8	4		3	3	10	216
96" E.LATHE		109	14	12	8	10		3			32	9	2	10		2	5	216
160' E.LATHE		72	34	13	14	6	2	1		4	21	3	1		2	6	37	216
11-1/2' PLNR		106	35	11	10	4				1	11	4	3		2	13	16	216
32' VBM		151	23	8	7	1			3	1	10	2	5	2		3		216
		1535	289	173	145	62	8	6	3	19	328	64	34	45	13	76	224	3024 =
%		50.7	9.6	5.9	4.8	2.1	.3	.2	.1	.6	10.8	2.1	1.1	1.5	.4	2.5	7.4	100% =

are assigned identifying numbers, and the analyst, at the time of the observation, merely checks the appropriate number in the space provided on his form. Every five minutes, the analyst rates the performance of the operator in the space provided. At the end of the study, a selective average rating factor is determined.

The normal minutes required for a given element of work may be expressed:

$$T_n = \frac{P \times N}{P_a}$$

where:

T_n = Normal minutes to perform the element
P = Selective average rating factor
N = Total observations for the given element
P_a = Total production for the period studied

For example, to determine the normal time for the element "write report heading on one BM report," the analyst may have prepared a work sampling study showing 84 observations during which the employee was actually engaged in performing this work element. During the study period, 12 reports were made, and it was determined that the selective average rating factor was 110 percent. The normal time would then equal:

$$\frac{(1.10)(84)}{12 \text{ reports}} = 7.7 \text{ minutes normal time per report heading}$$

Figure 21–16 illustrates the study form developed for taking a 40-minute study. The columns "I," "P," and "M" accommodate idle, personal delays and miscellaneous observations. Under the column headed "Lev." the performance factor of the operator is recorded every five minutes. The table of standard data shown on page 429 illustrates representative standard data developed by the technique just outlined.

A more generalized expression for determining the normal minutes required to perform the element under study where observations are taken at intervals of fixed duration would be:

$$T_n = \frac{\dfrac{\Sigma \text{ Ratings}}{100} \times I}{P_a}$$

where:

I = Established interval between observations

For example, let us assume that 30 observations are made at 0.50-minute intervals on a work assignment involving three elements, and that 12

FIGURE 21–16
Work sampling time study form for clerical operations

1. SIGHT–VERIFY REPORT TOTAL	11. OBTAIN INFORMATION (DISCUSS, PHONE, ETC.)
2. HEADINGS	12. OBTAIN SUPPLIES (PLUS TRIPS)
3. POSTING FROM REPORT	13. FILING
4. DELIVERY	14.
5. MANUAL REPORT PREPARATION	15.
6. EDITING –ASSIGN NO'S (RECAP NO'S ETC.)	16.
7. CORRECT ERROR CARDS	17.
8. ADDING	18. IDLE TIME
9. TYPING	19. PERSONAL TIME
10. TELEGRAMS –RECEIVING–SENDING	20. MISCELLANEOUS

TABULATING–CLERICAL

FROM __8:00__ TO __8:40__   DATE __1/12/ zupnaugk__

MI.	1	2	3	4	5	6	7	8	9	10	11	12	13	14	15	16	17	I	P	M	LEV.	REMARKS
1																			✓			NOT HERE
2																			✓			NOT HERE
3					✓																	
4						✓																
5						✓															110	
6						✓																
7				✓																		
8				✓																		
9					✓																	
10						✓															115	
11						✓																
12						✓																
13						✓																
14						✓																
15						✓															115	
16						✓																
17						✓																
18						✓																
19						✓																
20						✓															100	
21						✓																
22						✓																
23						✓																
24						✓																
25																			✓		105	SEARCH FOR STAMP
26																			✓			DATE STAMP
27							✓															
28							✓															
29							✓															
30																			✓		110	OBTAIN CLIPS
31							✓															
32							✓															
33							✓															
34							✓															
35							✓														110	
36							✓															
37							✓															
38							✓															
39							✓															
40							✓														115	
TOTAL				2		20	13													5		

units of product have been produced. The tabular data involved are shown in Table 21–5, and the normal minutes for the three elements would be as follows:

$$T_1 = \frac{8.60 \times 0.50}{12} = 0.358 \text{ minutes}$$

$$T_2 = \frac{7.05 \times 0.50}{12} = 0.293 \text{ minutes}$$

$$T_3 = \frac{11.80 \times 0.50}{12} = 0.490 \text{ minutes}$$

TABLE 21–5
Tabular data of a three-element work sampling study

Observation number	Performance rating observed			
	Element 1	Element 2	Element 3	Idle
1	90			
2				100
3		110		
4	95			
5	100			
6		100		
7			105	
8	90			
9			110	
10	85			
11			95	
12		90		
13			100	
14			95	
15	80			
16			110	
17		105		
18			90	
19	100			
20			85	
21			90	
22			90	
23	110			
24			100	
25		95		
26				100
27		105		
28		100		
29			110	
30	110			
Σ Rating	860	705	1180	100
$\dfrac{\Sigma \text{ Rating}}{100}$	8.60	7.05	11.80	1.00

This company's technique can be criticized in that random observations have not been taken. Consequently, biased results could occur. For example, some short-cyclic element may be completely omitted by observing at regular one-minute intervals. This would not happen if sufficient random observations were taken. The expression used for establishing standards on office work can be modified to be applicable on work sampling studies involving random observations rather than regular ones a minute apart. This may be expressed:

$$T_n = \frac{(n)(T)(P)}{(P_a)(N)}$$
$$T_a = T_n + \text{Allowances}$$

where:

T_n = Normal elemental time
T_a = Allowed elemental time
P = Performance rating factor
P_a = Total production for period studied
n = Total observations of element under study
N = Total observations of study
T = Total operator time represented by study

For example, assume that a standard was to be established on the maintenance operation of lubricating fractional horsepower motors. If a work sampling study of 120 hours revealed that, after 3,600 observations, lubrication of fractional horsepower motors on the facilities being studied was taking place in 392 cases and that a total of 180 facilities using fractional horsepower motors were maintained, and the average performance factor was 0.90, then the normal time for lubricating a fractional horsepower motor would be:

$$\frac{(392 \text{ observations})(7,200 \text{ minutes})(0.90 \text{ performance factor})}{(180 \text{ total production})(3,600 \text{ total observations})} =$$

$$\frac{(392)(7,200)(0.90)}{(180)(3,600)} =$$

3.92 minutes to lubricate one fractional horsepower motor

This tool has many applications. Machine downtime can be determined; the relative amount of setup and put-away elements for all classes of work can be made known; and the relative amount of manual, mental, and delay times for clerical work, direct labor, and administrative work can be studied as a basis for establishing the ideal work assignments and methods procedures.

Conclusion

The work sampling method is another tool that allows the time and methods study analyst to get the facts in an easier, faster way. Every person in the field of methods, time study, and wage payment should become familiar with the advantages, limitations, uses, and application of this technique. In summary, the following considerations should be kept in mind:

1. Explain and sell the work sampling method before using it.
2. Confine individual studies to similar groups of machines or operations.
3. Use as large a sample size as is practicable.
4. Take individual observations at random times so that observations will be recorded for all hours of the day.

5. Take the observations over a reasonably long period (two weeks or more).

TEXT QUESTIONS

1. Where was work sampling first used?
2. What advantages are claimed for the work sampling procedure?
3. In what areas does work sampling have application?
4. How many observations should be recorded in determining the allowance for personal delays in a forge shop if it is expected that a 5 percent personal allowance will suffice, and if this value is to remain between 4 and 6 percent 95 percent of the time?
5. How is it possible to determine the time of day to make the various observations so that biased results will not occur?
6. What considerations should be kept in mind when taking work sampling studies?
7. To get ±5 percent precision on work that is estimated to take 80 percent of the workers' time, how many random observations will be required at the 95 percent confidence level?
8. If the average handling activity during a 10-day study is 82 percent, and the number of daily observations is 48, how much tolerance can be allowed on each day's percentage activity?
9. If it should develop that four seconds of time would be desirable for each random observation, and if a sample size of 2,000 is necessary, how often would the analyst need to service the random activity analysis camera?
10. Over how long a period is it desirable to continue to acquire sampling data?
11. How biased can we expect work sampling data to be? Will this bias vary with the work situation? Explain.

GENERAL QUESTIONS

1. Is there an application for work sampling studies in determining fatigue? Explain.
2. How can the validity of work sampling be sold to the employee not familiar with probability and statistical procedure?
3. What are the pros and cons for using work sampling to establish standards of performance?

PROBLEMS

1. The analyst in the Dorben Reference Library decides to use the work sampling technique to establish standards. Twenty employees are involved. The operations include cataloging, charging books out, returning books to their proper location, cleaning books, record keeping, packing books for shipment, and handling correspondence.

A preliminary investigation resulted in the estimate that 30 percent of the time of the group was spent in cataloging. How many work sampling observations would be made if it were desirable to be 95 percent confident that the observed data were within a tolerance of plus or minus 10 percent of the population data? Describe how the random observations should be made.

The following table illustrates some of the data gathered from 6 of the 20 employees. From this data, determine a standard in hours per hundred for cataloging:

	Operators					
Item	*Smith*	*Apple*	*Brown*	*Green*	*Baird*	*Thomas*
Total hours worked.	78	80	80	65	72	75
Total observations (all elements)	152	170	181	114	143	158
Observations involving cataloging	50	55	48	29	40	45
Average rating.	90	95	105	85	90	100

The number of volumes cataloged equals 14,612.

Design a control chart based on three-sigma limits for the daily observations.

2. The work measurement analyst in the Dorben Company is planning to establish standards on indirect labor by using the work sampling technique. This study will provide the following information.

T = Total operator time represented by the study
N = Total number of observations involved in the study
n = Total observations of the element under study
P = Production for the period under study
R = Average performance rating factor during the study

With the above information, derive the equation for estimating the normal elemental time for an operation (t_n).

3. The analyst in the Dorben Company wishes to measure the percentage of downtime in the drop hammer section of the forge shop. The superintendent estimated the downtime to be about 30 percent. The desired results, using a work sampling study, are to be within plus or minus 5 percent of "p" with a level of significance of 0.95.

The analyst decides to take 300 random readings a day for a period of three weeks. Develop a "p" chart for "p" = 0.30 and subsample size N = 300. Explain the use of this "p" chart.

4. The Dorben Company is using the work sampling technique to establish standards in its typing pool section. This pool has varied responsibilities, including typing from tape recordings, filing, Kardex posting, and duplicating.

The pool has six typists who work a 40-hour week. Seventeen hundred random observations were made over a four-week period. During the period, 1,852 pages of routine typing were produced. Of the random observations, 1,225 showed that typing was taking place. Assuming a 20 percent P.D. & F. allowance and an adjusted performance rating factor of 0.85, calculate the hourly standard per page of typing.

SELECTED REFERENCES

Barnes, R. M. *Work Sampling,* 2d ed. New York: John Wiley, & Sons, Inc., 1957.

Heiland, R. E., and Richardson, W. J. *Work Sampling.* New York: McGraw-Hill Book Co., 1957.

22

Establishing standards on indirect and expense work

Since 1900 the percentage increase of indirect and expense workers has more than doubled that of direct labor workers. Groups usually classified as indirect labor include shipping and receiving, trucking, stores, inspection, material handling, toolroom, janitorial, and maintenance. Expense personnel include all positions not coming under direct or indirect—office clerical, accounting, sales, management, engineering, and so on.

The rapid growth in the number of office workers, maintenance workers, and other indirect and expense employees is due to several reasons. First, the increased mechanization of industry and the complete automation of many processes have decreased the need for craftsmen and even for operators. This trend toward mechanization has resulted in a greater demand for electricians, technicians, and other servicemen. Also, the design of complicated machines and controls has resulted in a greater demand for engineers, designers, and draftsmen.

The tremendous increase in paperwork brought on by federal, state, and local legislation is responsible, to a large extent, for the large number of clerical employees utilized today.

Another reason for the rise in the number of indirect and expense employees is that office work and maintenance work have not been subjected to the methods study and the technical advances that have been applied so effectively to industrial processes.

With a large share of most payrolls being earmarked for indirect and expense labor, progressive management is beginning to realize the opportunities for the application of methods and standards in this area.

Automation

The term *automation* may be defined as "increased mechanization." A completely automated manufacturing process is capable of operating for prolonged periods without human effort. Few industries or even processes within an industry have been completely automated. However, there has

542

been a pronounced tendency toward semiautomation in American, Japanese, and European industry. With the increasing demand for greater productivity, it is anticipated that American industry will make a continuing effort to automate.

A program of automation begins with the integration of a fully automatic machine, such as the automatic screw machine, with automatic transfer handling devices, so that a series of operations may be performed automatically.

To determine the extent of automation justified, two factors should be considered: (1) the quantity requirements of the product, and (2) the nature or design of the product itself.

If the quantity requirements of the product are large, the design engineer will endeavor to design the product so that it lends itself to automation. Frequently, by adding something to a part, such as a lug, fin, extension, or hole, it will be possible to provide means of mechanical handling to and from the work station. Such a redesign can accommodate indexing at a work station for successive production operations. For example, a holding fin was added to the die casting for producing a compressor piston. This permitted mechanical fingers to hold the work while it was being processed, as well as to automatically transport the work between production stations (see Figure 22–1).

Complete automation is feasible for the continuous processing of chem-

FIGURE 22–1
Die casting with a fin so that work can be handled in automated equipment

ical products, such as gasoline, oil, and detergents. Partial automation is now widespread in the making of many types of mass-produced items— food products (for example, cookies, cereals, and pretzels), light bulbs, radio tubes, automobile parts, cigarettes, and so forth.

Factors that encourage plant automation include:

1. The increasing cost of labor.
2. Increasing competition, especially from foreign manufacturers, which stimulates reductions in selling prices and a lessening of profits.
3. The prospects for market expansion through cost and price reduction.

Among the principal factors that may discourage automation in a particular plant are:

1. The large capital investment necessary which must, subsequently, be absorbed in operating profit.
2. A market potential presently inadequate to absorb the increased output.
3. An existing technology unable to provide automation to produce a particular design.
4. Opposition by production workers, and the possibility of having community relations affected for the worse by a reduction in the labor force.

However, there is no doubt that the present trend toward automation will continue.

As automation equipment is developed, the need for effective preventive maintenance becomes apparent. The failure of a single minor component may cause the shutdown of a complete process or even an entire plant. This fact, together with the complexity of automated equipment using pneumatic, hydraulic, and electronic controls, brings out the reasons for the growth of indirect workers not only in numbers but in diversification of occupation.

Methods improvements on indirect and expense work

The systematic approach to methods standards and wage payment outlined in Chapter 1 is just as applicable to indirect and expense areas as it is to direct labor. A program for establishing standards on indirect and expense labor should be preceded by careful fact-finding, analysis, development of the proposed method, presentation, installation, and development of job analysis. The methods analysis procedure in itself will result in the introduction of economies.

Work sampling is a good technique to determine the severity of the problem and the potential savings in the indirect and expense areas. It is not unusual to find that the work force is productively engaged only 40 to 50 percent of the time or even less. For example, in maintenance work, which represents a large share of the total indirect cost, the analyst may find that the following reasons are responsible for much of the time lost during the working day.

1. LOST TIME DUE TO INADEQUATE COMMUNICATION. It is quite common to find that incomplete and even incorrect job instructions are provided on the work order. This necessitates additional trips to the toolroom and supply room to obtain parts and tools that should have been available when the craftsman first started to work on the job. The work order that merely states "Repair leak in oil system" is indicative of poor planning and communication. The worker will need to know whether a new valve or a new pipe is needed, or whether a new gasket will do the job, or the valve will need repacking.

2. LOST TIME DUE TO THE UNAVAILABILITY OF PARTS, TOOLS, OR EQUIPMENT. If the craftsman does not have the facilities and parts to do the job, he will be obliged to improvise in a manner that will usually be wasteful of time and will frequently result in an inferior job. Proper planning will assure that the correct tools and equipment to perform the job are available and that suitable spares are supplied to the job site.

3. LOST TIME DUE TO THE INTERFERENCE OF PRODUCTION EMPLOYEES. With improper scheduling, maintenance employees may find that they are unable to begin a repair, service, or overhaul operation because the facility is still being used by production employees. This can result in the craftsman waiting idly by until the production department is ready to turn over the equipment.

4. LOST TIME DUE TO OVERMANNING THE MAINTENANCE JOB. In the experience of the writer, this has been one of the principal causes of lost time in maintenance work. Too often, a crew of three or four is supplied when two or three workers are really all that is needed.

5. LOST TIME BECAUSE THE WORK IS UNSATISFACTORY AND MUST BE REDONE. Poor planning will frequently result in an attitude of "this will get by" on the part of the mechanic. This will result in the repair work having to be redone.

6. LOST TIME DUE TO THE CRAFTSMAN HAVING TO WAIT FOR INSTRUCTIONS TO BEGIN THE NEXT JOB. With improper planning, the "wait for work" idle time can be significant. Good planning assures that there is ample work ahead of the maintenance man so that he seldom if ever will have to wait for the next job.

After good methods have been developed, they must be standardized and taught to the workers who will be using them. This seems obvious, but is too often neglected by management.

By giving as much attention to operation improvement in indirect work as has been given to operation improvement in direct work, industry can get good results.

Indirect labor standards

If standards are going to be established in indirect labor departments, such as clerical, maintenance, and toolmaking, it is essential that they be developed from standard data or formulas. Time would not permit using stopwatch techniques for each and every standard developed. However, both stopwatch studies and fundamental motion data will be needed in the development of indirect labor standards.

A time standard can be established on any operation or group of operations that can be quantified and measured. If the elements of the work performed by the electrician, blacksmith, boilermaker, sheet metal worker, painter, carpenter, millwright, welder, pipe fitter, material handler, toolmaker, and others doing indirect work, are broken down and studied, it will usually be possible to evolve equitable standards by summarizing the direct, transportation, and indirect elements. The tools used for establishing standards for indirect and expense work are identical with those used in establishing standards for direct work: time study, predetermined motion time standards, standard data, time formulas, and work sampling.

Thus, a standard can be established for such tasks as hanging a door, rewinding a one-horsepower motor, painting a centerless grinder, sweeping the chips from a department, or delivering a skid of 200 forgings. Standard times for each of these operations can be established by measuring the time required for the operator to perform the job, then performance rating the study and applying an appropriate allowance.

Careful study and analysis will reveal that crew balance and interference cause more unavoidable delays in indirect work than in direct work. Crew balance is the delay time encountered by one member of a crew while waiting for other members of the crew to perform elements of the job. Interference time is the time that a maintenance or other worker is delayed in waiting for other workers to do necessary work. Both crew balance and interference delays may be thought of as unavoidable delays; however, they are usually characteristic of indirect labor operations only, such as those performed by maintenance workers. Queuing theory is a useful tool for the best estimate of the magnitude of the waiting time.

Because of the high degree of variability characteristic of most maintenance and material handling operations, it will be necessary to conduct a sufficient number of independent time studies of each operation to assure completely that average conditions have been determined and that the resulting standard is representative of the time needed for the normal operator to do the job under those average conditions. For example, if a study indicates that 47 minutes are required to sweep a machine floor 60 feet wide by 80 feet long, it will be necessary to assure that average conditions prevailed when the study was taken. Obviously, the work of the sweeper would be considerably more time consuming if the shop were machining cast iron rather than an alloy steel. Not only is alloy steel much cleaner, but alloy steel chips are easier to handle, and the use of slower speeds and feed would result in fewer chips. If the standard of 47 minutes was established when the shop was working with alloy steel, it would be inadequate when the department was producing cast-iron parts. It would be necessary to take additional time studies to assure that average conditions had been determined and that the resulting standards were representative of those conditions.

In a similar manner, standards can be established for painting. Thus, a standard will be determined for overhead painting based on square footage. Likewise, vertical and floor painting can be measured, and time standards can be developed on a square-footage basis.

Just as the automotive industry has established time standards for common repair jobs, such as grinding valves, replacing piston rings, and adjusting brakes, so can standards be established for typical maintenance and repair operations performed on the machine tools within a plant. Thus, a set of time standards can be determined for rewinding fractional horsepower motors. A standard can be established for rewinding a $\frac{1}{8}$-HP motor, a $\frac{1}{4}$-HP motor, a $\frac{1}{2}$-HP motor, a $\frac{3}{4}$-HP motor, and so on.

Toolroom work is very similar to the work done in job shops. The method by which such tools as a drill jig, milling fixture, form tool, or die will be made can be predetermined quite closely. Consequently, it is possible for the analyst to use time study and/or predetermined elemental times to establish a sequence of elements and to measure the normal time required for each element. Work sampling provides an adequate tool to determine the allowances that must be added for fatigue, personal, unavoidable, and special delays. The standard elemental times thus developed can be tabularized in the form of standard data and can be used to design time formulas for pricing future work.

Factors affecting indirect and expense standards

All work that can be classified as indirect and/or expense may be thought of as being a combination of four divisions: (1) direct work, (2) transportation, (3) indirect work, (4) and unnecessary work and delays.

The direct work portion of indirect and/or expense labor is that segment of the operation which discernibly advances the progress of the work. For example, in the installation of a door, the work elements may include: cut door to rough size, plane to finish size, locate and mark hinge areas, chisel out hinge areas, mark for screws, install screws, mark for lock, drill out for lock, and install lock. Such direct work can be measured quite easily by using conventional techniques, such as stopwatch time study, standard data, or fundamental motion data.

Transportation refers to the work performed in movements during the course of the job or from job to job. Transportation may take place horizontally or vertically, or both. Typical transportation elements include: walk up and down stairs, ride elevator, walk, carry load, push truck, and ride on motor truck. Transportation elements similar to direct elements can easily be measured, and standard data established. For example, one company uses 0.50 minutes per 100-foot zone as its standard for horizontal travel time and 0.30 minutes as its standard time for 10 feet of vertical travel.

As a general rule, the indirect portion of indirect and/or expense labor cannot be evaluated by physical evidence in the completed job or at any stage during the work except by deductive inferences from certain characteristic features of the job. Indirect work elements may be separated into three division: (*a*) tooling, (*b*) material, and (*c*) planning.

Elements of the work that may be assigned to tooling are those having to do with the acquisition, disposition, and maintenance of all the tools needed to perform an operation. Typical elements under this category would include: getting and checking tools and equipment; returning tools and equipment at the completion of the job; cleaning tools; and repairing, adjusting, and sharpening tools. The tooling work elements are easily measured by conventional means, and statistical records can provide data on the frequency of their occurrence. Waiting line or queuing theory can provide information on the expected waiting time at supply centers.

The elements associated with material are those having to do with acquiring and checking the material used in an operation and disposing of scrap. Making minor repairs to materials, picking up and disposing of scrap, and getting and checking materials, are characteristic of the work elements in this division. As with tooling elements, the elements associated with material may be measured readily, and their frequency accurately determined through historical records. In acquiring material from storerooms, queuing theory provides the best estimate of waiting time.

The planning elements represent the most difficult area in which to establish standards. Here we have such elements as consultation with the foreman and planning work procedures, inspection, checking, and testing. Work sampling techniques provide a basis for determining the time required to perform the planning elements. The planning elements may also be measured through stopwatch procedures, but their frequency of occurrence is difficult to determine; consequently, work sampling is the more practical tool to use for this class of elements.

Unnecessary work and delays represent the portion of the cycle that should be eliminated through planning and methods improvement. It is not unusual to find that unnecessary work and delays represent as much as 40 percent of the indirect and expense payroll. Much of this wasted time is management oriented, and it is for this reason that the systematic procedure recommended in Chapter 1 should be carefully followed prior to establishing standards. In this work, the analyst should get and analyze the facts, and develop and install the method, before endeavoring to establish the standard.

Much of the delay time encountered in indirect and expense work is due to queues. Thus, the worker is obliged to stand in line at the tool crib, the storeroom, or the stockroom, waiting for a forklift truck, a desk calculator, or some other piece of office equipment. The analyst is frequently able to determine the optimum number of servicing units required in the given circumstances through the application of queuing theory. Such studies should be made to optimize the delay time in indirect and expense work.

Basic queuing theory equations

Queuing system problems may occur when a flow of arriving traffic (people, facilities, and so on) establishes a random demand for service at facilities (people, machines, and so on) of limited service capacity. The time interval between arrival and service will obviously vary inversely with the level of the service capacity. The greater the number of service stations and the faster the rate of servicing, the less the time interval between arrival and service.

The methods and work measurement analyst should endeavor to select an operating procedure that minimizes the total cost of operation. There must be an economic balance between waiting times and service capacity.

The following four characteristic define queuing problems:

1. THE PATTERN OF ARRIVAL RATES. The arrival rate (for example, a machine breaking down for repair) may be constant or random. If random, the pattern should be described in terms of a probability distribution of the various values of the intervals between successive arrivals. A probability distribution of the random pattern may be definable or nondefinable.

2. THE PATTERN OF THE SERVICING RATE. The servicing time may also be constant or random. If random, the analyst should endeavor to define the probability distribution that fits the service time random pattern.

3. THE NUMBER OF SERVICING UNITS. In general, multiple-service queuing problems are more complex than the problems of single-service systems. However, most problems are of the multiple-service type, such as the number of mechanics that should be employed to keep a battery of machines in operation.

4. THE PATTERN OF SELECTION FOR SERVICE. Service is usually on a "first-come, first-served" basis; however, in some cases, the selection may be completely random or according to priorities.

The methods of solution to queuing systems may be classified into two broad categories: analytic and simulation. The analytic category covers a wide range of problems for which mathematical probability and analytic techniques have provided equations representing systems with various assumptions about the above characteristics. One of the most common assumptions about the arrival pattern or arrivals per unit time interval is that they follow the Poisson probability distribution:

$$p(k) = \frac{a^k e^{-a}}{k!}$$

where a is the mean arrival rate, and k is the number of arrivals per time interval. A helpful graphic presentation of the cumulative Poisson probabilities is shown in Figure 22–2.

A further consequence of a Poisson-type arrival pattern is that the random variable time between arrivals obeys an exponential distribution with the same parameter a. Being a continuous distribution, the exponential distribution has a density function $f(x) = ae^{-ax}$. The exponential distribution has a

FIGURE 22–2

Poisson distribution of arrivals

$$P(c, a) = \sum_{x=c}^{\infty} \frac{e^{-a}a^x}{x!}$$

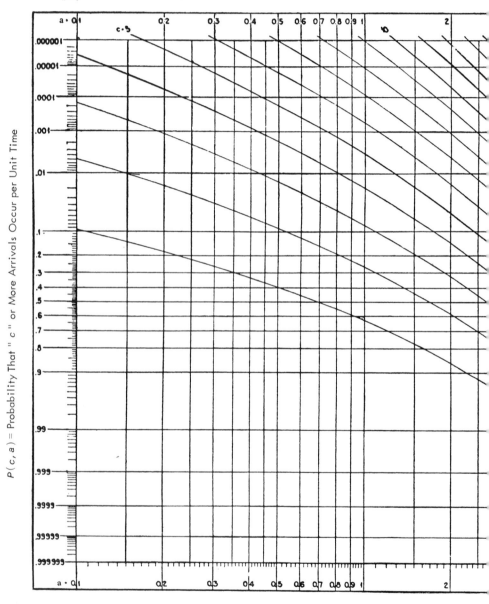

a = Average Number of Arrivals per Unit Time

;URE 22–2 (continued)

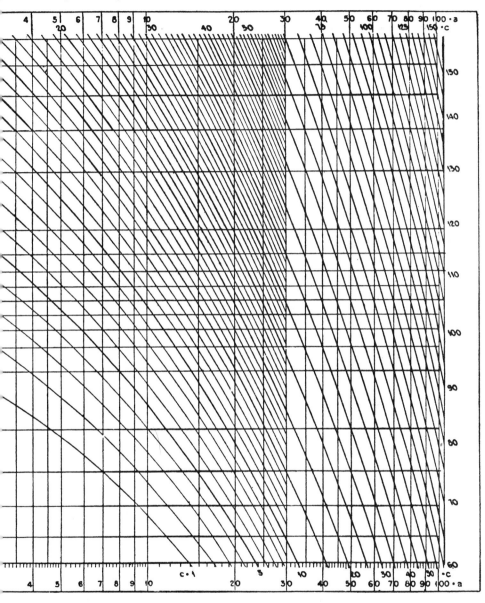

The Port of New York Authority

FIGURE 22–3
The exponential distribution

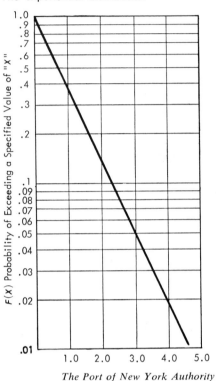

The Port of New York Authority

mean $\mu = \dfrac{1}{a}$, and a variance $= \dfrac{1}{a^2}$. μ can be recognized as the mean interval between arrivals. In some queuing systems the number of services per unit time may follow the Poisson pattern, and subsequently the service times follow the exponential distribution. Figure 22–3 illustrates the exponential curve $F(x) = e^{-x}$ which shows the probabilities of exceeding various multiples of any specified service time.

The basic equations governing the queue applicable to Poisson arrivals and service in arrival order may be classified into five categories. These are as follows:

1. Any service time distribution and a single server.
2. Exponential service time and a single server.
3. Exponential service time and finite servers.
4. Constant service time and a single server.
5. Constant service time and finite servers.

Equations have been developed for each of the above categories. These provide quantitative answers to such problems as mean delay time in the waiting line and mean number of arrivals in the waiting line.

Let us follow through on a typical example to see how queuing theory is applied. The analyst wishes to determine a standard for inspecting the hardness of large motor armatures. The time is composed of two distinct quantities: the time the inspector takes to make his Rockwell observations and the time the operator must wait until the next armature shaft is made available for inspection. The following assumptions will apply: (1) single server; (2) Poisson arrivals; (3) arbitrary servicing time; and (4) first-come, first-served discipline.

This is a situation which fits into the first category classified above. The equations which apply are:

$$a. \quad P > 0 = u$$

$$b. \quad w = \left[\frac{uh}{2(1 - u)}\right]\left[1 + \left(\frac{\sigma}{h}\right)^2\right]$$

$$c. \quad m = \frac{w}{P > 0} = \frac{w}{u}$$

where:

a = Average number of arrivals per unit of time
h = Mean servicing time
w = Mean waiting time of all arrivals
m = Mean waiting time of delayed arrivals
n = Number of arrivals present (both waiting and being served) at any given time.
s = Number of servers
u = Servers' occupancy ratio = $\dfrac{ah}{s}$
σ = Standard deviation of the servicing time
$P(n)$ = Probability of n arrivals being present at any random time
$P(\geq n)$ = Probability of at least n arrivals being present at any random time
t = Unit of time
$P > t/h$ = Probability of a delay greater than t/h multiples of the mean holding time
$P(d > 0)$ Probability of any delay (delay greater than 0)
L = Mean number of waiting individuals among all individuals

Through stopwatch time study, a normal time of 4.58 minutes per piece is established for the actual hardness testing. The standard deviation of the service time is 0.82 minutes, and 75 tests are made in every eight-hour working day. From the above data, we have:

$s = 1$

$a = \dfrac{75}{480} = 0.156$

$h = 4.58$

$\sigma = 0.82$

$u = (0.156)(4.58) = 0.714$

$w = \left[\dfrac{(0.714)(4.58)}{2(1 - 0.714)}\right]\left[1 + \left(\dfrac{0.82}{4.58}\right)^2\right]$

$= 5.95$ minutes mean waiting time of an arrival

Thus, the analyst is able to determine a total time of 10.53 minutes per shaft.

Many industrial problems fit into the second category; that is, exponential service and a single server with Poisson arrivals. The equations that apply here are:

a. $P > 0 = u$

b. $P > (t/h) = ue^{(u-1)(t/h)}$

c. $P(n) = (1 - u)(u)^n$

d. $P(\geqq n) = u^n$

e. $w = \dfrac{h(P > 0)}{(1 - u)} = \dfrac{uh}{1 - u}$

f. $m = \dfrac{w}{(P > 0)} = \dfrac{h}{(1 - u)}$

g. $L = \dfrac{m}{h} = \dfrac{1}{(1 - u)}$

For example, the arrivals at a tool crib are considered to be Poisson, with an average time of seven minutes between one arrival and the next. The length of the service time at the crib window is assumed to be distributed exponentially, with a mean of 2.52 minutes determined through stopwatch time study. The analyst wishes to determine the probability that a person arriving at the crib will have to wait, and the average length of the queues that form from time to time so that he can evaluate the practicality of opening a second tool-dispensing window. This information can be determined from the equations summarized above.

$a = 0.14$ average number of arrivals per minute

$h = 2.52$ minutes mean service time

$s = 1$ server

$P > 0 = u$

$u = \dfrac{ah}{s} = 0.35$ probability of a person arriving at crib having to wait

$L = \dfrac{1}{(1 - u)} = 1.52$ average length of the queue formed

Monte Carlo simulation

The method of simulation used for the solution of queuing system problems does not employ formal mathematical models, as the models may be too complex or the systems too unique in light of the mathematical models in existence. Here a model of a queuing system is created by a series of numerical statements. A sample set of input values drawn from specified arrival and service time distributions are introduced. These input data generate a sample output distribution of waiting line results for the period. This sampling procedure is referred to as "Monte Carlo."

The expected waiting time and service times can be estimated and an optimum solution developed through a proper balance of service stations, servicing rates, and arrival rates. This technique is most helpful to analyze the waiting line problem involved in the centralization-decentralization storage location of tools, supplies, and service facilities.

The Monte Carlo simulation technique can also be applied to problems with a Poisson arrival pattern. However, Monte Carlo is generally used for problems from which neither standard nor empirical formulas are obtainable.

For example, the analyst may wish to determine the number of operators to be used to set up machines in an automatic screw machine section to minimize total costs. An analysis of past records in the automatic department reveals the following probability distributions of the number of work stoppages per hour and the time required to service a machine.

Labor rate	$ 6.00 per hour
Machine rate.	$24.00 per hour
Number of operators	3
Number of automatics	15

Work stoppages per hour	Probability
0	0.108
1	0.193
2	0.361
3	0.186
4	0.082
5	0.040
6	0.018
7	0.009
8	0.003
	1.000

Hours to get machine into operation	Probability
0.100	0.111
0.200	0.254
0.300	0.009
0.400	0.007
0.500	0.005
0.600	0.008
0.700	0.105
0.800	0.122
0.900	0.170
1.000	0.131
1.100	0.075
1.200	0.003
	1.000

Note that the time to get a machine into operation results in a bimodal distribution and does not conform to any standard distribution.

The analyst now assigns blocks of three-digit numbers, in direct proportion to the probabilities associated with the data for both the arrival and the service rates.

So 1,000 three-digit numbers have been assigned to both the arrival and the service distributions. By using a table of random numbers, it is

Work stoppages per hour	Probability	Numbers assigned
0	0.108	000–107
1	0.193	108–300
2	0.361	301–661
3	0.186	662–847
4	0.082	848–929
5	0.040	930–969
6	0.018	970–987
7	0.009	988–996
8	0.003	997–999

Hours to get machine into operation	Probability	Numbers assigned
0.100.	0.111	000–110
0.200.	0.254	111–364
0.300.	0.009	365–373
0.400.	0.007	374–380
0.500.	0.005	381–385
0.600.	0.008	386–393
0.700.	0.105	394–498
0.800.	0.122	499–620
0.900.	0.170	621–790
1.000.	0.131	791–921
1.100.	0.075	922–996
1.200.	0.003	997–999

possible to simulate the behavior of what would be expected in the screw machine section over a period. Let us take 80 random observations, to simulate the number of work stoppages occurring during two weeks of activity on the floor. These 80 random numbers applied to work stoppages give the following:

Hour	Random number	Work stoppages	Hour	Random number	Work stoppages
1 221		1	14 369		2
2 . . . . : 193		1	15 188		1
3 167		1	16 885		4
4 784		3	17 097		0
5 032		0	18 . . . . : . 129		1
6 932		5	19 859		4
7 787		3	20 386		2
8 236		1	21 534		2
9 153		1	22 407		2
10 587		2	23 021		0
11 573		2	24 951		5
12 485		2	25 357		2
13 619		2	26 262		1

Hour	Random number	Work stoppages	Hour	Random number	Work stoppages
27 778		3	54 204		1
28 464		2	55 733		3
29 375		2	56 071		0
30 616		2	57 923		4
31 934		5	58 995		7
32 219		1	59 938		5
33 952		5	60 184		1
34 978		6	61 245		1
35 699		3	62 220		1
36 043		0	63 071		0
37 610		2	64 297		1
38 859		4	65 579		2
39 217		1	66 333		2
40 156		1	67 494		2
41 028		0	68 651		2
42 871		4	69 920		4
43 988		7	70 987		6
44 100		0	71 004		0
45 479		2	72 579		2
46 228		1	73 125		1
47 678		2	74 315		2
48 276		1	75 961		5
49 337		2	76 854		4
50 131		1	77 728		3
51 689		3	78 913		4
52 137		1	79 775		3
53 096		0	80 372		2

A different group of random numbers should be selected as input for each work stoppage to estimate the time required to get a machine into operation after it has stopped. These values indicate the following:

Hour	Random number	Hours to get machine into operation	Hour	Random number	Hours to get machine into operation
1 341		0.200	18 794		1.000
2 112		0.200	19 383		0.500
3 273		0.200	20 472		0.700
4 106		0.100	21 870		1.000
5 597		0.800	22 397		0.700
6 337		0.200	23 280		0.200
7 871		1.000	24 976		1.100
8 728		0.900	25 693		0.900
9 739		0.900	26 870		1.000
10 799		1.000	27 520		0.800
11 202		0.200	28 527		0.800
12 854		1.000	29 153		0.200
13 599		0.800	30 851		1.000
14 726		0.900	31 411		0.700
15 880		1.000	32 822		1.000
16 496		0.700	33 985		1.100
17 128		0.200	34 997		1.200

Hour	Random number	Hours to get machine into operation	Hour	Random number	Hours to get machine into operation
35	234	0.200	58	160	0.200
36	773	0.900	59	096	0.100
37	423	0.700	60	653	0.900
38	607	0.800	61	876	1.000
39	220	0.200	62	086	0.100
40	915	1.000	63	085	0.100
41	684	0.900	64	898	1.000
42	366	0.300	65	422	0.700
43	220	0.200	66	164	0.200
44	920	1.000	67	255	0.200
45	346	0.200	68	145	0.200
46	347	0.200	69	144	0.200
47	639	0.900	70	322	0.200
48	308	0.200	71	610	0.800
49	319	0.200	72	855	1.000
50	848	1.000	73	167	0.200
51	562	0.800	74	463	0.700
52	481	0.700	75	886	1.000
53	000	0.100	76	081	0.100
54	587	0.800	77	826	1.000
55	275	0.200	78	566	0.800
56	267	0.200	79	228	0.200
57	430	0.700	80	491	0.700

In light of the above-indicated outcomes for 80 hours of operation in our automatic screw machine department, we can look at the machine downtime predicted because of insufficient operators.

Hour	Random number	Work stoppages	Random number	Hours to get machine into operation	Operators available for next work stoppages	Downtime hours because of lack of operators for servicing
1	221	1	341	0.200	2	
2	193	1	112	0.200	2	
3	167	1	273	0.200	2	
4	784	3	106	0.100	2	
			597	0.800	1	
			337	0.200	0	
5	032	0	—	—	3	
6	932	5	871	1.000	2	
			728	0.900	1	
			739	0.900	0	
			799	1.000	0	0.9
			202	0.200	0	0.9
7	787	3	854	1.000	0	
			599	0.800	0	0.9
			726	0.900	0	0.1
8	236	1	880	1.000	1	
9	153	1	495	0.700	2	

Hour	Random number	Work stoppages	Random number	Hours to get machine into operation	Operators available for next work stoppages	Downtime hours because of lack of operators for servicing
10 587	2	128	0.200	2	
		794	1.000	1	
11 573	2	383	0.500	2	
		472	0.700	1	
12 485	2	870	1.000	2	
		397	0.700	1	
13 619	2	280	0.200	2	
		976	1.100	1	
14 369	2	693	0.900	2	
		870	1.000	1	
15 188	1	520	0.800	2	
16 885	4	527	0.800	2	
		153	0.200	1	
		851	1.000	0	
		411	0.700	0	0.200
17 097	0	–	–	3	
18 129	1	985	1.100	2	
19 859	4	997	1.200	1	
		234	0.200	1	0.100
		773	0.900	0	
		423	0.700	0	0.30
20 386	2	607	0.800	1	
		220	0.200	0	
21 534	2	915	1.000	2	
		684	0.900	1	
22 407	2	366	0.300	2	
		220	0.200	1	
23 021	0	–	–	3	
24 951	5	346	0.200	2	
		347	0.200	1	
		639	0.900	0	
		308	0.200	1	0.200
		319	0.200	0	0.200
25 357	2	898	1.000	2	
		562	0.800	1	
26 262	1	481	0.700	2	
27 778	3	000	0.100	2	
		587	0.800	1	
		275	0.200	0	
28 464	2	267	0.200	2	
		430	0.700	1	
29 375	2	160	0.200	2	
		096	0.100	1	
30 616	2	653	0.900	2	
		876	1.000	1	
31 934	5	086	0.100	2	
		085	0.100	1	
		898	1.000	0	
		422	0.700	0	0.100
		920	1.000	0	0.100
32 219	1	164	0.200	1	
33 952	5	255	0.200	2	
		145	0.200	1	

Hour	Random number	Work stoppages	Random number	Hours to get machine into operation	Operators available for next work stoppages	Downtime hours because of lack of operators for servicing
			144	0.200	0	
			322	0.200		0.200
			610	0.800		0.200
34 978		6	855	1.000	2	
			167	0.200	1	
			463	0.700	0	
			886	1.000	0	0.200
			081	0.100	0	0.700
			826	1.000	0	0.800
35 699		3	566	0.800	0	
			228	0.200	0	0.2
			491	0.700	0	0.8
36 043		0	623	0.900	1	
37 610		2	188	0.200	2	
			702	0.900	1	
38 859		4	323	0.200	2	
			354	0.200	1	
			464	0.700	0	
			806	1.000	0	0.200
39 217		1	494	0.700	1	
40 156		1	050	0.100	2	
41 871		0	–	–	3	
42 871		4	441	0.700	2	
			390	0.600	1	
			543	0.800	0	
			077	0.100	0	0.600
43 988		7	941	1.100	2	
			667	0.900	1	
			317	0.200	0	
			336	0.200	0	0.20
			650	0.900	0	0.40
			651	0.900	0	0.90
			370	0.300	0	1.40
44 100		0	–	–	0	
45 479		2	011	0.100	2	
			976	1.100	1	
46 228		1	229	0.200	1	
47 678		3	072	0.100	2	
			167	0.200	1	
			144	0.200	0	
48 276		1	843	1.000	2	
49 337		2	768	0.900	2	
			404	0.700	1	
50 131		1	636	0.900	2	
51 689		3	970	1.100	2	
			954	1.110	1	
			951	0.110	0	
52 137		1	562	0.800	0	0.100
53 096		0	–	–	0	
54 204		1	866	1.000	2	
55 733		3	348	0.200	2	
			327	0.200	1	
			995	1.100	0	

Hour	Random number	Work stoppages	Random number	Hours to get machine into operation	Operators available for next work stoppages	Downtime hours because of lack of operators for servicing
56 071	0	998	1.200	1	
		682	0.900	0	
		005	0.100	0	0.100
57 923	4	539	0.800	1	0.200
		616	0.800	0	
		181	0.200	0	
		647	0.900	0	0.200
58 995	7	273	0.200	2	
		683	0.900	1	
		623	0.900	0	
		018	0.100	0	0.200
		748	0.900	0	0.300
		408	0.700	0	0.900
		688	0.900	0	0.900
59 938	5	297	0.200	0	0.600
		985	1.100	0	0.600
		777	0.900	0	0.800
		294	0.200	0	0.800
		417	0.700	0	1.000
60 184	1	509	0.800	0	0.700
61 245	1	571	0.800	1	
62 220	1	925	1.100	2	
63 071	0	–	–	2	
64 297	1	753	0.900	2	
65 579	2	448	0.700	2	
		060	0.100	1	
66 333	2	899	1.000	2	
		238	0.200	1	
67 494	2	872	1.000	2	
		051	0.100	1	
68 651	2	123	0.200	2	
		886	1.000	1	
69 920	4	001	0.100	2	
		914	1.000	1	
		049	0.100	0	
		150	0.200	0	0.100
70 987	6	128	0.200	2	
		229	0.200	1	
		356	0.200	0	
		507	0.800	0	0.200
		114	0.200	0	0.200
		342	0.200	0	0.200
71 004	0	–	–		
72 579	2	513	0.800	2	
		361	0.200	1	
73 125	1	711	0.900	2	
74 315	2	619	0.800	2	
		558	0.800	1	
75 961	5	630	0.900	2	
		694	0.900	1	
		482	0.700	1	
		059	0.100	0	0.700
		321	0.200	0	0.800

				Hours to get	*Operators available for next*	*Downtime hours because of lack of operators*
Hour	*Random number*	*Work stoppages*	*Random number*	*machine into operation*	*work stoppages*	*for servicing*
76 854		4	027	0.100	2	
			426	0.700	1	
			161	0.200	0	
			524	0.800	0	0.100
77 728		3	572	0.800	2	
			780	0.900	1	
			298	0.200	0	
78 913		4	556	0.800	2	
			098	0.100	1	
			896	1.000	0	
			509	0.800	0	0.100
79 775		3	654	0.900	2	
			340	0.200	1	
			600	0.800	0	
80 372		2	996	1.100	2	
			651	1.100	1	
						Total 19.6

From the simulated model, we can expect to have 19.6 hours of machine downtime every two weeks because of the lack of an operator. At a machine rate of $24 per hour, this amounts to a weekly cost of $235.20. Since the cost of a fourth operator would result in an added weekly direct labor cost of $240 (40 × 6), the present setup of three men to service the 15 automatic screw machines appears to be the optimum solution.

Expense standards

More and more, management is recognizing its responsibility to determine accurately the appropriate office forces for a given volume of work. To control office payrolls, management must develop time standards, since they are the only reliable "yardsticks" for evaluating the size of any task.

As for other work, methods analysis should precede work measurement in all expense operations. The flow process chart is the ideal tool for presenting the facts of the present method. Once presented, the present method should be reviewed critically in all its details. Such factors as the purpose of the operation, design of forms, office layout, elimination of delays resulting from poor planning and scheduling, and adeqacy of existing equipment should be considered; the primary approaches to operation analysis should be used.

After the completion of a thorough methods program, standards development can commence. Many office jobs are repetitive; consequently, it is not particularly difficult to set fair standards. Central typing pools, billing groups, file clerks, addressograph operators, punch card operators, and tabulating machine operators are representative groups that readily lend themselves to work measurement either by stopwatch or motion-picture techniques. In studying office work, the analyst should carefully identify element end points, so that standard data may be established for pricing future work. For example, in typing production orders, the following elements of work normally occur on each page of each order typed:

1. Pick up production order from pile and position in typewriter.
2. Pick up sheet to be copied from, and place in Copy Right.
3. Read product order instructions.
4. Type heading on order:
    a. Date.
    b. Number of pieces.
    c. Material.
    d. Department.

Once standard data have been developed for most of the common elements used in an office, time standards can be achieved quite rapidly and economically. Of course, many clerical positions within the office are made up of a series of diversified activities that do not readily lend themselves to time study. Such work is not made up of a series of standard cycles that continually repeat themselves and, consequently, is more difficult to measure than are direct labor operations. Because of this characteristic of some office routines, it is necessary to take many time studies, each of which may be but one cycle in duration. Then, by calculating all the studies taken, the analyst develops a standard for typical or average conditions. Thus, a time standard based upon the page may be calculated for copy typing. Granted, some pages of technical typing requiring the use of symbols, radicals, fractions, formulas, and other special characters or spacing will take considerably longer than will typing a routine page. But if the technical typing is not representative of average conditions, it will not result in an unfair influence on the operator's performance over a period of time; simple typing and shorter than standard pages will tend to balance the extra time needed to type complex letters.

It is usually not practical to establish standards on office positions that require creative thinking. Thus, such jobs as tool designing and product designing should be carefully considered before a decision is made to establish time standards on the work done by the employees filling those jobs. If standards are established on work of this nature, they should be used for such purposes as scheduling, control, or labor budgeting and should not be used for purposes of incentive wage payment. Work demanding creative thinking can be retarded if pressure is put on the employee. The result may be inferior designs that can be more costly to the business than the amount saved through the greater productivity of the designer.

In setting standards for office workers, the analyst will usually find that the white-collar worker will not like the practice, since he is not accustomed to having his work measured. Therefore, it is important that the same observance of good human relations that is practiced in the shop be adhered to in the office.

Supervisory standards

By establishing standards for supervisory work, it is possible to determine equitable supervisor loads and to maintain a proper balance between supervision, facilities, clerical employees, and direct labor.

The work sampling technique is probably the most fruitful tool for the development of supervisory standards. The same information could also be attained through all-day stopwatch studies, but the cost of obtaining reliable data would usually be prohibitive. Supervisory standards can be expressed in terms of "effective machine running hours" or some other benchmark.

For example, one study of a manufacturer of vacuum tubes revealed that 0.223 supervisory hours were required per machine running hour in a given department (Figure 22–4). The work sampling study showed that, out of 616 observations, the supervisor was working with the grid machines, inspecting grids, doing desk work, supplying material, walking, or engaging in activities classified as miscellaneous allowances a total of 519 times. This figure converted to prorated hours revealed that 518 indirect hours were required while 2,461 machine running hours took place. When a 6 percent personal allowance was added, a standard of 0.223 supervisory hours per machine running hour was computed.

$$\frac{518}{(2,461)(1.00 - 0.06)} = 0.223$$

Thus, if one supervisor handled a department of a given type of facility operating 192 machine running hours a week, his efficiency would be:

$$\frac{192 \times 0.223}{40 \text{ (hrs./week)}} = 107 \text{ percent}$$

Standard data on indirect and expense labor

The development of standard data for establishing standards on indirect and expense labor operations is quite feasible. In fact, in view of the diversification of indirect labor operations, standard data are, if any-

GURE 22–4

				INDIRECT LABOR STANDARD				
JOB- SUPERVISION						DEPT.- GRID	DATE- 4-16	

Cost Center	Number of Obser- vations	Percent of Obser- vations	Prorated Hours	Base Indirect Hours	Effective Machine Running Hours (EMRH)	Direct Labor Hours		Base Indirect Hours per Machine Running Hour (includes 6% personal)
Grid Machines	129	21	130	130	2,461			0.223
Inspect Grids	161	26	160	160				
Desk Work	54	9	56	56				
Supply Material	18	3	18	18				
Misc. Allow.	150	24	148	148				
Walking	7	1	6	6				
Out of Dept.	11	2	12					
Idle	86	14	86					
Total	616	100	616	518	2,461			0.223

thing, more appropriate on office, maintenance, and other indirect work than on standardized production operations.

As individual time standards are calculated, it is wise to tabularize the elements and their respective allowed times for future reference. As the inventory of standard data is built up, the cost of developing new time standards will decline proportionally.

For example, tabularizing standard data for forklift truck operations can be based on only six different elements: travel, brake, raise fork, lower fork, tilt fork, and the manual elements required to operate the truck. Once standard data have been accumulated for each of these elements (through the required range), the standard time required to perform any fork truck operation may be determined by summarizing the applicable elements. In a similar manner, standard data can readily be established on janitorial work elements, such as sweep floor; wax and buff floor; dry mop; wet mop; vacuum rugs; clean, dust, and mop lounge.

The maintenance job of "inspecting seven fire doors in a plant and make minor adjustments" can readily be estimated from standard data. For example, the Department of the Navy developed this standard:

Operation	Unit time (hours)	No. units	Total time (hours)
Inspection of fire door, roll-up type (manual chain, crankshaft, or electrically operated) fusible link. Includes minor adjustment.	0.170	7	1.190
Walk 100 feet between each door, obstructed walking	0.00009	600	0.054
			1.24

The standard data time of 0.00009 hours per foot of obstructed walking was established from fundamental motion data time, and the inspection time of 0.170 hours per fire door from stopwatch time study.

Several manufacturers of material handling equipment have taken detailed studies of their product, and provide standard data applicable to their equipment when it is purchased. This information will save the analyst many hours in developing material handling time standards.

Universal indirect standards

Where maintenance and other indirect operations are numerous and diversified, the task of developing standard data and/or formulas to preprice all indirect operations may appear to be more costly than the expected savings brought about by introducing time standards. In order to reduce the number of different time standards for indirect operations, some engineers have sought to develop universal indirect standards.

The principle behind universal standards is assigning the major proportion of indirect operations (perhaps as much as 90 percent) to appropriate groups. Each group will have its own standard, which will be the average time for all indirect operations assigned to the group. For example, group A may include the indirect operations of replacing defective union, repairing door (replace two hinges), replacing limit switch, and replacing two sections (14 feet of 1-inch pipe). The standard time for any indirect operation performed in group A may be 48 minutes. This time will represent the mean ($\bar{x}$) of all jobs within the group, and the dispersion of the jobs within the group for $\pm 2\sigma$ will be some predetermined percentage of $\bar{x}$ (perhaps ± 10 percent).

Standards may be established for each group by developing many (several hundred) time standards using standard data, formulas, and stopwatch procedures for typical indirect operations. These standards will be arranged in ascending sequence and then be bracketed in sufficient divisions so that the range of any bracket does not exceed the desired accuracy of any operation included in the bracket. Similar operations that have not been studied, but have had the method standardized, may be assigned to the bracket that is representative of their work elements.

This procedure may utilize as few as 20 standards to cover many thousands of different indirect operations. Since the majority of operations will cluster about the mean of their respective brackets, the total average error introduced by such standards will be slight. This will be especially true when indirect employees perform a broad variety of assignments, which is usually the case. For example, the following jobs have been given a standard of 4.4 hours by the Department of the Navy in conjunction with its engineered performance standards program. These jobs, along with many others, have been assigned to a specific category:

a. Fabricate eight anchor plates, 1 × 10 × 6 inches. Cut off eight pieces from milled steel stock with power hacksaw. Drill four 1-inch holes in corners, using Cincinnati 21-inch vertical single-spindle drill press.

b. Fabricate one 8-inch flat belt pulley with 2-inch hole through center, ¼-inch keyway, and four 1½-inch stress holes. Material used: 9 inches diameter × 3 inches wide cast-iron blank. Machines used: engine lathe, horizontal mill, and radial drill.

c. Fabricate one shaft to dimensions from 4½ inches outside diameter × 25 inches milled steel. Turn four diameters; face, thread, mill keyway; and drill center pin holes. Machines used: power hacksaw, lathe, horizontal mill, and single-spindle vertical drill press.

When studying a new part to be made, the analyst will be able to fit the job to a category where similar jobs have been studied and standards established.

To determine the validity of the universal standard approach, the author recently took 270 maintenance standards developed from the Department of the Navy standard data. These were then arranged in order of magnitude—from 0.07 hours to 24.31 hours. They were then divided into 18 groups, based upon normal distribution slotting. Each group had its own mean. These groups and their means in hours were as follows:

Group 1	0.07	Group 10.	1.31
Group 2	0.08	Group 11.	2.04
Group 3	0.11	Group 12.	2.89
Group 4	0.15	Group 13.	4.08
Group 5	0.19	Group 14.	5.66
Group 6	0.28	Group 15.	7.76
Group 7	0.39	Group 16.	12.52
Group 8	0.57	Group 17.	16.50
Group 9	0.85	Group 18.	24.30

Using simulation, the author used these 18 standards to establish the standards for maintenance jobs for a worker over a period of 25 weeks. The weekly standard hours established with the 18 maintenance standards were then compared with the weekly figures for the same jobs, based upon the Navy standard data.

The difference between the two sums for one week's work would be the error representative of using universal maintenance standards. The magnitude of this error was as follows:

Week	Percent error	Week	Percent error
1	7.18	14	−7.55
2	−6.93	15	2.37
3	−6.42	16	−0.87
4	4.03	17	−2.88
5	1.67	18	5.21
6	0.47	19	−1.52
7	−6.08	20	2.29
8	−3.37	21	2.48
9	−5.34	22	8.31
10	6.45	23	6.72
11	0.32	24	0.45
12	1.75	25	1.01
13	4.24		

Increasing the pay period from one week (40 hours) to two weeks (80 hours) would markedly reduce the cumulative error per pay period. The magnitude of the error would also decrease as the number of groups is increased.

The above analysis should give the reader confidence in the application of universal maintenance standards. The application of universal indirect standards offers an opportunity to introduce standards for a majority of indirect operations at a moderate cost, and also minimizes the cost of maintaining the indirect standards system.

Advantages of work standards on indirect work

Standards on indirect work offer distinct advantages to both the employer and the employee. Some of these advantages are:

1. The installation of standards will lead to many operating improvements.
2. The mere fact that standards are being established will result in better performance.
3. Indirect labor costs can be related to the work load regardless of fluctuations in the overall work load.
4. The budgeting of labor loads can be achieved.
5. The efficiency of various indirect labor departments can be determined.
6. The costs of such items as specific repairs, reports, and documents can be easily determined. This frequently results in the elimination of needless reports and procedures.

7. System improvements can be evaluated prior to installation. Thus, it is possible to avoid costly mistakes by choosing the right procedure.
8. It is possible to install incentive wage payment plans on indirect work, thus allowing employees to increase their earnings.
9. All indirect labor work can be planned and scheduled more accurately, with the obvious resulting benefits of getting the job done on time.
10. Less supervision is required, as a program of work standards tends to enforce itself. As long as each employee knows what is required of him, he will not arbitrarily waste or "kill" time.

Conclusion

It is more difficult to study and determine representative standard times for the nonrepetitive task characteristic of many indirect labor operations than for the task that is performed over and over again. Since indirect labor operations are difficult to standardize and study, they have not been widely subjected to methods analysis. Consequently, this area usually offers a greater percentage potential for reducing costs and increasing profits through methods and time study than does any other.

After good methods have been introduced and operator training has taken place, it is both readily possible and practical to establish standards on indirect labor operations. The usual procedure is to take a sufficiently large sample of stopwatch time studies to assure the representation of average conditions, and then to tabularize the allowed elemental times in the form of standard data. Fundamental motion data also have a wide application for establishing standards on indirect work. Thus, allowed operation times may be established consistently, accurately, and economically.

Indirect elements involving waiting time can be estimated accurately by using waiting line or queuing theory. It is important that the analyst have an understanding of elementary waiting line theory so that he can establish the mathematical model that fits the parameters of the problem. Where the problem does not fit established waiting line equations, the analyst should consider Monte Carlo simulation as a tool for determining the extent of the waiting line problem in the work area being studied.

The reader should recognize that there are no automatic means for determining a standard job time for a maintenance or other type of indirect or expense job. The analyst must review each job request and determine both the labor and material requirements. He usually goes through the following steps:

1. He examines the work request in detail. He may consult with the initiator of the work request or even visit the jobsite in order to determine the exact requirements of the job.

2. He prepares a material estimate for the job.
3. He studies the job from the standpoint of the work elements to be performed.
4. From standard data tables or universal time standards, he selects the appropriate direct work or task times and the applicable tooling, material, and planning times.
5. If no specifically applicable times exist, he assigns an estimated time from similar work shown on the master table of universal standard times or spread sheet.
6. After selection of the appropriate times, he adds standard times to cover the transportation times for the work order being analyzed. To this sum is added the necessary allowance to cover unavoidable delays, personal delays, and fatigue.

TEXT QUESTIONS

1. What would be the expected waiting time per shipment if it were established through stopwatch time study that the normal time to prepare a shipment was 15.6 minutes? Twenty-one shipments are made every shift (eight hours). The standard deviation of the service time has been estimated as 1.75 minutes. It is assumed that the arrivals are Poisson distributed and that the servicing time is arbitrary.

2. Using Monte Carlo methods, what would be the expected downtime hours because of the lack of an operator for servicing if four operators were assigned to the work situation described in the text?

3. Differentiate between indirect labor and expense labor.

4. Explain what is meant by "queuing theory."

5. What four divisions constitute indirect and expense work?

6. How are standards established on the "unnecessary and delays" portion of indirect and expense work?

7. The arrivals at the company cafeteria are Poisson, with an average time between arrivals of 1.75 minutes during the lunch period. The average time for a customer to obtain his lunch is 2.81 minutes, and this service time is distributed exponentially. What is the probability that a person arriving at the cafeteria will have to wait, and how long can he expect to wait?

8. Why has there been a marked increase in the number of indirect workers employed?

9. Why do more unavoidable delays occur in maintenance operations than on production work?

10. What is mant by crew balance? By interference time?

11. Explain how time standards would be established on janitorial operations.

12. What office operations are readily time studied?

13. Why are standard data especially applicable to indirect labor operations?

14. Summarize the advantages of standards established on indirect work.

15. Why is work sampling the best technique for establishing supervisory standards?
16. Explain the application of "slotting" under a universal standards system.
17. Why will a universal standards system involving as few as 20 benchmark standards work in a large maintenance department where thousands of different jobs are performed each year?

PROBLEMS

1. Through work measurement procedures, an average time of 6.24 minutes per piece is established on the inspection of a complex forging. The standard deviation of the inspection time is 0.71 minutes. Usually, 60 forgings are delivered to the inspection station on the line every eight-hour turn. One operator is used to perform this inspection.

   Assuming that the castings arrive in Poisson fashion and that the service time is exponential, what would be the mean waiting time of a casting at the inspection station? What would be the average length of the casting queue?

2. In the tool and die room of the Dorben Company, the work measurement analyst wishes to determine a standard for the jig boring of holes on a variety of molds. The standard will be used to estimate mold costs only, and it will be based upon operator wait time for molds coming from a surface grinding section and upon operator machining time. The wait time is based upon: a single server, Poisson arrivals, exponential service time, and first-come, first-served discipline.

   A study revealed the average time between arrivals to be 58 minutes. The average jig-boring time was found to be 46 minutes. What is the possibility of a delay of a mold at the jig borer? What is the average number of molds in back of the jig borer?

SELECTED REFERENCES

Crossman, Richard M., and Nance, Harold, W. *Master Standard Data: The Economic Approach to Work Measurement.* Rev. ed. New York: McGraw-Hill Book Co., 1972.

Lewis, Bernard T. *Developing Maintenance Time Standards.* Boston: Cahners, 1967.

Nance, Harold W., and Nolan, Robert E. *Office Work Measurement.* New York: McGraw-Hill Book Co., 1971.

Newbrough, E. T. *Effective Maintenance Management.* New York: McGraw-Hill Book Co., 1967.

Pappas, Frank G., and Dimberg, Robert A. *Practical Work Standards.* New York: McGraw-Hill Book Co., 1962.

23

Follow-up method and uses of time standards

Follow-up is the last of the nine systematic steps in installing a methods improvement program. Although follow-up is as important as any of the other steps, it is the most frequently neglected step. There is a natural tendency for the analyst to consider the methods improvement program complete after he or she developed time standards. However, a methods installation should never be considered complete. The follow-up is made principally to be assured that the proposed method is being followed, that the established standards are being realized, and that the new method is being supported by labor, supervision, the union and management. Follow-up will usually result in additional benefits accruing from new ideas and new approaches that will eventually stimulate the desire to embark again on a methods engineering program on the existing design or process. The procedure is to repeat the methods improvement cycle shortly after it is completed, so that each process and each design is continually being scrutinized for possible further improvement.

Without follow-up, it is easy for the proposed methods to revert to the original procedures. The writer has made innumerable methods studies where the follow-up revealed that the method under study was slowly reverting to or had reverted to the original method. Humans are creatures of habit, and a work force must develop the habit of the proposed method if that method is to be preserved. Continual follow-up is the only way to be assured that the new method is maintained long enough to have all those associated with its details become completely familiar with its routines.

Making the follow-up

The initial follow-up should take place approximately one month after the development of time standards for production jobs. A second follow-up should be made 3 months after the development of standards, and a third follow-up 6 to 12 months after the development of standards. As brought out on page 400, the frequency of the audit should be based upon the expected hours of application per year.

On each follow-up the analyst should review his original method report and the development of the standard to be certain that all aspects of the proposed method are being followed. At times he will find that portions of the proposed method are being neglected and that the worker has reverted to his old ways. When this has happened, the analyst should contact the foreman immediately and determine why the unauthorized change has taken place. If no satisfactory reasons can be given for going back to the old method, the analyst should insist that the correct procedure be followed. Tact coupled with firmness is essential, and the analyst will need to display salesmanship and technical competence.

Not only should the method be followed up, but a follow-up should also be made on the performance of the operator. A check should be made on his daily efficiency. It should be verified that he is continually operating at a performance greater than standard. His performance should be evaluated with typical learning curves for the class of work. If the operator is not making the progress anticipated, a careful study should be made, including a conference with the operator to discover what unforeseen difficulties he may have encountered.

A thorough and regular follow-up system will help assure the expected benefits from the proposed method.

METHODS OF ESTABLISHING STANDARDS

We have found that time standards may be determined in several ways:

1. By estimate (see Chapter 12).
2. By performance records (see Chapter 11).
3. By stopwatch time study (see Chapter 14).
4. By standard data (see Chapter 18).
5. By time study formulas (see Chapter 20).
6. By work sampling studies (see Chapter 21).
7. By queuing theory (see Chapter 22).

Methods 3, 4, 5, 6, and 7 will give considerably more reliable results than either method 1 or 2. If standards are to be used for wage payment, it is essential that they be determined as accurately as possible. Consequently, standards determined by estimate and by performance records will not suffice. Of course, standards developed by performance records and estimates are better than no standards at all, and frequently can be used to exercise controls throughout an organization.

All of these methods have application under certain conditions, and all have limitations on accuracy and cost of installation. With regard to the more reliable methods of measurement; (that is, methods 3, 4, 5, 6, and 7), the following summary may prove helpful.

Stopwatch time study

A. Advantages.
1. Enables the analyst to observe the complete cycle, thereby giving him or her an opportunity to suggest and initiate methods improvements.
2. Is the only method that actually measures and records the actual time taken by the operator.
3. Is more likely to provide coverage of those elements that occur less than once per cycle.
4. Quickly provides accurate values for machine-controlled elements.
5. Is relatively simple to learn and explain.
B. Disadvantages.
1. Requires the performance rating of the worker's skill and effort.
2. Does not force a detailed record of the total method being employed, including workplace layout, motion patterns, condition of materials, tools, and so forth.
3. May not provide accurate evaluation of noncyclic elements.
4. Bases the standard upon a small sample since the standard is determined by one analyst studying one worker who is following one method.
5. Requires that the work be performed before the standard is established.

Predetermined motion time data systems

A. Advantages.
1. Force detailed and accurate description of the workplace layout; motion patterns; and shape, size, and fit of components and tools.
2. Encourage work simplification to reduce standard times.
3. Eliminate performance rating.
4. Permit establishing methods and standards in advance of actual production.
5. Permit easy and accurate adjustments of time standards in order to accommodate minor changes in method.
6. Provide more consistent standards.
B. Disadvantages.
1. Depend on complete and accurate descriptions of the required methods for the accuracy of the time standard.
2. Require more time for the training of competent analysts.
3. Are more difficult to explain to workers, supervisors, and union officials.
4. May require more man-hours to establish standards for long-cycle operations.

5. Must use stopwatch, standard data, or formulas for process-controlled and machine-controlled elements.

Standard data, formulas, and queuing methods

A. Advantages.
   1. Eliminate performance rating.
   2. Establish consistent standards.
   3. Permit establishing methods and standards in advance of actual production.
   4. Allow standards to be established rapidly and inexpensively.
   5. Permit an easy adjustment of time standards in order to accommodate minor changes in method.
B. Disadvantages.
   1. May not accommodate small variations in method.
   2. Complex formulas may require a more skilled technician.
   3. Are more difficult to explain to workers than is stopwatch procedure.
   4. May result in significant inaccuracies if extended beyond the scope of the data used in their development.

Work sampling

A. Advantages.
   1. Eliminates tension caused by constant observation of the worker (when using stopwatch time study).
   2. Represents typical or average conditions over a period of time where conditions change from hour to hour or from day to day.
   3. Permits the simultaneous development of standards for a variety of operations.
   4. Is ideally suited to studies of machine utilization, activity analysis, and unavoidable and personal delays.
   5. Can be used with performance rating to determine standard times.
B. Disadvantages.
   1. Assumes that the worker uses an acceptable and standard method.
   2. Requires that the observer be able to identify and classify a wide range of delays and work activities.
   3. Should be confined to constant populations.
   4. Makes it more difficult to apply a correct performance rating factor than does stopwatch time study.
   5. Accuracy of time standard is dependent upon the number of random observations made as well as upon the accuracy of the classification and recording of individual observations.

6. Requires accurate records of the hours worked and the number of units produced.

A further summary of the kinds of situations that are best suited to the various techniques may be of assistance to the reader.

A. Stopwatch time study.
    1. Where there are repetitive work cycles of from short to long duration.
    2. Where new operations can be performed without standards until a study is performed.
    3. Where a wide variety of dissimilar work is performed.
    4. Where process control elements constitute a part of the cycle.
B. Predetermined motion-time data systems.
    1. Where the work is predominantly operator controlled.
    2. Where there are repetitive work cycles of short to medium durations.
    3. Where it is necessary to plan work methods, including line balancing, in advance of production.
    4. Where there has been controversy over the performance rating procedure.
    5. Where there has been controversy over the consistency of standards.
C. Standard data, formulas, and queuing methods.
    1. Where there is similar work of from short to long duration.
    2. Where there has been controversy over the performance rating procedure.
    3. Where there has been controversy over the consistency of standards.
D. Work sampling.
    1. Where it is necessary to establish delay allowances for various processes or departments.
    2. Where there is considerable difference in the work content from cycle to cycle, as in certain shipping, material handling, and clerical activities.
    3. Where activity studies are needed to show machine or space utilization, or the percentage of time spent on various activities.
    4. Where standards are needed for crew activities which vary from cycle to cycle.
    5. Where there is objection to stopwatch time study.

PURPOSES OF STANDARDS

In the operation of any manufacturing enterprise or business, it is fundamental that we have time standards. Time is the one common denominator from which all elements of cost may be evolved. In fact, everyone uses time standards for practically everything he does or wants anyone else to do. A man rising in the morning will allow himself one hour to wash, shave, dress, eat breakfast, and get to work. This surely represents a time standard that has a bearing on the number of hours of sleep he gets. A football game involves but 60 minutes of playing time. This also represents a standard. The student reading this chapter will allow himself so many minutes to cover the assignment. This, too, is a standard that will affect the student's program for the remainder of the day.

We are particularly interested in the use of time standards as applied to the effective operation of a manufacturing enterprise or business. Several of the results of time study will be discussed.

A basis for wage incentive plans

Time standards are usually thought of in their relation to wage payment. As important as this is, there are many other uses of standards in the operation of an enterprise. However, the need for reliable and consistent standards is more pronounced in connection with wage payment than in any other area. Without equitable standards, no incentive plan that endeavors to compensate in proportion to output can possibly succeed. If we have no yardstick, how can we measure individual performance? With standardized methods and standard times, we have a yardstick that will form a basis for wage incentive application.

A common denominator in comparing various methods

Since time is a common measure for all jobs, time standards are a basis for comparing various methods of doing the same piece of work. If, for example, we thought that it might be advantageous to install broaching on a close tolerance inside diameter rather than to ream the part to size as is currently being done, we could make a sound decision on the practicality of the change only if time standards were available. Without reliable standards, we would be groping in the dark.

A means for securing an efficient layout of the available space

Time is the basis for determining how much of each kind of equipment is needed. By knowing the exact requirements for facilities, we can achieve the best possible utilization of space. If we know that we shall require 10 milling machines, 20 drill presses, 30 turret lathes, and 6 grinders in a particular machining department, then we can plan for the layout of this equipment to best advantage. Without time standards, we may have overprovided for one facility and underprovided for another, inefficiently utilizing the space available.

Storage and inventory areas are determined from the length of time a part is in storage and the demand for the part. Again, time standards are the basis for determining the size of such areas.

A means for determining plant capacity

Through the medium of time standards, not only is machine capacity determined, but also department and plant capacity. It is a matter of simple arithmetic to estimate product potential once we know the available facility hours and the time required to produce a unit of product. For example, if the bottleneck operation in the processing of a given product required 15 minutes per piece, and if 10 facilities for this operation existed, then the plant capacity based on a 40-hour per week operation on this product would be:

$$\frac{40 \text{ hours} \times 10}{0.25 \text{ hours}} = 1,600 \text{ pieces per week}$$

FIGURE 23–1
Time standards allow the determination of projected direct labor requirements

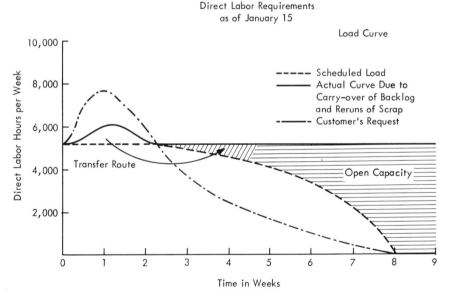

Figure 23–1 illustrates a weekly graphic analysis of the direct labor requirements for a specific industrial plant. Note how clearly this chart indicates when the plant will be in a position to produce new customer orders.

A basis for purchasing new equipment

Since time standards allow us to determine machine, department, and plant capacity, they also provide the necessary information for determining how many of what type of facilities must be provided for a given volume of production. Accurate comparative time standards also highlight the advantages of one facility over its competitors. For example, we may find it necessary to purchase three additional single-spindle, bench-type drill presses. By reviewing available standards, we shall be able to procure the style and design of drill press that will give the most favorable output per unit of time.

A basis for balancing the working force with the available work

By having concrete information on the required volume of production, as well as the amount of time needed to produce a unit of that production, we are able to determine the required labor force. For example, if our production load for a given week is evaluated as 4,420 hours, then we should need 4,420/40, or 111, operators. This use of standards is especially important in a retrenching market, where the volume of production is going down. When overall volume diminishes, if there is no yardstick to determine the actual number of people needed to perform the reduced load, then there will be a tendency for the entire working force to slow down so that the available work will last. Unless we balance our working force with the available volume of work, unit costs will rise progressively. It will only be a matter of time, under these circumstances, until production operations will be performed at a substantial loss, thus necessitating increased selling prices and further reductions in volume. The cycle will repeat until it becomes necessary to close the plant.

In an expanding market, it is equally important to be able to budget labor. As customer demands rise, necessitating a greater volume of manpower, it is essential that the exact number and type of personnel to be added to the payroll be known so that they can be recruited in sufficient time to meet customer schedules. If accurate time standards prevail, it is a matter of simple arithmetic to convert product requirements to departmental man-hours.

Figure 23–2 illustrates how overall plant capacity may be provided in an expanding market. Here the plant anticipates doubling its man-hour capacity in the period between January and November. This budget was based upon projecting the scheduled contracts in terms of man-hours and allowing a reasonable cushion (crosshatched section) for receiving additional orders.

In addition to budgeting plant labor requirements, time standards serve in budgeting the labor needs of specific departments. Figure 23–3 illustrates the budgeting of spindle hours in the multiple winding department

FIGURE 23–2
Chart illustrating actual projected man-hour load and budgeted man-hour load

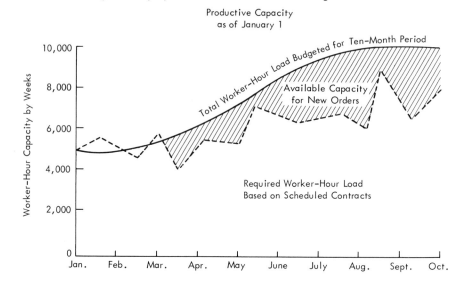

FIGURE 23–3
Chart illustrating the budgeting of the spindle hours of a specific department so as to best meet customer needs

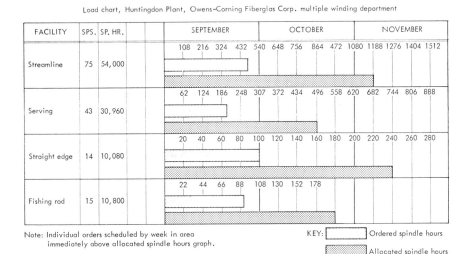

of one of the Owens-Corning Fiberglas plants. It will be noted that four product areas are served: "streamline," "serving," "straight edge," and "fishing rod." In view of present customer requirements, 75 spindles have been allotted for the multiple winding of "streamline" products, 42 spindles for "serving," 14 for "straight edge," and 15 for "fishing rod." Based upon this budgeting of spindles and man-hours, the load for "streamline" extends to the end of the first week in November, "serving" to the middle of October, "straight edge" to the middle of November, and "fishing rod" to the last week in October. Although customer requirements fluctuate, it will be possible through the use of time standards to adjust the labor budget of this department to best meet all customer requirements. The solid bar on this chart shows the material that has been allocated to the various product classes as of September 1.

With valid time standards, we are able to keep our working force in proportion to the volume of production required, thus controlling costs and maintaining operation in a competitive market.

Improving production control

Production control is the phase of an operation that schedules, routes, expedites, and follows up production orders so that operating economies are achieved and customer requirements are best satisfied. The whole function of controlling production is based upon determining where and when the work will be done. This obviously cannot be achieved unless we have a concrete idea of "how long."

Scheduling, one of the major functions of production control, is usually handled in three degrees of refinement: (1) long-range or master scheduling; (2) firm order scheduling; and (3) detailed operation scheduling, or machine loading.

Long-range scheduling is based upon the existing volume of production and the anticipated volume of production. In this case, specific orders are not given any particular sequence, but are merely lumped together and scheduled in appropriate time periods. Firm order scheduling involves scheduling existing orders to meet customer demands and still operate in an economical fashion. Here, degrees of priority are assigned to specific orders, and anticipated shipping promises are evolved from this schedule. Detailed operation scheduling, or machine loading, is assigning specific operations day by day to individual machines. This scheduling is planned to minimize setup time and machine downtime while meeting firm order schedules. Figure 23–4 illustrates the machine loading of a specific department for one week. Note that considerable capacity exists on milling machines, drill presses, and internal thread grinders.

No matter what the degree of refinement in the scheduling procedure, scheduling would be utterly impossible without time standards. The success of any schedule is in direct relation to the accuracy of the time values used in determining the schedule. If time standards do not exist, schedules formulated only on the basis of judgment cannot be expected to be reliable.

Through the use of time standards, the rate of flow of materials and of work in progress may be predetermined, thus forming a basis for accurate scheduling.

FIGURE 23–4
Machine loading of a machining department for one week (notice that several schedules depend on receiving additional raw material)

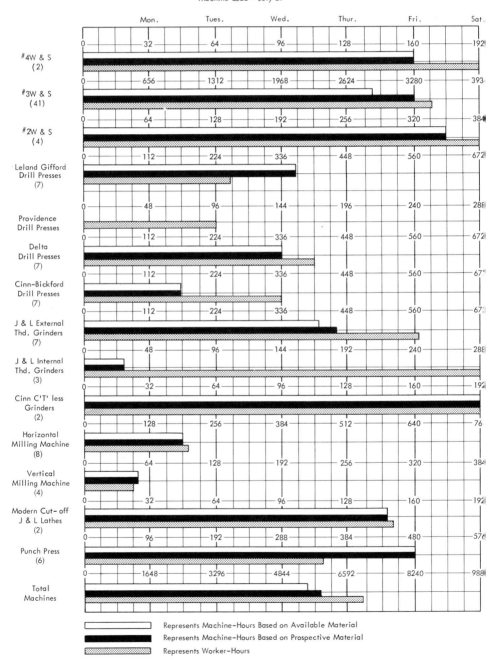

DORBEN MANUFACTURING COMPANY

Machine Load – July 29

The accurate control and determination of labor costs

With reliable time standards, a plant does not have to be on incentive wage payment to determine and control its labor costs. The ratio of departmental earned production hours to departmental clock hours provides information on the efficiency of the specific department. The reciprocal of the efficiency multiplied by the average hourly rate will give the hourly cost in terms of standard production. For example, the finishing department in a given plant using straight daywork may have had 812 clock hours of labor time, and in this period let us assume that it has earned 876 hours of production. The departmental efficiency would then be:

$$E = \frac{He}{Hc} = \frac{876}{812} = 108 \text{ percent}$$

If the average daywork hourly rate in the department was $6.40, then the hourly direct labor cost based on standard production would be:

$$\frac{1}{1.08} \times \$6.40 = \$5.92$$

Let us look at another example. We might assume that in another department the clock hours were 2,840 and that the hours of production earned for the period were only 2,760. In this case, the efficiency would be:

$$\frac{2,760}{2,840} = 97 \text{ percent}$$

and the hourly direct labor cost based on standard production with an average daywork rate of $6.40 would equal:

$$\frac{1}{0.97} \times \$6.40 = \$6.60$$

In the latter case, management would realize that its labor costs were running $0.20 per hour more than base rates and could take steps with supervision so that total labor costs would be brought into line. In the first example, labor costs were running less than standard, which would allow a downward price revision, increasing the volume of production, or making some other adjustment suitable to both management and labor. Figure 23–5 illustrates a direct labor variance report where departmental performance above and below standard is indicated.

Requisite for standard cost methods

Standard cost methods refer to the procedure of accurate cost determination in advance of production. The advantage of being able to

FIGURE 23–5

Weekly report illustrating departmental performance in a specific manufacturing plant (regular print indicates hours and percentage earned over standard; italics indicate to what degree standard has not been achieved)

DIRECT LABOR VARIANCE HOURS

Week Ending June 3

No.	Name	Allowed direct labor	Efficiency variance					Percent total variance over-under standards				
			Week ending			Weekly average		Week ending			4 weeks,	First
			$6/3$	$5/26$	$5/19$	First qtr.	Apr.	$6/5$	$5/26$	$5/19$	April	qtr.*
11	Machine shop	892	204	29	110	33	3	22.9	2.5	9.2	0.2	2.2
12	Wire brush	178	—	—	—	—	—	—	—	—	—	—
19	Punch press	41	18	8	—	6	3	43.9	9.5	—	4.5	9.1
20	Rubber milling	21	101	43	124	21	51	481.0	18.1	172.2	37.4	13.8
31	Rubber fabricating	1,183	36	29	12	116	59	3.0	1.5	0.7	3.2	5.2
35	Pilot plant	53										
39F	Finishing	339	60	107	27	42	50	17.7	18.5	5.8	10.0	6.6
39P	Paint	23	1	9	12	8	3	4.3	12.7	23.1	11.0	26.4
40	Assembly	13	1	15	15	14	4	7.7	28.3	25.9	6.0	19.9
50	Reclaim	20										
65	Toolroom	—	—	—	—	—	—	—	—	—	—	—
	Total—This week	2,763	217	148	192	104	59	7.9%	3.7%	4.5%	1.2%	1.9%
	Last week	4,462										

Note: The latest planning and method changes are reflected in all groups.
* First quarter includes the 13 weeks from January 1 through March 31.

predetermine cost is obvious. It is really necessary in many instances today to compute costs and contract work at the predetermined price. By having time standards on direct labor operations, it is possible to preprice those elements entering into the prime cost of the product. (Prime cost is usually thought of as the sum of the direct material and direct labor costs.)

As a basis for budgetary control

Budgeting is the establishment of a course of procedure: the majority of budgets are based on the allocation of money for a specific center or area of work. Thus, for a given period we may establish a sales budget, a production budget, and so forth. Since money and time are definitely related, any budget is a result of standard times regardless of how the standards were determined.

As a basis for the supervisory bonus

Wage incentives will be discussed at length in later chapters of this text. At this time, it will suffice to point out that any type of supervisory bonus keyed to productivity will depend directly on having equitable time standards. And since workers receive more and better supervisory attention under a plan where the supervisory bonus is related to output, the majority of supervisory plans give consideration to worker productivity as the principal criterion for the supervisory bonus. Other factors that are usually considered in the supervisory bonus are indirect labor costs, scrap cost, product quality, and methods improvements.

Quality standards are enforced

The establishing of time standards forces the maintenance of quality requirements. Since production standards are based upon the quantity of acceptable pieces produced in a unit of time, and since no credit is given for defective work turned out, there will be a constant intense effort by all workers to produce only good parts. If an incentive wage payment plan is in effect, operators are compensated for good parts only, and to keep their earnings up they will keep their scrap down.

If some of the pieces produced are found to be defective through subsequent inspection, either the operator who produced the parts will be held responsible for their salvage or else his earnings will be adjusted so that he is compensated only for the satisfactory parts that he produced.

Personnel standards are raised

Where standards are used, there will be a natural tendency to "put the right man on the right job," so that the standards established will be either met or exceeded. Placing employees on work for which they are best suited goes a long way toward keeping them satisfied. Workers tend to be motivated when they know the goals that have been established and how these goals dovetail with organizational objectives.

Problems of management are simplified

With time standards go many control measures that it would be otherwise impossible to exercise, such as scheduling, routing, material control, budgeting, forecasting, planning, and standard costs. With controls on practically every phase of an enterprise, including production, engineering, sales, and cost, the problems of management are minimized. By exercising the "exception principle," in which attention is given only to the items deviating from the planned course of events, management will be able to confine its efforts to only a small segment of the total activity of the enterprise.

Governmental operations have found time standards extremely helpful. Added emphasis was given to the need for a standards program in all governmental agencies when President Truman, by Executive Order 10072, in July 1949, and the 81st Congress, by Public Law 429, passed the same year, emphasized the need for the continuous examination and review of governmental operations to ensure the achievement of planned programs in each department. Title X of Public Law 429 makes specific provisions for establishing an efficiency awards system in each governmental agency. Time standards provide a means for evaluating the individual or group of individuals submitting proposals, and also provide a basis for enabling those submitting entries to examine the potential of their ideas.

Service to customers is bettered

With the use of time standards, up-to-date production control procedures can be introduced, with the resulting advantage to the customer of getting his merchandise when he wants and needs it. Also, time standards tend to make any company more time- and cost-conscious; this usually results in lower selling prices to the customer. As has been explained, quality will be maintained under a work standards plan, thus assuring the customer of more parts made to required specifications.

Conclusion

Thorough and regular follow-up will help assure the expected benefits from the proposed method.

Several of the uses of time standards have been briefly summarized. There are, of course, many more applications in all areas of any enterprise. Probably the most significant result of time standards is the maintenance of overall plant efficiency. If efficiency cannot be measured, it cannot be controlled, and without control it will diminish markedly.

Once efficiency goes down, labor costs rapidly rise, and the result is eventual loss of competitive position in the market. Figure 23–6 illus-

FIGURE 23–6
Relationship of labor costs to efficiency

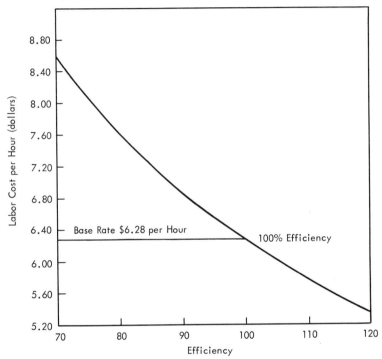

Developed from data furnished by AC Spark Plug Division, General Motors Corp.

trates the relationship of labor costs to efficiency in one leading manufacturer's automobile accessories business. By establishing and maintaining effective standards, a business can standardize direct labor costs and control overall costs.

TEXT QUESTIONS

1. Explain the use of learning curves in the "follow-up" step of the systematic approach to motion and time study.
2. In what different ways may time standards be determined?
3. How can valid time standards help develop an ideal plant layout?
4. Explain the relationship between time standards and plant capacity.
5. In what way are time standards used for effective production control?

6. How do time standards allow the accurate determination of labor costs?
7. How does developing time standards help maintain the quality of product?
8. In what way is customer service bettered through valid time standards?
9. What is the relationship between labor cost and efficiency?
10. If a daywork shop was paying an average rate of $6.60 per hour and had 250 direct labor employees working, what would be the true direct labor cost per hour if during a normal month 40,000 hours of work were produced?
11. How are management problems simplified through the application of time standards?
12. In what way did President Truman provide emphasis for the use of time standards in governmental operations?
13. Explain how inventory and storage areas can be predicted accurately.

GENERAL QUESTIONS

1. What other uses of time standards not mentioned in this chapter can be realized?
2. What is the relationship between the accuracy of time standards and production control? Does the law of diminishing returns apply?
3. How does work measurement improve the selection and placement of personnel?
4. When is it no longer necessary to follow up the installed method?
5. Do you believe that the typical worker today is motivated if the company goals and objectives are clearly presented to him? Discuss and give examples.

PROBLEMS

1. In the XYZ Manufacturing Company, direct labor cost is based on the efficiency relationship illustrated in Figure 23–6. For a given product line, the selling price was based upon the company's running at 95 percent efficiency. How much additional profit did the company realize if the actual efficiency turned out to be 110 percent? Profit was originally estimated at 10 percent of the total cost. On this product line, the total overhead was estimated to be 100 percent of prime cost (direct labor plus direct material). Material cost averaged $5 per unit of output, and direct labor averaged 0.50 hours per unit of output.
2. What would be the costing rate in the finishing department of the XYZ Company, where 32 employees work? Twenty-seven of these employees are on standard and have been averaging 1,310 hours earned per standard 40-hour week. The daywork hourly rate for these workers is $5.05 per hour. The remaining five workers are line supervisors who are not on standard. Their daywork hourly rate is $5.80 per hour.
3. In the XYZ Company, management is considering going from two 8-hour shifts per day to three 8-hour shifts per day or two 10-hour shifts per day in order to increase capacity.

Management realizes that shift start-up results in a loss of productivity that averages 0.5 hour per employee. The premium for the third shift is 15 percent per hour. Time over eight hours worked per day gives the operator 50 percent more pay. In order to meet projected demands, it is necessary to increase the man-hours of production by 25 percent. In view of insufficient space and capital equipment, this increase cannot be accommodated by increasing the number of employees on either the first or the second shift.

How should management proceed?

SELECTED REFERENCES

Buffa, Elwood S. *Modern Production Management,* 4th ed. New York: John Wiley & Sons, Inc., 1973.

Graham, C. F. *Work Measurement and Cost Control.* Oxford, England: Pergamon Press, 1965.

Greene, J. H. *Production and Inventory Control: Systems and Decisions,* Rev. ed. Homewood, Ill.: Richard D. Irwin, Inc., 1974.

Magee, J. F., and Boodman, D. M. *Production Planning and Inventory Control,* 2d ed. New York: McGraw-Hill Book Co., 1967.

Moore, F. G., and Jablonski, R. *Production Control.* 3d ed. New York: McGraw-Hill Book Co., 1969.

Mundel, Marvin E. *A Conceptual Framework for the Management Sciences.* New York: McGraw-Hill Book Co., 1967.

24

Methods and standards automation

If data processing equipment is to be utilized in connection with methods and standards work, it is axiomatic that standard data are being tabularized from either, or both, stopwatch time studies and fundamental motion data. With these data available, the digital computer offers a means of storing and recovering detailed standard data and elemental standard data for application on jobs that are being performed for the first time and for the development of proposals.

This chapter outlines the content of the basic files necessary for the implementation of an automated methods and work measurement system, and the flowcharts and procedures that would be typically utilized when operating such a system.[1]

Data processing and the establishment of standards

The conventional procedure for the establishment of a new standard without data processing equipment, but utilizing standard data, is illustrated in Figure 24–1.

Here the methods analyst develops a work station layout and motion pattern based upon his knowledge of motion economy and shop operations. From this proposed method, he makes an elemental breakdown and looks up the appropriate standard data times. If fundamental motion data are being used, the elemental breakdown is further subdivided so that basic motion times may be assigned. The time standard for the operation is developed by extending elemental time values by their frequency, totaling the times for each element, applying the correct allowance, and finally summarizing the allowed elemental times in order to determine the allowed operation time.

---

[1] The author is indebted to the IBM Corporation for assisting in the preparation of this chapter.

FIGURE 24–1
The development of time standards through the application of standard data

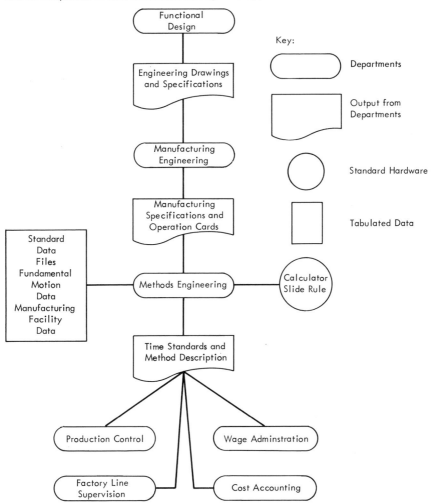

It is apparent that much of the above procedure represents burdensome clerical activities that are prone to human error. An automated methods and standards data processing system can minimize this clerical and computational work in the establishment of a time standard.

Such a data processing system would operate as follows:

1. Methods engineering develops a work station layout and motion pattern.
2. The proposed method is identified in detail by an elemental breakdown.

3. Through the utilization of data processing equipment, the description of each element is retrieved, the normal elemental times are identified, the elemental times are extended by frequencies, allowances are applied, and the allowed time for the operation is computed.
4. All associated reports are prepared by the system.
5. The operation time and description are entered into a permanent file for future use and maintenance.

The reader should understand that the only way the data processing system can accomplish step 3, as enumerated above, is by standard data coding. The coding can take any form a company may wish to use: numeric, alphabetic, or mnemonic. Generally, coding similar to what is used in the fundamental motion data systems (MTM, Work-Factor, BMT) is appropriate.

Advantages of methods and standards automation

The principal advantages of methods and standards automation include increased coverage, more accurate standards, and improved standards maintenance. Since standards may be developed much more rapidly through the use of data processing, it follows that from the standpoint of both time and money, data processing makes it feasible to increase the plant coverage of measured work. The less the amount of unmeasured work, the greater the opportunity for effective control and efficient operation.

Standards that are developed through the medium of data processing are inevitably more error-free, since manual arithmetic and lookup techniques are subject to human error.

Manual techniques prohibit accurate standard maintenance for no other reason than the avalanche of clerical work involved. For example, if basic research revealed that 2.2 TMUs was a more valid value for a GIA grasp than 2.0 TMU's, think of the clerical work involved to adjust every existing standard in the plant by increasing all GIA grasps by 10 percent! With a carefully installed data processing system, such a change could be economically effected within a short period of time.

Approach to automation through data processing

In order to use data processing equipment, it is necessary to develop files of all the existing standard data, which must be identified by an acceptable code. For example, a standard data element characteristic of such a file might be:

Code	Element description	Standard time
PSP3A	Pick up small piece, place in three-jaw air chuck and clamp	0.062

Here the code "PSP3A" represents the input to the system and the element description "pick up small piece, place in three-jaw air chuck and clamp," along with the standard time of 0.062, the output.

It is possible to program the computer so that greater amounts of detail can appear on a specific output document, such as a detailed instruction sheet. This is done by placing special symbols at appropriate locations within the element description.

For example, the lozenge ($\diamondsuit$) symbol might be inserted between the words "piece" and "place" as follows: "pick up small piece $\diamondsuit$ place in three-jaw air chuck and clamp." One output document could show the element as stated, and another could show additional detail, such as "pick up small piece, casting #437106, place in three-jaw air chuck and clamp."

The flow of information that takes place in the development of a detailed instruction sheet, operation card, and operation standard is illustrated in Figure 24–2. Process information—such as cutting feeds, cutting speeds, depths of cuts, and so on, along with a listing of the codes of the elements required, with their frequencies—is keypunched and entered into the system. The element description and its standard time are retrieved from the standard data element file for each element involved in the operation. Records involved in the development of the operation standard are stored in a temporary file until final review of the output has been completed by the analyst. Upon completion of the operation review, the operation records are transferred to a permanent operation file, which may be in either tape or disk form.

The output may take many forms. Typically included in output are the operation standard, which lists the applicable elements and their standard times; an operation card, which gives a brief description of the operation and its standard in minutes per piece and pieces per hour (this can serve as a move card, pay voucher, and so on); and an instruction sheet, which may be used by operating personnel.

With an automated data processing system, it is relatively easy to introduce a change that affects one or several operations. With details of operations stored in a file of magnetic tapes or disks, retrieval techniques allow changes to be made through "add" and/or "delete" instructions entered for the sections of the operation to which the changes should be applied. When the operation file is initially generated, each element of an operation is assigned a line number identification. Space signifying line

FIGURE 24–2
Automated work measurement systems for processing new operations and changing existing operations (dotted line applies to change operations only, and solid lines apply to both change and new operations)

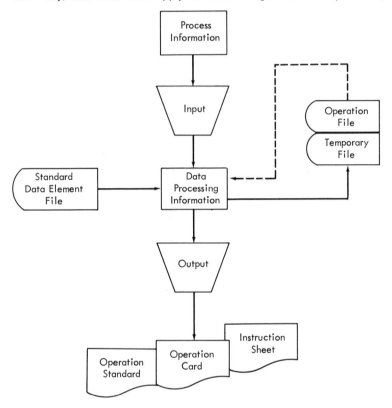

numbers is provided for possible element additions. When the deletion of a part of an operation is required, reference is made to the line number. When an element is to be added, it is assigned a line number in the proper sequence that had been reserved for possible future additions of elements.

The work measurement system

The general flow of information through a typical work measurement system is illustrated in Figure 24–3.

By the application of motion study (either visual or micromotion), time study, and process information, a methods study is made that results in the development of a work station utilizing the basic principles of motion economy and operation analysis. The proposed method is then described in appropriate element coding on the input document. Subsequently, data cards are punched and verified for each line entry on the input document. The data cards are now processed, and their information is

GURE 24–3

e flow of information in a typical work measurement system

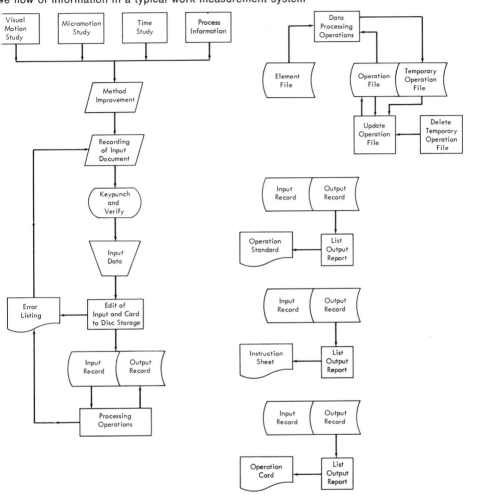

transferred to disk storage. Simultaneously, the input records are edited for any obvious errors. If errors have occurred, the recording of the corrected input document is made, and corrected data cards are punched and verified.

Now, while input records are being processed, information is retrieved from the major input files, that is, the element file and the operation file.

A temporary file is produced for all output records, including operations, and any error conditions are listed.

The output records are processed, and the various reports are printed.

After the analyst has reviewed the output reports, the operation file is updated, using the temporary file.

In the installation of a work measurement system, two major input files are needed: the element file and the operation file. The element file is more basic than the operation file, since it contains the description and the normal time of all existing elements. This information is retrieved when a new operation is processed by the work measurement system. After the new operation has been processed, its records are stored in the operation file.

In order to create an operation record on an existing operation, an analysis input sheet is keypunched and entered into the system. The computer program then retrieves element record data from its element file, applies the data to each recorded element, performs the required frequency time multiplications and accumulations, applies the appropriate allowances, and stores the entire operation in a temporary storage tape or disk location. Upon approval of the operation standard, the temporary operation records are transferred to the operation file.

The operation record is normally composed of two sections: (1) heading information, where the operation is described and its standard time is recorded; and (2) element information, where the elements involved in the operation are described and assigned a time value. Each element in an operation is usually given a "line item" number automatically by the computer in order to provide for an easy method of updating a particular element.

It is apparent that both the element and operation files must be adequately maintained if the work measurement system is to be reliable. It should be emphasized that all operations must be processed by using the most recent elemental descriptions and time values.

A typical work measurement input document is illustrated in Figure 24–4. The computer program compares the element code with its element master file, retrieves element times, and computes the time standard based upon the element time value, the element's frequency, and the appropriate allowance.

Changes in element description and/or time may have an appreciable effect on existing standards and operating instructions. Under a manual system, it would not be feasible to review all existing standards in order to introduce every elemental change. However, an automated methods and standards system can readily incorporate element changes that would result in a significant alteration of the operation standard.

This can be handled by generating a special file to obtain the impact of standard data changes. Those operations that have a significant change can then be reprocessed and a new standard developed.

In addition to the rapid development and maintenance of time standards, an automated work measurement system is capable of rendering several helpful reports. For example, it is possible to develop a report that shows the old and new standard for the operation, the variance, and the percentage of variance. Thus, the effect of change by area or depart-

Work measurement input document

GEAR	STUDY NO.	OPER. PARA. STEP	PART NO.	E.C. NO.	M.E. NO.	MC. NO.	DEPT. & MGR.	PFO %	DESCRIPTION	PAGE NO. 1 OF 1
		ELEMENT CODE	FREQUENCY NUMBER DENOM						DESCRIPTION	NAME J.J. JONES
	0 1 MTMST 12		0 3 4 7 5 5 0	0 0 0 0 0 0 0	0 0 0 1 6	0 5 1 4 4	0 9 1 1	1 5	ASSEMBLY OF TRIGGERS	ADJUSTMENT
									START ELEMENT GRID CODE OR ELEMENT DESCRIPTION	

CODE	ELEMENT CODE	DESCRIPTION
1	0010	ASSEMBLE AND HOT UPSET 34755/1/23146/1 -THIS OPERATION TO BE RUN
		ON MACHINE WITH CHILL AND TEMPER CONTROL ONLY
	001	ASSEMBLE PN/PO34755/ TRIGGER WITH PN/PO231461 USING TN/TO13 87,39
		ANVIL AND PLACE IN FIN. PARTS CONTR A70
	01	
2	R 6 C	/ / WITH RIGHT HAND AND
	G 4 B	/ / TRIGGER.
	R I C	/ / WITH LEFT HAND AND
	G 4 B	/ / STUD
	M 6 C	/ / AND
	P 2 SD	/ / STUD IN ANVIL
	G 2	/ / TO ORIENT,
	M 2 C	/ / AND
	P 2 SD	/ / TRIGGER ON STUD.
	A P 2	/ / TO HOLD SQUARE.

IBM Corporation, Data Processing Division

ment is readily available. Another helpful report is the "where-used index." This summary can identify where specific components, tools, and so on are used. For example, a particular drill jig may be used for the production of several components. The "where-used" index can show the study numbers, as well as the specific lines on which the drill jig is employed.

It is possible to generate a workplace layout that can be used by line supervision to help assure that the prescribed method is being employed. Figures 24–5 and 24–6 represent typical input and output documents.

The automated work measurement system can be extremely helpful in the rapid development of time estimates and quotations. Many businesses today are experiencing substantial costs in the preparation of quotations, only a small fraction of which result in contracts. Let us review how the system would function in a small job shop automatic screw machine business that prepares approximately 100 quotations per week.

The request for the quotation, along with a drawing of the part, will be studied by a methods analyst, who will prepare a planning sheet that includes the following information:

1. Alloy of material required, size (diameter and so on), material weight, cost per pound of material.
2. Length of material required per piece (length of part plus cutoff).
3. Recommended cutting speed in surface feet per minute.
4. Recommended feed in inches per revolution.
5. Money rate of the automatic to be employed.
6. Coding of the setup elements.
7. Dollar rate of setup man.
8. Percent efficiency of the production machine employed.
9. Any miscellaneous costs and one-time charges.

The planning sheet, along with the request for the quotation, will then be routed to the computer operator who will prepare the quotation. The computer operator's responsibilities are to:

1. Place a magnetic ledger card into the magnetic ledger card reader which will read and print out on the primary printer the customer's name and address on the quotation as shown on the request for the quotation. Other pertinent information, such as the part name, part number, or customer request number, can be printed out from the magnetic ledger card or else be typed manually on the quotation.
2. Key in as a column heading the quantities which are being quoted on, as read from the request for the quotation.
3. Type in the size, shape, and kind of material required, as shown on the planning sheet.
4. Key in part length and cutoff length.
5. Key in the material weight (pounds per foot).

FIGURE 24-5
Input document calling for a workplace layout

GROUP	STUDY NO.	PART NO.	E.C. NO.	M.E. NO.	M.C. NO.	DEPT. & MGR.	PFD %	DESCRIPTION	PAGE NO. 1	NAME A.W.
01	MTMST02	0347550	0000000000	00016	05144	0911	15	ASSEMBLY OF TRIGGERS		

CODE	OPER	PARA	STEP	ELEMENT CODE	FREQUENCY NUMER/DENOM	START ELEMENT GRID CODE OR ELEMENT DESCRIPTION	ADJUSTMENT
	0010					ASSEMBLE AND HOT UPSET 347551/123I461/1-THIS OPERATION	
						ON MACHINE WITH CHILL AND TEMPER CONTROL ONLY	
WP	WORK ARE M	A31	12 TO	138739		ANVIL/1/STD.ELECTRODE/1/	00000
WP	OPER SEAT	B31	00 80	000123		OPERATOR SEAT	00000
WP	PART LOC	A21	18 PO	231461		STUDS	00500
WP	PART LOC	A41	06 PO	347551		TRIGGERS	00100
WP	PART LOC	A42	06 PO	347550		FIN. PARTS IN SMALL CNTR.	00050
WP	CNTR 405 S	A11	18 PO	231461		STUD SUPPLY	00500
WP	CNTR 705	B11	18 PO	347551		TRIGGER SUPPLY	00500
WP	CNTR 150	B41	12 PO	347550		FIN. PARTS IN CNTR.A70	00200
	001					ASSEMBLE PN/PO347551 TRIGGER WITH PN/PO231461 USING TN/TO	
						ANVIL AND PLACE IN FIN PARTS CNTR A70	
		01					
2	R6C					WITH RIGHT HAND AND	
	G4B					TRIGGER	
	R1C					WITH LEFT HAND AND	
	G4B					STUD	
	M6C					AND	
	P2SD					STUD IN ANVIL	
	G2					TO ORIENT.	
	M2C					AND	
	P2SD					TRIGGER ON STUD.	
	AP2					TO HOLD SQUARE.	

IBM Corporation, Data Processing Division

FIGURE 24–6
Output document showing workplace layout

GROUP 01   ASSEMBLY OF TRIGGERS   PART NUMBER 0347550
STUDY MTMST02   MAY 20, 19--
EC 00000000 ME 00016   OCN 05-20-4 DEPT 091-1

0010

```
XXXXXXXXXXXXXXXXXXXXXXXXXXXXXXXXXXXXXXXXXXXXXXXXXXXXXX
X                 X              X           X             X
X  VENDORS   X    PART      X           X  PART LOC 2   X
X  CONTAINER X              X    WORK   X             X
X  --------  X    LOCATION  X           XXXXXXXXXXXXXXX
X                 X              X    AREA   X             X
X  SEALED    X    1         X           X  PART LOC 1   X
X                 X              X           X             X
XXXXX -A1-XXXXXXXXX -A2- XXXXXXXXX-A3- XXXXXXXXX-A4- XXXXX
XXXXXXXXXXXXXXX                              XXXXXXXXXXXXXXX
X                 X                          X             X
X                 X              XXXXXXX  X             X
X   705      X                  X    X  X   PLASTIC    X
X   STACK    X                  X  X  X   TOTE BOX    X
X   BIN      X                  XXX     X  WITH PROT.  X
X                 X                  OPERATOR X             X
X                 X                          X             X
XXXXX -B1- XXXXX                    -B3-    XXXXX -B4- XXXXX
```

PARTS LOCATION TABLE

FROM LOC.	TO LOC.	PART/TOOL NUMBER	DIST	PART QTY.	PART/TOOL DESCRIPTION	REP. QTY.
A1-1		P0231461	18	0500	STUD SUPPLY	
A2-1		P0231461	18	0500	STUDS	
A3-1		T0138739	12	0000	ANVIL/1/STD. ELECTRODE/1/	
A4-1		P0347551	06	0100	TRIGGERS	
A4-2		P0000000	06	0050	FIN. PARTS IN SMALL CONTR.	
B1-1		P0347551	18	0500	TRIGGER SUPPLY	
B3-1			00			
B4-1			12	0200	FIN. PARTS IN CONTR. A70	

IBM Corporation, Data Processing Division

6. Key in the cost per hundredweight according to the price, as read from the planning sheet.

7. Key in the major diameter and recommended surface feet, as read from the planning sheet.

8. Key in the throw and feed for each operation.

9. Key in the percent efficiency.

The computer will calculate the amount of material needed per hundred pieces (or some other unit of production) and will print this information on the quotation. Also, such information as the pounds of material per hundred pieces, weight per quantity quoted on, cost per hundredweight, gross and net production, and spindle speed required for each operation will be automatically computed.

The computer operator will type in the description of the operation as read from the planning sheet, and key in the gross and net production, and the dollar rate of the facility. The computer will automatically calculate the dollar amount per hundred pieces and will print out this information on the quotation.

This procedure will be repeated for each line of charges on the quotation.

To total the quotation, the operator will depress a command key, and the total price per 100 pieces will print out in the appropriate quantity breakdown column on the quotation.

Such information as miscellaneous or nonrecurring tooling charges, delivery schedule, and so on will now be typed in.

Upon completion of the quotation preparation, the computer will automatically tab to the quotation journal in the primary printer, and post a recap of the quotation for future reference purposes. This recap would contain such information as the customer name, customer request number, quotation number, prices per quantity quoted on, and delivery time. As this journal was printing, the printer inside the magnetic ledger card reader would post the same information to the customer account card so that it would provide a reference of quotations outstanding, by customer.

The procedure outlined would permit processing about 10 quotations per hour. Figure 24–7 illustrates the flow of information on this estimating and quotation writing procedure.

A computer program for indirect labor standards

In order to utilize data processing equipment for standard development through measurement, the time study analyst must record his data in a language acceptable to the computation equipment being utilized. Figure 24–8 illustrates a time study observation sheet designed for use with the IBM 360 system and for establishing standards on indirect labor work.[2] Pertinent information is transferred from this observation sheet to the IBM keypunched input card (see Figure 24–9) by the keypunch operator. It should be noted that one card is utilized for each line of information on the time study form and that the information from the form is transferred to the identical column on the keypunch card. The time study form as designed permits 25 elements or readings per page.

The following information has been included in this example:

Heading
The date, employee, job or operation description, job location, and name of the analyst are included. This information is compatible with that provided on most time study forms.

Date
Columns 01, 02, and 03 are reserved for the recording of the date the study

---

[2] This method and program were developed by Kenneth Kopp of The Pennsylvania State University.

FIGURE 24–7
Estimating and quotation-writing procedure

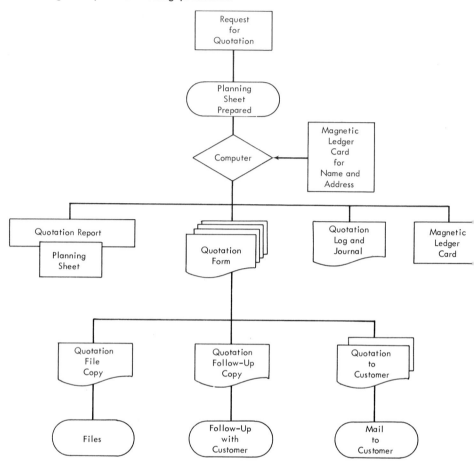

was taken. A number is assigned to each calendar day of the year. Thus, the number 41 would indicate February 10.

Study Number

Two columns, 04 and 05, are used to identify the study number. Thus, space is provided for up to 99 studies per day. The first study taken each day is assigned the number 01.

Cost Center

Columns 06, 07, and 08 are used to identify the specific cost center applicable to the study being taken.

Position or Location

Columns 09 through 18 are used to identify the building and room where the worker is engaged during the course of the study.

FIGURE 24–8
Time study observation sheet for use with the IBM 360 system

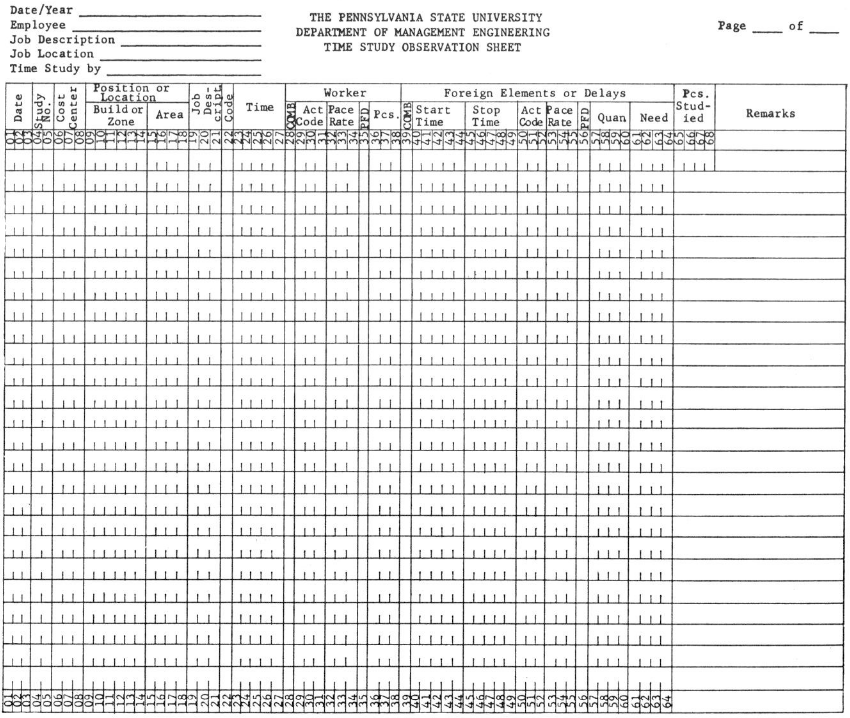

Date/Year _____
Employee _____
Job Description _____
Job Location _____
Time Study by _____

THE PENNSYLVANIA STATE UNIVERSITY
DEPARTMENT OF MANAGEMENT ENGINEERING
TIME STUDY OBSERVATION SHEET

Page ____ of ____

FIGURE 24–9
IBM keypunched input card

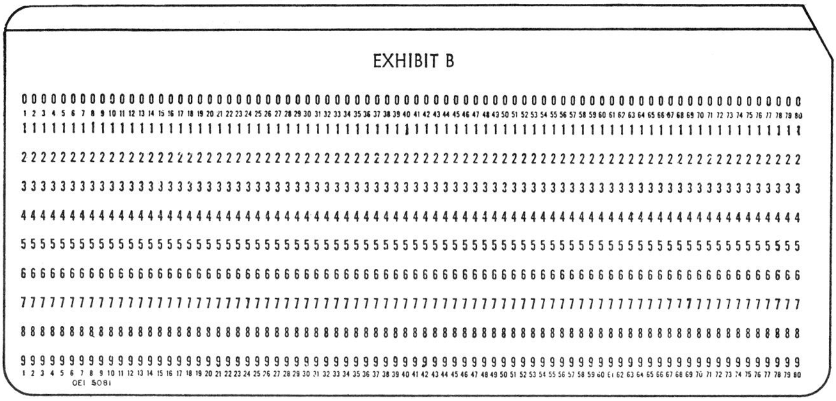

EXHIBIT B

FIGURE 24–10

Time study of an office cleaning operation

Date/Year 9/20/--
Employee Jack Keller
Job Description Custodian
Job Location Willard Building
Time Study by K. Kopp

THE PENNSYLVANIA STATE UNIVERSITY
DEPARTMENT OF MANAGEMENT ENGINEERING
TIME STUDY OBSERVATION SHEET

Page 1 of 2

Date	Study No.	Cost Center	Position of Location: Build or Zone	Position of Location: Area	Job Description	Code	Time	Worker: COMB	Worker: Act Code	Worker: Pace Race	Worker: PFD	Worker: Pcs.	Foreign Elements or Delays: COMB	Foreign Elements or Delays: Start Time	Foreign Elements or Delays: Stop Time	Foreign Elements or Delays: Act Code	Foreign Elements or Delays: Pace Race	Foreign Elements or Delays: PFD	Foreign Elements or Delays: Quan	Foreign Elements or Delays: Need	Pcs. Studied	Remarks
01 02 03	04 05	06 07 08	09 10 11 12 13 14	15 16 17 18	19 20 21	22	23 24 25 26 27	28	29 30 31	32 33 34	35	36 37 38	39	40 41 42 43 44	45 46 47 48 49	50 51 52	53 54 55	56	57 58 59 60	61 62 63 64	65 66 67 68	
263	05	290	184	3010	293	0	10															
				303			280		100	100	2			64	187	201	100	2				Start Room 300
				303			426		200	105												100 Walk
				315			440		100	100												201 Instructions from Foreman
							453		101	105												200 Get Supplies & Cart
							485		106													101 Unlock Door Turn on Light
							511		106													106 Empty Pencil Sharpener
							542		110													110 Empty Waste Basket
							579		110													107 Dust Desk Top
							655	0	107													108 Dust Telephone
							690		108													103 Dust Chairs
							700		107													104 Dust Tops of Files
							713		103													113 Dust Table Top
							736															301 Talk
							749															105 Wash Chalk Board (5' x 10')
							765															206 Get Bucket of Water
							790		104			3										207 Get Supplies (Erasers)
							910		107													111 Dry Mop Floor
							1232		113					962	1205	301						202 Change Mop Head
						1	0															222 Miscellaneous (Close Window)
							1050	1	105	100	2		0	300	480	206	100	2		2		167 Adjust Shades
							1130	1						1070	1102	207		2	30	4		115 Turn off Lights, Lock Doors
							1345	1					0	1138	1300	206		2				
							1464	2	111	105												
							1576		202	100												12.32 Went For Lunch
01 02 03	04 05	06 07 08	09 10 11 12 13 14	15 16 17 18	19 20 21	22	23 24 25 26 27	28	29 30 31	32 33 34	35	36 37 38	39	40 41 42 43 44	45 46 47 48 49	50 51 52	53 54 55	56	57 58 59 60	61 62 63 64		

FIGURE 24-10 (continued)

Date/Year 9/20/—
Employee Jack Keller
Job Description Custodian
Job Location Willard Building
Time Study by K. Kopp

THE PENNSYLVANIA STATE UNIVERSITY
DEPARTMENT OF MANAGEMENT ENGINEERING
TIME STUDY OBSERVATION SHEET

Remarks

Pcs. Studied

Foreign Elements or Delays

Need | Quan | Pace Race | PFD | Act Code | Stop Time | Start Time | COMB

Worker

Pcs. | PFD | Pace Race | Act Code | COMB | Time

Code | Job Description

Position of Location: Area | Build or Zone

Cost Center | Study No. | Date

Job Description

A three-digit number is used to identify each job description. The three-digit number associated with the study being made is recorded in columns 19, 20, and 21.

Time

Five columns (23 through 27) are used to record the elapsed time for each element.

Code

Column 22 has been introduced for verification purposes. This column provides the means to ascertain that each keypunched card is in its proper location within the deck of cards before the run on the computer starts. Column 22 provides for as many as nine interruptions during the course of one study. The first is identified as 1, the second as 2, and so on. In order to put the keypunched cards in their proper location within the deck, a check on column 22 is made. If no entry is shown, the cards are placed in numerical sequence, based upon the time entries shown in columns 23 through 27. If a recording is shown in column 22, the sorter will put all the 0's together, the 1's next, the 2's next, and so on, before going to the time column.

Combine

When an element is interrupted by a foreign element prior to its completion, it is necessary to combine that portion of the elemental time prior to the interruption with that portion of the elemental time required to complete the element after the interruption. By placing like digits (0 through 9) in column 28 to the rear of the two stopwatch readings, the computer will add these two values to arrive at the total elemental elapsed time.

Activity Code

Columns 29, 30, and 31 are used to identify standard elemental descriptions through the use of a three-digit number. For example, the three-digit number 111 might signify "dry mop floor."

Pace Rate

The performance rating factor is shown in columns 32, 33, and 34.

P–F–D

The allowance for personal, fatigue, and delay time is shown in column 33. The number 1 indicates a 10 percent allowance, the number 2 a 15 percent allowance, the number 3 a 20 percent allowance, the number 4 a 25 percent allowance, and so on.

Pieces

The number of units completed during each element will be entered in columns 36, 37, and 38. This permits the computer to compute an average elemental time per piece. The absence of an entry in the piece columns is interpreted as meaning that one unit is being processed.

Foreign Elements or Delays

Columns 39 through 64 are used to record foreign elements or delays as they occur.

FIGURE 24–11

Printout of a time study taken of an office cleaning operation

					THE PENNSYLVANIA STATE UNIVERSITY MANAGEMENT ENGINEERING MOTION AND TIME STUDY DEC 8, 19--				PAGE	1	
STUDY NUMBER 263-05											
CTR	LOCATN	AREA	DSC	TIME	ACT. TIME	RATING	EXTENSION	PFD	EXTENSION	CODE	DELAYS
290	184	300	293	00000							
			0303	00280	1.57	100	1.5700	115	1.8055	100	
				00187	1.23	100	1.2300	115	1.4145	201	
				00426	1.46	105	1.5330	115	1.7630	200	
			0315	00440	.14	100	.1400	115	.1610	100	
				00453	.13	105	.1365	115	.1570	101	
				00485	.32	105	.3360	115	.3864	106	
				00511	.26	105	.2730	115	.3140	106	
				00542	.31	105	.3255	115	.3743	110	
				00579	.37	105	.3885	115	.4468	110	
				00655	.76	105	.7980	115	.9177	107	
				00690	.35	105	.3675	115	.4226	108	
				00700	.10	105	.1050	115	.1208	107	
				00713	.13	105	.1365	115	.1570	103	
				00736	.23	105	.2415	115	.2777	103	
				00749	.13	105	.1365	115	.1570	103	
				00765	.16	105	.1680	115	.1932	103	
				00790	.25	105	.2625	115	.3019	104	
				00910	1.20	105	1.2600	115	1.4490	107	
				01232	.79	105	.8295	115	.9539	113	2.43
				00000							
				01050	8.70	100	8.7000	115	10.0050	105	
				00480	1.80	100	1.8000	115	2.0700	206	
				01130	.48	100	.4800	115	.5520	105	
				01102	.04	100	.0400	115	.0460	207	
				01345	.53	100	.5300	115	.6095	105	
				01300	1.62	100	1.6200	115	1.8630	206	
				01464	1.19	105	1.2495	115	1.4369	111	
				01576	1.12	100	1.1200	115	1.2880	202	
				01610	.34	105	.3570	115	.4106	111	
				01698	.70	105	.7350	115	.8453	167	
				01643	.03	100	.0300	115	.0345	222	
				01723	.25	105	.2625	115	.3019	115	
					26.69	1.03	27.1615	1.15	31.2360		
					UNITS	1	AVERAGE		31.24		

Combine

Column 39 is used to combine entries made under the foreign elements or delays section.

Start Time and Stop Time

Columns 40 through 49 provide space for the recording of the start and stop time for each foreign element or delay.

Activity Code, Pace Rate, P–F–D

Columns 50 through 56 are used in conjunction with foreign elements so as to identify their activity code, pace rating factor, and applicable allowance.

Quantity

Columns 57 through 60 are used to identify the quantity involved in conjunction with the foreign element taking place.

Need

The quantity of materials needed to complete the study is shown in columns 61 through 63. The fraction of the quantity processed (as shown in columns 57 through 60) in order to meet the need is shown in column 64.

Pieces Studied

The total number of units studied is entered in columns 65 through 68.

FIGURE 24–12

DEPARTMENT OF MANAGEMENT ENGINEERING
INVENTORY OF ROOM FURNITURE

Building __301__   Room __104__   Square Feet __324__

Code	Item & Description	Fre	Pcs	Code	Item & Description	Fre	Pcs	Code	Item & Description	Fre	Pcs
01A	Prepare	D			CHAIRS				BKCASE (glass)		
02A	Clean up	D		15A	Dust T A M, F.	M		22A	Dust top each tier	W	5
03A	Nonscheduled time	D		15B	Dust office	M	3	22B	S C each glass	M	22
04A	Instruct & inspect	D		15C	W & P or V	A	3	22C	Dust each unit	M	22
05A	Walk	D	1		SILL, OFFICE			22D	W & P each unit	A	22
06A	Lock	D	1	17A	Dust	W			BKCASE (no glass)		
07A	Unlock	D	1	17B	Damp wipe	M		23A	Dust top each unit	W	
09A	Extra walk 500	D			ASHTRAYS			23B	Dust comp. each row	M	
10A	Water	M		18A	Empty	D	1	23C	W & P each unit	A	
10B	Water	Tu		18B	Wipe	W	1		PENCIL SHARPENER		
10C	Water	Wed			TELEPHONE			24A	Empty	D	
10D	Water	T		19A	Dust	W			CUBICLE GLASS		
10E	Water	F		19B	Wash	M		28A	Wash sq. ft. of 1 side	Q	
11A	Extra walk 5A,6A,7A	A			FILES (2,3,4 drws.)				WALL SHELVES		
	DESK			20A	Dust top	W	6	29A	Dust (lin. ft.)	W	
12A	Dust top	W	1	20B	Dust comp.	M	6		SHADES, ROLLER		
12B	Dust comp.	M	1		BASKET, OFFICE			30A	Dust 25'/1	SA	
12C	W & P	A	1	21A	Empty	Tu&F+1			VENETIAN BLINDS		
	TABLES (large)			21B	Wipe	M		31A	Dust 25'/1	SA	3
13A	Dust top	W	1	21C	W & P	SA			RADIATORS, OFFICE		
13B	Dust comp.	M	1		COAT TREE			32A	Dust top (lin. ft.)	W	10
13C	W & P	A	1	25A	Dust	M		32B	Dust comp. (lin. ft.)	M	10
	TABLES (small)				COATRACK 10'/1			32C	Wash (lin. ft.)	A	10
14A	Dust top	W		26A	Dust	M			STOR-CAB. (closed)		
14B	Dust comp.	M			COUNTERS			33A	Dust top	W	6
14C	W & P	A		27A	Dust top (lin. ft.)	W		33B	Dust comp.	M	6
	CHALKBOARD			27B	Dust comp. (30 sq.'/1)	M			MIRRORS (5 sq. ft. = 1)		
16A	Wash 100% C, 15%/O	Tu&T		27C	W & P (lin. ft.)	A		40A	Wash	D	

Code	Item & Description	Fre	Pcs	Code	Item & Description	Fre	Pcs	Code	Item & Description	Fre	Pcs
	FLOORS (sq. ft.)				RADIATOR (RR) (lin. ft.)				DOORS		
34A	Dry mop (offices)	W	324	41A	Damp wipe	W		49A	S C half clear	W	
35B	Dry mop (classrooms)	MWF		41B	Wash	A		49B	S C half cloud	Q	
35C	Dry mop (labs–library)	Tu&T			BASKETS (RR)			49C	S C full clear	W	
35F	Dry mop (special)	D		43A	Empty	D		49D	S C Wood	M	1
35K	Dry mop (open areas)	W		43B	Wipe	W		49E	Dust door	M	1
35L	Dry mop (open areas)	Tu&T		43C	W & P	SA		49F	Dust jamb	Q	1
35M	Dry mop (open areas)	MWF			DISPENSER			49G	W & P	A	1
35A	Align (movable)	Tu&T		44A	Fill	D		49H	Kick damp	M	
35E	Align (fixed)	MWF		44B	Damp wipe	W		49I	Kick W & P	M	
35G	Align (movable)	MWF		44C	Wash Kotex	SA		49J	Push damp	M	
35H	Align (fixed)	Tu&T			WALLS(RR) (sq. ft.)			49K	Push W & P	Q	
	FLOOR (RR)			45A	Damp wipe	M		49L	K & P damp	M	
36A	Damp mop	Tu		45B	Wash	A		49M	K & P W & P	Q	
36B	Wet mop	T		45C	Water	M			STAIRS (per floor)		
36C	S C	MWF		45D	Water	A		50A	Dust mop	MWF	
	FLOOR (corr.) (sq. ft.)				URINALS			50B	Wet mop	W	
37A	Dry mop	MWF		46A	Wash inside	D		50C	Wet mop	M	
37B	Damp mop	W		46B	Wash comp.	W		50D	Water	M	
37C	Wet mop	M			WATER CLOSET			50E	Water	M	
37D	Water	W		47A	Wash inside	D		50F	Wash	MWF	
37E	Water	M		47B	Wash comp.	W			HANDRAILS		
	RUGS (sq. ft.)				WASHBOWL			51A	Dust	W	
38A	Vac. comp.	Tu		48A	Wash inside	D		51B	Wash	SA	
38B	Spot vac.	T		48B	Wash comp.	W			URNS		
38C	Vac. hook up	Tu&T			WALLS (sq. ft.)			52A	Empty & wipe	D	
	RUNNERS (sq. ft.)			53A	Dust	SA			FOUNTAINS		
39A	Vac.	MWF		53B	Walk	SA		55A	Wash	D	
39B	Vac. hook up	MWF			COOLERS				MISC.		
	SILLS (RR) (lin. ft.)			54A	Wash	D		56A	Dust	M	3
42A	Damp wipe	W		54B	Polish	M					

Remarks Column

This column is for the analyst's use as required.

Figure 24–10 illustrates a time study taken on the cleaning of an office, and Figure 24–11 shows the printout of the information supplied by the data processing department.

BLDG NO. 00301	ROOM NO. 0104		SQ. FEET 324							DATE 3-15-		PAGE 371	
CDE	ITEM AND DESCRIPTION	FREQ	STD	PCS	PERWK	MON	TUES	WED	THUR	FRI	WEEKLY	MONTHLY	PROJECT
05A	WALK	D			1.75	.35	.35	.35	.35	.35			
06A	LOCK	D			1.25	.25	.25	.25	.25	.25			
07A	UNLOCK	D			1.15	.23	.23	.23	.23	.23			
12A	DESK, DUST TOP	W			.42						.42		
12B	DESK, DUST COMP	M			.20							.20	
12C	DESK, W + P	A			.19								.19
13A	TABLES (L) DUST TOP	W			.42						.42		
13B	TABLES (L) DUST COMP	M			.05							.05	
13C	TABLES (L) W + P	A			.08								.08
15B	CHAIRS, DUST OFFICE	M			.30							.30	
15C	CHAIRS, W + P OR V	A			.14								.14
18A	ASHTRAYS EMPTY	D			1.25	.25	.25	.25	.25	.25			
18B	ASHTRAYS WIPE	W			.20						.20		
20A	FILES DUST TOP	W			1.20						1.20		
20B	FILES DUST COMP	M			.30							.30	
22A	BKCASE (G) DUST TOP	W			1.15						1.15		
22B	BKCASE (G) S C GLASS	M			1.54							1.54	
22C	BKCASE (G) DUST COMP	M			.55							.55	
22D	BKCASE (G) W + P	A			1.12								1.12
31A	V BLINDS DUST	SA			.20								.20
32A	RADIATOR (O) DST TOP	W			.60						.60		
32B	RADIATOR (O) DST COM	M			.05							.05	
32C	RADIATOR (O) WASH	A			.10								.10
33A	STOR CAB DUST TOP	W			2.40						2.40		
33B	STOR CAB DUST COMP	M			.39							.39	
34A	FLOOR (OFF) DRY MOP	W			5.51						5.51		
49D	DOORS SC WOOD	M			.16							.16	
49E	DOORS DUST DOOR	M			.09							.09	
49F	DOORS DUST JAMB	Q			.06							.06	
49G	DOORS W + P	A			.04								.04
53A	WALLS DUST	SA			.21								.21
56A	MISC DUST	M			.30							.30	
	TOTAL				23.37	1.08	1.08	1.08	1.08	1.08	11.90	3.93	2.14

To prevent misinterpretation and possible incorrect use, the standards and pieces have been omitted.

Here the stopwatch readings are reproduced as they appear on the time study. Also recorded are the elemental rating factor and the P–F–D allowances. The normal and allowed elemental times are computed and coded for identification purposes. The mean elemental rating factor and the allowance are also computed, along with the operation standard.

Using the computer to calculate standards from standard data

Once standard data have been developed, they can readily be stored in the digital computer, and recalled as necessary. The rapidity and accuracy of this recall, computation, and printout are such that standards may be economically developed on almost all kinds of work. Through the use of the computer, the application of standards in indirect work, including janitorial, repair, maintenance, toolmaking, tool design, nursing, clerical

FIGURE 24–14

6 7 8	20

3916
GENERALIZED AUTOMATIC STANDARDS PROGRAM (GASP) Date:

Standard 01--For shearing flat stock, straight cut--up to & including 7" wide, $1\frac{1}{4}$" thick.
Standard 02--For shearing round stock--up to & including $2\frac{1}{4}$" diameter (EXCLUDE 3/4" diameter ABOVE C1025 H.R. & C1040 C.R.)
Standard 03--For combination cuts (1 R.E.)-(2 R.E.)--up to & including $2\frac{1}{4}$" wide & $\frac{1}{2}$" thick.
Standard 04--For cut on curved knife (1 R.E.)--up to & including 3 3/4" wide & 7/8" thick.

STD. NO. (4)	PART NUMBER (6)	OPER. NO. (21)	ORIGINAL BAR: THICK. -or- GAUGE (27)	ORIGINAL BAR: WIDTH (32)	ORIGINAL BAR: LENGTH (37)	ORIGINAL BAR: FIN. PCS. (43)	SHEARED PIECE: LENGTH (46)	SHEARED PIECE: FIN. PCS. (52)	SHEAR	ADD LABOR	OLD PRICE 2d Hlp (67)	OLD PRICE 1st Hlp (71)	OLD PRICE Oper. (75)	RSN CODE (79)	ACCT
01	467335 R1	010	0.7500	6.0000	144.000	012	11.5000	001					0237	A	5
01	518036 R1	010	1.2500	2.5000	192.000	007	27.2500	001				0259	0267	F	8
		010													
		010													
		010													
		010													
		010													
		010													
		010													
		010													
		010													
		010													
		010													
		010													

Col. 23 12 zone=cards only
11 zone=forms only
no zone=cards & forms

If gauge is given, col 27 must contain # followed by gauge. Leave col 30 & 31 blank.

Equal Length=E
Extra stroke for C/K = 1
2 R.E. (C/C) = 2

REASON CODE
10 A= A Ex M =E
10 B = B Prc Adq =F
9 C= C Mtl Ch =G
10 D= D Mtd Ch =H
Eqp Ch =I

501=5
501=8

ALL DIMENSIONS MUST BE GIVEN IN INCHES AND DECIMAL PARTS OF AN INCH. (ORIGINAL BAR LENGTH MUST BE ROUNDED TO 3 DEC.)

Repairs = R Package= P Exports= E
Machine= M Interworks= I Special Labor= X

(including stenographic), office routine, and warehousing, has become a reality.

Continuing with the janitorial example cited earlier will give the reader a feeling of confidence in the use of computers in developing such standards. It has already been explained how janitorial methods were standardized and how the computer was used to assist in the development of time standards.

The next step was for the methods analyst to determine the frequency of the various cleaning tasks in connection with the custodial work characteristic of a given building or work center. This was done, and each janitorial operation in a work center was coded "D" for daily, "W" for weekly, "M" for monthly, and so on. Figure 24–12 illustrates an inventory of a work center, giving information as to the frequency of each operation

FIGURE 24–15

PIECEWORK PRICE AUTHORIZATION

DP NO. 20	ST NO. 01	EFFECTV. DY MN YR 03 09 64	CHANGE REASON 10-A A	TAG # 4737	GROUP TYPE SHR.PRS.	NO. 1	MACHINE TAG # TYPE	PART NUMBER 467335 R1	M/P 01	USED	ON MODEL NO.

OPER NO.010	METHOD DESCRIPTION	SHEAR 12 PCS./BAR

MATERIAL STEEL	FORM FLAT BAR	IN PER 1	OUT CYC 12	FIN IN 12	PCS/ OUT 1	DIMENSIONS				SET ANALYSIS BY G&P CANTN 062664
						.7500 THICKNS	6.0000 WIDTH	144.000 LENGTH	11.5000 SHR-LNG	STND. MINUTES PER

CL. NO.	ELEMENT DESCRIPTION	FREQ/CY	CYCLE	FIN. PCE.
V 1	HANDLE ORIGINAL BAR/S TO SHEAR	1.000	.3214	
V 2	SHIFT BAR/S TO END GAGE	10.000	.5693	
V 3	SHIFT BAR/S TO END GAUGE- END OF BAR/S ON TABLE	1.000	.0579	
V 4	BACK GAUGE TO MARK	1.000	.0693	
V 6	LAST PIECE/S AWAY	1.000	.0314	
V 8	LEVEL LOAD.	1.000	.1450	
V 9	GAUGE ONE PIECE	.060	.0110	
V10	TRUCKING	.060	.2400	
V11	BUNDLE CHANGE	.024	.0734	
	DOWNTIME PERSONAL 24 CLEAN-UP 3 OTHER - TOTAL 27 MIN/DAY	TOTAL	1.5189	

POSTED BY DEPT		ACCT #	OC. CLS	LG	O	H	STANDARD PER HUNDRED PIECES			PRODUCTION		CLASS	SUB TOTAL
20	COST STND.	501	SX050B	06	1	0	$ .2370	$ .2350	$	STD. TOT.	473.933 447.427	VARIABLE MACHINE CONSTANT	
NEW PIECEWORK PRICE APPROVED BY				OLD PRICE	NEW PRICE	EXCS. COST	PIECES/HOUR					IDLE	
STANDARDS ENGINEER		FOREMAN		NEW PRICE RETROACTIVE TO EFFECTIVE DATE AS PER LABOR CONTRACT ARTICLE XII, SEC. 10 VALID ONLY WHEN PROPERLY APPROVED								STANDARD	.1266
												DOWN	.0075
												TOTAL	.1341
												HRS.CPS.	.211

and the number of work units (chairs, desks, bookcases, radiators, and so on) located in that work center.

Once the inventory of a building has been completed and the work frequencies have been assigned, the inventory sheets can be summarized and work standards developed by the computer. Figure 24–13 illustrates a printout. The 23.37-minute total which is shown in the last line is the time needed each week by the janitor to clean a given room (work area).

Some typical computer programs

One company has developed a modular program capable of processing a sequence of the known elements necessary to perform an operation. The input involves a description of the work location, equipment, raw material or stock, and desired piece dimensions.

FIGURE 24–16

PIECEWORK PRICE AUTHORIZATION

DP NO. 20	ST NO. 01	EFFECTV. DY MN YR 03 | 09 | 64	CHANGE REASON PRCADQ F	TAG # 4230	GROUP TYPE SHR.PRS.	NO. 1	MACHINE TAG # TYPE	PART NUMBER |518036| R1	M/P 01	USED	ON MODEL NO.

OPER. NO.010	METHOD DESCRIPTION	SHEAR 7 PCS./BAR

MATERIAL STEEL	FORM FLAT BAR	IN PER 1	OUT CYC 7	FIN IN 7	PCS/ OUT 1	DIMENSIONS 1.2500 THICKNS	2.5000 WIDTH	192.000 LENGTH	27.2500 SHR-LNG	SET ANALYSIS BY G&P CANTN 062664 STND. MINUTES PER

CL. NO.	ELEMENT DESCRIPTION	FREQ/CY	CYCLE	FIN. PCE.
V 1	HANDLE ORIGINAL BAR/S TO SHEAR	1.000	.2961	
V 2	SHIFT BAR/S TO END GAGE	6.000	.3906	
V 4	BACK GAUGE TO MARK	1.000	.0693	
V 7	LAST PIECE THRU KNIFE TO HELPER	1.000	.0310	
V 8	TRUCKING	.049	.1997	
V 9	BUNDLE CHANGE	.022	.0679	

	DOWNTIME PERSONAL 24 CLEAN-UP 3 OTHER – TOTAL 27 MIN/DAY	TOTAL	1.0548	

POSTED BY DEPT 20	COST	STND.	ACCT # 801	OC. CLS. SX050A SX245	LG 05 04	O 1 0	H 0 1	STANDARD PER HUNDRED PIECES $ .2670 $ .2690 .2590 .2610		PRODUCTION STD. 398.142 TOT. 375.704	CLASS VARIABLE MACHINE CONSTANT	SUB TOTAL

NEW PIECEWORK PRICE APPROVED BY	OLD PRICE	NEW PRICE	EXCS. COST	PIECES PER HR.	IDLE		
STANDARDS ENGINEER	FOREMAN	NEW PRICE RETROACTIVE TO EFFECTIVE DATE AS PER LABOR CONTRACT (ARTICLE XII, SEC. 10) VALID ONLY WHEN PROPERLY APPROVED			STANDARD DOWN TOTAL HRS/CPS.		.1507 .0090 .1597 .251

An example of the input data for the development of a shearing standard is illustrated in Figure 24–14. Figures 24–15 and 24–16 reflect the program output.

Where group technology is practiced, the use of the computer may be especially timesaving in the development of work standards. A computer program can describe all the operations for making a part characteristic of a product group. For example, gears of similar types may be classified by a family or group number. The input information for a specific gear would include the pertinent dimensions, the part name and number, the material, and other descriptive details. The computer output includes a tabulation providing a sequence of the operations; the tooling involved; the feeds, speeds, and depths of cuts; and the standard times.

The United California Bank at Los Angeles stores its standards in a computer file. These standards are regularly recalled and used to evaluate work centers and determine manpower policies.

TEXT QUESTIONS

1. Explain in detail how an automated methods and standards data processing system can minimize standards clerical and computational work.
2. Outline how a change that affects several operations is introduced where an automated data system is employed.
3. What two major input files are needed in the installation of a work measurement system?
4. What helpful management reports can be readily generated if an automated work measurement system is being used?

GENERAL QUESTIONS

1. What disadvantages, if any, do you visualize can exist if an automated work measurement system is installed?
2. What percentage time do you feel the work measurement analyst would save using the computer technique illustrated in Figure 24–8? Will the keypunch operator make as many input errors as the time study man's arithmetic errors? Why or why not?

PROBLEMS

1. Write a computer program to provide output data for the painting of a line of castings from the equation:

$$T = [C_1 + 0.162S + n(v_1 + 0.0820]1 + \alpha$$

where:

T = Time in minutes per casting
C_1 = 0.16 or 0.25 or 0.38, depending upon the grade of paint used

S = Surface area of casting in square inches
n = Number of holes that need to be plugged prior to painting
V_1 = 0.04 or 0.07 or 0.13, depending upon the plug diameter
α_1 = 0.12 or 0.15 or 0.18, depending upon which paint booth is used

2. Prepare a flow diagram and write a computer program to estimate standard times for producing a variety of jigs and fixtures from the equation:

$$T = \sqrt{2.5 + 0.14W^2} + 0.75N + 1.82Q + 0.51S + 2.42B + 6.14 + t$$

where:

T = Time in hours to produce one jig or fixture
W = Gross weight of jig or fixture in pounds
N = Number of holes jig bored in the tool
Q = Number of clamps contained on the tool
S = Number of stops contained on the tool
B = Number of bushings contained on the tool
t = 0.25 when tolerances are closer than ± 0.001 inch
  = 0.10 when tolerances are between ± 0.001 and ± 0.005
  = 0 when tolerances are greater than ± 0.005

SELECTED REFERENCES

Bedworth, David D. *Industrial Systems: Planning, Analysis, Control.* New York: Ronald Press Co., 1973.

Lientz, Bennet P. *Computer Applications in Operations Analysis.* Englewood Cliffs, N.J.: Prentice-Hall, Inc., 1975.

Maisel, Herbert, and Gnugnoli, Givliano. *Simulation of Discrete Stochastic Systems.* Chicago: Science Research Associates, Inc., 1972.

25

Wage payment

Experience has proven that workers will not give extra or sustained effort unless some incentive, either direct or indirect, is in the offing. In one form or another, incentives have been used in business and industry for many years. Today, with the increasing need for American business and industry to improve productivity in order to halt spiraling inflation and maintain or improve their position in the world market, the advantages of wage incentives should not be overlooked. Only about 25 percent of manufacturing employees are now on incentive. If this figure were doubled in the next decade with sound installations, the improvement in our national productivity could be phenomenal.

With fringe benefits becoming increasingly significant (today, they average about 30 percent of direct labor), it is important that these costs be spread over more units of output. At present, fringe benefits average approximately 19 vacation days a year, 9 paid holidays a year, $11,000 in life insurance, disability insurance at 72 percent of salary, and sick pay up to 85 days.

In a broad sense, all incentive plans that tend to increase the employee's production will fall under one of the following three classes: (1) direct financial plans, (2) indirect financial plans, and (3) plans other than financial.

Before the analyst designs a wage payment plan for a specific plant, it is wise for him or her to review the strengths and weaknesses of past plans. For this reason, the more important historical plans will be reviewed in this chapter, even though most of them are no longer being used as they were originally designed.

Direct financial plans

Direct financial plans include all plans in which the worker's compensation is commensurate with his output. In this category are included both individual incentive plans and group plans. In the individual type of plan, each employee's compensation is governed by his own performance for the period in question. Group plans are applicable to two or more persons who are working as a team on operations that tend to be dependent on one another. In these plans, each employee's compensation within the group is based upon his or her base rate and on the performance of the group for the period in question.

The incentive for high or prolonged individual effort is not nearly as great in group plans as in individual plans. Hence, there has been a tendency for industry to favor individual incentive methods. In addition to lower overall productivity, group plans have other drawbacks: (1) personnel problems brought about by nonuniformity of production coupled with uniformity of pay, and (2) difficulties in justifying base rate differentials for the various opportunities within the group.

Of course, group plans do offer some decided advantages over individual incentive, of which these are noteworthy: (1) ease of installation brought about through ease of measuring group rather than individual output; and (2) reduction of cost in administration through the reduced amount of paperwork, less verification of inventory in process, and less in-process inspection.

In general, higher rates of production and lower unit cost of the product can be expected under individual incentive plans. If practical to install, the individual incentive plan should be given preference over group systems. On the other hand, the group approach has more application where it is difficult to measure individual output and where individual work is variable and is frequently performed in cooperation with another employee on a team basis. For example, where four men are jointly working together in the operation of an extrusion press for extruding brass rods, it would be virtually impossible to install an individual incentive system, but a group plan would be applicable.

Indirect financial plans

Company policies that tend to stimulate employee morale and result in increased productivity, yet have not been designed to bring about a direct relation between the amount of compensation and the amount of production would fall in the indirect financial classification. Such overall company policies as fair and relatively high base rates, equitable promotion practices, sound suggestion systems, a guaranteed annual wage, and relatively high fringe benefits will lead to building healthy employee attitudes which stimulate and increase productivity. Thus, they are classified as indirect financial plans.

All indirect incentive methods have the weakness of allowing too broad a gap between employee benefits and productivity. After a period of time, the employee takes for granted the benefits bestowed upon him and fails to realize that the means for their continuance must result entirely from his productivity. The theories, philosophies, and techniques of indirect incentives are beyond the scope of this text; for added information in this area, the student is referred to books on personnel administration.

Plans other than financial

Nonfinancial incentives include any rewards that have no relation to pay, and yet improve the spirit of the employee to such an extent that added effort will be evidenced. Under this category would come such company policies as periodic shop conferences, frequent talks between the supervisor and the employee, proper employee placement, job enrichment, job enlargement, nonfinancial suggestion plans, the maintenance of ideal working conditions, the posting of individual production records, and many other techniques utilized by effective supervisors and capable, conscientious managers.

Nonfinancial incentives represent an area that is thoroughly covered in texts on industrial relations and will not be further discussed in this book.

Classification of direct wage financial plans

The majority of individual plans may be placed within two classifications of direct wage financial plans. These two classifications, with representative specific plans, are as follows:

1. The employee participates in all the gain above standard.
   a. Piecework.
   b. The standard hour plan.
   c. The Taylor differential piece rate.
   d. The Merrick multiple piece rate.
   e. Measured daywork.
2. The employee shares the gains with the employer.
   a. The Halsey plan.
   b. The Bedaux point system.
   c. The Rowan plan.
   d. The Emerson plan.
   e. Cost savings sharing plans.
   f. Profit sharing.

There are some plans in which the relationship between output and earnings has been established empirically, and at different levels of performance, these plans may allow the employee either all the gains or a share of the gains. For purposes of simplicity, plans with these characteristics have been included in the group of "gain sharing with the employer."

Plans where the employee participates in all the gain above standard

PIECEWORK. Piecework implies that all standards are expressed in terms of money and that the operator is rewarded in direct proportion to his output. Under piecework, the day rate is not guaranteed. Today, piecework is not used, since federal law requires a minimum guaranteed hourly rate. Prior to World War II, piecework was used more extensively than any other incentive plan. The reasons for the popularity of piecework are that it is easily understood by the worker, easily applied, and one of the oldest types of wage incentive plans.

FIGURE 25–1
Operator earnings and unit direct labor cost under piecework

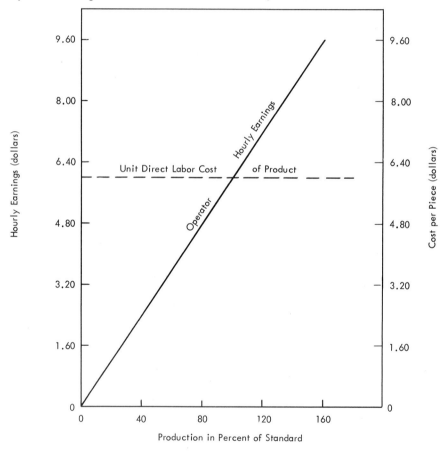

Figure 25–1 illustrates graphically the relationship between the operator's earnings and unit direct labor costs under a piecework plan.

THE STANDARD HOUR PLAN. The fundamental difference between the standard hour plan and the piecework plan is that under the former, standards are expressed in terms of time rather than money. The operator is rewarded throughout in direct proportion to his output.

Graphically, the relationship between operator earnings and unit direct labor cost, when plotted against production and money, would be identical to piecework.

For example, a standard may be expressed as 2.142 hours per 100 pieces. It is an easy operation to calculate either the money rate or the operator's earnings, once the base rate of the operator is known. If the operator had a base rate of $3, then the money rate of this job would be: $(3.00)(2.142) = 6.42 per hundred, or $0.0642 per piece. Let us assume that an operator produced 412 pieces in an 8-hour working day; his earnings for the day would be: $($3.00)(2.142)(4.12) = 26.48, and his hourly earnings would be: $26.48/8, or $3.310. The operator's efficiency for the day, in this case, would then be: $(2.142)(4.12)/8$, or 110 percent.

The standard hour plan offers all the advantages of piecework and eliminates the major disadvantages. However, it is somewhat more difficult for the worker to compute his or her earnings under this plan than if standards were expressed in terms of money. The principal advantage is, of course, that standards are not changed when base rates are altered. Thus, clerical work is reduced over a period of time when compared to the piecework plan. Moreover, the term *standard hour* is more palatable to the worker than is the term *piecework,* and with standards expressed in time, the amount of money earned by the worker is not quite so closely linked with time study practice. For these reasons, there has been a marked growth in the popularity of standard hour plans where the base rate is guaranteed.

THE TAYLOR DIFFERENTIAL PIECE RATE. The Taylor differential piece rate plan is not in use today. In this plan, two piece rates were established and expressed in terms of money. The lower rate compensated in direct proportion to output until the operator's performance reached standard. Once standard performance was achieved or exceeded, the high piece rate went into effect. Thus, the worker was encouraged not only to reach standard, but, since he was paid in direct proportion to output beyond standard, he was also encouraged to execute his maximum performance. Under the Taylor plan, standards were set "tight" enough so that only the proficient employees could exceed it. On so doing, they were generously compensated. It is also to be noted that the poor worker was penalized.

Under this plan, for example, the piece rates for machining a part may be $0.40 each up to a standard of seven pieces per hour. Daily production averaging seven or more pieces per hour may be paid $0.50 each. Thus, if an operator turned out 56 pieces during a day, his earnings would be $28. However, if he produced only 55 pieces, he would be paid at the lower piece rate and would have earned $22 for the day. Figure 25–2 illustrates the unit cost and earnings relationship of this plan.

FIGURE 25–2
Operator earnings and unit direct labor cost under the Taylor differential piece
rate plan

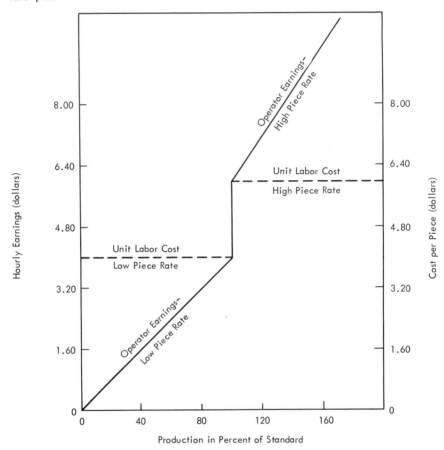

THE MERRICK MULTIPLE PIECE RATE. Under the Merrick multiple
piece rate plan, three graded piece rates were established rather than two,
as advocated by Taylor. One piece rate was established for beginners, one
for average employees, and one for superior workers. This plan endeavored
to correct the low rate paid to operators performing below task, and so
overcome the defects of Taylor's plan, which had caused severe criticism
by employees. Merrick chose 83 percent of task as the point for his first
step, which involved a piece rate with a 10 percent bonus. His high piece
rate beginning at standard included an additional 10 percent bonus. The
Merrick plan is not in use today.

MEASURED DAYWORK. During the early 1930s, shortly after the era of
the "efficiency expert," there was an effort by organized labor to get away
from time study practice and, in particular, piece rates. At this time,
measured daywork was introduced as an incentive system that broadened

the gap between the establishment of a standard and the worker's earnings. Many modifications of measured daywork installations are in operation today, and the majority of them follow a specific pattern. First, base rates are established by job evaluation for all opportunities falling under the plan. Then, standards are determined for all operations by means of some form of work measurement. A progressive record of each employee's efficiency is maintained for a period of time, usually one to three months. This efficiency, multiplied by his base rate, forms the basis of his guaranteed base rate for the next period. For example, the base rate of a given operator may be $3.30 per hour. Let us assume that the governing performance period is one month, or 173 working hours. If, during the month, the operator earned 190 standard hours, his efficiency for the period would be 190/173, or 110 percent. Then, in view of the performance, the operator would receive a base rate of (1.10)(3.30), or $3.63, for every hour worked during the next period, regardless of his performance. However, his achievement during this period would govern his base rate for the succeeding period.

In all measured daywork plans today, the base rate is guaranteed; thus an operator falling below standard (100 percent) for any given period would receive his base rate for the following period.

The length of time used in determining performance usually runs three months so as to diminish the clerical work of calculating and installing new guaranteed base rates. Of course, the longer the period, the less incentive effort can be expected. When the spread between performance and realization is too great, the effect of incentive performance is diminished.

The principal advantage of measured daywork is that it takes the immediate pressure off the worker. He knows what his base rate is, and realizes that, regardless of his performance, he will receive that amount for the period.

The limitations of measured daywork are apparent. First, because of the length of the performance period, the incentive feature is not particularly strong. Then, in order to be effective, it places a heavy responsibility on supervisors for the maintenance of production above standard. Otherwise, the employee's performance will drop, thus lowering his base rate for the following period and causing employee dissatisfaction. The keeping of detailed rate records and the making of periodic adjustments is costly in all base rates. In fact, as much clerical work is involved under measured daywork as under any straight incentive plan where the employee is rewarded according to his output.

Although we have classified measured daywork as a type of wage payment plan in which the employee participates in all the gain above standard, it can be seen that this method of wage payment is really a hybrid of the other "one-for-one" plans. As the worker's seniority increases, his total earnings will approximate the amount he would have earned if he had been paid in direct proportion to his output. Therefore, the plan is, in effect, similar to other types discussed that fall under this category.

Plans where the employee shares the gains with the employer

THE HALSEY PLAN. About 1890, Frederick Halsey developed a wage incentive plan that was one of the first to deviate from piecework. In Halsey's method, standards were established from past records, as time study had not yet come into use. The resulting standards, as could be expected, were loose.

However, if the worker failed to meet the standard, he was paid his regular wages; thus, this early proposal guaranteed the base rate, which is a requisite of an effective wage payment plan today. In Halsey's original plan, he compensated the operator for performance above standard so that the worker received one third of the time saved. The amount of premium in subsequent installations varied with individual plants but tended to be set at 50 percent. Using this percentage, it was easy to present the plan to the employees as an equitable arrangement in which the employee and the company each received one half of the time saved.

Another characteristic of the Halsey method that has sound application today was that the standards were expressed in terms of time rather than money.

The Halsey plan was established as a curb to runaway piece rates that were characteristic of the time because scientific methods of work measurement had not as yet been developed and methods improvements were the prerogative of the employee. Since the employee was assured his base rate and was given up to half the share of the earnings above task, he was usually attracted to the plan. Figure 25–3 illustrates the curves depicting the operator's earnings and unit direct labor cost for a modified Halsey system. Inspection of these curves will show that a varying unit labor cost is provided. This is one distinct disadvantage of the Halsey plan and all plans where the employee shares the gain with his employer. With variable unit labor costs, it is difficult to establish the true overall costs and budgets which are so necessary in the efficient operation of any business.

Since the Halsey plan does not reward the operator in direct proportion to his output, it would not be accepted by labor today. Labor's philosophy is that, if management is willing to pay a given amount per piece at the task point, it should be willing to pay the worker a like amount for production beyond task. Since total unit cost declines with increased productivity because overhead is spread over more pieces, management can afford to compensate the worker in direct proportion to his output.

THE BEDAUX POINT SYSTEM. The Bedaux point system, as intro-

FIGURE 25–3
The operator's earnings and unit direct labor cost under the Halsey plan

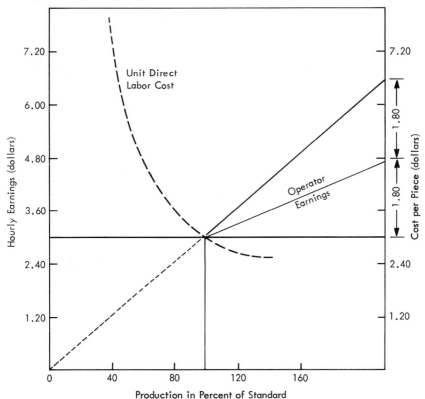

duced in 1916 by Charles E. Bedaux, is similar in many respects to the Halsey plan. The hourly rate is guaranteed up to task or standard, and beyond this point a constant sharing of time saved takes place. Bedaux expressed his standards in terms of "B" which was defined as one minute "composed of relative proportions of work and rest as indicated by the whole job." A normal operator was expected to perform 60 B's every hour he worked. The number of B's comprised by any job was determined by time study practice. Under the original Bedaux plan, the worker participated in 75 percent of the B's earned above standard. The remaining 25 percent of B's earned above standard was used to compensate indirect labor and supervision.

For example, if an operator earned 520 B's during the working day, his efficiency would be: 520/480, or 108.3 percent.

Of the 40 B's above standard earned, the operator would be compensated for 75 percent, or 30. Let us assume an hourly rate of $3.60. Then each B would have a value of $0.06, and the incentive earned on the above job would be $1.80. Graphically, the operator earnings curve and unit cost curves for the Bedaux point system would be identical to those for the Halsey plan except for the slopes above the task point. Since this plan does not reward the worker in direct proportion to output, it too is not acceptable to organized labor today.

THE ROWAN PLAN. In 1898, James Rowan proposed a sharing plan in which the incentive was determined by the ratio of the time saved to the standard time.[1] The base rate was guaranteed, and the premium earning curve began at 62½ percent of standard. Since it is impossible to save all the standard time, it would be impossible for any operator to earn 200 percent of his base wage. The fundamental purpose of Rowan's plan was to protect the employer from "runaway rates" that might be established from past performance records, and still provide sufficient incentive to the operator for high continuous effort. Although this system accomplished what was intended, the fact that large gains on the part of the operator were virtually impossible discouraged high production. The fact that the worker was compensated on an incentive basis at an early stage of the earnings curve made the plan more attractive to workers than the Halsey plan if tight rates prevailed and high production performance was impossible.

Operator earnings under this system can be expressed as follows:

$$E_a = R_a T + \frac{S_t R_a T}{T_a}$$

where:

E_a = Earnings
R_a = Rate per hour
T = Time spent on the work
S_t = Time saved
T_a = Allowed time

For example, if an operator at a base rate of $3 per hour spent three hours on a job which had a standard of 3.5 hours, his earnings for the job would be:

$$E_a = (3.00)(3) + \frac{(0.5)(3.00)(3)}{3.5} = \$10.28$$

His hourly rate for this assignment would amount to $3.43.

Figure 25–4 illustrates operator earnings and unit labor costs under the Rowan plan. This system has several disadvantages that limit its use today. First, it limits earnings and consequently restricts production. Then, this plan is more complex, making it difficult for the worker to understand, and causing more clerical work in the calculation of wages earned.

For example, let us assume a base rate of $4 per hour and a standard

[1] "A Premium System of Remunerating Labor," *Proceedings, Mechanical Engineers* (British), 1901, p. 865.

of one hour per piece for a given job. The task point, taken at 62.5 percent of standard, would require 1.6 hours to do the job. As performance increased, the resulting pay, labor cost, and rates of production would be as shown in Figure 25–4 and the accompanying table.

FIGURE 25–4
Operator earnings and unit cost curve under the Rowan plan

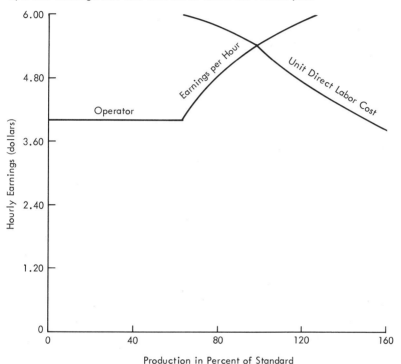

Percent of standard	Time taken (hours)	Time saved (hours)	Ratio of time saved of time/ allowed	Premium pay	Pay for job	Hourly rate
62.5.	1.6	0	0	0	$6.40	$4.00
80.	1.25	0.350	0.219	$1.09	6.09	4.87
100.	1.00	0.600	0.375	1.50	5.50	5.50
120.	0.833	0.767	0.480	1.60	4.93	5.92
140.	0.714	0.886	0.554	1.58	4.44	6.22
160.	0.625	0.975	0.610	1.52	4.02	6.43
200.	0.500	1.100	0.686	1.38	3.38	6.75

THE EMERSON PLAN. In many respects, the wage payment plan advocated by Harrington Emerson is not unlike the Halsey system. Emerson assured the employee his base rate, and established standards based on careful study of all the details entering into production. However, the incentive portion of his plan differed somewhat from that of other plans. At 66⅔ percent of a standard, he established a small incentive which increased as the performance increased until the task point was reached. Beyond this point, he established a straight-line earning curve which compensated the operator in direct proportion to his output plus 20 percent. The premium paid from two thirds of task to task was empirically determined and is tabularized in Table 25–1.

TABLE 25–1

Efficiency in percent	Bonus expressed as decimal proportion of base rate	Efficiency in percent	Bonus expressed as decimal proportion of base rate
67	0.0001	84	0.0553
68	0.0004	85	0.0617
69	0.0011	86	0.0684
70	0.0022	87	0.0756
71	0.0037	88	0.0832
72	0.0055	89	0.0911
73	0.0076	90	0.0991
74	0.0102	91	0.1074
75	0.0131	92	0.1162
76	0.0164	93	0.1256
77	0.0199	94	0.1352
78	0.0238	95	0.1453
79	0.0280	96	0.1557
80	0.0327	97	0.1662
81	0.0378	98	0.1770
82	0.0433	99	0.1881
83	0.0492	100	0.2000

Under Emerson's plan, the operator's earnings below two thirds of task may be computed from the expression:

$$E_a = R_a T$$

and between two thirds of task and task:

$$E_a = R_a T + F_t(R_a T)$$

and above task:

$$E_a = R_a T + S_t R_a + 0.20 R_a T$$

where:

F_t = Factor taken from table
R_a = Rate per hour
S_t = Time saved in hours
T = Time taken in hours

In adopting his plan, Emerson advocated the calculation of efficiency for a pay period of either a week or a month, thus tending to equalize the very low and the very high daily efficiencies. For example, if during a month involving 173 working hours an operator earned 180 hours for which he was paid a base rate of $3, his earnings for the period would be:

$$E_a = (3.00)(173) + (7)(3.00) + (0.20)(3.00)(173) = \$643.80$$

COST SAVING SHARING PLANS. Most progressive managements accept the principle of rewarding employees for improvements in productivity and/or cost no matter whether the improvements are due to performance above normal or to improvements in work methods. Proponents of cost saving sharing plans point out that, from the motivational viewpoint, direct incentives miss many improvement opportunities, including savings in material, both direct and factory supplies, and the myriad of methods changes that can result in improvements. Most of these plans (for example, the Rucker plan) tie the incentive to a "production value" above standard. Production value is the monetary difference between sales and purchases, and consequently represents the value that all employees have contributed. Previous performance for a period of at least a year is considered standard. Future "value added" can be improved by:

1. Savings in raw materials, purchased parts, supplies, fuel, and power.
2. A reduction in scrap.
3. A reduction in customer allowances.
4. An increase in the volume of output.

Under plans of this type, incentives are computed on a monthly basis. Customarily, only two thirds of the incentive earned in a given pay period is distributed. The remaining third is placed in a reserve fund to be used any month that performance falls below standard.

Cost saving sharing plans are relatively simple to establish since they do not require the development of time standards. However, only a limited number of successful applications are in effect today.

PROFIT SHARING. The Council of Profit Sharing Industries has defined profit sharing as "any procedure under which an employer pays to all employees, in addition to good rates of regular pay, special current or deferred sums based not only upon individual or group performance, but on the prosperity of the business as a whole."[2]

No one specific type of profit sharing has received general industrial acceptance. In fact, just about every installation has certain "tailor-made" features which distinguish it from others. However, the majority of profit sharing systems can be placed under one of the following broad categories: (1) cash plans, (2) deferred plans, and (3) combined plans.

As the name implies, the straight cash plan involves the periodic distribution of money from the profits of the business to the employees. The payment is not included with the regular pay envelope, but is made

---

[2] Council of Profit Sharing Industries, *Profit Sharing Manual* (First National Tower, Akron, Ohio, 1949), p. 3.

separately, so as to identify it as an extra reward, brought about by the individual and combined efforts of the entire operating force. The amount of the cash distribution is based upon the degree of financial success of the enterprise for the bonus period. The period varies in different companies. In some installations it is as short as one month, and in others it is one year. There is general agreement, however, that the shorter the period, the closer will be the connection between effort and financial reward to the employees. Longer periods are selected because they will reflect more truly the status of the business over that particular length of time. For example, if a period of one month is used, the employees might give excellent performance for the entire period in anticipation of a sizable bonus. However, due to conditions beyond their control, such as high inventories or slow-moving products, there may be no profit for the period. With no additional reward, the employees would quickly lose confidence in the plan and reduce their efforts appreciably.

Another factor of importance in the selection of the period for the distribution of profits is the amount of cash that will be shared. If the period is too short, the amount may be so insignificant in size that the plan itself may backfire. But, if a longer period, such as six months or a year, is used, the amount will be adequate to accomplish the purpose of the profit sharing system.

Deferred profit sharing plans are characterized by the periodic investment of portions of the profits of the business for the employees, so that upon retirement or separation from the company, they will have a source of income at a time when their needs may be more pronounced,

The deferred type of profit sharing plan obviously does not provide the incentive stimulus to the degree that cash plans do. However, deferred profit sharing plans do offer the advantage of being easier to install and administer. Also, plans of this type offer more security than do the cash reward plans. This makes them especially appealing to the stable worker.

Combined plans arrange to have some of the profits invested for retirement and similar benefits, and some distributed in the form of cash rewards. This class of plans can realize the advantages of both the deferred plans and those employing the straight cash system. A representative installation might provide for sharing half the profits with the employees. Of this amount, one third may be distributed to the employees in the form of an extra bonus check, one third may be held in reserve to be given out during a less successful financial period, and the remaining third may be placed with a trustee for deferred distribution.

Methods of distribution under profit sharing. There are three methods in common use for determining the amount of money to be given individual employees from the company's profits. The first and least used is the "share and share alike" plan. Here each employee, regardless of his job class, participates in an equal amount of the profits, once he has attained the prescribed amount of company service. Proponents of this method believe that individual base rates have already taken care of the relative importance of the different workers to the company. The share and share alike plan supplies a feeling of teamwork and importance to each employee, no matter what his or her position in the plant may be. Just as a complex mechanism is made inoperative by the removal of an insignificant pin, so an enterprise is dependent upon each and every employee for its efficient operation.

The most commonly used method of distribution under profit sharing is on the basis of the regular compensation paid to the workers. The theory is that the employee who was paid the most during the period contributed to the greatest extent to the company's profits and consequently should share in them to the greatest extent. For example, a toolmaker earning $10,000 during a six-month period would receive a greater share of the company's profits than a chip-hauler who was paid $4,000 during the same period.

Another popular means of profit distribution is through the allocation of points. Points are given for each year of seniority and each $100 of pay. Some point plans also endeavor to evaluate such factors as attendance, the worker's cooperation, and standards of production. The number of points accumulated by each employee for the period then determines his share of the profits. This method endeavors to take into consideration the factors that influence the company's profits and then to distribute the profits on a basis that gives each employee his just share. Perhaps the principal disadvantage of the point method is the difficulty of maintenance and administration brought about through complex and detailed records.

Criteria for a successful profit sharing plan. The Council for Profit Sharing Industries has summarized 10 principles which it feels is fundamental if a profit sharing plan is to succeed. These are:

1. There must be a compelling desire on the part of management to install the plan in order to enhance the team spirit of the organization. If there is a union, its cooperation is essential.
2. The plan should be generous enough to forestall any feeling on the part of the employees that the lion's share of the results of extra effort will go to management and the stockholders.
3. The employees must understand that there is no benevolence involved, that they are merely receiving their fair share of the profits they have helped to create.
4. The emphasis should be placed on partnership, not on the amount of money involved. If financial return is emphasized, it will be more difficult to retain the employees' loyalty and interest in loss years.
5. The employees must be made to feel that it is their plan as well as management's and not something that is being done for them by

management. The employees should be well represented on any committee set up to administer the plan.

6. Profit sharing is inconsistent with arbitrary management—it functions best in companies operating under a democratic system. This does not mean that management relinquishes its right, and, in fact, its obligation, to manage, but rather that management functions by leadership instead of arbitrary command.

7. Under no circumstances can profit sharing be used as an excuse for paying lower than prevailing wages.

8. Whatever its technical details, the plan must be adapted to the particular situation, and should be simple enough so that all can readily understand it.

9. The plan should be "dynamic," both as to its technical details and as to its administration. Both management and employees—and in profit sharing companies it is often difficult to find the demarcation line—should give constant thought to ways of improving it.

10. Management should recognize the fact that profit sharing is no panacea. No policy or plan in the industrial relations field can succeed unless it is well adapted and unless it evidences the faith of management in the importance, dignity, and response of the human individual.

Attitude toward profit sharing. For the most part, union officials have not supported profit sharing. Their principal criticism is that the technique is a "hard-times, wage cutting method."[3] There can be no doubt that profit sharing, when practiced with perfect harmony between labor and management, minimizes the necessity of a union in the eyes of the employees. Therefore, union leaders can hardly be expected to uphold and propagate something that will diminish their personal prestige, power, and income.

Where profit sharing has been practiced by a sincere, fair, and competent management team, individual workers are enthusiastic, and wholeheartedly support the plan. Attempts to unionize plants where profit sharing is flourishing, such as occurred in the Lincoln Electric Company, have resulted in violent worker reaction against the organizers.

A successful profit sharing program depends on the profits of the company, which frequently are not under the control of the direct labor force. In periods of low profits or of losses, the plan may actually weaken rather than strengthen employee morale. Also, unless the team spirit is enthusiastically supported by *all* members of supervision, the plan does not provide the incentive to continuous effort characteristic of wage incentive installations. This is a result of the length of time between performance and reward; most companies wait 12 months and use the year-end to declare their profits. In large companies, it becomes virtually im-

---

[3] Charles E. Britton, *Incentives in Industry* (Esso Standard Oil Co., 1953), p. 30.

possible to instill the team idea throughout the plant, and unless this is done generally, the plan will not succeed completely.

Perhaps the greatest objection to profit sharing that the writer has observed is the taking for granted that an "extra" check will be received at the end of the year. The employee expects to receive it, and even makes substantial purchases on a time basis in anticipation of the added remuneration. If the company has experienced a lean year and the worker gets no extra remuneration, he does not see any relation between his productivity for the past year and the fact that no bonus was received. All he feels is that he has been cheated. Thus, the whole profit sharing plan fails to engender the spirit intended.

For these reasons, any employer should be very cautious before embarking on a profit sharing program. On the other hand, many companies today are experiencing high worker efficiency, decreased costs, reduction of scrap, and better worker morale as a result of profit sharing installations. James F. Lincoln, past president and founder of the Lincoln Electric Company, who developed one of the most successful profit sharing installations, offers the following suggestions to companies which are contemplating the installation of some type of profit sharing system:

1. Determine that the system is going to be adopted and decide that whatever needs to be done to install it will be done.

2. Determine what plan and products the company will make that will carry out the philosophy of "more and more for less and less."

3. Get the complete acceptance of the board of directors and all management involved in the plan, together with their assurance that they will continue to take whatever steps are necessary for a successful application of it.

4. Arrange a means whereby management can talk to the men and the men can talk back. That means full discussion by all.

5. Make sure of co-operative action on the agreed plan of operation. This will include the plan for progressively better manufacturing by all people in the organization and the proper distribution of the savings that result from it.

6. Set your sights high enough. Do not try to get just a little better efficiency with the expectation that such gain will be to the good and expect to leave the matter there.

7. Remember, this plan for industry is a fundamental change in philosophy. From it new satisfactions will flow to all involved. There is not only more money for all concerned, there is also the much more important reward—the satisfaction of doing a better job in the world. There is that greatest of all satisfactions, the becoming a more useful man.[4]

The unions' attitudes toward wage incentives

The subject of wage incentives has always been controversial to employees, unions, and management. Industries that have assured a good

---

[4] James F. Lincoln, *Lincoln's Incentive System* (New York: McGraw-Hill Book Co.), pp. 171–72.

living wage and then applied incentive earnings which can easily be calculated for extra or prolonged effort, will find that their employees will be receptive to wage incentives. In fact, where a successful installation of this nature has been made, considerable labor unrest would result if any attempt were made to do away with the plan. On the other hand, in industries or businesses where the worker finds it necessary to work at an incentive pace to earn the necessities of life, he or she can hardly be expected to be enthusiastic about any form of wage incentive payment.

The majority of union officials with whom the writer has been in contact oppose the installation of incentive wage payment in plants where incentives do not exist. However, where incentives do exist, most unions will not only insist upon their continuance but will endeavor to expand the coverage of their membership under incentives. For example, the United Steelworkers of America summary of the August 1, 1969, incentive arbitration award stated:

The first issue tackled at the arbitration hearing was the Union's demand that a rational system of coverage be provided in place of the present chaos. It was the Union's position that all jobs should be covered by incentives. The Companies, on the other hand, argued unsuccessfully that in many instances already too many jobs are covered by incentives, and that the Companies should be given the right to remove coverage from many jobs which now enjoy incentives.

On the other hand, in a recent AFL-CIO Collective Bargaining Report that is still being used to state union attitude toward incentives, President George Meany states:

Wage incentive plans, that is, plans which offer more wages for more production, present a host of special problems which normally far outweigh any possible benefits.

With few exceptions, unions are opposed to wage incentive systems, both because of the damaging past experience with the abuses under such plans, and because of the difficulties and ill effects inherent in incentive plans.

Unions which have accepted them or permitted them to continue have usually done so only with reluctance and misgivings. It simply has not always been practical or expedient actively to oppose or to eliminate such plans.

A few unions, primarily in the rubber and needle trade industries where wage incentives are most firmly entrenched, have at least temporarily accepted incentives as part of their collective bargaining programs.

Many industrial engineers agree that wage incentives have been abused "in the past," but maintain that this is no longer true. In fact, although some of the worst abuses have been toned down, primarily through union action, a variety of ill effects and strains on workers are still very much the rule.

The presence of an incentive system invariably means special problems of education, representation, and protection of workers. It puts a strain on the entire collective bargaining process, making it more difficult, more complex, and more costly.

The value of any such plan is also highly questionable because, even though it may initially yield increased earnings, it inevitably requires a speedup of work efforts, creates friction between workers, and produces continual wrangling over production standards.

The unions' principal objections to incentive plans, in the opinion of the author, arise from a fear that a reduction in personnel will be brought about by high effort, given only a fixed production of goods and services, and that incentive installations will deemphasize the necessity of unions since the unions' major role is to achieve higher wages, which the incentive plan will automatically do. Union officials have stated that they object to incentives because these "pit worker against worker." They state that when one worker makes high earnings and another low, a feeling of distrust and suspicion permeates the working group, which disrupts the partnership relations among workers.

In the Grand Lodge constitution of the International Association of Machinists (Section 6 of Article J), the policy of this labor group toward incentives is clearly expressed in the statement that "any member guilty of advocating or encouraging any of these systems where they are not in existence is liable to expulsion." Carl Huhndorff, as director of research of this labor group, stated that the aforementioned clause had never been invoked against an individual member to the best of his knowledge; however, the Grand Lodge policy "has always opposed these [incentive] plans whenever possible as we do not feel that they work to the best interests of working men and women."

There are, of course, unions which approve of incentives. In fact, the late Philip Murray, a past president of the CIO, was known to look with favor upon incentive wage payment and had expressed his belief that practically any sound system of wage payment can be made to work when a harmonious relationship prevails between labor and management.

In summary, most unions will fight to keep incentives where they already exist. Where incentives do not exist, unions will resist any attempt to install them.

Prerequisites for a sound wage incentive plan

Certainly the majority of companies that have incentive installations favor their continuance and believe that their plans are (1) increasing the rate of production, (2) lowering overall unit costs, (3) reducing supervision costs, and (4) promoting increased earnings of their employees.

However, in a survey of 160 companies where incentive wage payment was practiced, 84 replies from managers implied that they felt their plans to be only fair, and that additional improvement could be made.[5] Five plant managers felt that their incentive systems were poor, and that some changes must be made to warrant continuance.

Before installing a wage incentive program, management should sur-

---

[5] National Metal Trades Association.

vey its plant to be sure that the plant is ready for an incentive plan. First, a policy of methods standardization must be introduced so that valid work measurement can be accomplished. If each operator follows his own pattern of performing his work, and the sequence of elements has not been standardized, then the organization is not yet ready for the installation of wage incentives.

The scheduling of work must be handled so that there is always a backlog of orders for each operator, and so that the chances of his running out of work are held to a minimum. Of course, this implies that adequate inventories of material are available, and that machines and tools are properly maintained. Also, established base rates should be fair, and should provide for a sufficient spread between job classes to recognize the positions that demand more skill, effort, and responsibility. Preferably, base rates should have been established through a sound job evaluation program.

Lastly, fair standards of performance must be developed before wage incentive installation can take place. In no case should these rates be set by judgment or past performance records. In order to be sure that they are correct, some form of work measurement based upon the time study or work sampling procedure should be used.

Once these prerequisites have been completed and management is fully sold on incentive wage payment, the company will be in a position to design the system.

Design for a sound wage incentive plan

To be successful, an incentive plan must be fair to both the company and the operator. The plan should give the operator the opportunity to earn approximately 20 to 35 percent above his base rate if he is normally skilled and executes high effort continuously. Management will benefit through the added productivity by being able to prorate fixed costs over a greater number of pieces, thus reducing total cost.

Perhaps next to fairness, the most important qualification of a good incentive plan is simplicity. In order to be successful, the plan must be completely sold to the employee, the union, and to management itself. The simpler the plan is, the more easily it will be understood by all parties; and with understanding the chances for approval are enhanced. Individual incentive plans are more easily understood and will work the best—if individual output can be measured.

The plan should guarantee the basic hourly rate set by job evaluation; the rate should be a good living wage comparable to the prevailing wage rate of the area for each job in question.

There should be a range of pay rates for each job, and these base rates should be related to total performance. By total performance, we mean giving consideration to quality, reliability, safety, and attendance, as well as to output. At periodic intervals, such as every six months or every year, the employee's rate step should be reviewed in relation to his total performance. For performance greater than standard, the operator should be compensated in direct proportion to his output, thus discouraging any restriction of production.

In order for the employee to associate his effort with compensation, the paycheck stub should clearly show both the regular and the incentive earnings. It is also advisable to indicate on a separate form, placed in the pay envelope, the efficiency of the operator for the past pay period. This is calculated as the ratio of the standard hours produced during the period to the hours worked during the period.

Provision must be made in an efficient manner for unavoidable loss of time not included in the standard. Also, techniques must be established for ensuring accurate piece counts of the required quality requirements. The plan should entail the control of indirect labor through measuring the productivity of those involved. Indirect labor should not be given a bonus equivalent to that of the direct workers unless some yardstick has been determined to justify their incentive earnings. Once the plan has been installed, management must accept the responsibility for maintaining it. Administration of the plan calls for keen judgment in making decisions, and for close analysis of all grievances submitted. Management must exercise its right to change the standards when the methods and equipment are changed. Employees must be guaranteed an opportunity to present their suggestions, but the advisability of their requests must be proven before any change is made. Compromising on standards must be avoided. A liberalizing of standards already based on facts will lead to complete failure of the plan. No incentive plan will continue to operate unless it is effectively maintained by a watchful, competent management team.

The motivation for incentive effort

The methods analyst should recognize that unless operators perform at good effort, the most favorably designed work station will not result in levels of productivity synonymous with company objectives. In order to achieve high levels of productivity, in addition to good physical work center design, the conditions surrounding the work should be such that all employees will do their best to help realize the stated objectives of the enterprise.

Some hypotheses related to motivation that the reader should be cognizant of include:

1. Work is and always has been a natural human activity. Today, as always, most people basically want to work and to achieve.
2. Almost all people want to be involved in the achievement of the goals established by their group or organization.
3. Most people will perform better if given both independence and control in their work situation.

4. Workers expect to see a relationship between their contributions and the magnitude of the resulting rewards.

To introduce an effective wage incentive plan, all four of the above hypotheses should be observed. It is insufficient to provide a one-for-one–type incentive plan based on fair standards. When people feel equitably treated as to compensation, there is no guarantee of high performance. A climate of motivation must accompany any formal incentive plan.

The reader may rightly ask, "What is meant by a climate of motivation?" Perhaps the first requirement in establishing the proper climate is to provide a management style which emphasizes a supporting role rather than a directive role. Thus, the goal should be to have all the workers feel that it is their responsibility to meet the objectives of the business and that it is the supervisors' responsibility to assist the workers as best they can.

Second, the goals of the enterprise should be clearly established, and should be broken down into division, department, work center, and individual goals. It is important that established goals be realistic, that they emphasize not just quantity but also quality, reliability, and any other characteristic essential to the success of the business. Every worker should understand the objectives of the company and the goals related to his work. These goals should be quantified in such a manner that every worker is cognizant of his achievement in relation to established goals.

Third, there should be regular feedback to all employees. Timely reporting should give all workers confirmation as to the results of their efforts and the impact these efforts have had on the established goals.

Fourth, every work situation should be designed so that the operator is in a position to control to a large extent the assignment he has been given. The reader will recognize that there is a considerable variation in different jobs from the standpoint of the amount of control that can be given to the operator. However, experience has verified that positive results are achieved where management has adopted the attitude that "job enrichment is important." A sense of responsibility is an important source of motivation, as is recognition for achievement.

Reasons for incentive plan failures

An incentive plan may be classified as a failure when it costs more for its maintenance than it actually saves, and thus must be discontinued. Usually it is not possible to put a finger on the precise immediate cause of failure for a given incentive installation. If the facts were completely known, numerous reasons would be found for the plan's lack of success.

In one survey, the principal causes of failure were shown to be poor employee attitude; excessive cost; an insufficiently stable product resulting in high standards costs; and too liberal allowances, resulting in too costly a program.

Certainly, any plan should be discontinued when its cost of maintenance exceeds the benefits derived through its use. The reasons given in the aforementioned survey are, for the most part, symptoms of a sick plan, of a plan doomed to failure—they are not really causes. The actual cause of the failure of any plan is incompetence in management: a management that permits the installation of a plan with poor scheduling, unsatisfactory methods, a lack of standardization or loose standards, and the compromising of standards, has no one but itself to blame for the plan's failure.

Of course, all the requisites of a sound incentive system may be met and the plan may still be unsatisfactory because of failure to promote good industrial relations relative to the program. The complete cooperation of the employees, the union, and supervision must be won in order to get the team spirit which is so necessary to attain ultimate success from the incentive installation.

Bruce Payne, president of Bruce Payne & Associates, Inc., reported on 246 companies where incentive plans had either failed or had developed weaknesses that necessitated revision. Three factors were enumerated as being responsible for the failures:[6]

		Percent
1.	Fundamental deficiencies in the plan	41.5
2.	Inept human relations	32.5
3.	Poor technical administration	26.0

The reasons falling under each of these divisions and their relative importance were summarized as follows:

Fundamental deficiencies		*Percent*
a.	Poor standards	11.0
b.	Low incentive coverage of direct productive work	8.6
c.	Ceiling on earnings	7.0
d.	No indirect incentives	6.8
e.	No supervisory incentives	6.1
f.	Complicated pay formula	2.0
Inept human relations		
a.	Insufficient supervisor training	6.9
b.	No guarantee of standards	5.7
c.	A fair day's work not required	5.0
d.	Standards negotiated with the union	4.8
e.	Plan not understood	4.1
f.	Lack of top-management support	3.6
g.	Poorly trained operators	2.4
Poor technical administration		
a.	Method changes not coordinated with standards	7.8
b.	Faulty base rates	5.1
c.	Poor administration, i.e., poor grievance procedure	4.9
d.	Poor production planning	3.2
e.	Large group on incentive	2.8
f.	Poor quality control	2.2

[6] Britton, *Incentives*, p. 60.

Administration of the wage incentive system

As has been pointed out, in order to be successful, an incentive system must be adequately maintained; it cannot maintain itself. To maintain a plan effectively, management must keep all employees aware of how the plan works and of any changes that may be introduced into the plan. One technique frequently used is to distribute to all employees an "Operating Instruction" manual outlining in detail not only company policy relative to the plan, but also all its working details with sample examples. The basis of job classifications, time standards, performance rating procedure, allowances, and grievance procedure should be thoroughly explained. The technique of handling any unusual situation should be described. Finally, the objectives of the organization and the role of each employee in the fulfillment of those objectives should be presented.

In the administration of the plan, a daily check should be made of low performance and excessively high performance in an effort to determine their causes. Low performance is not only costly to management in view of the guaranteed hourly rate, but it will lead to employee unrest and dissatisfaction. Unduly high performance is a symptom of loose standards, or the introduction of a methods change for which no standard revision has been made. In any case, a loose rate will lead to dissatisfaction on the part of employees in the immediate vicinity of the operator who is working on the job carrying the low standard. A sufficient number of such poor standards can cause the whole incentive plan to fail. Frequently, the operator who has the loose rate will restrict his daily production in fear that management will adjust the standard. This restriction of output is costly to the operator and the company, and results in dissatisfaction among neighboring workers who see a fellow employee on a soft job.

There should be a continuing effort to include a greater share of the employees in the incentive plan. When only a portion of the plant is on standard, there will be a lack of harmony among operating personnel because of significant differentials in take-home pay. The reader should be cautioned, however, that work generally should not be put on incentive unless:

1. It can readily be measured.
2. The volume of available work is sufficient to economically justify an incentive installation.
3. The cost of measuring the output is not excessive.

Periodic reviews of old standards should be made to assure their validity. On standards that have proven to be satisfactory, elemental values should be recapped for standard data purposes, so that even greater utilization may be made of the time values. Thus, greater coverage of the plant relative to the use of standards can be achieved.

Fundamental in the administration of any wage incentive plan that is keyed to production is the constant adjustment of standards to changes in the work. No matter how insignificant a methods change may be, it is well to review the standard for possible adjustment. Several minor methods improvements, in aggregate, can mount to a sufficient time differential to bring about a loose rate if the standard is not changed. When revising time standards due to methods changes, it is necessary to study only those elements affected by the changes.

So as to keep the incentive plan healthy, the company should arrange periodic meetings with operating supervisors to discuss fundamental weaknesses of the plan, and possible improvements in the installation. At these meetings, departmental performance should be compared, and specific standards that appear unsatisfactory should be brought to light and discussed. Employees expect and should receive an equitable working climate that assures a relationship between their contributions and their rewards.

Progress reports should be maintained, showing such pertinent information as departmental efficiency, overall plant efficiency, the number of workers not achieving standard performance, and the highest individual performance. These reports provide information on areas that need attention as well as areas where the plan is working satisfactorily.

For effective administration of the plan, it is essential that there be a continuing effort to minimize the nonproductive hours of direct labor. This nonproductive time for which allowance must be given to the operator represents lost time due to machine breakdowns, material shortages, tool difficulties, and long interruptions of any sort not covered in the allowances applied to the individual time standards. This time, frequently referred to as "blue ticket time" or "extra allowance time," must be carefully watched, or it will destroy the purpose of the entire plan.

For example, let us assume that on a certain job a production rate of 10 pieces per hour is averaged, and that an hourly rate of $3 is in effect under a straight daywork operation. Thus, we would have a unit direct labor cost of $0.30. Now this shop changes over to incentive wage payment where the day rate of $3 per hour is guaranteed, and, above task, the operator is compensated in direct proportion to his output. Let us assume that the standard developed through time study is 12 pieces per hour, and that for the first five hours of the working day, a certain operator averages 14 pieces per hour. His earnings for this period would then be:

$$(\$3.00)(5)\left(\frac{14}{12}\right) = \$17.48$$

Now let us assume that for the remainder of the working day, due to a material shortage, the operator could not be productively engaged in work. He would then expect at least his base rate or:

$$(3)(\$3.00) = \$9.00$$

which would give him earnings for the day of

$$\$17.48 + \$9.00 = \$26.48$$

This would result in a unit direct labor cost of:

$$\frac{\$26.48}{70} = \$0.378$$

Under daywork, even with the low performance, the operator would have produced the 70 pieces in less than the working day. Here his earnings would have been: $8 \times \$3.00$, or \$24.00, and the unit direct labor cost would have been: \$24.00/70, or \$0.342.

Under incentive effort, production performance will be considerably higher than under daywork operation, and with the shorter accompanying in-process time of materials, there will need to be very careful inventory control to prevent material shortages. Likewise, a program of preventive maintenance should be introduced so as to assure the continuous operation of all machine tools. Equally important to material control is the control of all nondurable tools, so that shortages with resulting operator delays do not develop.

An effective technique often employed to control the "extra allowance time" is to key the supervisor's bonus to the amount of this nonproductive time credited to the operator. The more of this time turned in for the pay period, the less would be the supervisor's compensation. Since the foreman is in an ideal position to watch schedules and material inventories, and to maintain facilities, he can control nonproductive downtime better than anyone else in the plant.

In addition to controlling the "extra allowance" or daywork time, it is essential that exact piece counts be recorded at each work station. The piece count which determines the operator's earnings is usually done by the operator himself. To prevent the operator from falsifying his production output, controls must be established.

Where the work is small (several pieces can be held in one hand), the operator makes a "weigh" count of his production at the end of the day or the end of the production run, whichever period is shorter. This "weigh" count is verified by his immediate supervisor, who initials his production report.

On larger work, one technique that is frequently employed is to have a tray or box with built-in compartments to hold the work. The tote box will hold round numbers of the work, such as 10, 20, or 50. Thus, at the end of the shift, it is a simple matter for the operator's supervisor to authenticate the operator's production report by merely counting the number of boxes and multiplying by 10, 20, 50, or whatever number each box holds.

Basically, wage incentive plans are established to increase productivity. In a sound and properly maintained installation, the percentage of incentive earnings of those workers on incentive would remain relatively constant over time. If analysis shows that incentive earnings continue to rise over a period of years, then one can be sure that the installation has problems that will ultimately erode the effectiveness of the plan. If, for example, the average incentive earnings increased from 17 percent to 40 percent in a period of 10 years, it would be suspected that the 23 percent rise was not due to a proportionate increase in productivity, but was probably due to a creeping looseness in standards.

Conclusion

The only acceptable wage incentive plan applied to individual workers today is the standard hour plan with a guaranteed day rate. Similarly, group plans must guarantee their respective day rates to all members of the group, as well as reward members of the group in direct proportion to their productivity once standard performance has been achieved.

Profit sharing or related types of cost improvement savings plans have met with success in many cases. In general, they tend to be more effective when they are installed in addition to, rather than instead of a, direct incentive installation.

Incentive principles have been applied in both job shops and production shops; in the manufacture of both hard goods and soft goods; in both manufacturing and service industries; and in both direct and indirect labor operations. Incentives have been used to increase productivity, improve the quality and reliability of the product, reduce waste, improve safety, and stimulate good working habits, such as punctuality and regularity of attendance.

Soundly administered incentive systems possess important advantages, both for workers and management. The chief benefit to employees is that these plans make it possible for them to increase their total wages, not at some indefinite time in the future, but immediately—in their next paycheck. Management obtains a greater output and, assuming that some profit is being made on each unit produced, a greater volume of profits. Normally, profits increase, not in proportion to production, but when a higher rate of production takes place, so that overhead costs per unit decrease. Then, too, the higher wages that result from incentive plans improve employee morale and tend to reduce labor turnover, absenteeism, and tardiness.

Since the proper functioning of incentive systems implies the existence of many prerequisites, such as good methods, good standards, good scheduling, and good management practices, the installation of incentives normally results in important improvements in production and supervisory methods. The reader should recognize that the activities that bring about these improvements should be performed even though incentives are not introduced; the improvements, therefore, are not necessarily attributable to the employment of the incentive plan or plans.

In general, the harder the work is to measure, the more difficult it will be to install a successful wage incentive plan. Usually, it is not advantageous to install incentives unless the work is subject to reasonably accurate measurement. Furthermore, it is usually not advantageous to introduce incentives if the availability of work is limited to less than 120 percent of normal.

The following 15 fundamental principles are recommended as a guide for sound practice in wage incentive installation and administration:

1. *Agreement of General Principles.* Management and labor should be in real agreement on the general principles involved in the relationship between work and wages.

2. *A Foundation of Job Evaluation.* There should be a sound wage rate structure, based upon an evaluation of the skill, responsibility, and working conditions inherent in the various jobs.

3. *Individual, Group, or Plant-wide Incentives.* It is generally conceded that standards, applied to individuals or to small integrated groups, are most effective. Such standards need to be set with the utmost care, and undoubtedly tend toward the lowest unit cost. At times, due to difficulties in recording individual production, or due to the possibilities of teamwork, group standards may be advisable. The larger the group, the less the individual response. With plant-wide incentives, some of the jealousies and transfer difficulties often inherent in group plans are eliminated, but without an unusual degree of leadership and co-operation, the incentive effect is greatly diluted.

4. *The Production-Incentive Relationship.* When production standards are properly set, and based upon well-engineered conditions, good practice has demonstrated the desirability of adopting an incentive payment in which earnings above the established standard are in direct proportion to the increased production.

5. *Simplicity.* The plan should be as simple as possible, without causing inequities. Workers should be able to understand the effect of their own efforts on their earnings.

6. *Quality Control and Improvement.* The desirable and economical degree of quality should be determined and maintained, tied in with bonus payment where advisable.

7. *Improved Methods and Procedures.* To secure the lowest costs and to prevent uneven standards and inequitable earnings, which lead to poor labor relations, the establishment of production standards should be preceded by basic engineering improvements in design, equipment, methods, scheduling, and material handling.

8. *Based on Detailed Time Studies.* Standards should be developed from detailed time studies. A permanent record of standard elemental times for each unit of an operation eliminates the occasion for many arguments. A table of

basic standard times prepares the way for proper introduction of technological improvements.

9. *Based on Normal Operation under Normal Conditions.* In general, the production standard should be established by management setting up the amount of work performed per unit of time by a normal qualified operator under normal conditions.

10. *Changes in Standards.* The plan should provide for the changing of production standards whenever changes in methods, materials, equipment, or other controlling conditions are made in the operations represented by the standards. In order to avoid misunderstandings, the nature of such changes and the logic of making them should be made clear to labor or its representatives who should have the opportunity to appeal through the grievance machinery.

11. *Considerations in Changing Standards.* Except to correspond properly with changed conditions, production standards once established should not be altered unless by mutual agreement between management and labor representatives.

12. *Keep Temporary Standards at Minimum.* The practice of establishing temporary standards on new operations should be kept at a minimum. It should, in any event, be made clear to all that the standards are for a reasonably short period only.

13. *Guarantee of Hourly Rates.* Under ordinary circumstances, the employees' basic hourly rates should become guaranteed rates.

14. *Incentives for Indirect Workers.* Effective standards may be established for most indirect jobs in the same manner as for direct jobs. If the exigencies of a situation demand that some form of incentive payment be applied to indirect workers as a whole, or in groups, then the indirect man-hours should be correlated to some measurable unit, such as production or direct employee hours, so that indirect labor cost may be kept under control.

15. *Thorough Understanding of Human Relations Involved.* Finally, it should be emphasized that unless management is prepared to work on the problem with a thorough understanding of the human relations involved, it had better have no incentive plan. Whereas such a plan may be a progressively constructive force for increased production, it may also be a means of disrupting labor relations and of actually lowering production. While necessarily retaining its functions, management should take into account labor's point of view. It should impart to labor a complete understanding of the plan and patiently consider grievances, in whatever manner may be agreed upon.[7]

Well-planned and -administered incentives will increase production and decrease total unit cost. Usually, they will more than compensate for the price paid in increased costs of industrial engineering, quality control, and timekeeping which may have resulted from their use.

TEXT QUESTIONS

1. Under what three general classes may the majority of wage incentive plans be classified?

---

[7] From a talk by John W. Nickerson, "The Importance of Incentives," given at the Second Annual Time Study and Methods Conference, New York.

2. Differentiate between individual wage payment plans and group-type plans.

3. What is meant by the term *fringe benefits?*

4. What company policies are included under nonfinancial incentives?

5. What are the characteristics of piecework? Plot the unit cost curve and operator earning curve for daywork and piecework on the same set of coordinates.

6. Why did measured daywork become popular in the 1930s?

7. Define "profit sharing."

8. What specific type of profit sharing plan has received general acceptance?

9. What three broad categories cover the majority of profit sharing installations?

10. Upon what does the amount of money distributed depend under the cash plan?

11. What determines the length of the period between bonus payments under the cash plan? Why is it poor practice to have the period too long? What disadvantages are there to the short period?

12. What are the characteristic features of the deferred profit sharing plan?

13. Why is the "share and share alike" method of distribution not particularly common? On what basis is this technique advocated by its proponents?

14. What 10 principles summarized by the Council for Profit Sharing Industries are fundamental for a successful profit sharing plan?

15. Why have many unions shown an antagonistic attitude toward profit sharing?

16. What suggestions has James Lincoln offered for those embarking on an incentive installation?

17. How does job enrichment differ from job enlargement?

18. Why is it usually advisable to have a range of pay rates that are applicable to each job?

19. What four important hypotheses related to motivation should the work measurement analyst be cognizant of?

20. Why do many union officials prefer straight daywork as a form of wage payment?

21. What are the fundamental prerequisites of a successful wage incentive plan?

22. What are the requisites of a sound wage incentive system?

23. What is extra allowance or "blue ticket" time? How may it be controlled?

24. Why is it fundamental to keep time standards up-to-date if a wage incentive plan is to succeed?

25. What does unduly high performance indicate?

26. What responsibilities of the foreman, time study department, quality control department, production department, accounting department, and industrial relations department are essential to the effective administration of a sound wage incentive system?

27. What 15 fundamental principles should be followed in order to assure a successful wage payment plan?
28. When would it be inadvisable to put an indirect labor activity on incentive?
29. How would you go about establishing a climate for worker motivation?

GENERAL QUESTIONS

1. Why do many unions object to management posting incentive earnings?
2. Should indirect labor that is not on incentive carry a higher base rate than direct labor on incentive when both jobs carry the same evaluation? Why or why not?
3. What is the major reason that causes many union officials to disfavor any form of incentive wage payment?
4. How can the piece count be controlled within the plant?
5. How can a yardstick of performance be established in the following jobs: tool crib attendant, tool grinder, floor sweeper, chip-hauler?
6. Explain how you would enrich the position of "punch press operator."

PROBLEMS

1. An allowed time of 0.0125 hours/piece is established for machining a small component. A setup time of 0.32 hour is established also, as the operator performs the necessary setup work on "incentive." Compute:
    a. Total time allowed to complete an order of 860 pieces.
    b. Operator efficiency, if job is completed in an eight-hour day.
    c. Efficiency of the operator if he requires 12 hours to complete the job.
2. A "one-for-one" or 100 percent time premium plan for incentive payment is in operation. The operator base rate for this class of work is $5.20. The base rate is guaranteed. Compute:
    a. Total earnings for the job at the efficiency determined in problem 1(*b*).
    b. Hourly earnings, from above.
    c. Total earnings for job at the efficiency determined in problem 1(*c*).
    d. Direct labor cost per piece from (*a*), excluding setup.
    e. Direct labor cost per piece from (*c*), excluding setup.
3. A forging operation is studied, and a rate of 0.42 minute per piece is set. The operator works on the job for a full eight-hour day and produces 1,500 pieces.
    a. How many standard hours does the operator earn?
    b. What is his efficiency for the day?
    c. If his base rate is $4.90 per hour, compute his earnings for the day. (Use a 100 percent time premium plan.)
    d. What is the direct labor cost per piece at this efficiency?
    e. What would be the proper piece rate (rate expressed in money) for this job, assuming that the above time standard is correct?
4. A 60–40 gain sharing plan is in operation in a plant. The established time value on a certain job is 0.75 minute, and the base rate is $4.40. What is the direct labor cost per piece when the operator efficiency is:

a. 50 percent of standard?
b. 80 percent of standard?
c. 100 percent of standard?
d. 120 percent of standard?
e. 160 percent of standard?

5. A worker is employed in a plant where all the rates are set on a money basis (piece rates). He is regularly employed at a job where the guaranteed base rate is $4.40. His regular earnings are in excess of $44 per day. Due to the pressure of work, he is asked to help out on another job, classified so that it pays $5 per hour. He works three days on this job and earns $40 each day.
   a. How much should the operator be paid for each day's work on this new job? Why?
   b. Would it make any difference if he had worked on a new job where the base rate was $4 per hour and had earned $36? Explain.

6. An incentive plan employing a "low-rate high-rate" differential is in use. A certain class of work has the guaranteed "low-rate" of $3 per hour and the "high-rate" for work on standard of $4.60 per hour. A job is studied and a rate of 0.036 hours per piece is set. What is the direct labor cost per piece at the following efficiencies:
   a. 50 percent?
   b. 80 percent?
   c. 98 percent?
   d. 105 percent?
   e. 150 percent?

SELECTED REFERENCES

Ells, Ralph W. *Salary and Wage Administration.* New York: McGraw-Hill Book Co., 1945.

Lincoln, James F. *Lincoln's Incentive System.* New York: McGraw-Hill Book Co., 1946.

Zollitsch, Herbert G., and Langsner, Adolph. *Wage and Salary Administration.* 2d ed. Cincinnati: South-Western Publishing Co., 1970.

26

Training and research for methods, time study, and wage payment

In a survey conducted by Ralph E. Balycat on subjects found in industrial engineering curriculas and reported in the May 1954 issue of the *Journal of Industrial Engineering,* educators ranked motion and time study first in order of importance in a listing of 41 subject areas. A somewhat similar survey made 10 years later and reported in the January 1964 issue of *Factory* magazine found that industrial engineers still spent most of their time on work measurement. This information was based on the response of 250 of the larger U.S. manufacturing companies and was indicative of more than 8,700 nonclerical employees in industrial engineering. Table 26–1 presents the percent of the industrial engineer's time spent on major functions.

Today, the majority of the work of the practicing industrial engineer in industry is still centered on methods, standards, and wage payment.

Independent surveys made by the author within the past year indicate that most businesses and industries expect at least half of the industrial engineering effort to be directed toward work methods and standards. It is true that the utilization of industrial engineering techniques is spreading to all areas of the modern business, including marketing, finance, sales, and top management, yet the data input for almost all optimization techniques includes standard times. The importance of work measurement in such indirect areas as office activities, maintenance, shipping and receiving, sales work, inspection, and the toolroom will continue to grow.

To meet the demand, and to reap more quickly the benefits of training in this field, many industries have embarked on education programs of their own, conducted in their own plants on company time. For example, an extensive 1974 survey of over 5,300 U.S. companies revealed that 80 percent of the companies surveyed were providing formal training programs for first-line supervisors, and that 42 percent of these training programs dealt with work simplification.

TABLE 26–1
Percent of industrial engineer's time spent on major functions

		Percent
1.	Work measurement	33.4
2.	Work methods	21.1
3.	Production engineering	13.0
4.	Manufacturing analysis and control	9.9
5.	Facilities planning	8.6
6.	Wage administration	5.6
7.	Safety	2.6
8.	Production and inventory planning	2.0
9.	Quality control	1.1
10.	Other	2.7
		100.0

One might ask, "What will it cost and what are the potential savings if a methods and standards program is introduced?" The following has been the experience of the author with reference to the aforementioned question. Production under a typical daywork environment is about 80 percent of normal productivity. Thus, by establishing standards after methods have been studied, improved, and standardized under daywork operation, an improvement of 25 percent in productivity can be expected.

Of course, several items of cost need to be considered in the introduction of such a program. If the annual wages of 1 operator represents 1 unit of direct labor payroll, 100 direct labor employees will represent 100 units. Usually one capable methods and standards analyst will be able to establish good methods and maintain reliable standards for approximately 100 employees. The cost of the methods and standards analyst can be estimated as three units of cost. In order to keep production records, make routine calculations, assist with the taking of studies, and so forth, a junior standards analyst engineering technician may be required at a salary of two units of cost. Let us estimate that inspection costs rise by two units in view of the added productivity. (This often is not the case—in fact, inspection costs frequently lessen because of methods improvement). Secretarial help will require one unit of cost.

Thus, the total extra labor cost to introduce a methods and standards program per 100 direct labor employees is estimated at:

	Units
Methods and standards analyst	3
Junior standards analyst engineering technician	2
Secretarial assistance	1
Increased inspection	2
Total	8

The above cost estimate should be adequate for most jobbing shops with a wide variety of operations and frequent changes of products. It can be expected that costs will be considerably lower for well-standardized processes or mass-production plants.

With gross direct labor savings of 20 units and an additional administrative expense of 8 units, a good methods and work measurement program should effect a net savings of 12 units or 12 percent of a daywork direct labor payroll.

The savings in a fixed factory burden resulting from the introduction of a work measurement program may be even more significant. Since it has been estimated that a methods and work measurement program would increase productivity by 25 percent, then each dollar of fixed burden is distributed over 125 units, where formerly it was distributed over only 100 units. Thus, the fixed burden cost per unit of output is reduced to 100/125, or 80 percent, of the former rate. For example, if a company's fixed burden rate is 100 percent of the direct labor cost (this is a low estimate), the reduction in burden cost is equivalent to 20 percent of the direct labor payroll.

Individual plant methods training programs

When properly undertaken, any methods training program will prove to be self-supporting. In the experience of the writer, several concerns that have introduced and gone ahead with training in methods have almost immediately realized substantial savings in all areas of their plants. One manufacturer of farm machinery gave a 64-hour course in methods analysis over a period of 21 weeks. In attendance were 42 foremen, assistant foremen, time study analysts, and other key operating personnel. At the termination of the course, 28 methods projects were turned in, representing methods improvement solutions that could be introduced in the plant. These ideas included machine coupling, redesign of tools and product, paperwork simplification, improved layout, better means of material handling, improved dies, jigs, and fixtures, elimination of operations, adjusting of tolerances and specifications, and many other usable improvements that resulted in the saving of thousands of dollars annually.

In addition to the immediate gains mentioned, the training in operation analysis and work simplification developed an analytic approach on the part of the operating personnel so that in the future they will be continually on the alert to find a "better way." They developed an appreciation of cost of manufacture, and, at the completion of the course, were more cognizant of the relationship between output and selling price.

This company, by providing the means for employee "on-the-job" training in the field of methods and related areas, has gone a long way toward assuring its place in an extremely competitive market.

Training in methods and time study

The lack of success of some time and methods study programs is in part due to a lack of understanding of the techniques by both management and the operating personnel. One of the easiest ways to assure the success of any practical innovation is to inform all affected parties as to how and why it operates. When the theories, techniques, and economic necessity of methods, work measurement, and employee motivation are understood by all parties, little difficulty will be encountered in their application. In unionized shops, it is especially important that the officers of the union at the local level understand the need for and the steps involved in establishing standards of performance. Training in the areas of performance rating, application of allowances, standard data methods, and job evaluation are especially important. In the experience of the writer, the companies that have provided training in the elements of time study for union officials and stewards, as well as representatives of management, have had harmonious relationships in the field of methods, standards, and wage payment.

In addition to informative training to acquaint the various operating and supervisory members of the plant with the philosophies and techniques of time and motion study, it is wise for industry to provide training for those of its personnel who plan to make this field of endeavor their lifework. Not only is it necessary to train the neophyte, but the experienced analyst should be continually checked to make certain that his conception of normal is not deviating from standard. As new developments, which are constantly being made, are recognized, the information should be circulated throughout the personnel of the methods, time study, and wage payment sections. This should be done through the medium of formalized training.

It is recommended that any company that has, or plans to have, a program of work simplification or methods analysis, time study, or work measurement, and incentive wage payment, include as part of its installation a continuing training program. The devotion of a two-hour period once a week for training foremen, union stewards, direct labor, and management, will be well worth the time and money spent. The periodic verification of the rating ability of the staff of the time study department is fundamental.

Developing creativity

Creative work is not confined to a particular field or to a few individuals, but is carried on in varying degrees by people in many occupations: the artist sketches, the newspaper writer promotes an idea, the teacher encourages student development, the scientist experiments with a theory, the methods and time study analyst develops improved methods of doing work.

Creativity implies newness, but it is as often concerned with the improvement of old products as it is with the creation of new ones. A "how to produce something better" attitude, tempered with good judgment, is an important characteristic of the effective methods and time study analyst.

Developing creativity in the practicing methods and time study analyst is a continuing problem. Knowledge of the fundamental principles of physics, chemistry, mathematics, and engineering subjects is a good foundation for creative thinking. If the practicing analyst does not have this basic background, he should acquire it either by education programs or else through study alone. Of course, knowledge is only a basis for creative thinking, and does not necessarily stimulate it.

The inherent personal characteristics of curiosity, intuition, perception, ingenuity, initiative, and persistence produce an effective creative thinker. Curiosity seems to stimulate more ideas than does any other personal characteristic. One aid to the development or restoration of curiosity is to train oneself to be observant. The methods and time study analyst should get into the habit of asking himself how a particular object was made, of what materials it was constructed, why it was designed to be of a particular size and shape, why and how it was finished as it was, and how much it cost. If he isn't able to answer these questions, he should find the answers, either through analysis or by referring to source materials and consulting others. These observations lead the creative thinker to see ways in which products or processes can be improved through cost reduction, quality improvement, ease of maintenance, or improved aesthetic appeal.

One significant creative idea usually opens up fields of activities that lead to many new ideas. Frequently one idea that has application on a given product or process will have equal application on other products and similar processes.

Decision-making methods and processes

The methods and time study analyst should become trained in the various decision-making processes. Judgment alone will not always provide the best answer. The analyst is continually confronted with such questions as: "Would it be economically wise to go ahead with this process?" "Should this tool be redesigned with the probability of 0.50 that production requirements will be increased by between 500 and 1,000 pieces?" "Will it be necessary to take 5,000 work sampling observations if we are seeking an element that occurs approximately 12 percent of the time and if tolerance limits are so much?" These are typical questions that must be answered, and analytic means provide a much better method than does mere judgment.

To be able to handle problems similar to these, the analyst should have a working knowledge of engineering economy, statistical analysis, and the theory of probability. He should also become familiar with the various decision-making processes for the evaluation of alternatives. For example, let us assume that the analyst has four alternatives to consider (a_1, a_2, a_3, a_4) which would be applied to four possible states of the product or market (S_1, S_2, S_3, S_4). Let us also assume that he estimates the following outcomes for the various alternatives and states of the market.

Alternatives	States of product or market			
	S_1	S_2	S_3	S_4
a_1	0.30	0.15	0.10	0.06
a_2	0.10	0.14	0.18	0.20
a_3	0.05	0.12	0.20	0.25
a_4	0.01	0.12	0.35	0.25

If the outcomes represent profits or returns, and the analyst is certain that the state of the market will be S_2, then he would definitely decide on alternative a_1. If the outcomes represent scrap or some other factor that he wishes to minimize, then alternative a_3 would be chosen. (Although a_4 also has an outcome of 0.12, he chooses a_3, since under this alternative there is less variability as to outcome than with a_4.)

Seldom should decisions be made under an assumed certainty. Usually, some risk is involved in predicting the future state of the market. Let us assume that the analyst is able to estimate the following probability values associated with each of the four states of the market:

$$
\begin{aligned}
S_1 &\ldots\ldots\ldots\ldots\ldots\ldots 0.10 \\
S_2 &\ldots\ldots\ldots\ldots\ldots\ldots 0.70 \\
S_3 &\ldots\ldots\ldots\ldots\ldots\ldots 0.15 \\
S_4 &\ldots\ldots\ldots\ldots\ldots\ldots \underline{0.05} \\
& 1.00
\end{aligned}
$$

A logical decision-making strategy would be to calculate the expected return under each decision alternative and then select the largest value if we are maximizing or the smallest if we are minimizing. Here

$$ E(a) = \sum_{j=1}^{n} P_j C_{ij} $$

$$ E(a_1) = 0.153 $$
$$ E(a_2) = 0.145 $$
$$ E(a_3) = 0.132 $$
$$ E(a_4) = 0.15 $$

Thus, alternative a_1 would be selected if we are maximizing.

A different decision-making strategy would be to consider the state of the market that has the greatest chance of occurring. This state of the market would, of course, be S_2, since S_2 carries a probability value of 0.70. The choice again, based upon the most probable future, would be alternative a_1, with a 0.153 return.

A third decision-making strategy under risk would be based upon a "level of aspiration." Here, we assign an outcome value (C_{ij}) which represents the consequence of what we are willing to settle for if we are reasonably sure that we will get at least this consequence most of the time. This assigned value may be referred to as representing a level of aspiration which we shall denote as "A." We now determine the probability for each a_j that the C_{ij} in connection with each decision alternative is greater than or equal to "A." The alternative with the greatest $P(C_{ij} \gtreqless A)$ is selected.

For example, if we assign the consequence of 0.10 to A, we would have the following:

$$(C_{ij} \gtreqless 0.10)$$
$$a_1 = 0.95$$
$$a_2 = 1.00$$
$$a_3 = 0.90$$
$$a_4 = 0.90$$

Since decision alternative a_2 has the greatest $P(C_{ij} \gtreqless A)$, it would be recommended.

The analyst may be unable to assign with confidence probability values to the various states of the market and will therefore want to consider any one of them as being equally likely.

A decision-making strategy that may be used under these circumstances is based upon the "principle of insufficient reason," since there is no reason to expect that any state is more likely than any other state. Here we would compute the various expected values based upon:

$$E(a) = \frac{\sum\limits_{j=1}^{n} C_{ij}}{n}$$

In our example, this would result in:

$$E(a_1) = 0.153$$
$$E(a_2) = 0.155$$
$$E(a_3) = 0.155$$
$$E(a_4) = 0.183$$

and, based upon this choice, alternative 4 would be proposed.

A second decision-making strategy that the analyst may consider when making decisions under uncertainty is based upon the criterion of pessimism. When one is pessimistic, the worst is anticipated. Under a problem of maximization, the minimum consequence is selected for each decision alternative. These minimum values are compared, and the alternative that has the maximum of these minimum values is selected. Thus, in our example:

Alternative	Min. C_{ij}
a_1	0.06
a_2	0.10
a_3	0.05
a_4	0.01

Here, alternative a_2 would be recommended since its minimum value of 0.10 is a maximum when compared with the minimum values of the other alternatives.

The plunger criterion is a third decision-making strategy that the analyst may want to consider. This criterion is based upon an optimistic approach. If one is optimistic, the best is expected, regardless of the alternative chosen. In a maximizing problem, the maximum C_{ij} is selected for each alternative, and then the alternative having the largest of these maximum values is selected. Thus:

Alternative	Max. C_{ij}
a_1	0.30
a_2	0.20
a_3	0.25
a_4	0.35

Here, decision alternative a_4 would be recommended because of its maximum value of 0.35.

Most decision makers will be neither completely optimistic nor completely pessimistic. Instead, a coefficient of optimism, X, is established where:

$$0 \leq X \leq 1$$

Then, a Q_i is determined for each alternative, where:

$$Q_i = (X)(\text{Max. } C_{ij}) + (1 - X)(\text{Min. } C_{ij})$$

The alternative recommended is the one that is associated with the maximum Q_i in cases of maximization and with the minimum Q_i in cases of minimization.

A final decision-making approach based upon uncertainty is the minimax regret criterion. This criterion involves the calculation of a regret matrix. For each alternative, based upon a state of the market, a regret value is calculated. This regret value is the difference between the payoff actually received and the payoff that could have been received if the decision maker had known the state of the market that was going to occur.

To construct the regret matrix, select the maximum C_{ij} for each state

S_j and then subtract from this maximum the C_{ij} value of each alternative associated with that state. In our example, the regret matrix would be:

Alter-natives	States			
	S_1	S_2	S_3	S_4
a_1	0	0	0.25	0.19
a_2	0.20	0.01	0.17	0.05
a_3	0.25	0.03	0.15	0
a_4	0.29	0.03	0	0

We now select the alternative associated with the minimum of the maximum regrets (minimax).

Alternative	*Max. r_{ij}*
a_1	0.25
a_2	0.20
a_3	0.25
a_4	0.29

Based upon the minimax regret criterion, a_2 would be selected in view of its minimum regret of 0.20.

It is apparent that the different decision-making processes may very well give different answers. The analyst should become familiar with the various processes and make use of those that are most appropriate to his organization.

Labor relations and work measurement

The importance of harmonious labor-management relations is recognized by every business owner. Sound work measurement philosophies and practices can do a great deal to promote good relations between labor and management; lack of consideration of the human element in work measurement procedures can cause sufficient turmoil to make the profitable operation of a business impossible. Management should identify and implement those conditions that are most likely to enable its employees to achieve the organization's objectives.

Union objectives

To understand the relationship between work measurement and labor relations, it is important to understand the objectives of the typical labor union. Briefly, the principal objectives of the typical union are to secure for its members higher wage levels, decreased working hours per work-week, increased social and fringe benefits, improved working conditions, and job security. The philosophy underlying the union movement has in the past had much to do with the opposition of organized labor to incentive systems. Unions formerly looked upon themselves primarily as fighting units, and endeavored to unite workers by seeking ends common to all members. It was not to the advantage of the early unions to emphasize differences in workers' abilities and interests; to do this would increase rivalries and jealousies among their members and potential members. Consequently, organized labor usually sought percentage wage increases for all members of a group rather than means by which remuneration would be adjusted to the worth of individual workers. The work of the methods standards and wage payment analyst came to be looked upon as means by which management sought to destroy the solidarity of workers by stressing the differences in their capabilities.

The enactment of government legislation, however, has changed the status of labor unions. Management today recognizes the union as a bargaining agent for its employees. Unions, therefore, are less fighting units than bodies concerned with the orderly negotiation of wage contracts for all their members. It is also true that many of their members will not remain content with the type of wage negotiation that concerns itself only with obtaining high minimum wages for the entire group, and leaves the determination of extra rewards for the more valuable workers entirely up to management. To satisfy great numbers of their members, unions must take on the function of obtaining equitable wages (recognizing workers' different skills and qualities) as well as high wages for all; in fact, unions have already done this in many instances. Methods, time standards, job evaluation, merit rating, and incentive systems are tools for assuring equitable wages and good working conditions and may soon be as important to organized labor as to management.

Labor executives, in bargaining over wage contracts, are today in an excellent position to obtain safeguards for the proper handling of the development of work methods and standards and for fair wage payment practices. They may, for example, demand that provisions be inserted in contracts: (*a*) forbidding the reduction of standard times without a methods change; (*b*) requiring minimum hourly rates and payment for time lost due to no fault of the workers; (*c*) establishing grievance procedures for handling all workers' complaints arising out of the functioning of job evaluation, merit rating, and wage incentives; and (*d*) even giving labor the right to participate in job- and worker-rating activities, the setting of standard times, and the determination of piece rates.

Many unions today train their own time study men. In most instances, however, these time study men are employed to check standard times and to explain them to workers, but do not take part in their initial establishment. However, the training the union time study men receive usually treats the concepts, philosophies, and techniques of methods, work measurement, and wage payment from a different point of view than that of management's time study men. For example, in a recent collective bargaining report published by the AFL–CIO for the training of union stewards in motion and time study, the subject of time study is introduced as follows:

Time study, which is widely used for determining work loads and wage incentive standards, is an imprecise tool and lends itself to easy abuse. Unions confronted with it must be consistently on guard against the use of arbitrary, unreasonable and unrealistic time study results.

Of the whole field of so-called "scientific management," time study is the area in which most of labor's distrust and suspicions are centered. Ever since its introduction in the 1880's most unions have opposed the use of stopwatch time study.

Labor's distrust stems from its practical experiences with time study. Problems arise both because of the inherent shortcomings of the time study process itself as well as the application of the technique in industry.

Time study is usually represented to unions as "scientific." But time study produces results which are simply judgments. They are not, and cannot be, scientific or accurate: at best they represent approximations, and at worst they are no better than wild guesses.

Time study is supposed to be a method of determining the time which should be allowed for a worker to perform a defined job according to a specific method and under prescribed conditions.

Time studies are usually made by timing workers with a stopwatch while they are doing a certain job. This time is then adjusted for such factors as delays, fatigue, personal needs and incentive factors. The result is frequently called a standard. The job or time standard may be in terms of units per hour, standard hours per 100 units or time per unit. If an incentive plan exists, the standard may also be expressed in monetary terms, such as 1 cent per piece produced.

Since most time studies are taken of an individual worker or a small group of workers, disagreements over time studies are usually expressed as grievances on specific job standards at the local union level.

While international unions can and do provide expert assistance and information to their locals, the investigation and processing of time study disputes remain primarily a local union problem.

Based on their own experiences, local union have devised approaches to time study which fall into four general categories:

1. Some locals prevent the use of time study altogether.
2. Some locals allow management to use any method of setting job standards it desires, but the locals reserve the right to bargain on the results.
3. Some locals participate directly with management in making time studies and in setting standards.
4. A majority of locals allow management to make time studies but insist on bargaining on both the methods used and their applications.

Unions faced with time study must be certain that they have complete information on how it is used by their company. In order to provide essential protection to members, this information must not be limited in any respect. It must include not only the results of time study, that is, the individual job standards, but also the plan in use by the company and the exact procedures followed.

That this information is essential to collective bargaining is apparent. Without it, a union would be unable to sensibly process grievances or discuss pertinent contract clauses.

The union's legal right to such information has been definitely established by decisions of arbitrators and the National Labor Relations Board.

In spite of this, some unions are still finding it difficult to secure all the necessary data. Some managements still claim that such information is confidential. When union pressure forces compliance, some managements attempt to restrict the information they will provide or the means of providing it.

To insure immediate availability of time study data, without question or limitation, most unions insist on a contract clause such as:

The company shall furnish to the union a copy of the time study plan presently being used. It shall make available for inspection by the union any and all records pertaining to time study and the setting of production standards including original time study observation sheets. Upon request by any shop steward or union officer, the company will furnish copies of any of the above information including copies of time studies.

Some companies, by contract, have been required to furnish the union with photostatic copies of the time study observation sheet each time a study is made.

Accuracy of Time Studies

We do not bargain standards!! is a frequent assertion by management when faced with union time study grievances.

Management claims that time studies produce facts and, of course, facts are not subject to bargaining or compromise.

If time study did result in facts, then this management position might be sound and the union might be able to bargain only on whether or not time study should be used in the plant.

But such is not the case with time study. At best, standards developed from time study are only approximations. They involve the use of considerable judgment by the time study man at every step of the time study procedure.

The actual recording of the time involves the least judgment of all, but even here a 10 percent error is expected and recognized by time study experts.

Among some of the variable factors which will significantly affect a time study result are:

1. The selection of the worker to be studied.
2. The conditions under which the work is performed during the time study.
3. The manner in which the operation is broken down into parts or elements.
4. The method of reading the stopwatch.
5. The duration of the time study.

6. The rating of the worker's performance.
7. The amount of allowances for personal needs, fatigue and delays.
8. The method of applying the allowances.
9. The method of computing the job standard from the timing data.

Inaccuracies of "Rating" Process

After a time study man has completed the stopwatch timing, he has obtained a figure which represents the time, on the average, that it took the particular worker he observed to perform the job.

This time could be used to set a standard only if the worker who has been time studied could be considered a qualified, average worker, working at a normal pace exhibiting a normal amount of skill.

If the worker observed does not meet these specifications, then the observed time must be adjusted to make it conform to the normal time. This adjusting procedure most frequently is called "rating," but is sometimes called "leveling" or "normalizing."

Rating has been defined as the process whereby the time study man compared the actual performance which he observed in a concept or idea of what normal work performance would be on the job being studied. This concept of "normal" must by its nature be carried around in the time study man's head.

During the rating process, the time study man must make two distinctly personal judgments. First, he must formulate in his mind the concept of what a normal performance should be on the job being studied and second, he must numerically compare the observed performance with his mentally conceived normal performance.

The very nature of rating opens it to abuse. By manipulating this rating factor, it is easy for the time study man to end up with practically any result he chooses.

In fact, as many unionists know, the rating factor is often used to enable the time study man to end up with a standard determined before the time study is taken. In other words, time study is often used to "prove" to the workers that a workload or standard set by the company is fair.

Time Study Men "Adjust" Findings

For example, a company may decide, without measurement of any kind, that a certain item should be produced at the rate of 60 pieces an hour or one piece per minute. Its time study man then studies the job and finds that the average time it took the operator to make one piece was 1¼ minutes. This means that only 48 pieces would be produced in an hour. Obviously, this falls far short of management's set goal of 60 pieces per hour.

So the time study man simply decides that the operator was working below "normal" during the time study and that a "normal" operator would have worked faster. The time study man then adjusts his findings; he cuts the 1¼ minutes that it took the "slow" operator by 20 percent—to come up with the standard that the company wanted of 1 piece per minute or 60 pieces per hour.

It is practically impossible for a union to prove this kind of deliberate deceit since the "normal" operator is only a figment of the time study man's imagination.

So it is easy to see that unscrupulous time study men can readily manipulate time studies.

But the situation unfortunately is not much better even when management and its time study men are trying to be completely honest and objective.

Test of Time Study Men's Estimates

Most time study men claim that they can judge worker pace within 5 percent; some have even claimed that with experience this could be reduced to an average error of 2 percent. But these claims of precision have been proved false by many studies by university researchers and management groups to determine the ability of time study men to rate workers' performance.

These studies show that time study men will, in more than half their ratings, misjudge by more than 10 percent variations in work pace. Such errors of as much as 40 percent are not at all uncommon among qualified and experienced time study men.

Specifically, the Society for the Advancement of Management, the largest national organization of industrial engineers and supervisors, recently conducted one of the most extensive rating studies ever undertaken.

The study used films of workers performing a variety of industrial operations. Time study men rated different paces of these operations. (From the calibration of the film, it was possible to get an accurate measurement of variations in pace.)

Trade unionists were not surprised at the published results of the SAM study. They showed, for example, that for 599 time study men:

1.   The average error in estimating variations in work pace was 10.57 percent.
2.   59 percent of the time study men had average errors larger than 10 percent; 41 percent averaged less than 10 percent error and less than 12 percent had errors averaging below 5 percent.

One man's errors averaged as high as 22 percent. These, of course, are average errors. Some individual ratings were lower and some were higher than the averages. The SAM did not publish the range of errors, but similar studies have shown errors higher than 40 percent.

Burden on Grievance Procedure

Time study has become a major issue in union-management relations. Unions have found that grievance problems are greatly increased in plants where time study exists. Not only are there more grievances, but a greater amount of time must be spent in investigating and processing time study grievances.

One local union recently reported that in the first five months of this year, it had carried 254 grievances to the fourth step of the grievance procedure. (That's when the grievance committee meets with the plant superintendent.) Of these 254 grievances, 221 or 87 percent were time study cases.

Local Union Representatives Can Handle Time Study Problems

Some unionists have felt that time study grievances could not be handled in the same manner as other grievances. Some have felt that time study grievances could not be handled by local union representatives but that "outside" experts were needed.

While local unions may seek help in special cases, the majority of time study grievances not only should, but can be successfully handled by shop stewards. See the listing starting below under "Union Handling of Time Study Grievances," for points to be considered in handling disputes over production standards.

The most important factor to remember is that no time study is accurate. Union judgment is on a par with management's. The worker on the job is as accurate a judge as anyone of the propriety of any job standards.

Because management does not like to bargain on standards and because there is so little that is factual about time study, a large proportion of time study grievances are taken to arbitration.

Unfortunately, while many arbitrators may not be biased in favor of the company or union, they all too often accept time study as being scientific and mathematically precise. They do not recognize the shortcomings of time study and therefore are more likely to accept its results.

Despite their difficulties and shortcomings, local unions can do a good job of protecting workers against unfair management time studies. To do an effective job, however, locals need informed and alert officers and stewards. Above all union representatives should not be "snowed under" by the so-called scientific procedures and arguments of management time study men.

In locals where trouble with time study persists, the international union should be requested to give advice and assistance. Various international unions have staff members with specialized experience in handling time study difficulties.

Union Handling of Time Study Grievances

Time study grievances should be handled the same as any other grievances. The most important factor is that of getting facts.

While some knowledge of time study is helpful, it is not necessary for the shop steward or other union representative to be a time study man in order to process a time study grievance.

The union representative should:

1. Secure a copy of the company's record of the operation in dispute.
2. Make certain that the records of job conditions and job description are complete. If either is incomplete, it will be impossible to reproduce the job as it was when the time study was made, and therefore the company's time study cannot be checked. This alone is grounds for rejecting the study.
3. If the time study sheet does contain sufficient information as to how and under what conditions the job was being performed when the time study was made, then it is necessary to determine whether the job is still being performed in exactly the same way now.

   Check to see if there has been any change in machines, materials, tools, equipment, tolerances, job layout, etc. Any change should be checked to see if it affects the ability of the operator to produce at the same rate as that achieved during the time study.
4. Usually the total operation or job cycle is broken down for timing purposes into parts which are called elements. Check the elemental break-

down of the job on the time study sheet. See that the beginning and ending point of each element is clearly defined. If it is not, any attempt to measure elements and give them a time value was pure guesswork.

5.   Check the descriptions of each element to see if they describe what the operator is presently required to do. Any change invalidates the original study.

6.   Make sure that everything the operator is required to do as part of the job has been recorded and timed on the time study sheet. Watch for tasks which are not part of every cycle.

   Such things as getting stock, adjusting machines, changing tools, reading blueprints, and waiting for materials are examples of items most frequently missed.

7.   Check for strike-outs. A time study finding is based on a number of different timings of the same job. The time study man may discard some of his timings as abnormal.

   If the time study man has discarded any of his recorded times, he must record his reasons for doing so. This enables the union intelligently to determine if the strike-out is valid. The fact that a particular time is larger or smaller than other times for the same element is not sufficient reason for discarding it.

8.   Determine if the time study was long enough to accurately reflect all of the variations and conditions which the operator can be expected to face during the job. Was it a proper sample of the whole job? If not, the time study should be rejected, as its results are meaningless.

9.   See that only a simple average, and not the median, mode or other arithmetic device was used in calculating the elemental times. The average is the only proper method for time study purposes.

10.   Check the rating factor on the time study sheet. Try to find out if the time study man recorded his rating factor before leaving the job or after he computed the observed times. Ask the operator who was time studied if he feels the rating factor is a proper one.

   Watch the worker who was timed work at the pace he considers proper and then at the pace required to produce the company's workload. The judgment of the worker and steward are as valid as that of the time study man.

11.   Make sure that allowances for personal time, fatigue and delays have been provided in proper amounts.

12.   And finally check all the arithmetic for errors.

A final note:

Most union representatives have found it unwise to take additional time studies themselves as a check, except as a last resort. It is more effective to show the errors in management's study than to try to prove a new union time study is a proper one.

A time study taken by a union is still only the result of judgment. Even when proper methods are used, they tend merely to reduce the inconsistencies of time study, not eliminate them.[1]

---

[1] Collective Bargaining Report prepared by Department of Research, AFL-CIO, *American Federationist*, November 1965.

Company training of union time study representatives has been quite successful, in many instances, as a means of promoting a more cooperative atmosphere in installing and maintaining methods, standards, and wage payment systems. This procedure provides for the joint training of company and union personnel. Having received this training, union representatives are much more qualified to evelute the fairness and accuracy of the technique and to discuss any technical points relative to a specific case.

Employee reactions

In addition to having an understanding of the unions' objectives and its attitudes toward the methods, standards, and wage payment approach, it is important that the practitioner have a vivid understanding of the psychological and sociological reactions of the operator.

There are three points that should always be recognized. These are that: (1) most people do not respond favorably to change; (2) job security is uppermost in most men's minds; and (3) people are influenced by the group to which they belong.

One may ask, "Why is it that most people, regardless of their positions, have an inherent resistance to changing anything associated with their work patterns or work centers?"

The answer stems from several psychological factors. First, change indicates dissatisfaction with the present situation; there is a natural tendency to defend the present way since it is associated intimately with the individual. No one likes others to be dissatisfied with one's work; if a change is even suggested, there is immediate reaction to expound on why the proposed change will not work.

Then, too, people tend to be creatures of habit. Once a habit is acquired, not only is it difficult to give up, but there is resentment if someone endeavors to alter the habit. If one gets in the habit of eating at a certain place, he is reluctant to change to another restaurant even though the food may be better and less expensive.

It is only natural that people desire security. The desire to feel secure in one's position is just as basic as the feeling for self-preservation. In fact, security and self-preservation are related. Most workers prefer job security over high wages when choosing a place to work.

To the worker, all work relating to methods and standards will result in either a change or an effort to increase productivity. The immediate and understandable reaction is to believe that if production goes up, the demand will be filled in a shorter period; and that without demand, there will be fewer jobs.

The solution to the problem of job security lies principally in the sincerity of the management leadership. When methods improvement results in job displacement, management has the responsibility to make an honest effort to relocate those who have been displaced. This may include provision for retraining. Some companies have gone so far as to guarantee that no one will lose his employment as a result of methods improvement. Since the labor turnover rate is usually greater than the improvement rate, the natural attrition through resignation and retirement can usually absorb any displacement of people as a result of improvement.

The sociological impact of "behaving as the group wants him to behave" also has an influence on change. Frequently the worker, as a union member, feels that he is expected to resist any change that has been instituted by management; accordingly, he is reluctant to cooperate with any contemplated changes resulting from methods and standards work. Another factor is the resistance toward anybody who is not part of one's own group. Just as any parent would take offense to a neighbor suggesting that his son part his hair on the other side, so does a member of a group take offense when a member of another group suggests changes. A company represents a "group" which has several groups within its major boundaries. These individual groups respond to basic sociological laws. Change proposed by someone outside one's own group is often received with open hostility. The worker associates himself with a different group than that of the practitioner of methods and standards, and so will tend to resist any effort the analyst makes that might interfere with the usual performance within his group.

The human approach

The method of approach and the behavior of the practitioners of methods engineering and work measurement are the real key to good management-labor relations and the success of the whole production system. Too many of the men practicing in this field of work today do not have an adequate appreciation of the importance of the human approach. When doing methods work, the analyst should take time to talk to the operator and get his ideas and reactions. The work progresses so much more smoothly and effectively if the operator becomes a "part of the team." However, this he cannot do and will not do unless he is asked to "join the team." The operator is closer to the individual operations than anyone else and usually has more specific knowledge relative to operation details than anyone else. It is important that his knowledge be realized, respected, and utilized. Operator suggestions should be gratefully received; if they are practical and worthwhile, they should be introduced as soon as possible. If they are used, the operator should be appropriately rewarded. If they cannot be used at present, a complete explanation should be given as to why they cannot be used.

The analyst should take time to explain all facets of the methods-standards work being undertaken, and point out how this work improves the competitive situation of the company. It can be brought out that both the security of a business and the job security of its employees depend on the business' earning a profit.

As the work progresses, the analyst should keep the operator informed on the probable results of the analysis; he should take time to answer questions related to any personal problems the operator may have in relation to his work or to contemplated changes in his work.

The analyst should at all times think of himself as being in the place of the worker and then use the approach he would like to have used toward himself if he were the operator. Friendliness, courtesy, cheerfulness, and respect, tempered with firmness, are the human characteristics that must be practiced if one is going to be successful in this work. In short the "golden rule" must be applied.

To summarize the "how" of the human approach, it would be well to consider Dale Carnegie's advice on handling people, making people like you, influencing the thinking of people, and changing people. Mr. Carnegie summarizes as follows:

A. Fundamental techniques in handling people.
    1. Instead of condemning (criticizing) people, try to understand them.
    2. Remember that all people *need* to feel important; therefore, try to figure out the other man's good points. Forget flattery; give honest, sincere appreciation.
    3. Remember that all people are interested in *their* own needs; therefore, talk about what *he* wants and show him how to get it.
B. Six ways to make people like you.
    1. Become genuinely interested in other people.
    2. Smile.
    3. Remember that a man's name is to him the sweetest and most important sound in the English language.
    4. Be a good listener. Encourage others to talk about themselves.
    5. Talk in terms of the other man's interest.
    6. Make the other person feel important—and do it sincerely.
C. Twelve ways to win people to your way of thinking.
    1. The only way to get the best of an argument is to avoid it.
    2. Show respect for the other man's opinions. Never tell a man he is wrong.
    3. If you are wrong, admit it quickly and emphatically.
    4. Begin in a friendly way.
    5. Get the other person saying yes, immediately.
    6. Let the other man feel that the idea is his.
    7. Let the other man do a great deal of the talking.
    8. Try honestly to see things from the other person's point of view.
    9. Be sympathetic with the other person's ideas and desires.
    10. Appeal to the nobler motives.
    11. Dramatize your ideas.
    12. Throw down a challenge.

D. Nine ways to change people without giving offense or arousing resentment.
    1. Begin with praise and honest appreciation.
    2. Call attention to people's mistakes indirectly.
    3. Talk about your own mistakes before criticizing the other person.
    4. Ask questions instead of giving direct orders.
    5. Let the other man save his face.
    6. Praise the slightest improvement and praise every improvement. Be hearty in your approbation and lavish in your praise.
    7. Give the other person a fine reputation to live up to.
    8. Use encouragement. Make the fault seem easy to correct.
    9. Make the other person happy about doing the thing you suggest.

Research in methods, time study, and wage payment

The reader of this text should have recognized that many of the techniques presented are not based upon a scientific method, but upon judgment, experience, and iterative methods. This is not because of choice on the part of the industrial engineer associated with methods, time study, and wage payment, but because the technology does not exist today to provide a scientific method for many aspects of this work. The techniques described in this text have been proven in countless instances; they can and do produce acceptable solutions, and they have contributed significantly to the economy of the country. However, there is no doubt that many of the aforementioned techniques can be improved. In particular, the following areas are in need of more research so that objective, quantitative methods can be employed, thus assuring that judgment, bias, and favoritism cannot influence solutions to problems in such areas:

Methods

1. Man and machine relationships where arrival and service times are not characteristic of typical distributions.
2. Production and assembly line balancing.
3. Human capabilities—mental and physical. The effect of age, sex, physical structure, intelligence, and manual dexterity on worker performance.
4. Human behavior patterns. The effect of education, home life, social groups, and the like on worker output.
5. Environmental factors. The effect of working conditions—such as noise, temperature, humidity—on worker output.

Time Study

1. Allowances:
    a. Fatigue. The influence of environmental factors and different work situations.
2. The appropriate time to take the time study.

3. Learning curves. Their shape for different work situations.
4. Performance rating. Development of an objective method that gives consistent, reliable results.
5. Fundamental motion data:
   a. The additivity of fundamental data.
   b. The time required for combined motions.
   c. More data on body motions.

Wage Payment

1. Job evaluation. More objective means of determining job worth.
2. Nonfinancial incentives.
   a. Human wants, reasons for motivation, and so on.

Methods

Although simulation techniques, such as "Monte Carlo," provide a solution to man and machine systems studies of systems that are too unique in light of existing mathematical models, the solutions may not be optimal. Past records are used as a basis for probability distributions, and this procedure alone can introduce data whose reliability may be questioned. The assignment of a random number to a block of time can also be questioned. The size of the block of time and the point within the block of time in which the event (work stoppage, delivery, and so forth) occurs are points of concern.

The classical assembly line balancing problem does not provide usable solutions in most real-life situations. Some of our most advanced industries still sort decks of cards, each card representing a standard for an operation, until a satisfactory balance has been determined by manual means. The General Electric Company, in its publication "Assembly Line Balancing" states: "There is no question but what a capable methods and time standards technician can improve on any computer-generated assignment by a few percentage points. Our goal should be to give him a very good base, much better on the average than he could do manually, from which he can make intelligent and effective improvements." This statement speaks for itself as to the need for more analytic work in this area.

The standardization of the established method advocated by practitioners of methods, standards, and wage payment is certainly not in agreement with the views held by many psychologists. A question arises concerning the extent to which the operations of the man-machine system can be divided, and the job resynthesized and standardized for use by every individual.

The motivation of the worker, as affected by the sociological system of the community and the plant, may influence a constant cause system.

Much research needs to be done in job placement, where job require-ments are matched with the inherent capabilities of employees. We all know of excellent milers who would come in last in a respectable hun-dred-yard dash. We also know first-class baseball players who would be most awkward on the basketball court. Many industrial assignments bear no information as to what human capabilities are suited for the work.

Time study

With the technology available today, it is not feasible to develop a practical, deterministic, mathematical structure of the work measurement problem. The complexity of the mathematical relationships, involving variables as yet unmeasured, is so great that mathematics has not yet developed the power to treat a phenomenon as complex as work measurement.

In William Gomberg's text *A Trade Union's Analysis of Time Study,* four sources of variation that influence the output of the operator are mentioned: mechanical, physiological, psychological, and sociological. It is Gomberg's contention that these sources of variation result in a variable chance cause system which cannot be stabilized sufficiently to be mea-sured over an extended period. Since present-day work measurement is a sampling approach in which the time for a unit of work is assumed to be a random variable, it may well be that the effect of all sources of variation results in a variable chance cause system which cannot be stabilized suf-ficiently to be measured accurately over an extended period.

For example, such mechanical effects as temperature, light, heat, hu-midity, and noise are suspected of affecting the fatigue accumulation of the worker, but little is known of the effect of fatigue on the constant chance cause system. Fatigue, a primary source of physiological varia-tion, has been a stumbling block, as no satisfactory method of measuring it has yet been found. Much experimentation has been done in an effort to relate fatigue to physical output, but with incomplete success. Some researchers have tried to relate fatigue with Kefosteriod excretion, with some degree of success.[2] Some manufacturing concerns (Du Pont and Eli Lilly) are endeavoring to correlate the fatigue of different jobs through the measurement of energy consumed. This is done by measuring the heartbeat, and correlating it with oxygen consumption. Studies have indi-cated that the average man should expend no more than 2,500 calories during an eight-hour working day.

In practice, a fatigue allowance is often set by collective bargaining for the development of plantwide standards. This practice can result in realistic allowances in only a portion of the developed standards, in that fatigue undoubtedly varies with the type of operation and the conditions

---

[2] Hoagling and Pincas.

surrounding it. Surely, a different fatigue allowance should be applied for different classes of work.

Selecting the appropriate time to take a time study is a realistic problem. Ambruzzi recommends that a criterion of "local and grand stability" be satisfied before an operation is ready for measurement.[3] He suggests using the Shewhart control chart technique for this purpose. The local stability criterion covers continuous observations taken over a few hours during a day, while the grand stability criterion includes small samples of observations taken over an extended time interval considered to be under homogeneous conditions. In applying Ambruzzi's criterion, consideration should be given to the use of the learning curve. This curve is generally conceded to be exponential in form and, consequently, it would be most appropriate to apply the stability criterion after the knee of the curve has been reached.

Assuming the validity of the statistical formulation, the analyst should be sure to recognize that an unknown mean performance time for productive work is to be estimated by work measurement. In general, measurement systems have three major components: a sensing mechanism, a translation mechanism, and a readout system. Each of these sources is capable of contributing bias error (positive or negative) and a variable error to the measurements, resulting in the total error for the system.

Probably the most fertile field for additional research in the area of work measurement lies in performance rating. All performance rating systems endeavor to establish a "within company" concept of a "normal" rate of work. Arbitrary times for benchmark jobs, rated motion-picture films of representative jobs, and the concept of normal embodied in predetermined time systems or in the views of experienced time study technicians may be bases for adopting the concept of normal for a company. Research needs to be undertaken to develop an objective means of rating that eliminates or minimizes the need of judgment on the part of the analyst. Perhaps the synthetic rating method using the nomogram suggested by the author can point the way toward more valid performance rating. In this connection, research needs to be accomplished to determine the variability in performance of successive elements within an operation.

The development of fundamental motion data, certainly, represents one of the most significant contributions of industrial engineering. MTM, Work-Factor, and other systems provide means for establishing realistic standards in a short period of time by semitechnical personnel after a relatively short (three to six months) training period. However, much research still needs to be done in this area.

The fact that fundamental motion data is not additive on certain motion patterns has been established in several research papers. Thus,

---

[3] *Work Measurement* (New York: Columbia University Press, 1952).

more studies need to be made on combined motion patterns, and data need to be established for these patterns.

Existing fundamental motion data on body motions is inadequate. For example, data need to be established for body motions (walking and other) under load and when performed at different angles of inclination.

Job evaluation

In the establishment of equitable base rates, more research needs to be done. Job evaluation removed some of the bias and favoritism that were characteristic of many wage payment programs prior to the inception of job evaluation techniques. However, even with job evaluation, job descriptions can be manipulated so that more or less points are assigned and, consequently, the base rate is affected. There is still a real need to develop a completely fair method for determining base rates.

There is much need of research in the whole area of nonfinancial incentives. What do most people really want out of life after a satisfactory wage has been earned? How can these objectives be incorporated as part of every job? What barriers are inadvertently introduced in every job through poor supervision, poor communication, and poor leadership?

Summary

The work of the motion and time study analyst influences to a large extent the labor relations within an enterprise. It is important that the analyst understands the objectives of the union that represents his plant. It is also important that the analyst knows the nature of the training that officers in his company's local are receiving. With this information he will be better able to understand the attitudes and problems of the worker on the bench.

At all times the analyst must be cognizant of the necessity of using the human approach. He must always ask for and develop standards and methods that are fair to both the company and the operator.

The analyst must recognize that much of the work relating to methods, standards, and wage payment is not based on science but on judgment and experience. Consequently, he should continually be alert for the output of research taking place in the field as it is published so that he will be able to introduce quantitative techniques to supplement or supplant those that can result in inequity. He should take an active part in professional societies that are continually exploring the field in order to develop more objective techniques.

TEXT QUESTIONS

1. Outline the objectives of the typical labor union.
2. Why have unions in the past sought flat "across-the-board" wage increases for their members?

3. Explain why the union training program says that time study is an "imprecise tool and lends itself to easy abuse."

4. What are the four approaches to time study suggested by the typical union?

5. What three points related to the psychological and sociological reactions of the operator should be recognized by the analyst?

6. What do we mean by "the human approach"?

7. In what 12 ways can you get people to agree with your ideas?

8. Is methods training conducted within a plant self-supporting?

9. Why is plantwide training in the areas of methods and time study a healthy management step?

10. Why should training in time study be looked upon as a continuing project?

11. Why is American industry more receptive to the time study analyst today than it was prior to World War II?

12. What order of importance does diversified industry place upon time study in the development of an industrial engineering curriculum?

13. What intangible benefits can be achieved through a methods training program on the foreman level?

14. Why should the experienced analyst be continually checked on his ability to performance rate?

15. How can a person develop creative ability?

16. What are some of the areas where research needs to be done in methods? In standards? In job evaluation?

17. What is meant by "local and grand stability"?

GENERAL QUESTIONS

1. Who should handle in-plant training in methods, time study, and wage payment?

2. Is it advisable to provide training in the area of methods, time study, and wage payment down to the operator level? Why or why not?

3. To what opportunities within an industry can the experienced analyst in methods and time study expect to be promoted?

4. What is the relation between work measurement and operations research?

5. Will "automation" diminish the need for the methods and time study analyst?

6. What government legislation has changed the status of labor unions?

7. Why do unions often train their own time study men?

8. Would a young engineer find the work in the time study division of the CIO a challenging opportunity? Why or why not?

PROBLEMS

1. Based upon the cost relationships presented in this chapter, what would be the total annual dollar savings (estimated) of a company having an annual

direct labor payroll of $2,500,000 and a fixed burden rate of 150 percent of direct labor if it initiated a methods and standards program?

2. A company employing straight daywork as a method of wage payment is compensating its employees an average of $5.10 per hour. In addition, the cost of fringe benefits is running 30 percent of direct labor. Overhead in this company is 125 percent of direct labor. A methods, standards, and incentive plan is being contemplated in which the average incentive earnings have been estimated as equaling 20 percent of base wages. What payoff will the proposed plan yield?

SELECTED REFERENCES

Abruzzi, A. *Work Measurement.* New York: Columbia University Press, 1952.

Krick, Edward V. *Methods Engineering.* New York: John Wiley & Sons, Inc., 1962.

Appendix 1

Glossary of terms used in methods, time study, and wage payment

Abnormal time. Elemental time values taken during the course of a time study that are either considerably higher or lower than the mean of the majority of observations. Synonym for "wild value."

Activity sampling. (See Work sampling.

Actual time. The average elemental time actually taken by the operator during a time study.

Address. A designation indicating where a specific piece of information can be found in the memory storage.

Aerobic efficiency. The efficiency of the body during moderate work where the oxygen intake is adequate.

Allowance. An amount of time added to the normal time to provide for personal delays, unavoidable delays, and fatigue.

Allowed time. The time utilized by the normal operator to perform an operation while working at a standard rate of performance with due allowance for personal and unavoidable delays and fatigue.

Anaerobic efficiency. The efficiency (ratio of work done in calories to net energy used in calories) of the body during heavy work.

Arithmetic unit. That part of the computer processing section that does the adding, subtracting, multiplying, dividing, and computing.

Assemble. The act of bringing two mating parts together.

Assignable cause. A source of variation in a process or operation which can be isolated.

Automation. Increased mechanization to produce goods and services.

Average cycle time. The sum of all average elemental times divided by the number of cycle observations.

Average elemental time. The mean elemental time taken by the operator to perform the task during a time study.

Average hourly earnings. The mean dollar-and-cent moneys paid to an operator on an hourly basis. They are determined by dividing the hours worked per period into the total wages paid for the period.

Avoidable delay. A cessation of productive work entirely due to the operator and not occurring in the regular work cycle.

Balanced motion pattern. A sequence of motions made simultaneously by both the right and the left hand in directions which facilitate rhythm and coordination.

Balancing delay. The cessation of productive work by one body member as a result of orienting another body member in the process of useful work.

Base time. The time required to perform a task by a normal operator working at a standard pace with no allowance for personal delays, unavoidable delays, or fatigue.

Base wage rate. The hourly money rate paid for a given work assignment performed at a standard pace by a normal operator.

Basic motion. A fundamental motion related to primary physiological and/or biomechanical performance capabilities of body members.

Bedaux plan. A constant sharing wage incentive plan with a high task and with the provision that the bonus for incentive effort be distributed between the employee and management.

Benchmark. A standard that is identified with characteristics in sufficient detail so that other classifications can be compared as being above, below, or comparable to the identified standard.

Binary digits. The numbers 1 and 0 (on and off) as used internally by computers.

Binomial distribution. A discrete probability distribution with mean $= np$ and variance $= np\ (1 - p)$ having a probability function $= C_{n,k}\ p^k\ (1 - p)^{n-k}$.

Bonus earnings. Those moneys paid in addition to the regular wage or salary.

Breakpoint. A readily distinguishable point in the work cycle that is selected as the boundary between the completion of one element and the beginning of another element.

Change direction. A basic division of accomplishment characterized by a slight hesitation when the hand alters its directional course while reaching or moving.

Check study. A review of a job with either a stopwatch or a regular wristwatch to determine the appropriateness of a standard.

Chronocyclegraph. A photographic record of body motion that may be used to determine both the speed and the direction of body motion patterns.

Chronograph. A time-recording device that operates by marking a tape driven at a constant speed. Time is determined by measuring the distance between successive markings on the tape.

Coding. A system of symbols and rules that sequence the operation of the computer.

Constant element. An element whose performance time does not vary significantly when changes in the process or dimensional changes in the product occur.

Continuous timing method. A method of studying an operation in which the stopwatch is kept running continuously during the course of the study and is not snapped back at elemental termination.

Controlled time. Elapsed elemental time that depends entirely on the facility or process.

Coverage. The number of jobs that have been assigned a standard during the reporting period or the number of personnel whose jobs have been assigned a standard during the reporting period.

Curve. A graphic representation of the relation between two factors, one of which is usually time.

Cycle. A series of elements that occur in regular order and make an operation possible. These elements repeat themselves as the operation is repeated.

Cycle timing. The measurement of the time for a complete work cycle rather than for the individual elements of the cycle.

Cyclegraph. A photographic record of body motion.

Daywork. Any work for which the operator is compensated on the basis of time rather than output.

Decimal hour stopwatch. A stop watch used for work measurement, the dial of which is graduated in 0.0001 of an hour.

Decimal minute stopwatch. A stop watch used for work measurement, the dial of which is graduated in 0.01 of a minute.

Delays. Any cessation in the work routine that does not occur in the typical work cycle.

Differential piecework. Compensation of labor in which the money rate per piece is based on the total pieces produced during the period (usually one day).

Differential timing. Timing an element by combining it with preceding and/or succeeding elements and then determining the elemental times by solving the simultaneous collective time equations.

Direct labor. Labor performed on each piece that advances the piece toward its ultimate specifications.

Disassemble. The basic division of accomplishment that takes place when two mating parts are separated.

Downtime. A period of time that is represented by operation cessation due to machine or tool breakdown, lack of material, and so on.

Drop delivery. The disposal of a part by dropping it on a conveyor or a gravity chute, thus minimizing move and position therbligs.

Earned hours. The standard hours credited to a worker or a work force as a result of the completion of a job or a group of jobs.

Efficiency. The ratio of actual output to standard output.

Effort. The will to perform either mental or manual productive work.

Effort time. The portion of the cycle time that depends on the skill and effort of the operator.

Elapsed time. The actual time that has transpired during the course of a study or an operation.

Element. A division of work that can be measured with stopwatch equipment and that has readily identified terminal points.

Ergonomics. The analysis of a work situation in order to optimize the physiological cost of performing an operation.

Exponential distribution. A continuous probability distribution with mean $= \frac{1}{a}$ and variance $= \frac{1}{a^2}$, and having a density function $= ae^{-ax}$.

External time. The time required to perform elements of work when the machine or process is not in operation.

Extra allowance. An added allowance to compensate for required work in addition to that which is specified in the standard method.

Fair day's work. The amount of work performed by an operator or a group of operators that is fair to both the company and the operator, considering the wages paid. It is the "amount of work that can be produced by a qualified employee when working at a normal pace and effectively utilizing his time where work is not restricted by process limitations."

Fatigue. A lessening in the capacity for the will to work.

Fatigue allowance. An amount of time added to the normal time to compensate for fatigue.

Feed. The speed at which the cutting tool is moved into the work, as in drilling and turning, or the rate that the work is moved past the cutting tool.

Film analysis. The frame-by-frame observation and study of a film of an operation or process with the objective of improvement.

First-piece time. The time allowed to produce the first piece of an order. This time has been adjusted to allow for the operator's unfamiliarity with the method and to accommodate for minor delays resulting from the newness of the work. It does not include time for setting up the work station.

Fixture. A tool that is usually clamped to the work station and that holds the material being worked upon.

Flow diagram. A pictorial representation of the layout of a process showing the location of all activities appearing on the flow process chart and the paths of travel of the work.

Flow process chart. A graphic representation of all operations, transportations, inspections, delays, and storages occurring during a process or procedure. The chart includes information considered desirable for analysis, such as the time required and the distance moved.

Foreign element. An interruption in the regular work cycle.

Frame. The space occupied by a single picture on a motion-picture film.

Frame counter. A device that automatically tabulates the number of frames that have passed the lens of the projector.

Frequency function. The complete listing of the values of a random variable together with their probabilities of occurrence.

Fringe benefits. The portion of tangible compensation given by the employer to employees that is not paid in the form of wages, salaries, or bonuses. These include insurance, retirement funds, and other employee services. They exclude benefits paid for by employees by pay deductions, such as their participating portions of insurance premiums and retirement funds.

Gain sharing. Any method of wage payment in which the worker participates in all or a portion of the added earnings resulting from his production above standard.

Gantt chart. A series of graphs consisting of horizontal lines or bars in positions and lengths that show schedules or quotas and progress plotted on a common time scale.

Get. The act of picking up and gaining control of an object. It consists of the therbligs reach and grasp, and move; it also sometimes includes search and select.

Grasp. The elemental hand motion of closing the fingers around a part in an operation.

Gravity feed. Conveying materials either to the work station or away from the work station by using the force of gravity.

Hand time. That part of the work cycle that is controlled by manual elements exclusive of power or mechanized pacing elements.

Human factors. Those axioms and postulates that concern themselves with the physical, mental, and emotional constraints that affect operator performance.

Idle time. Time during which the worker is not working.

Incentive. Reward, financial or other, that compensates the worker for high and/or continued performance above standard.

Incentive pace. A performance that is above normal or standard.

Indirect labor. Labor that does not directly enter into transforming the material used in making the product, but that is necessary to support the manufacture of the product.

Input storage. The temporary storage of a group of facts by a computing machine until the time that this group of facts should be processed.

Instruction. A coded program step that tells the computer what to do for a single operation in a program.

Interference time. Idle machine time due to insufficient operator time to service one or more machines that need servicing, because of operator engagement on other assigned work.

Internal work. Work that is performed by the operator during the operation of the machine or equipment.

Jig. A tool that may or may not be clamped to the work station and that is used both to hold the work and to guide the tool.

Job analysis. A procedure for making a careful appraisal of each job and then recording the details of the work so that it can be equitably evaluated.

Job evaluation. A procedure for determining the relative worth of various work assignments.

Key job. A job that is representative of similar jobs or classes of work in the same plant or industry.

Learning curve. A graphic presentation of the progress in production effectiveness as time passes.

Leveling. A term used synonymously with performance rating. It is the assignment of a percentage to the operator's average observed time to adjust his time to the observer's conception of normal.

Loose rate. An established allowed time permitting the normal operator to achieve standard performance with poorer than average effort.

Machine attention time. That time during the work cycle in which the operator must devote his attention to the machine or process.

Machine cycle time. The time required for the machine in process to complete one cycle.

Machine downtime. That time when the machine or process is inoperative because of some breakdown or because of a material shortage.

Machine idle time. That time when the machine or process is inoperative.

Man-hour. The standard amount of work performed by one man in one hour.

Man-machine process chart. A chart showing the exact relationship in time between the working cycle of the operator and the operating cycle of his machine or machines.

Manufacturing progress function. The progress in production effectiveness with the passing of time.

Marsto-chron. A time study instrument that records elapsed time on a tape driven by a synchronous motor. Elapsed time is determined by measuring the distance on the tape between parallel markings recorded at element terminal points.

Maximum performance. That performance which will result in the highest obtainable production.

Maximum working area. The area that is readily reached by the operator when the arms are fully extended, while he is situated at his normal working position.

Mean of x. The expected value of x.

Measured daywork. Work for which performance standards have been established but where the operator is compensated on an hourly basis with no provision for incentive earnings. (First choice.)
An incentive system in which hourly rates are periodically adjusted on the basis of operator performance during the previous period. (Second choice.)

Memomotion study. The division of a work assignment into elements by analyzing motion pictures taken at speeds of 50, 60, or 100 frames per minute, and then improving the operation.

Memory storage. The section of the computer that files or holds facts.

Merit rating. A method of evaluating an employee's worth to a company in terms of such factors as quantity of work, quality of work, dependability, and general contribution to the company.

Merrick differential piece rate. An incentive wage payment plan in which three different piece rates are established on the basis of operator performance.

Method. A term used to signify the technique employed for performing an operation.

Methods study. Analysis of an operation to increase the production per unit of time and consequently reduce the unit cost.

Microchronometer. A specially designed clock devised by Frank B. Gilbreth that is capable of measuring elapsed time in "winks" (0.0005 minute).

Micromotion study. The division of a work assignment into therbligs by analyzing motion pictures frame by frame and then improving the operation by eliminating unnecessary movements and simplifying the necessary movements.

Minimum time. The least amount of time taken by the operator to perform a given element during a time study.

Modal time. The elapsed elemental time value that occurs most frequently during a time study. Occasionally used in preference to the average elemental time.

Motion study. The analysis and study of the motions constituting an operation to improve the motion pattern by eliminating ineffective motions and shortening the effective motions.

Move. The term used to signify a hand movement with a load.

MTM. A procedure analyzing any manual operation or method into the basic motions required to perform it, and assigning to each motion a predetermined time standard based upon the nature of the motion and the conditions under which it is made.

Normal distribution. A continuous probability distribution with mean $= m$ and variance $= \sigma^2$ and having a density function $=$

$$\frac{1}{\sigma \sqrt{2\pi}} \exp\left[\frac{-(x - m)^2}{2\sigma^2}\right]$$

Normal operator. An operator who can achieve the established standard of performance when following the prescribed method and working at an average pace.

Normal performance. The performance expected from the average trained operator when he is following the prescribed method and working at an average pace.

Normal time. The time required for the standard operator to perform the operation when working at a standard pace without delay for personal reasons or unavoidable circumstances.

Normal working area. That space at the work area which can be reached by either the left hand or the right hand when both elbows are pivoted on the edge of the work station.

Observation. The gathering and recording of the time required to perform an element, or one watch reading.

Observation board. A convenient board used to support the stopwatch and hold the observation form during a time study.

Observation form. A form designed to accommodate the elements of a given time study with space for recording their duration.

Observer. The analyst taking a time study of a given operation.

Occupational physiology. Scientific study of the worker and his environment.

Occurrence. An incident or event happening during a time study.

Operation. The intentional changing of a part toward its ultimate desired shape, size, form, and characteristics.

Operation analysis. An investigative process dealing with operations in factory or office work. Usually the process leading to operation standardization, including motion and time study.

Operation card. A form outlining the sequence of operations, the time allowed, and the special tools required in manufacturing a part.

Operation process chart. A graphic representation of an operation showing all operations, inspections, time allowances, and materials used in a manufacturing process.

Operator process chart. A graphic representation of all movements and delays made by both the right and the left hand, and of the relationship between the relative basic divisions of accomplishment performed by the two hands.

Output. The total production of a machine, process, or worker for a specified unit of time.

Output. (applied to computer). Computer results, such as answers to mathematical problems.

Output devices. Converting results of a problem solved on the computer into the desired form, such as punched paper tape, magnetic tape, and punched and printed cards.

Overall study. Recording cycle time as a verification of a developed time study standard.

Pareto's distribution. A distribution that reflects the fact that the major part of an activity (usually 80–85 percent is accounted for by a minority (usually 15–20 percent). For example, 15 percent of the employees account for 85 percent of the absenteeism.

Performance. The ratio of the operator's actual production to the standard production.

Performance rating. The assignment of a percentage to the operator's average observed time, based on the actual performance of the operator compared to the observer's conception of normal.

Personal allowance. A percentage added to the normal time to accommodate the personal needs of the operator.

Piece rate. A standard of performance expressed in terms of money per unit of production.

Plan. A basic division of accomplishment involving the mental process of determining the next action.

Point. A unit of output identified as the production of one standard operator in one minute. Used as a basis for establishing standards under the Bedaux system.

Point system. A method of job evaluation in which the relative worth of different jobs is determined by totaling the number of points assigned to the various factors applicable to the different jobs.

Poisson distribution. A discrete probability distribution with mean $= \lambda$ and variance $= \lambda$, and having a probability function $= \dfrac{\lambda^k e^{-\lambda}}{k!}$

Position. An element of work which consists of locating an object so that it will be properly oriented in a specific location.

Pre-position. An element of work which consists of positioning an object in a predetermined place so that it may be grasped in the position in which it is to be held when needed.

Process. A series of operations that advance the product toward its ultimate size, shape, and specifications.

Process chart. A graphic representation of a manufacturing process.

Productive time. Any time spent in advancing the progress of a product toward its ultimate specifications.

Profit sharing. Any procedure by which an employer pays to all employees, in addition to good rates of regular pay, special current or deferred sums, based not only upon individual or group performance but also upon the prosperity of the business as a whole.

Program. A set of instructions providing the information needed by the computer to handle a complete program.

Qualified operator. An employee who has had sufficient training and education and has demonstrated an adequate level of skill and effort so that he can be expected to perform at an acceptable level with respect to both quantity and quality.

Queuing theory. (See Waiting line theory.)

Random variable. A chance number resulting from a trial from among the set of numbers x_1, x_2, and so on.

Rate. A standard expressed in dollars and cents.

Rate setting. The act of establishing money rates or time values on any operation.

Ratio-delay study. A work sampling study in which a large number of observations are taken at random intervals.

Selected time. An elemental time value that is chosen as being representative of the expected performance of the operator being studied.

Setup. The preparation of a work station or a work center to accomplish an operation or a series of operation.

Skill. Proficiency at following a prescribed method.

Standby time. That time in which the worker is not actively engaged, but is prepared to take action when needed.

Synthetic basic motion times. A collection of time standards assigned to fundamental motions and groups of motions.

Temporary standard. A standard established to apply for a limited number of pieces or a limited period of time so as to account for the newness of the work or some unusual job condition.

Therblig. An abbreviated segment of a work element that describes the sensorimotor activities.

Tight rate. A time standard that allows less time than that required by a normal operator working at a normal pace to do the work.

Transmission. Facility of transmission of nerve impulses across the motor end plate in the muscle fiber.

Unavoidable delay. An interruption in the continuity of an operation beyond the control of an operator.

Use. An objective therblig occurring when either or both hands have control of an object during that part of the cycle when productive work is being performed.

Variable element. An element whose time is affected by one or more characteristics, such as size, shape, hardness, or tolerance, so that as these conditions change, the time required to perform the element changes.

Variance of x. A measure of the expected dispersion of the values of x about its mean.

Wage incentive. Providing a financial inducement for effort above normal.

Wage rate. The money rate expressed in dollars and cents paid to the employee per hour.

Waiting line theory. Mathematical analysis of the laws governing arrivals, service times, and the order in which arriving units are taken into service.

Waiting time. That time when the operator is unable to do useful work because of the nature of the process or because of the immediate lack of material.

Wild value. An elemental time value taken during a time study that is either considerably higher or lower than the mean of the majority of observations. Synonym for "abnormal time."

Wink. One division on the microchronometer equal to $\frac{1}{2,000}$ (0.0005) minute.

Wink counter. A mechanically or electrically driven timing device that records elapsed time in winks.

Work factor. Index of the additional time required over and above the basic time as established by the Work-Factor system of synthetic basic motion times.

Work physiology. The specification of the physiological and psychological factors characteristic of a work environment.

Work sampling. A method of analyzing work by taking a large number of observations at random intervals, to establish standards and improve methods.

Work station. The area where the worker performs the elements of work in a specific operation.

Appendix 2

Collection of helpful formulas

(1) *Quadratic*

$$Ax^2 + Bx + C = 0$$

$$x = \frac{-B \pm \sqrt{B^2 - 4AC}}{2A}$$

(2) *Logarithms*

$$\log ab = \log a + \log b$$

$$\log \frac{a}{b} = \log a - \log b$$

$$\log a^n = n \log a$$

$$\log \sqrt[n]{a} = \frac{1}{n} \log a$$

$$\log 1 = 0$$
$$\log_a a = 1$$

(3) *Binomial theorem*

$$(a + b)^n = a^n + na^{n-1}b + \frac{n(n-1)}{2} a^{n-2}b^2$$

$$+ \frac{n(n-1)(n-2)}{3} a^{n-3}b^3 + \cdots$$

(4) *Circle*

$$\text{Circumference} = 2\pi r$$
$$\text{Area} = \pi r^2$$

(5) *Prism*

$$\text{Volume} = Ba$$

(6)   *Pyramid*

$$\text{Volume} = \tfrac{1}{3}Ba$$

(7)   *Right circular cylinder*

$$\text{Volume} = \pi r^2 a$$
$$\text{Lateral surface} = 2\pi r a$$
$$\text{Total surface} = 2\pi r(r + a)$$

(8)   *Right circular cone*

$$\text{Volume} = \tfrac{1}{3}\pi r^2 a$$
$$\text{Lateral surface} = \pi r s$$
$$\text{Total surface} = \pi r(r + s)$$

(9)   *Sphere*

$$\text{Volume} = \tfrac{4}{3}\pi r^3$$
$$\text{Surface} = 4\pi r^2$$

(10)   *Frustrum of a right circular cone*

$$\text{Volume} = \tfrac{1}{3}\pi a(R^2 + r^2 + Rr)$$
$$\text{Lateral surface} = \pi s(R + r)$$

(11)   *Measurement of angles*

$$1 \text{ degree} = \frac{\pi}{180} = .0174 \text{ radians}$$
$$1 \text{ radian} = 57.29 \text{ degrees}$$

(12)   *Trigonometric functions*

Right triangles:

The sine of the angle A is the quotient of the opposite side divided by the hypotenuse. $\text{Sin } A = \dfrac{a}{c}$

The tangent of the angle A is the quotient of the opposite side divided by the adjacent side. $\text{Tan } A = \dfrac{a}{b}.$

The secant of the angle A is the quotient of the hypotenuse divided by the adjacent side. $\text{Sec } A = \dfrac{c}{b}.$

The cosine, cotangent, and cosecant of an angle are, respectively, the sine, tangent, and secant of the complement of that angle.

Law of sines:

$$\frac{a}{\sin A} = \frac{b}{\sin B} = \frac{c}{\sin C}$$

Law of cosines:

$$a^2 = b^2 + c^2 - 2bc \cos A$$

(13) *Equations of straight lines*
Slope—intercept form

$$y = mx + b$$

intercept form

$$\frac{x}{a} + \frac{y}{b} = 1$$

Appendix 3

Special tables

TABLE A3–1

Natural sines and tangents

Angle	Sin	Tan	Cot	Cos	
0	0.0000	0.0000	∞	1.0000	90
1	0.0175	0.0175	57.2900	0.9998	89
2	0.0349	0.0349	28.6363	0.9994	88
3	0.0523	0.0524	19.0811	0.9986	87
4	0.0698	0.0699	14.3007	0.9976	86
5	0.0872	0.0875	11.4301	0.9962	85
6	0.1045	0.1051	9.5144	0.9945	84
7	0.1219	0.1228	8.1443	0.9925	83
8	0.1392	0.1405	7.1154	0.9903	82
9	0.1564	0.1584	6.3138	0.9877	81
10	0.1736	0.1763	5.6713	0.9848	80
11	0.1908	0.1944	5.1446	0.9816	79
12	0.2079	0.2126	4.7046	0.9781	78
13	0.2250	0.2309	4.3315	0.9744	77
14	0.2419	0.2493	4.0108	0.9703	76
15	0.2588	0.2679	3.7321	0.9659	75
16	0.2756	0.2867	3.4874	0.9613	74
17	0.2924	0.3057	3.2709	0.9563	73
18	0.3090	0.3249	3.0777	0.9511	72
19	0.3256	0.3443	2.9042	0.9455	71
20	0.3420	0.3640	2.7475	0.9397	70
21	0.3584	0.3839	2.6051	0.9336	69
22	0.3746	0.4040	2.4751	0.9272	68
23	0.3907	0.4245	2.3559	0.9205	67
24	0.4067	0.4452	2.2460	0.9135	66
25	0.4226	0.4663	2.1445	0.9063	65
26	0.4384	0.4877	2.0503	0.8988	64
27	0.4540	0.5095	1.9626	0.8910	63
28	0.4695	0.5317	1.8807	0.8829	62
29	0.4848	0.5543	1.8040	0.8746	61
30	0.5000	0.5774	1.7321	0.8660	60
31	0.5150	0.6009	1.6643	0.8572	59
32	0.5299	0.6249	1.6003	0.8480	58
33	0.5446	0.6494	1.5399	0.8387	57
34	0.5592	0.6745	1.4826	0.8290	56
35	0.5736	0.7002	1.4281	0.8192	55
36	0.5878	0.7265	1.3764	0.8090	54
37	0.6018	0.7536	1.3270	0.7986	53
38	0.6157	0.7813	1.2799	0.7880	52
39	0.6293	0.8098	1.2349	0.7771	51
40	0.6428	0.8391	1.1918	0.7660	50
41	0.6561	0.8693	1.1504	0.7547	49
42	0.6691	0.9004	1.1106	0.7431	48
43	0.6820	0.9325	1 0724	0.7314	47
44	0.6947	0.9657	1.0355	0.7193	46
45	0.7071	1.0000	1.0000	0.7071	45
	Cos	Cot	Tan	Sin	Angle

TABLE A3–2
Areas of the normal curve

z	Area	z	Area
−3.0	.0013	0.1	.5398
−2.9	.0019	0.2	.5793
−2.8	.0026	0.3	.6179
−2.7	.0035	0.4	.6554
−2.6	.0047	0.5	.6915
−2.5	.0062	0.6	.7257
−2.4	.0082	0.7	.7580
−2.3	.0107	0.8	.7881
−2.2	.0139	0.9	.8159
−2.1	.0179	1.0	.8413
−2.0	.0228	1.1	.8643
−1.9	.0287	1.2	.8849
−1.8	.0359	1.3	.9032
−1.7	.0446	1.4	.9192
−1.6	.0548	1.5	.9332
−1.5	.0668	1.6	.9452
−1.4	.0808	1.7	.9554
−1.3	.0968	1.8	.9641
−1.2	.1151	1.9	.9713
−1.1	.1357	2.0	.9772
−1.0	.1587	2.1	.9821
−0.9	.1841	2.2	.9861
−0.8	.2119	2.3	.9893
−0.7	.2420	2.4	.9918
−0.6	.2741	2.5	.9938
−0.5	.3085	2.6	.9953
−0.4	.3446	2.7	.9965
−0.3	.3821	2.8	.9974
−0.2	.4207	2.9	.9981
−0.1	.4602	3.0	.9987
0.0	.5000		

Above table tabularizes $P(z)$ where

$$P(z) = \int_{-\infty}^{z} \frac{1}{\sqrt{2\pi}} e^{-\frac{z^2}{2}} \, dz$$

$$\text{and } z = \frac{x - u}{\sigma}$$

TABLE A3–3

Percentage points of the *t* distribution* (probabilities refer to the sum of the two tail areas; for a single tail, divide the probability by 2)

Probability (*P*).

n	·9	·8	·7	·6	·5	·4	·3	·2	·1	·05	·02	·01	·001
1	·158	·325	·510	·727	1·000	1·376	1·963	3·078	6·314	12·706	31·821	63·657	636·619
2	·142	·289	·445	·617	·816	1·061	1·386	1·886	2·920	4·303	6·965	9·925	31·598
3	·137	·277	·424	·584	·765	·978	1·250	1·638	2·353	3·182	4·541	5·841	12·941
4	·134	·271	·414	·569	·741	·941	1·190	1·533	2·132	2·776	3·747	4·604	8·610
5	·132	·267	·408	·559	·727	·920	1·156	1·476	2·015	2·571	3·365	4·032	6·859
6	·131	·265	·404	·553	·718	·906	1·134	1·440	1·943	2·447	3·143	3·707	5·959
7	·130	·263	·402	·549	·711	·896	1·119	1·415	1·895	2·365	2·998	3·499	5·405
8	·130	·262	·399	·546	·706	·889	1·108	1·397	1·860	2·306	2·896	3·355	5·041
9	·129	·261	·398	·543	·703	·883	1·100	1·383	1·833	2·262	2·821	3·250	4·781
10	·129	·260	·397	·542	·700	·879	1·093	1·372	1·812	2·228	2·764	3·169	4·587
11	·129	·260	·396	·540	·697	·876	1·088	1·363	1·796	2·201	2·718	3·106	4·437
12	·128	·259	·395	·539	·695	·873	1·083	1·356	1·782	2·179	2·681	3·055	4·318
13	·128	·259	·394	·538	·694	·870	1·079	1·350	1·771	2·160	2·650	3·012	4·221
14	·128	·258	·393	·537	·692	·868	1·076	1·345	1·761	2·145	2·624	2·977	4·140
15	·128	·258	·393	·536	·691	·866	1·074	1·341	1·753	2·131	2·602	2·947	4·073
16	·128	·258	·392	·535	·690	·865	1·071	1·337	1·746	2·120	2·583	2·921	4·015
17	·128	·257	·392	·534	·689	·863	1·069	1·333	1·740	2·110	2·567	2·898	3·965
18	·127	·257	·392	·534	·688	·862	1·067	1·330	1·734	2·101	2·552	2·878	3·922
19	·127	·257	·391	·533	·688	·861	1·066	1·328	1·729	2·093	2·539	2·861	3·883
20	·127	·257	·391	·533	·687	·860	1·064	1·325	1·725	2·086	2·528	2·845	3·850
21	·127	·257	·391	·532	·686	·859	1·063	1·323	1·721	2·080	2·518	2·831	3·819
22	·127	·256	·390	·532	·686	·858	1·061	1·321	1·717	2·074	2·508	2·819	3·792
23	·127	·256	·390	·532	·685	·858	1·060	1·319	1·714	2·069	2·500	2·807	3·767
24	·127	·256	·390	·531	·685	·857	1·059	1·318	1·711	2·064	2·492	2·797	3·745
25	·127	·256	·390	·531	·684	·856	1·058	1·316	1·708	2·060	2·485	2·787	3·725
26	·127	·256	·390	·531	·684	·856	1·058	1·315	1·706	2·056	2·479	2·779	3·707
27	·127	·256	·389	·531	·684	·855	1·057	1·314	1·703	2·052	2·473	2·771	3·690
28	·127	·256	·389	·530	·683	·855	1·056	1·313	1·701	2·048	2·467	2·763	3·674
29	·127	·256	·389	·530	·683	·854	1·055	1·311	1·699	2·045	2·462	2·756	3·659
30	·127	·256	·389	·530	·683	·854	1·055	1·310	1·697	2·042	2·457	2·750	3·646
40	·126	·255	·388	·529	·681	·851	1·050	1·303	1·684	2·021	2·423	2·704	3·551
60	·126	·254	·387	·527	·679	·848	1·046	1·296	1·671	2·000	2·390	2·660	3·460
120	·126	·254	·386	·526	·677	·845	1·041	1·289	1·658	1·980	2·358	2·617	3·373
∞	·126	·253	·385	·524	·674	·842	1·036	1·282	1·645	1·960	2·326	2·576	3·291

* Reprinted from Table III of R. A. Fisher and F. Yates, *Statistical Tables for Biological, Agricultural, and Medical Research* (Edinburgh: Oliver & Boyd, Ltd.), by permission of the authors and publishers.

TABLE A3–4
Random numbers III*

22 17 68 65 84	68 95 23 92 35	87 02 22 57 51	61 09 43 95 06	58 24 82 03 47
19 36 27 59 46	13 79 93 37 55	39 77 32 77 09	85 52 05 30 62	47 83 51 62 74
16 77 23 02 77	09 61 87 25 21	28 06 24 25 93	16 71 13 59 78	23 05 47 47 25
78 43 76 71 61	20 44 90 32 64	97 67 63 99 61	46 38 03 93 22	69 81 21 99 21
03 28 28 26 08	73 37 32 04 05	69 30 16 09 05	88 69 58 28 99	35 07 44 75 47
93 22 53 64 39	07 10 63 76 35	87 03 04 79 88	08 13 13 85 51	55 34 57 72 69
78 76 58 54 74	92 38 70 96 92	52 06 79 79 45	82 63 18 27 44	69 66 92 19 09
23 68 35 26 00	99 53 93 61 28	52 70 05 48 34	56 65 05 61 86	90 92 10 70 80
15 39 25 70 99	93 86 52 77 65	15 33 59 05 28	22 87 26 07 47	86 96 98 29 06
58 71 96 30 24	18 46 23 34 27	85 13 99 24 44	49 18 09 79 49	74 16 32 23 02
57 35 27 33 72	24 53 63 94 09	41 10 76 47 91	44 04 95 49 66	39 60 04 59 81
48 50 86 54 48	22 06 34 72 52	82 21 15 65 20	33 29 94 71 11	15 91 29 12 03
61 96 48 95 03	07 16 39 33 66	98 56 10 56 79	77 21 30 27 12	90 49 22 23 62
36 93 89 41 26	29 70 83 63 51	99 74 20 52 36	87 09 41 15 09	98 60 16 03 03
18 87 00 42 31	57 90 12 02 07	23 47 37 17 31	54 08 01 88 63	39 41 88 92 10
88 56 53 27 59	33 35 72 67 47	77 34 55 45 70	08 18 27 38 90	16 95 86 70 75
09 72 95 84 29	49 41 31 06 70	42 38 06 45 18	64 84 73 31 65	52 53 37 97 15
12 96 88 17 31	65 19 69 02 83	60 75 86 90 68	24 64 19 35 51	56 61 87 39 12
85 94 57 24 16	92 09 84 38 76	22 00 27 69 85	29 81 94 78 70	21 94 47 90 12
38 64 43 59 98	98 77 87 68 07	91 51 67 62 44	40 98 05 93 78	23 32 65 41 18
53 44 09 42 72	00 41 86 79 79	68 47 22 00 20	35 55 31 51 51	00 83 63 22 55
40 76 66 26 84	57 99 99 90 37	36 63 32 08 58	37 40 13 68 97	87 64 81 07 83
02 17 79 18 05	12 59 52 57 02	22 07 90 47 03	28 14 11 30 79	20 69 22 40 98
95 17 82 06 53	31 51 10 96 46	92 06 88 07 77	56 11 50 81 69	40 23 72 51 39
35 76 22 42 92	96 11 83 44 80	34 68 35 48 77	33 42 40 90 60	73 96 53 97 86
26 29 13 56 41	85 47 04 66 08	34 72 57 59 13	82 43 80 46 15	38 26 61 70 04
77 80 20 75 82	72 82 32 99 90	63 95 73 76 63	89 73 44 99 05	48 67 26 43 18
46 40 66 44 52	91 36 74 43 53	30 82 13 54 00	78 45 63 98 35	55 03 36 67 68
37 56 08 18 09	77 53 84 46 47	31 91 18 95 58	24 16 74 11 53	44 10 13 85 57
61 65 61 68 66	37 27 47 39 19	84 83 70 07 48	53 21 40 06 71	95 06 79 88 54
93 43 69 64 07	34 18 04 52 35	56 27 09 24 86	61 85 53 83 45	19 90 70 99 00
21 96 60 12 99	11 20 99 45 18	48 13 93 55 34	18 37 79 49 90	65 97 38 20 46
95 20 47 97 97	27 37 83 28 71	00 06 41 41 74	45 89 09 39 84	51 67 11 52 49
97 86 21 78 73	10 65 81 92 59	58 76 17 14 97	04 76 62 16 17	17 95 70 45 80
69 92 06 34 13	59 71 74 17 32	27 55 10 24 19	23 71 82 13 74	63 52 52 01 41
04 31 17 21 56	33 73 99 19 87	26 72 39 27 67	53 77 57 68 93	60 61 97 22 61
61 06 98 03 91	87 14 77 43 96	43 00 65 98 50	45 60 33 01 07	98 99 46 50 47
85 93 85 86 88	72 87 08 62 40	16 06 10 89 20	23 21 34 74 97	76 38 03 29 63
21 74 32 47 45	73 96 07 94 52	09 65 90 77 47	25 76 16 19 33	53 05 70 53 30
15 69 53 82 80	79 96 23 53 10	65 39 07 16 29	45 33 02 43 70	02 87 40 41 45
02 89 08 04 49	20 21 14 68 86	87 63 93 95 17	11 29 01 95 80	35 14 97 35 33
87 18 15 89 79	85 43 01 72 73	08 61 74 51 69	89 74 39 82 15	94 51 33 41 67
98 83 71 94 22	59 97 50 99 52	08 52 85 08 40	87 80 61 65 31	91 51 80 32 44
10 08 58 21 66	72 68 49 29 31	89 85 84 46 06	59 73 19 85 23	65 09 29 75 63
47 90 56 10 08	88 02 84 27 83	42 29 72 23 19	66 56 45 65 79	20 71 53 20 25
22 85 61 68 90	49 64 92 85 44	16 40 12 89 88	50 14 49 81 06	01 82 77 45 12
67 80 43 79 33	12 83 11 41 16	25 58 19 68 70	77 02 54 00 52	53 43 37 15 26
27 62 50 96 72	79 44 61 40 15	14 53 40 65 39	27 31 58 50 28	11 39 03 34 25
33 78 80 87 15	38 30 06 38 21	14 47 47 07 26	54 96 87 53 32	40 36 40 96 76
13 13 92 66 99	47 24 49 57 74	32 25 43 62 17	10 97 11 69 84	99 63 22 32 98

* Reprinted with permission from Random Numbers III of Table XXXIII of R. A. Fisher and F. Yates, *Statistical Tables for Biological, Agricultural and Medical Research* (Edinburgh: Oliver & Boyd, Ltd.).

TABLE A3–4 (continued)
Random numbers IV*

10 27 53 96 23	71 50 54 36 23	54 31 04 82 98	04 14 12 15 09	26 78 25 47 47
28 41 50 61 88	64 85 27 20 18	83 36 36 05 56	39 71 65 09 62	94 76 62 11 89
34 21 42 57 02	59 19 18 97 48	80 30 03 30 98	05 24 67 70 07	84 97 50 87 46
61 81 77 23 23	82 82 11 54 08	53 28 70 58 96	44 07 39 55 43	42 34 43 39 28
61 15 18 13 54	16 86 20 26 88	90 74 80 55 09	14 53 90 51 17	52 01 63 01 59
91 76 21 64 64	44 91 13 32 97	75 31 62 66 54	84 80 32 75 77	56 08 25 70 29
00 97 79 08 06	37 30 28 59 85	53 56 68 53 40	01 74 39 59 73	30 19 99 85 48
36 46 18 34 94	75 20 80 27 77	78 91 69 16 00	08 43 18 73 68	67 69 61 34 25
88 98 99 60 50	65 95 79 42 94	93 62 40 89 96	43 56 47 71 66	46 76 29 67 02
04 37 59 87 21	05 02 03 24 17	47 97 81 56 51	92 34 86 01 82	55 51 33 12 91
63 62 06 34 41	94 21 78 55 09	72 76 45 16 94	29 95 81 83 83	79 88 01 97 30
78 47 23 53 90	34 41 92 45 71	09 23 70 70 07	12 38 92 79 43	14 85 11 47 23
87 68 62 15 43	53 14 36 59 25	54 47 33 70 15	59 24 48 40 35	50 03 42 99 36
47 60 92 10 77	88 59 53 11 52	66 25 69 07 04	48 68 64 71 06	61 65 70 22 12
56 88 87 59 41	65 28 04 67 53	95 79 88 37 31	50 41 06 94 76	81 83 17 16 33
02 57 45 86 67	73 43 07 34 48	44 26 87 93 29	77 09 61 67 84	06 69 44 77 75
31 54 14 13 17	48 62 11 90 60	68 12 93 64 28	46 24 79 16 76	14 60 25 51 01
28 50 16 43 36	28 97 85 58 99	67 22 52 76 23	24 70 36 54 54	59 28 61 71 96
63 29 62 66 50	02 63 45 52 38	67 63 47 54 75	83 24 78 43 20	92 63 13 47 48
45 65 58 26 51	76 96 59 38 72	86 57 45 71 46	44 67 76 14 55	44 88 01 62 12
39 65 36 63 70	77 45 85 50 51	74 13 39 35 22	30 53 36 02 95	49 34 88 73 61
73 71 98 16 04	29 18 94 51 23	76 51 94 84 86	79 93 96 38 63	08 58 25 58 94
72 20 56 20 11	72 65 71 08 86	79 57 95 13 91	97 48 72 66 48	09 71 17 24 89
75 17 26 99 76	89 37 20 70 01	77 31 61 95 46	26 97 05 73 51	53 33 18 72 87
37 48 60 82 29	81 30 15 39 14	48 38 75 93 29	06 87 37 78 48	45 56 00 84 47
68 08 02 80 72	83 71 46 30 49	89 17 95 88 29	02 39 56 03 46	97 74 06 56 17
14 23 98 61 67	70 52 85 01 50	01 84 02 78 43	10 62 98 19 41	18 83 99 47 99
49 08 96 21 44	25 27 99 41 28	07 41 08 34 66	19 42 74 39 91	41 96 53 78 72
78 37 06 08 43	63 61 62 42 29	39 68 95 10 96	09 24 23 00 62	56 12 80 73 16
37 21 34 17 68	68 96 83 23 56	32 84 60 15 31	44 73 67 34 77	91 15 79 74 58
14 29 09 34 04	87 83 07 55 07	76 58 30 83 64	87 29 25 58 84	86 50 60 00 25
58 43 28 06 36	49 52 83 51 14	47 56 91 29 34	05 87 31 06 95	12 45 57 09 09
10 43 67 29 70	80 62 80 03 42	10 80 21 38 84	90 56 35 03 09	43 12 74 49 14
44 38 88 39 54	86 97 37 44 22	00 95 01 31 76	17 16 29 56 63	38 78 94 49 81
90 69 59 19 51	85 39 52 85 13	07 28 37 07 61	11 16 36 27 03	78 86 72 04 95
41 47 10 25 62	97 05 31 03 61	20 26 36 31 62	68 69 86 95 44	84 95 48 46 45
91 94 14 63 19	75 89 11 47 11	31 56 34 19 09	79 57 92 36 59	14 93 87 81 40
80 06 54 18 66	09 18 94 06 19	98 40 07 17 81	22 45 44 84 11	24 62 20 42 31
67 72 77 63 48	84 08 31 55 58	24 33 45 77 58	80 45 67 93 82	75 70 16 08 24
59 40 24 13 27	79 26 88 86 30	01 31 60 10 39	53 58 47 70 93	85 81 56 39 38
05 90 35 89 95	01 61 16 96 94	50 78 13 69 36	37 68 53 37 31	71 26 35 03 71
44 43 80 69 98	46 68 05 14 82	90 78 50 05 62	77 79 13 57 44	59 60 10 39 66
61 81 31 96 82	00 57 25 60 59	46 72 60 18 77	55 66 12 62 11	08 99 55 64 57
42 88 07 10 05	24 98 65 63 21	47 21 61 88 32	27 80 30 21 60	10 92 35 36 12
77 94 30 05 39	28 10 99 00 27	12 73 73 99 12	49 99 57 94 82	96 88 57 17 91
78 83 19 76 16	94 11 68 84 26	23 54 20 86 85	23 86 66 99 07	36 37 34 92 09
87 76 59 61 81	43 63 64 61 61	65 76 36 95 90	18 48 27 45 68	27 23 65 30 72
91 43 05 96 47	55 78 99 95 24	37 55 85 78 78	01 48 41 19 10	35 19 54 07 73
84 97 77 72 73	09 62 06 65 72	87 12 49 03 60	41 15 20 76 27	50 47 02 29 16
87 41 60 76 83	44 88 96 07 80	83 05 83 38 96	73 70 66 81 90	30 56 10 48 59

* Reprinted with permission from Random Numbers IV of Table XXXIII of R. A. Fisher and F. Yates, *Statistical Tables for Biological, Agricultural, and Medical Research* (Edinburgh: Oliver & Boyd, Ltd.).

TABLE A3–5

Useful Information

To find the circumference of a circle, multiply the diameter by 3.1416.

To find the diameter of a circle, multiply the circumference by .31831.

To find the area of a circle, multiply the square of the diameter by .7854.

The radius of a circle $\times$ 6.283185 = the circumference.

The square of the circumference of a circle $\times$ .07958 = the area.

Half the circumference of a circle $\times$ half its diameter = the area.

The circumference of a circle $\times$ .159155 = the radius.

The square root of the area of a circle $\times$ .56419 = the radius.

The square root of the area of a circle $\times$ 1.12838 = the diameter.

To find the diameter of a circle equal in area to a given square, multiply a side of the square by 1.12838.

To find the side of a square equal in area to a given circle, multiply the diameter by .8862.

To find the side of a square inscribed in a circle, multiply the diameter by .7071.

To find the side of a hexagon inscribed in a circle, multiply the diameter of the circle by .500.

To find the diameter of a circle inscribed in a hexagon, multiply a side of the hexagon by 1.7321.

To find the side of an equilateral triangle inscribed in a circle, multiply the diameter of the circle by .866.

To find the diameter of a circle inscribed in an equilateral triangle, multiply a side of the triangle by .57735.

To find the area of the surface of a ball (sphere), multiply the square of the diameter by 3.1416.

To find the volume of a ball (sphere), multiply the cube of the diameter by .5236.

Doubling the diameter of a pipe increases its capacity four times.

To find the pressure in pounds per square inch at the base of a column of water, multiply the height of the column in feet by .433.

A gallon of water (U. S. Standard) weighs 8.336 pounds and contains 231 cubic inches. A cubic foot of water contains $7\frac{1}{2}$ gallons, 1728 cubic inches, and weighs 62.425 pounds at a temperature of about 39° F.

These weights change slightly above and below this temperature.

TABLE A3–6

Decimal and Millimeter Equivalents
of Fractional Parts of an Inch

Inches		Inches	mm	Inches		Inches	mm
	1-64	.01563	.397		33-64	.51563	13.097
1-32		.03125	.794	17-32		.53125	13.494
	3-64	.04688	1.191		35-64	.54688	13.890
1-16		.0625	1.587	9-16		.5625	14.287
	5-64	.07813	1.984		37-64	.57813	14.684
3-32		.09375	2.381	19-32		.59375	15.081
	7-64	.10938	2.778		39-64	.60938	15.478
1-8		.125	3.175	5-8		.625	15.875
	9-64	.14063	3.572		41-64	.64063	16.272
5-32		.15625	3.969	21-32		.65625	16.669
	11-64	.17188	4.366		43-64	.67188	17.065
3-16		.1875	4.762	11-16		.6875	17.462
	13-64	.20313	5.159		45-64	.70313	17.859
7-32		.21875	5.556	23-32		.71875	18.256
	15-64	.23438	5.953		47-64	.73438	18.653
1-4		.25	6.350	3-4		.75	19.050
	17-64	.26563	6.747		49-64	.76563	19.447
9-32		.28125	7.144	25-32		.78125	19.844
	19-64	.29688	7.541		51-64	.79688	20.240
5-16		.3125	7.937	13-16		.8125	20.637
	21-64	.32813	8.334		53-64	.82813	21.034
11-32		.34375	8.731	27-32		.84375	21.431
	23-64	.35938	9.128		55-64	.85938	21.828
3-8		.375	9.525	7-8		.875	22.225
	25-64	.39063	9.922		57-64	.89063	22.622
13-32		.40625	10.319	29-32		.90625	23.019
	27-64	.42188	10.716		59-64	.92188	23.415
7-16		.4375	11.113	15-16		.9375	23.812
	29-64	.45313	11.509		61-64	.95313	24.209
15-32		.46875	11.906	31-32		.96875	24.606
	31-64	.48438	12.303		63-64	.98438	25.003
1-2		.5	12.700	1		1.00000	25.400

HOURLY PRODUCTION TABLE

Showing 60% to 80% Efficiency

Sec per Piece	Gross Prod. Per Hr	60%	65%	70%	75%	80%
1/2	7200	4320	4680	5040	5400	5760
5/8	5760	3456	3744	4032	4320	4608
3/4	4800	2880	3120	3360	3600	3840
7/8	4114	2468	2674	2880	3086	3291
1	3600	2160	2340	2520	2700	2880
1 1/4	2880	1728	1872	2016	2160	2304
1 1/2	2400	1440	1560	1680	1800	1920
1 3/4	2057	1234	1337	1440	1543	1646
2	1800	1080	1170	1260	1350	1440
2 1/4	1600	960	1040	1120	1200	1280
2 1/2	1440	864	936	1008	1080	1152
2 3/4	1309	785	851	916	982	1047
3	1200	720	780	840	900	960
3 1/4	1107	664	720	775	830	886
3 1/2	1028	617	668	720	771	822
3 3/4	960	576	624	672	720	768
4	900	540	585	630	675	720
4 1/4	847	508	551	593	635	678
4 1/2	800	480	520	560	600	640

Sec per Piece	Gross Prod. Per Hr	60%	65%	70%	75%	80%
12 1/2	288	173	187	202	216	230
13	276	166	179	193	207	221
13 1/2	267	160	174	187	200	214
14	257	154	167	180	193	206
14 1/2	248	149	161	174	186	198
15	240	144	156	168	180	192
15 1/2	232	139	151	162	174	186
16	225	135	146	158	169	180
16 1/2	218	131	142	153	164	174
17	212	127	138	148	159	170
17 1/2	206	124	134	144	155	165
18	200	120	130	140	150	160
18 1/2	195	117	127	137	146	156
19	189	113	123	132	142	151
19 1/2	185	111	120	130	139	148
20	180	108	117	126	135	144
21	171	103	111	120	128	137
22	164	98	107	115	123	131
23	156	94	101	109	117	125

Sec per Piece	Gross Prod. Per Hr	60%	65%	70%	75%	80%
50	72	43	47	50	54	58
52	69	41	45	48	52	55
54	67	40	44	47	50	54
56	64	38	42	45	48	51
58	62	37	40	43	47	50
60	60	36	39	42	45	48
62	58	35	38	41	44	46
64	56	34	36	39	42	45
66	54	32	35	38	41	43
68	53	32	34	37	40	42
70	51	31	33	36	38	41
72	50	30	33	35	38	40
74	49	29	32	34	37	39
76	47	28	31	33	35	38
78	46	28	30	32	35	37
80	45	27	29	32	34	36
82	44	26	29	31	33	35
84	43	26	28	30	33	34
86	42	25	27	29	32	34

Size (in.)																				
4 3/4	757	454	492	530	568	606	24	150	90	98	105	113	120	88	41	25	27	29	31	33
5	720	432	468	504	540	576	25	144	86	94	101	108	115	90	40	24	26	28	30	32
5 1/4	686	412	446	480	515	549	26	138	83	90	97	104	110	92	39	23	25	27	29	31
5 1/2	654	392	425	458	491	523	27	133	80	86	93	100	106	94	38	23	25	27	29	30
5 3/4	626	376	407	438	470	501	28	128	77	83	90	96	102	96	37	22	24	26	28	30
6	600	360	390	420	450	480	29	124	74	81	87	93	99	98	37	22	23	26	28	30
6 1/4	576	346	374	403	432	461	30	120	72	78	84	90	96	100	36	20	22	24	26	29
6 1/2	553	332	359	387	415	442	31	116	70	75	81	87	93	105	34	20	21	23	25	27
6 3/4	533	320	346	373	400	426	32	112	67	73	78	84	90	110	33	19	20	21	23	26
7	514	308	334	360	386	411	33	109	65	71	76	82	87	115	31	18	20	20	23	25
7 1/4	497	298	323	348	373	398	34	106	64	69	74	80	85	120	30	17	19	20	21	24
7 1/2	480	288	312	336	360	384	35	103	62	67	72	77	82	125	29	17	18	19	20	23
7 3/4	465	279	302	326	349	372	36	100	60	65	70	75	80	130	28	16	18	18	20	22
8	450	270	293	315	338	360	37	97	58	63	68	73	78	135	27	16	17	18	19	21
8 1/4	436	262	283	305	327	349	38	95	57	62	67	71	76	140	26	15	16	17	18	20
8 1/2	423	254	275	296	317	338	39	92	55	60	64	69	74	145	25	14	16	16	17	19
8 3/4	411	247	267	288	308	329	40	90	54	59	63	68	72	150	24	14	15	15	17	18
9	400	240	260	280	300	320	41	88	53	57	62	66	70	155	23	13	14	15	16	18
9 1/4	389	233	253	272	292	311	42	86	52	56	60	65	69	160	22	13	14	15	16	17
9 1/2	379	227	246	265	284	305	43	84	50	55	59	63	67	165	21	12	14	14	15	17
9 3/4	369	221	240	258	277	295	44	82	49	53	57	62	66	170	21	12	13	13	15	16
10	360	216	234	252	270	288	45	80	48	52	56	60	64	175	20	12	13	13	14	15
10 1/2	342	205	223	239	257	274	46	78	47	51	55	59	62	180	20	11	12	13	14	14
11	327	196	213	229	245	262	47	77	46	50	54	58	62	185	19	11	12	12	14	
11 1/2	313	188	203	219	235	250	48	75	45	49	53	56	60	190	19		12	12	13	
12	300	180	195	210	225	240	49	73	44	47	51	55	58	195	18		11			

Source: National Twist Drill & Tool Co.

TABLE A3–8

Speed and feed calculations for milling cutters and other rotating tools

Ft. per Min.	30	40	50	60	70	80	90	100	110	120	130	140	150
Diam. In.						Revolutions per Minute							
$\frac{1}{16}$	1833	2445	3056	3667	4278	4889	....	....	....	....	....	....	....
$\frac{1}{8}$	917	1222	1528	1833	2139	2445	2750	3056	3361	3667	3973	4278	4584
$\frac{3}{16}$	611	815	1019	1222	1426	1630	1833	2037	2241	2445	2648	2852	3056
$\frac{1}{4}$	458	611	764	917	1070	1222	1375	1528	1681	1833	1986	2139	2292
$\frac{5}{16}$	367	489	611	733	856	978	1100	1222	1345	1467	1589	1711	1833
$\frac{3}{8}$	306	407	509	611	713	815	917	1019	1120	1222	1324	1426	1528
$\frac{7}{16}$	262	349	437	524	611	698	786	873	960	1048	1135	1222	1310
$\frac{1}{2}$	229	306	382	458	535	611	688	764	840	917	993	1070	1146
$\frac{5}{8}$	183	244	306	367	428	489	550	611	672	733	794	856	917
$\frac{3}{4}$	153	204	255	306	357	407	458	509	560	611	662	713	764
$\frac{7}{8}$	131	175	218	262	306	349	393	437	480	524	568	611	655
1	115	153	191	229	267	306	344	382	420	458	497	535	573
1$\frac{1}{8}$	102	136	170	204	238	272	306	340	373	407	441	475	509
1$\frac{1}{4}$	91.7	122	153	183	214	244	275	306	336	367	397	428	458
1$\frac{3}{8}$	83.3	111	139	167	194	222	250	278	306	333	361	389	417
1$\frac{1}{2}$	76.4	102	127	153	178	204	229	255	280	306	331	357	382
1$\frac{5}{8}$	70.5	94.0	118	141	165	188	212	235	259	282	306	329	353
1$\frac{3}{4}$	65.5	87.3	109	131	153	175	196	218	240	262	284	306	327
1$\frac{7}{8}$	61.1	81.5	102	122	143	163	183	204	224	244	265	285	306
2	57.3	76.4	95.5	115	134	153	172	191	210	229	248	267	287
2$\frac{1}{4}$	50.9	67.9	84.9	102	119	136	153	170	187	204	221	238	255
2$\frac{1}{2}$	45.8	61.1	76.4	91.7	107	122	138	153	168	183	199	214	229
2$\frac{3}{4}$	41.7	55.6	69.5	83.3	97.2	111	125	139	153	167	181	194	208
3	38.2	50.9	63.7	76.4	89.1	102	115	127	140	153	166	178	191
3$\frac{1}{4}$	35.3	47.0	58.8	70.5	82.3	94.0	106	118	129	141	153	165	176
3$\frac{1}{2}$	32.7	43.7	54.6	65.5	76.4	87.3	98.2	109	120	131	142	153	164
3$\frac{3}{4}$	30.6	40.7	50.9	61.1	71.3	81.5	91.7	102	112	122	132	143	153
4	28.7	38.2	47.7	57.3	66.8	76.4	85.9	95.5	105	115	124	134	143
4$\frac{1}{2}$	25.5	34.0	42.4	50.9	59.4	67.9	76.4	84.9	93.4	102	110	119	127
5	22.9	30.6	38.2	45.8	53.5	61.1	68.8	76.4	84.0	91.7	99.3	107	115
5$\frac{1}{2}$	20.8	27.8	34.7	41.7	48.6	55.6	62.5	69.5	76.4	83.3	90.3	97.2	104
6	19.1	25.5	31.8	38.2	44.6	50.9	57.3	63.7	70.0	76.4	82.8	89.1	95.5
6$\frac{1}{2}$	17.6	23.5	29.4	35.3	41.1	47.0	52.9	58.8	64.6	70.5	76.4	82.3	88.2
7	16.4	21.8	27.3	32.7	38.2	43.7	49.1	54.6	60.0	65.5	70.9	76.4	81.9
7$\frac{1}{2}$	15.3	20.4	25.5	30.6	35.7	40.7	45.8	50.9	56.0	61.1	66.2	71.3	76.4
8	14.3	19.1	23.9	28.7	33.4	38.2	43.0	47.7	52.5	57.3	62.1	66.8	71.6
8$\frac{1}{2}$	13.5	18.0	22.5	27.0	31.5	36.0	40.4	44.9	49.4	53.9	58.4	62.9	67.4
9	12.7	17.0	21.2	25.5	29.7	34.0	38.2	42.4	46.7	50.9	55.2	59.4	63.6
9$\frac{1}{2}$	12.1	16.1	20.1	24.1	28.2	32.2	36.2	40.2	44.2	48.3	52.3	56.3	60.3
10	11.5	15.3	19.1	22.9	26.7	30.6	34.4	38.2	42.0	45.8	49.7	53.5	57.3
11	10.4	13.9	17.4	20:8	24.3	27.8	31.3	34.7	38.2	41.7	45.1	48.6	52.1
12	9.5	12.7	15.9	19.1	22.3	25.5	28.6	31.8	35.0	38.2	41.4	44.6	47.8
Ft. per Min.	30	40	50	60	70	80	90	100	110	120	130	140	150

Source: National Twist Drill & Tool Co.

TABLE A3–9
Table of cutting speeds for fractional sizes

TO FIND	HAVING	FORMULA
Surface (or Periphery) Speed in Feet per Minute = S.F.M.	Diameter of Tool in Inches $= D$ and Revolutions per Minute $= R.P.M.$	$S.F.M. = \dfrac{D \times 3.1416 \times R.P.M.}{12}$
Revolutions per Minute = R.P.M.	Surface Speed In Feet per Minute = S.F.M. and Diameter of Tool in Inches $= D$	$R.P.M. = \dfrac{S.F.M. \times 12}{D \times 3.1416}$
Feed per Revolution in Inches $=$ F.R.	Feed in Inches per Minute $= F.M.$ and Revolutions per Minute $= R.P.M.$	$F.R. = \dfrac{F.M.}{R.P.M.}$
Feed in Inches per Minute = F.M.	Feed per Revolution in Inches = F.R. and Revolutions per Minute $= R.P.M.$	$F.M. = F.R. \times R.P.M.$
Number of Cutting Teeth per Minute $=$ T.M.	Number of Teeth in Tool $= T$ and Revolutions per Minute $= R.P.M.$	$T.M. = T \times R.P.M.$
Feed per Tooth $=$ F.T.	Number of Teeth in Tool $= T$ and Feed per Revolution in Inches $= F.R.$	$F.T. = \dfrac{F.R.}{T}$
Feed per Tooth $=$ F.T.	Number of Teeth in Tool $= T$ Feed in Inches per Minute $= F.M.$ and Speed in Revolutions per Minute $= R.P.M.$	$F.T. = \dfrac{F.M.}{T \times R.P.M.}$

Source: National Twist Drill & Tool Co.

TABLE A3–10

Comparative Weights of Steel and Brass Bars

Steel—Weights cover hot worked steel about .50% carbon. One cubic inch weighs .2833 lbs. High speed steel 10% heavier.
Brass—One cubic inch weighs .3074 lbs.
Actual weight of stock may be expected to vary somewhat from these figures because of variations in manufacturing processes.

Size, Inches	Weight of Bar One Foot Long, Lbs.					
	Steel ○	Steel □	Steel ⬡	Brass ○	Brass □	Brass ⬡
1/16	.0104	.013	.0115	.0113	.0144	.0125
1/8	.042	.05	.046	.045	.058	.050
3/16	.09	.12	.10	.102	.130	.112
1/4	.17	.21	.19	.18	.23	.20
5/16	.26	.33	.29	.28	.36	.31
3/8	.38	.48	.42	.41	.52	.45
7/16	.51	.65	.56	.55	.71	.61
1/2	.67	.85	.74	.72	.92	.80
9/16	.85	1.08	.94	.92	1.17	1.01
5/8	1.04	1.33	1.15	1.13	1.44	1.25
11/16	1.27	1.61	1.40	1.37	1.74	1.51
3/4	1.50	1.92	1.66	1.63	2.07	1.80
13/16	1.76	2.24	1.94	1.91	2.43	2.11
7/8	2.04	2.60	2.25	2.22	2.82	2.45
15/16	2.35	2.99	2.59	2.55	3.24	2.81
1	2.67	3.40	2.94	2.90	3.69	3.19
1 1/16	3.01	3.84	3.32	3.27	4.16	3.61
1 1/8	3.38	4.30	3.73	3.67	4.67	4.04
1 3/16	3.77	4.80	4.16	4.08	5.20	4.51
1 1/4	4.17	5.31	4.60	4.53	5.76	4.99
1 5/16	4.60	5.86	5.07	4.99	6.35	5.50
1 3/8	5.04	6.43	5.56	5.48	6.97	6.04
1 7/16	5.52	7.03	6.08	5.99	7.62	6.60
1 1/2	6.01	7.65	6.63	6.52	8.30	7.19
1 9/16	6.52	8.30	7.19	7.07	9.01	7.80
1 5/8	7.05	8.98	7.77	7.65	9.74	8.44
1 11/16	7.60	9.68	8.38	8.25	10.51	9.10
1 3/4	8.18	10.41	9.02	8.87	11.30	9.78
1 13/16	8.77	11.17	9.67	9.52	12.12	10.49
1 7/8	9.39	11.95	10.35	10.19	12.97	11.24
1 15/16	10.02	12.76	11.05	10.88	13.85	12.00
2	10.68	13.60	11.78	11.59	14.76	12.78
2 1/16	11.36	14.46	12.53	12.33	15.69	13.60
2 1/8	12.06	15.35	13.30	13.08	16.66	14.42
2 3/16	12.78	16.27	14.09	13.87	17.65	15.29
2 1/4	13.52	17.22	14.91	14.67	18.68	16.17
2 5/16	14.28	18.19	15.75	15.50	19.73	17.09
2 3/8	15.06	19.18	16.62	16.34	20.81	18.02

Source: Brown & Sharpe Manufacturing Co.

TABLE A3–11

S. A. E. Standard Specifications for Steels

S. A. E. STEEL NUMBERING SYSTEM

A numerical index system is used to identify compositions of S. A. E. steels, which makes it possible to use numerals that are partially descriptive of the composition of materials covered by such numbers. The first digit indicates the type to which the steel belongs. The second digit, in the case of the simple alloy steels, generally indicates the approximate percentage of the predominant alloying element and the last two or three digits indicate the average carbon content in points, or hundredths of 1 per cent. Thus, 2340 indicates a nickel steel of approximately 3 per cent nickel (3.25 to 3.75) and 0.40 per cent carbon (0.35 to 0.45).

In some instances, it is necessary to use the second and third digits of the number to identify the approximate alloy composition of a steel. An instance of such departure is the steel numbers selected for several of the High Speed Steels and corrosion and heat resisting alloys. Thus, 71360 indicates a Tungsten Steel of about 13 per cent Tungsten (12 to 15) and 0.60 per cent carbon (0.50 to 0.70).

The basic numerals for the various types of S. A. E. steel are listed below:

Type of Steel	*Numerals (and Digits)*
Carbon Steels	1xxx
Plain Carbon	10xx
Free Cutting, (Screw Stock)	11xx
Free Cutting, Manganese	X13xx
High Manganese	T13xx
Nickel Steels	2xxx
0.50 Per Cent Nickel	20xx
1.50 Per Cent Nickel	21xx
3.50 Per Cent Nickel	23xx
5.00 Per Cent Nickel	25xx
Nickel Chromium Steels	3xxx
1.25 Per Cent Nickel, 0.60 Per Cent Chromium	31xx
1.75 Per Cent Nickel, 1.00 Per Cent Chromium	32xx
3.50 Per Cent Nickel, 1.50 Per Cent Chromium	33xx
3.00 Per Cent Nickel, 0.80 Per Cent Chromium	34xx
Corrosion and Heat Resisting Steels	30xxx
Molybdenum Steels	4xxx
Chromium	41xx
Chromium Nickel	43xx
Nickel	46xx and 48xx
Chromium Steels	5xxx
Low Chromium	51xx
Medium Chromium	52xxx
Corrosion and Heat Resisting	51xxx
Chromium Vanadium Steels	6xxx
Tungsten Steels	7xxx and 7xxxx
Silicon Manganese Steels	9xxx

Source: Brown & Sharpe Manufacturing Co.

HORSEPOWER REQUIREMENTS

FOR TURNING

When metal is cut in a lathe there is a downward pressure on the tool. This pressure, called chip pressure, depends on the material cut, shape and sharpness of the tool, and the size and shape of the chip.

For average conditions, a simple formula which will give sufficiently accurate results for power estimating purposes is as follows:

$P = CA$, where

A = cross sectional area of the chip in square inches, which is the product of depth of cut and feed per revolution of the work.

C = a constant depending on material being cut.

P = chip pressure on the tool in pounds.

Values of C

MATERIAL CUT	CONSTANT C
Low alloy steel	270,000
High alloy steel	350,000
High carbon steel	340,000
Medium carbon steel	300,000
Mild Steel	270,000
Cast iron—soft	132,000
Wrought iron	198,000
Malleable iron	170,000
Brass and bronze	110,000

Horsepower may be figured by using the following formula:

$$H.P. = \frac{P \times S}{33,000} \text{ where}$$

Example 2: Assuming that SAE 4140 is heat treated so that its strength is 100,000 pounds per square inch, the horsepower necessary to cut it, when other conditions remain the same as in Example 1, is:

$$H.P. = \frac{3.25 \times 100,000 \times \frac{3.14 \times 4}{12} \times 200}{33,000} = 7.8 \text{ as before.}$$

Multiplying the result by 1.25, we get 10 horsepower.

FOR MILLING

Generally accepted approximate values of power for steel cutting is one horsepower per $\frac{3}{4}$ cubic inches of material removed per minute, although 1 cubic inch can be used for rapid estimating purposes. The horsepower is figured using the following formula:

$H.P. = KdfNnw$, in which

  d = depth of cut taken in inches

  f = feed per tooth in inches

$H.P.$ = horsepower necessary to cut

  K = a constant depending on material cut

  n = number of teeth in the cutter

  N = number of revolutions per minute the cutter makes

  w = the width of cut in inches

For estimating the horsepower, approximate values of constant K are given below:

MATERIAL CUT	CONSTANT K
Bakelite	0.2
Brass	0.4
Cast Iron, soft	0.5
Cast Iron, medium hard	0.7
Cast Iron, hard	1.0
Steel: 120 Brinell	1.2
150 Brinell	1.4

H.P. = the horsepower necessary to revolve the work against the cutting pressure.

P = pressure on the tool in pounds.

S = cutting speed in feet per minute, and is equal to $\dfrac{3.14 \times D}{12}$

in which D is the diameter of the work.

Example 1: Determine the horsepower necessary to take a cut ¼ inch deep, with feed of 1/64 inch per revolution, on SAE 4140 steel bar 4 inches in diameter turning 200 times per minute.

Solution: Using C = 325,000

P = 325,000 x ¼ x 1/64 = 1220 pounds

The H.P. $= \dfrac{1220 \times \dfrac{3.14 \times 4}{12} \times 200}{33,000} = 7.8$

This should be multiplied by 1.25 to allow for the efficiency of the machine, thus:

H.P. = 7.8 x 1.25 = 9.7 or 10.

When the tensile strength of the material cut is known, the following formula may be used for computing the horsepower necessary to cut the material:

H.P. $= \dfrac{3.25 \text{ ATS}}{33,000}$ where

A = the cross section area of the chip in square inches, and is equal to the product of depth of cut and feed per revolution.

H.P. = the horsepower necessary to cut the metal.

S = the cutting speed in f.p.m.

T = the ultimate strength of the material cut.

Source: Vascoloy-Ramet Corp.

175 Brinell	1.5
250 Brinell	1.7
300 Brinell	1.9
400 Brinell	-2.0
500 Brinell	2.3
600 Brinell	2.5

It is to be noted that for a given material cut, fixed width of cut, and fixed number of teeth, the horsepower will vary with the depth of cut, the feed per tooth and the r.p.m.

Example: Assuming a width of cut 2 inches, the depth ⅛ inch, the feed 0.004" per tooth, what horsepower will be required to mill with a 3-inch 6-tooth cutter running at 600 r.p.m. and cutting steel 250 Brinell hardness.

Solution: K = 1.7 from Table; d = ⅛" or 0.125"; f = 0.004"; n = 6; N = 600, and w = 2".

Substituting these values in the formula, we get:

H.P. = 1.7 x 0.125 x 0.004 x 6 x 600 x 2 = 6.14

If the machine was powered with a 5 horsepower motor, we could reduce the r.p.m. and come within the capacity of the machine, using formula: N = $\dfrac{\text{H.P.}}{\text{Kdfnw}}$ in which the symbols have the same meaning as before.

Substituting the known values in this formula, we get:

N $= \dfrac{5}{1.7 \times 0.125 \times 0.004 \times 6 \times 2} = 490$ (approx.)

The speed of the machine should not be allowed to drop more than 50 per cent from that recommended on page 25, since this would impair the performance of the cutter. 490 r.p.m. for a 3-inch cutter will give us a cutting speed

S $= \dfrac{3.14 \times 3 \times 490}{12} = 385$ f.p.m. (approx.)

TABLE A3–13

+00000000 01	+00000000 01	+40000000 51	+10000000 51	+78539750 50	+31415900 51	+00000000 00	0001 000
+00000000 40	+00000003 14	+15000000 51	+42500000 50	+00000000 01	+00000000 00	+00000000 00	0002 000
+40000000 51	+10000000 51	+00000000 01	+13166432 51	+00000000 00	+00000000 00	+00000000 00	0003 015
+00000000 01	+00000000 02	+40000000 51	+12500000 51	+12271836 51	+49087344 51	+00000000 00	0004 000
+00000000 32	+00000004 90	+12500000 51 ·	+42500000 50	+00000000 01	+00000000 00	+00000000 00	0005 000
+40000000 51	+12500000 51	+00000000 01	+12186924 51	+00000000 00	+00000000 00	+00000000 00	0006 015
+00000000 01	+00000000 03	+40000000 51	+15000000 51	+17671444 51	+70685776 51	+00000000 00	0007 000
+00000000 26	+00000000 06	+12500000 51	+47500000 50	+00000000 01	+00000000 00	+00000000 00	0008 000
+40000000 51	+15000000 51	+00000000 01	+12920935 51	+00000000 00	+00000000 00	+00000000 00	0009 015
+00000000 01	+00000000 04	+40000000 51	+17500000 51	+24052798 51	+96211192 51	+00000000 00	0010 000
+00000000 22	+00000009 62	+10000000 51	+50000000 50	+00000000 01	+00000000 00	+00000000 00	0011 000
+40000000 51	+17500000 51	+00000000 01	+12122121 51	+00000000 00	+00000000 00	+00000000 00	0012 015
+00000000 01	+00000000 05	+40000000 51	+20000000 51	+31415900 51	+12566360 52	+00000000 00	0013 000
+00000000 20	+00000012 56	+10000000 51	+55000000 50	+00000000 01	+00000000 00	+00000000 00	0014 000
+40000000 51	+20000000 51	+00000000 01	+12793782 51	+00000000 00	+00000000 00	+00000000 00	0015 015
+00000000 02	+00000000 01	+41250000 51	+10000000 51	+78539750 50	+32397647 51	+00000000 00	0016 000
+00000000 41	+00000003 23	+15000000 51	+42500000 50	+00000000 00	+00000000 00	+00000000 00	0017 000
+41250000 51	+10000000 51	+00000000 01	+13193287 51	+00000000 00	+00000000 00	+00000000 00	0018 015
+00000000 02	+00000000 02	+41250000 51	+12500000 51	+12271836 51	+50621324 51	+00000000 00	0019 000
+00000000 33	+00000005 06	+12500000 51	+47500000 50	+00000000 01	+00000000 00	+00000000 00	0020 000
+41250000 51	+12500000 51	+00000000 01	+12839626 51	+00000000 00	+00000000 00	+00000000 00	0021 015
+00000000 02	+00000000 03	+41250000 51	+15000000 51	+17671444 51	+72894707 51	+00000000 00	0022 000
+00000000 27	+00000007 28 ·	+12500000 51	+47500000 50	+00000000 01	+00000000 00	+00000000 00	0023 000
+41250000 51	+15000000 51	+00000000 01	+12950110 51	+00000000 00	+00000000 00	+00000000 00	0024 015
+00000000 02	+00000000 04	+41250000 51	+17500000 51	+24052798 51	+99217792 51	+00000000 00	0025 000
+00000000 23	+00000009 92	+10000000 51	+50000000 50	+00000000 01	+00000000 00	+00000000 00	0026 000

23AUG

This is a FOR TRANSIT coding of the oil-sand mix core problem specifiying a single clamp for lengths 9 inches or less and two clamps for longer lengths. The following definitions are listed:

CLO: Initial length

DL: Increment length

NL: Number of lengths to be used in calculation

CDO: Initial diameter

DD: Increment diameter

ND: Number of diameters to be used in calculation

$Y = f\dfrac{L}{D}$ was stored in tables as whole numbers and handled as subroutine function ADJYF, which was separately programmed for "table lookup."

CT was stored in tables as whole numbers and handled as subroutine function CORTF, which was separately programmed for "table lookup."

TABLE A3–13 (continued)

```
------------------------------------------------------------------------------

------------------------------------------------------------------------------

0000001500+     6164718866+      9595916572+      ADJYF       551EK
0000001500+     6376798366+      9595926572+      CORTF       552EK
0000001500+     8278798366+      9190916572+      SQRTF       101EK

           C          CYLINDRICAL OIL-SAND CORES
           C          FORMULA M-11-NO.15 - NIEBEL
           C          AUGUST 24, 19
           C
           1          READ, CLO, DL, NL
           2          READ, CDO, DD, ND
           3          PI = 3.14159
           4          DO 15 I=1,NL
           5          DO 15 J=1,ND
           6          CL = CLO + (I-1) * DL
           7          CD = CDO + (J-1) * DD
           8          A = PI * CD**2 / 4.0
           9          V = A * CL
          10          LD = (CL / CD) * 10.0
          11          KV = V * 100.0
          12          Y = ADJYF(LD) / 100.0
          13          CT = CORTF(KV) / 1000.0
                      PUNCH, I,J,CL,CD,A,V,LD,KV,
                     1 Y,CT
          14          AT = 0.173*V + 0.021*CL
                     1 + 0.0157*A
                     2 + SQRTF(0.0067 + 16.0E=6*V**2)
                     3 + Y*CT + 0.327
          15          PUNCH, CL, CD, AT
          16          STOP
          17          END
   08/21/
```

TABLE A3–14

Tables of waiting time and machine availability for selected servicing constants*†
(values expressed as percentages of total time, where $T_1 + T_2 + T_3 = 100$ percent)

Block 1

n	(a)		(b)	
	T_3	T_1	T_3	T_1
k = 0.01				
1	0.0	99.0	0.0	99.0
10	0.1	99.0	0.1	98.9
20	0.1	98.9	0.2	98.8
30	0.2	98.8	0.4	98.6
40			0.6	98.4
50			0.9	98.1
60			1.3	97.8
70			1.8	97.2
80			2.7	96.3
85			3.4	95.7
90			4.2	94.9
95			5.2	93.8
100			6.7	92.4
105			8.5	90.6
110			10.7	88.4
115			13.4	85.8
120			16.3	82.9
121			16.9	82.3
122			17.5	81.7
123			18.1	81.1
124			18.8	80.4
125			19.4	79.8
126			20.0	79.2
127			20.6	78.6
128			21.2	78.1
129			21.8	77.5
130			22.4	76.9
131			22.9	76.3
132			23.5	75.7
133			24.1	75.2
134			24.6	74.6
135			25.2	74.1
136			25.7	73.5
137			26.3	73.0
138			26.8	72.5
139			27.3	71.9
140			27.9	71.4
141			28.4	70.9
142			28.9	70.4
143			29.4	69.9
144			29.9	69.4
k = 0.02				
1	0.0	98.0	0.0	98.0
5	0.1	98.0	0.2	97.9
10	0.2	97.8	0.4	97.6
15	0.4	97.7	0.7	97.4
20	0.6	97.5	1.1	97.0
25	0.8	97.2	1.6	96.5
30	1.2	96.9	2.2	95.9
35			3.1	95.0
40			4.3	93.8
45			6.1	92.0
50			8.7	89.5

Block 2

n	(a)		(b)	
	T_3	T_1	T_3	T_1
k = 0.02 (cont.)				
51			9.3	88.9
52			10.0	88.3
53			10.7	87.6
54			11.5	86.8
55			12.3	86.0
56			13.1	85.2
57			14.0	84.3
58			14.9	83.4
59			15.9	82.5
60			16.8	81.5
61			17.9	80.5
62			18.9	79.5
63			19.9	78.5
64			21.0	77.5
65			22.0	76.4
66			23.1	75.4
67			24.2	74.4
68			25.2	73.3
69			26.2	72.3
70			27.2	71.3
71			28.2	70.4
72			29.2	69.4
k = 0.03				
1	0.0	97.1	0.0	97.1
5	0.2	96.9	0.4	96.7
10	0.5	96.6	1.0	96.2
15	1.0	96.2	1.8	95.4
20	1.6	95.5	3.0	94.2
25	2.8	94.4	4.7	92.5
26	3.1	94.1	5.2	92.1
27	3.4	93.7	5.7	91.6
28	3.8	93.4	6.2	91.1
29	4.3	92.9	6.8	90.5
30	4.8	92.4	7.4	89.9
31			8.1	89.2
32			8.9	88.5
33			9.7	87.7
34			10.6	86.8
35			11.6	85.9
36			12.6	84.9
37			13.7	83.8
38			14.9	82.6
39			16.1	81.4
40			17.4	80.2
41			18.8	78.9
42			20.1	77.5
43			21.6	76.2
44			23.0	74.8
45			24.4	73.4
46			25.9	72.0
47			27.3	70.6
48			28.7	69.2

Block 3

n	(a)		(b)	
	T_3	T_1	T_3	T_1
k = 0.04				
1	0.0	96.2	0.0	96.2
2	0.1	96.1	0.2	96.0
3	0.2	96.0	0.3	95.9
4	0.2	95.9	0.5	95.7
5	0.3	95.8	0.7	95.5
6	0.5	95.7	0.9	95.3
7	0.6	95.6	1.1	95.1
8	0.7	95.5	1.3	94.9
9	0.8	95.4	1.5	94.7
10	1.0	95.2	1.8	94.4
11	1.1	95.1	2.1	94.1
12	1.3	94.9	2.4	93.8
13	1.5	94.7	2.8	93.5
14	1.8	94.5	3.2	93.1
15	2.0	94.2	3.6	92.7
16	2.3	94.0	4.0	92.3
17	2.6	93.6	4.5	91.8
18	3.0	93.3	5.1	91.3
19	3.4	92.9	5.7	90.7
20	3.9	92.4	6.4	90.0
21	4.5	91.8	7.1	89.3
22	5.2	91.2	8.0	88.5
23	6.0	90.4	8.9	87.6
24	6.8	89.6	9.9	86.7
25	7.9	88.6	11.0	85.6
26	9.0	87.5	12.2	84.5
27	10.4	86.2	13.4	83.2
28	11.9	84.7	14.8	81.9
29	13.6	83.0	16.3	80.5
30	15.5	81.3	17.9	79.0
31			19.6	77.4
32			21.3	75.7
33			23.0	74.0
34			24.8	72.3
35			26.6	70.6
36			28.4	68.9
37			30.1	67.2
k = 0.05				
1	0.0	95.2	0.0	95.2
2	0.1	95.1	0.2	95.0
3	0.2	95.0	0.5	94.8
4	0.4	94.9	0.7	94.5
5	0.5	94.7	1.0	94.3
6	0.7	94.6	1.4	94.0
7	0.9	94.4	1.7	93.6
8	1.1	94.2	2.1	93.3
9	1.4	93.9	2.5	92.9
10	1.6	93.7	3.0	92.4
11	2.0	93.4	3.5	91.9
12	2.3	93.0	4.1	91.4
13	2.7	92.6	4.7	90.8
14	3.2	92.2	5.4	90.1
15	3.8	91.7	6.2	89.3

Block 4

n	(a)		(b)	
	T_3	T_1	T_3	T_1
k = 0·05 (cont.)				
16	4.4	91.0	7.1	88.5
17	5.2	90.3	8.1	87.6
18	6.1	89.5	9.1	86.5
19	7.1	88.5	10.4	85.4
20	8.4	87.3	11.7	84.1
21	9.8	85.9	13.1	82.7
22	11.5	84.3	14.7	81.2
23	13.4	82.5	16.5	79.6
24	15.5	80.5	18.3	77.8
25	17.8	78.2	20.2	76.0
26	20.3	75.9	22.2	74.1
27	22.8	73.6	24.3	72.1
28	25.3	71.2	26.4	70.1
29	27.9	68.8	28.5	68.1
k = 0.06				
1	0.0	94.3	0.0	94.3
2	0.2	94.2	0.3	94.0
3	0.4	94.0	0.7	93.7
4	0.6	93.8	1.1	93.3
5	0.8	93.6	1.5	92.9
6	1.1	93.3	2.0	92.5
7	1.4	93.1	2.5	92.0
8	1.7	92.7	3.1	91.4
9	2.1	92.4	3.7	90.8
10	2.6	91.9	4.5	90.1
11	3.1	91.4	5.3	89.4
12	3.8	90.8	6.2	88.5
13	4.5	90.1	7.3	87.5
14	5.4	89.2	8.4	86.4
15	6.5	88.2	9.7	85.2
16	7.8	87.0	11.2	83.8
17	9.3	85.6	12.8	82.3
18	11.1	83.9	14.6	80.6
19	13.2	81.9	16.5	78.8
20	15.6	79.7	18.6	76.8
21			20.8	74.7
22			23.1	72.5
23			25.5	70.3
24			27.9	68.0
25			30.3	65.8
k = 0.07				
1	0.0	93.5	0.0	93.5
2	0.2	93.2	0.4	93.1
3	0.5	93.0	0.9	92.6
4	0.8	92.7	1.4	92.1
5	1.1	92.4	2.0	91.6
6	1.5	92.1	2.7	91.0
7	1.9	91.7	3.4	90.3
8	2.4	91.2	4.3	89.5
9	3.1	90.6	5.2	88.6
10	3.8	89.9	6.3	87.6

* All tables assume random calls for service. Column (a) is for constant servicing time and column (b) for an exponential distribution of servicing times. It is hoped that the missing values in column (a) can be secured by approximation in the near future.

† Where no entry appears in column, the figures were not available.

TABLE A3–14 (continued)

n	(a) T_2	T_1	(b) T_2	T_1
	$k = 0.07$ (cont.)			
11	4.7	89.1	7.5	86.4
12	5.7	88.1	8.9	85.1
13	7.0	86.9	10.4	83.7
14	8.6	85.4	12.2	82.1
15	10.4	83.7	14.1	80.3
16	12.6	81.6	16.2	78.3
17	15.2	79.3	18.5	76.2
18	18.1	76.6	21.0	73.9
19	21.1	73.7	23.5	71.5
20	.24.4	70.7	26.2	69.0
21			28.9	66.5
	$k = 0.08$			
1	0.0	92.6	0.0	92.6
2	0.3	92.3	0.5	92.1
3	0.6	92.0	1.2	91.5
4	1.0	91.7	1.9	90.9
5	1.4	91.2	2.7	90.1
6	2.0	90.8	3.5	89.3
7	2.6	90.2	4.5	88.4
8	3.4	89.5	5.7	87.3
9	4.3	88.6	7.0	86.1
10	5.4	87.6	8.5	84.8
11	6.7	86.4	10.1	83.2
12	8.4	84.8	12.0	81.4

n	(a) T_2	T_1	(b) T_2	T_1
	$k = 0.08$ (cont.)			
13	10.4	83.0	14.2	79.5
14	12.8	80.8	16.5	77.3
15	15.6	78.2	19.0	75.0
16	18.8	75.2	21.8	72.4
17	22.2	72.0	24.6	69.8
18	25.7	68.8	27.6	67.1
19	28.2	66.5	30.5	64.4
	$k = 0.09$			
1	0.0	91.5	0.0	91.7
2	0.4	91.4	0.7	91.1
3	0.8	91.0	1.4	90.4
4	1.3	90.6	2.3	89.6
5	1.9	90.0	3.3	88.7
6	2.6	89.4	4.5	87.7
7	3.4	88.6	5.8	86.5
8	4.5	87.6	7.3	85.1
9	5.7	86.5	9.0	83.5
10	7.3	85.0	10.9	81.7
11	9.3	83.2	13.1	79.7
12	11.7	81.0	15.6	77.5
13	14.5	78.4	18.3	75.0
14	17.8	75.4	21.2	72.3
15	21.5	72.0	24.2	69.5
16	25.3	68.5	27.4	66.6
17	29.2	65.0	30.6	63.7

n	(a) T_2	T_1	(b) T_2	T_1
	$k = 0.10$			
1	0.0	90.9	0.0	90.9
2	0.4	90.5	0.8	90.2
3	1.0	90.0	1.8	89.3
4	1.6	89.5	2.8	88.3
5	2.3	88.8	4.1	87.2
6	2.2	88.0	5.5	85.9
7	4.4	86.9	7.1	84.4
8	5.8	85.7	9.0	82.7
9	7.5	84.1	11.2	80.8
10	9.7	82.1	13.6	78.5
11	12.4	79.8	16.3	76.1
12	15.6	76.8	19.3	73.4
13	19.2	73.4	22.5	70.4
14	23.3	69.8	25.9	67.4
15	27.4	66.0	29.4	64.2
16	31.5	62.0		
	$k = 0.15$			
1	0.0	87.0	0.0	87.0
2	0.9	86.2	1.7	85.5
3	2.1	85.1	3.6	83.8
4	3.9	83.8	6.0	81.8
5	5.5	82.2	8.7	79.4
6	8.0	80.0	11.8	76.7
7	11.2	72.2	15.4	73.5
8	15.2	73.7	19.5	70.0
9	20.1	69.5	23.8	66.2
10	25.5	64.8	28.4	62.3
11	31.0	60.0		

n	(a) T_2	T_1	(b) T_2	T_1
	$k = 0.20$			
1	0.0	83.3	0.0	83.3
2	1.5	82.0	2.7	81.1
3	3.6	80.4	5.9	78.4
4	6.3	78.1	9.8	75.2
5	10.0	75.0	14.2	71.5
6	14.7	71.1	19.2	67.4
7	20.6	66.2	24.6	62.8
8	27.3	60.6	30.3	58.1
9	32.6	56.1		
	$k = 0.30$			
1	0.0	76.9	0.0	76.9
2	3.0	74.6	5.1	73.0
3	7.4	71.3	11.1	68.4
4	13.3	66.7	18.0	63.1
5	21.1	60.7	25.4	57.4
6	29.9	53.9	33.0	51.6
	$k = 0.40$			
1	0.0	71.4	0.0	71.4
2	4.8	68.0	7.5	66.0
3	11.8	63.0	16.3	59.8
4	21.2	56.3	25.6	53.1
5	31.9	48.6	34.9	46.5

TABLE A3–15
Metric system conversion chart*

LENGTH

U.S.   METRIC

1 inch = 25.4 millimeters
1 foot = 30.48 centimeters
1 yard = 0.914 meter

METRIC   U.S.

1 millimeter = 0.039 inch
1 centimeter = 0.394 inch
1 meter = 39.37 inches

METRIC (cm, meters)

ENGLISH (in., ft., yds.)

THICKNESS

1 mil = .025 millimeter

1 millimeter = 39.37 mils

MM

MILS

AREA

1 sq. foot = 929.03 sq. centimeters

1 sq. centimeter = 0.155 sq. inch

CM²

SQ. FT.

VOLUME

1 gallon = 3.785 liters

1 liter = 0.264 gallon

LITERS

GALLONS

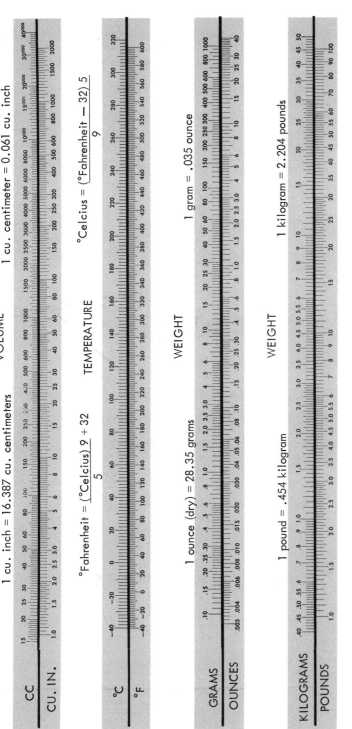

VOLUME

1 cu. centimeter = 0.061 cu. inch

1 cu. inch = 16.387 cu. centimeters

CC

CU. IN.

TEMPERATURE

$$°Celcius = \frac{(°Fahrenheit - 32)\,5}{9}$$

$$°Fahrenheit = \frac{(°Celcius)\,9}{5} + 32$$

°C

°F

WEIGHT

1 gram = .035 ounce

1 ounce (dry) = 28.35 grams

GRAMS

OUNCES

WEIGHT

1 kilogram = 2.204 pounds

1 pound = .454 kilogram

KILOGRAMS

POUNDS

PRESSURE

1 kilogram/sq. centimeter = 14.22 pounds/sq. inch

1 kilogram/sq. inch = 0.703 kilogram/sq. centimeter

1 pound/sq. inch = 0.703 kilogram/sq. centimeter

KG/CM²

LBS./SQ. IN.

* With the compliments of McGraw-Hill Book Company, College Division. Publishing for the engineer's diversity. 1221 Avenue of the Americas, New York, N.Y. 10020.

Index

This book has been set in 10 point and 9 point Times Roman, leaded 2 points. Chapter numbers are 36 point Bodoni and chapter titles are 18 point Times Roman. The size of the type page is 27 by 46½ picas.